RUBBER RECYCLING

RUBBER RECYCLING

Sadhan K. De • Avraam I. Isayev
Klementina Khait

CRC Press is an imprint of the
Taylor & Francis Group, an **informa** business
A TAYLOR & FRANCIS BOOK

CRC Press
Taylor & Francis Group
6000 Broken Sound Parkway NW, Suite 300
Boca Raton, FL 33487-2742

First issued in paperback 2019

ISBN-13: 978-0-8493-1527-5 (hbk)
ISBN-13: 978-0-367-39265-9 (pbk)
Library of Congress Card Number 2004064945

Library of Congress Cataloging-in-Publication Data

Rubber recycling / edited by Sadhan K. De, Avraam I. Isayev, Klementina Khait.
p. cm.
Includes bibliographical references and index.
ISBN 0-8493-1527-1 (alk. paper)
1. Rubber—Recycling. I. De, Sadhan K. (Sadhan Kumar) II. Isayev, Avraam I., 1942– III. Khait, Klementina. IV. Title.

TS1892.R7725 2005
678'.29—dc22 2004064945

Visit the Taylor & Francis Web site at
http://www.taylorandfrancis.com

and the CRC Press Web site at
http://www.crcpress.com

Preface

The safe disposal and reuse of industrial and postconsumer rubber wastes are serious challenges to environmental safety and public health. Rubber wastes are chemically crosslinked rubbers and are among the most difficult materials to recycle, since they will not dissolve or melt. Generally, recycling of rubber can be achieved by the following techniques:

- Size reduction and production of powders
- Devulcanization and reclaiming
- Recovery of chemicals and energy

This book highlights the problems involved with recycling vulcanized rubbers and presents probable solutions. Each chapter has been written by experts in a particular area; the present status of and future outlook for each area are discussed.

An issue to keep in mind is the strong resistance from industrial users to the use of crumb rubber materials (CRMs) from scrap tires. Their concern is that quality cannot be met on a consistent basis. Such hurdles and mindsets can be overcome if the tire-derived materials can be consistently demonstrated to be viable raw material components in the final compound for high-end applications, mainly new tire manufacture.

In the first chapter, Michael Rouse explains the strategy, practices, and philosophy required to provide high-quality raw materials derived from scrap tires.

The tire is designed to last and is probably one of the best engineered products developed by man. Although it is nearly indestructible to normal mechanical fracturing mechanisms, if the tire could be broken down into its separate components of fiber, steel, and rubber, it could be made into a very viable compounding ingredient for new tire manufacture. This subject is discussed by Michael Rouse in the second chapter. Rouse explains the application areas of CRMs and describes the methods used to produce high-grade CRMs for today's markets.

The third chapter, by Wilma K. Dierkes, deals with the grinding technologies: cutting and shearing, and grinding by impact. Applications for untreated, particulated rubber; surface treatment of the rubber powder; and application areas of the surface-treated rubber powder are included. Surface treatment includes chemical activation, physical activation, mechanical activation, and microbial treatment.

An unconventional polymer recycling technology involves mechanochemical processing. A polymer is subjected to simultaneous actions of compression and shear, resulting in both size reduction and changes in polymer structure. In the fourth chapter, Klementina Khait describes tire rubber recycling by the mechanochemical processing technology, also called solid-state shear pulverization (S^3P), which utilizes a modified single- or twin-screw extruder. The product of this process is powder of controlled particle size, unique elongated shape, and well-developed surface. Among its areas of application, recycled tire rubber powder could be used in large amounts to replace virgin rubber.

In the fifth chapter, Richard J. Farris, Drew E. Williams, and Amiya R. Tripathy explain the technique of high-pressure, high-temperature sintering for recycling chemically crosslinked rubber. The rubbers are first ground into a powder via cryogenic grinding, then sintered back together under high pressure and temperature. The authors describe the effects of polymer structure, molding temperature, and organic acid additives on the sintering mechanisms.

Ceni Jacob and S.K. De discuss utilization of powdered rubber waste in rubber compounds in the sixth chapter. The results of investigations on fine rubber powders obtained by the abrading technique show that, in rubber compounds such as EPDM, the rubber powder concentration can be as high as 300 phr without adversely affecting the properties of the final rubber vulcanizates. Moreover, incorporation of fine rubber powders increases the smoothness of the extrudate surface during processing and improves the final properties.

The seventh chapter, by D. Mangaraj, deals with composites of ground rubber with both thermoplastics and thermosets, with special reference to compatibilization, surface modification, and functionalization of ground rubber. Production of thermoplastic vulcanizates (TPVs) using waste rubber and thermoplastics is also included in this chapter, and the future potential for recycling waste rubber by blending it with waste plastics is discussed.

In the eighth chapter, D. Raghavan explains the strategies for reuse of rubber tires, with special emphasis on the use of tire particles as replacement aggregates for low-strength concrete material. It is believed that the understanding of interface chemistry between rubber and cement is fundamental to improved performance of rubber-filled cementitious systems.

Ultrasonic devulcanization of used tires and waste rubbers is presented in the ninth chapter by A.I. Isayev and Sayata Ghose. The chapter provides an up-to-date account of ultrasonic devulcanization of rubbers, including the developmental history of the technology and currently available ultrasonic devulcanization reactors. The chapter also deals with degradation mechanisms of the rubber network by ultrasonic waves, and various properties of blends of devulcanized and virgin rubbers.

The tenth chapter by Marvin Myhre deals with the nonultrasonic devulcanization processes, with special reference to different chemicals, mechanisms of devulcanization, and equipment used in the devulcanization process. Application of biotechnology in achieving devulcanization is also discussed. Finally, the results of reclaiming and devulcanization processes are explained, and the property values obtained upon recuring the material are shown.

In the eleventh chapter, C. Roy and his colleagues describe the development of the vacuum pyrolysis process and the characterization and end uses of the pyrolytic carbon black and oil. Data on the process feasibility and suitability of the pyrolysis products for different applications are also presented in the chapter.

The last chapter, by Tjaart P. Venter, covers the markets for scrap tires and recycled rubber. According to the author, "Do not first produce a product and then go and look for a market. Start with a market."

The editors are thankful to CRC Press/Taylor & Francis Group for undertaking the project and cooperating with the authors of the chapters at all stages of preparation.

Last but not least, the editors would like to acknowledge the financial support from the following organizations: the All India Council of Technical Education for providing support to S.K. De; the Illinois Department of Economic Opportunities (formerly the Department of Energy and Recycling) for providing support to K. Khait; and the DMI Division of the National Science Foundation for providing support to Avraam I. Isayev.

S.K. De
A.I. Isayev
K. Khait

About the Editors

Sadhan Kumar De has been Emeritus Fellow at the Government College of Engineering and Leather Technology, Calcutta, India, since 2003. Prior to 2003, he was Professor in the Rubber Technology Center at the Indian Institute of Technology, Kharagpur, India. He was the head of the Rubber Technology Center during both its formative stage (1982–1987) and later (1995–1999). He was also dean of postgraduate studies at the institute during 1987–1990. In 2002, he was the Harold A. Morton Distinguished Visiting Professor in the polymer engineering department at the University of Akron in Ohio.

In 2004, Dr. De was Visiting Professor at Mahidol University, Bangkok, Thailand. For his excellence in teaching and research, he received the George Whitby Award from the Rubber Division of the American Chemical Society. He is the co-author of five books in rubber science and technology and has published about 300 research papers in international journals.

Avraam I. Isayev is Distinguished Professor of Polymer Engineering and Director of the Molding Technology Center at the Institute of Polymer Engineering at the University of Akron. He is a native of Azerbaijan. He received M.Sc. degrees from the Azerbaijan Institute of Oil and Chemistry in chemical engineering and the Moscow Institute of Electronic Machine Building in applied mathematics. He also received a Ph.D. in polymer engineering and science from the Institute of Petrochemical Synthesis of the USSR Academy of Sciences, Moscow. Before joining the University of Akron in 1983 as associate professor, he was Senior Research Associate at Cornell University, Senior Research Fellow at Technion, and Research Associate at the Institute of Petrochemical Synthesis of the USSR Academy of Sciences.

Dr. Isayev's research interests are plastics, rubber, and composite processing, especially molding processes; rheology, rheo-optics, constitutive equations and process modeling; self-reinforced or *in-situ* composites based on liquid crystalline polymers; and processing of polymers, continuous decrosslinking of thermosets and rubbers, and *in-situ* copolymerization of polymer blends with the aid of high-power ultrasound.

Dr. Isayev has edited three books and has published over 200 papers, book chapters, and encyclopedia articles. He has been awarded 21 patents. His biography is listed in many *Who's Who* publications. His numerous honors include: Laureate of Young Scientists, Moscow, USSR, 1970; recognition from the Society of Plastics Engineers for significant contribution to the society and plastics industry, 1994; Distinguished Corporate Inventor, American Society of Patent Holders, 1995; Silver Medal, Institute of Materials, London,

England, 1997; Melvin Mooney Distinguished Technology Award of the Rubber Division of the American Chemical Society, 1999; and the Vinogradov Prize of Society of Rheology, Moscow, Russia, 2000.

Klementina Khait holds a Ph.D. in polymer chemistry and an M.S. in chemical engineering from the Technological Institute, St. Petersburg, Russia. She has 13 years of experience in applied research projects and group management in the United States with Borg-Warner Chemicals, Inc. (now part of General Electric) and Quantum Chemical Corp. (now Equistar LP), as well as 20 years of extensive experience in the former USSR. Her research interests include modification of polymers, polymer blends and alloys, polymer processing and recycling, and property–structure relationships.

Dr. Khait's recent research has focused on the development and commercialization of a novel solid-state shear pulverization (S^3P) technology for producing polymeric powders, for creating new polymer blends and plastic/rubber composites, and for recycling of rubber and unsorted plastic waste.

Dr. Khait has authored several U.S. and Canadian patents. She has also authored two USSR patents and numerous technical publications. In 2001, she co-authored the book *Solid-State Shear Pulverization: A New Polymer Processing and Powder Technology,* published by Technomic Publishing Co. (Lancaster, Pennsylvania).

Active in the Society of Plastic Engineers (SPE), Dr. Khait currently serves on the board of directors of the Plastics Environmental Division. She recently retired as Research Associate Professor and Director of the Polymer Technology Center in the Department of Chemical Engineering at Northwestern University in Evanston, Illinois.

Contributors

B. de Caumia
Département de Génie Chimique
Université Laval
Québec, Canada

A. Chaala
Département de Génie Chimique
Université Laval
Québec, Canada

H. Darmstadt
Département de Génie Chimique
Université Laval
Québec, Canada

S.K. De
Rubber Technology Center
Indian Institute of Technology
Kharagpur, India

Wilma Dierkes
Faculty of Chemical Technology
Rubber Technology Group
University of Twente
Enschede, The Netherlands

Richard J. Farris
Polymer Science and Engineering Department
University of Massachusetts Amherst
Amherst, Massachusetts

Sayata Ghose
Institute of Polymer Engineering
The University of Akron
Akron, Ohio

A.I. Isayev
Institute of Polymer Engineering
The University of Akron
Akron, Ohio

Ceni Jacob
Rubber Technology Center
Indian Institute of Technology
Kharagpur, India

Klementina Khait
Chemical Engineering Department (retired)
Polymer Technology Center
Northwestern University
Evanston, Illinois

D. Mangaraj
Innovative Polymer Solutions
Dublin, Ohio
METSS Corporation
Westerville, Ohio

Marvin Myhre
Canadian Rubber Testing and Development, Limited
St. Catharines, Ontario, Canada

H. Pakdel
Département de Génie Chimique
Université Laval
Québec, Canada

D. Raghavan
Polymer Group
Department of Chemistry
Howard University
Washington, DC

Michael W. Rouse
Star Trends, Inc.
Sunriver, Oregon

C. Roy
Département de Génie Chimique
Université Laval
Québec, Canada

Amiya R. Tripathy
Polymer Science and Engineering Department
University of Massachusetts Amherst
Amherst, Massachusetts

Tjaart P. Venter
Gloss Recycling and Chemicals (Pty) Ltd.
Johannesburg, South Africa

Drew E. Williams
Polymer Science and Engineering Department
University of Massachusetts Amherst
Amherst, Massachusetts

J. Yang
Département de Génie Chimique
Université Laval
Québec, Canada

Contents

1

Manufacturing Practices for the Development of Crumb Rubber Materials from Whole Tires

Michael W. Rouse

CONTENTS

1.1 Abstract

The processing or reduction of whole tires into useful particle sizes has been a challenge to the recycler since Charles Goodyear first discovered vulcanization. This chapter discusses the development of converting tires into a useful particle form, for reuse in many applications. Materials made from tires are called tire-derived materials (TDMs) and include a higher refined portion, crumb rubber materials (CRMs), which can be reused in the manufacture of tire compounds and for many other applications.

The tire is probably one of the best-engineered, longest-lasting products developed by man. It is designed for performance, aesthetics and comfort, control, and safety. Scrap tires are a valuable and viable national resource; the challenge today is to find the best way to utilize them in a wide range of applications.

1.2 Background

The tire is nearly indestructible to normal mechanical fracturing mechanisms. However, if the tire can be broken down into its separate major components — fiber, steel, and rubber — it can be made into very viable compounding ingredients for new tires. This rubber material or TDM can also be used in manufactured products, as an asphalt modifier for enhancing plastics, and for energy conversion and soil amendments. The TDM can be made into many gradations of crumb rubber material, or CRM. With the use of CRM, the potential for a wider variety of applications becomes even more viable. These higher and better uses include industrial, commercial, governmental, agricultural, and sports applications. Depending on the final particle size and gradation of the CRM derived from TDM using a particular tire type, the CRM can be an excellent raw material for many other applications. TDM can be used as a supplemental fuel, with or without the inherent wire reinforcements, or for civil engineering applications in road construction.

In general, CRM cannot be produced in a one-step process, such as punching a casing to make a part for a child's swing or a punched part for a muffler hanger, and cannot be used as one-step shredding for civil engineering applications. The production of CRM requires a number of key processing steps. This chapter discusses the methods and methodology to produce high-grade CRMs for today's markets. The production of TDM and CRM for specific uses encompasses many options and variables. The availability and cost of appropriate equipment, plus the expertise needed to operate the equipment for desired end products, present a challenge for the processor. Only the highlights can be covered in this chapter. It is hoped that the ideas presented herein can be a guide to initiate or improve the successful development of TDM in a safe and environmentally sound manner.[1]

Approximately 290 million tires[2] are produced and 23 million passenger, truck/bus, and off-the-road tires retreaded annually in the U. S.[3] Eventually, the tires reach the end of their useful life and are introduced into the scrap tire stream. This production value represents what is entering our environment.[4] Today, less than 10% of the tires removed from vehicles are recovered for recapping or resale as used tires. The remainder of this scrap tire stream is increasingly finding new markets or higher economic value other than disposal. To recycle means to put or pass through a cycle again, as for further treatment or different application. Recycling is the major focus of processing to convert TDM to its highest use.

The use of TDM as a fuel is not a form of recycling, but rather, a method of disposal. What is intended is for the processor to take advantage of the tires' inherent properties and develop unique markets for the processor's efforts. However, if the production of tire-derived fuels leads eventually to higher and better uses of the scrap tire ingredients, then this stepped approach is a major move in the right direction of utilizing the scrap tire to its fullest potential.

Tires are an oil derivative; its components of construction, such as the steel and cord fibers, depend on energy for production. With the world price of crude oils continually escalating, the value of TDM will only increase. This price escalation will spur on the necessary processing developments to utilize the scrap tire to its maximum value.

The polymer composition of a tire makes it very unique and valuable for other applications. However, scrap tires from a landfill have no economic value and present health, safety, environmental, and handling problems. Tires do not biochemically degrade sufficiently when buried. They may resurface in landfills, providing an excellent breeding ground for vermin and mosquitoes. The exception to this case is tire monofills that have been properly operated for future recovery of whole tires or some shredded form of them.

The purpose of this chapter is to discuss various alternatives to disposal of waste tires, so they can become a resource. Worn tires can be sold as "used tires" or recapped for continued use as tires, but eventually tires lose their total recyclability and must find another use. It is true that a small portion of the scrap tires entering the environment can find a recyclability life (use over many more applications without loss of integrity of the tire rubber compound or its components). Eventually, the useful life as a tire will end, but the rubber polymers can be successfully recovered and used again in a new tire compound as a filler or extender. Tire compound applications do require very stringent CRM quality requirements if CRM is to be used for this purpose. Worn tires can be used as artificial reefs, highway crash barriers, bank reinforcements, children's swings, and industrial and commercial parts to luggage items, just to mention a few uses. Whole or shredded tires can be burned directly (pyrolyzed) or gasified for energy recovery. If oil or carbon char is produced from tires as a building block of a new raw material ingredient, it could be argued that this activity is recycling, although the major composition of the tire has been altered due to depolymerization.

The focus herein is the reduction or commutation techniques for reducing whole tires of all types into a tire-derived material form, TDM. This includes the manufacture of tire-derived fuel (TDF). TDM can be further refined into CRM. CRM is not used in TDF due to cost restrictions. An equivalent cost competitiveness of CRM can be made to a basic fossil fuel such as gas, oil, or coal to wood wastes; refuse-derived fuels (RDF); and other energy sources. To illustrate this point, assume that No. 5 fuel-grade oil is selling for $30 (US) per barrel and is being used in a multi-fuel furnace for power generation at the same conversion efficiency. This would equate to $148 (US) per ton or roughly 7.4 cents per pound if a minus 2-inch wire free TDF was used. If the equivalent fuel price for both fuels were used, this would leave no room for return on investment to the energy user additional metering equipment and initial testing by the time the CRM was produced and shipped to the energy user. The economics under these conditions is not viable. Therefore, if a 30% discount price for TDF was used or $103 (US) per ton compared to oil, this may be a starting point for the energy user to consider using TDF as a supplemental fuel to justify the costs of testing, metering equipment installation, TDF handling, and the possibility of new monitoring controls for stack emissions.

Table 1.1 compares the various fuel values of TDF and TCF to other fuels on an "as received" basis. Normally, a range of high and low value range reflects fuel values. Care should be taken to identify the particular fuel value being used. This table can assist in obtaining a quick idea of a particular fuel's equivalent value.

Figure 1.1, showing the tire flow diagram, depicts the life or ecosystem of the tire from its initial manufacture, to its useful life as a tire, to the capture of its qualities for other uses when it is removed from the transport mode. This figure does not detail the many opportunities that are available today, but is intended as an overview to the endless possibilities for the scrap tire after its useful life. TDM has only come into real existence and acceptance within the last 20 years. Some of the branching opportunities noted under the TDM section of Figure 1.1 illustrate many new interrelationships and the potential for TDM if the scrap tire can be processed into useful products before going to landfills or being removed or diverted from these sites. New market uses have become economically viable only due to TDM processors' market efforts and developments, and their ability to overcome negative perceptions and replace them with positive acceptance of these new materials.

1.3 The Tire

The tire is designed to last and has evolved over the past 16 decades. If we were to design and build a tire today, the tire compounder would find this to be a Herculean task. Figure 1.2, indicating tire functions, depicts the complex set of forces that act on a tire and that must be considered in its design and construction. Esthetics and comfort comprise one component that is expanding to minimize noise migration. The lateral direction means cornering and center of gravity control as it is tied to the automobile's performance and safety. The tire designer is constantly challenged to design

TABLE 1.1

Comparison of Various Fuels on an "As Received" Basis[1]

Fuel	Grade	Average Heat Value
Gas	Natural	1,000 BTU/Std. Cubic Foot
Oil	No. 6 Fuel Oil "Bunker C"	151,000 BTU/Gal (42 Gal/Bbl.)
TDF	Tire-Derived-Fuel	15,500 BTU/Lb.
TCF	Tire-Chip-Fuel	14,200 BTU/Lb.
Coke	Petroleum	13,700 BTU/Lb.
Coal	Lignite	7,665 BTU/Lb.
Coal	Subbituminous	10,500 BTU/Lb.
Coal	Bituminous	12.750 BTU/Lb.
Wood	Wet Wood (50% Moisture) — Hog Fuel	4.375 BTU/Lb.

Gas $/mm BTU	Oil $/Bbl	TDF $/Ton	TCF $/Ton	Coke $/Ton	Subbitum. $/Ton	Bitum. $/Ton	Lignite $/Ton	Wood $/Ton
0.75	4.76	23.25	21.30	20.55	15.75	19.13	11.50	6.56
0.90	5.71	27.90	25.56	24.66	18.90	22.95	13.80	6.56
1.05	6.66	32.55	29.82	28.77	22.05	26.78	16.10	9.19
1.20	7.61	37.20	34.08	32.88	25.20	30.60	18.40	10.50
1.35	8.56	41.85	38.34	36.99	28.35	34.43	20.70	11.81
1.50	9.51	46.50	42.60	41.10	31.50	38.25	23.00	13.13
1.65	10.46	51.15	46.86	45.21	34.65	42.08	25.29	14.44
1.80	11.42	55.80	51.12	49.32	37.80	45.90	27.59	15.75
1.95	23.37	60.45	55.38	53.43	40.95	49.73	29.89	17.06
2.10	13.32	65.10	59.64	57.54	44.10	53.55	32.19	18.38
2.25	14.27	69.75	63.90	61.65	47.25	57.38	34.49	19.69
2.40	15.22	74.40	68.16	65.76	50.40	61.20	36.79	21.00
2.55	16.17	79.05	72.42	69.87	53.55	65.03	39.09	22.31
2.70	17.12	83.70	76.68	73.98	56.70	68.85	41.39	23.63
2.85	18.07	88.35	80.94	78.09	59.85	72.68	43.69	24.94
3.00	19.03	93.00	85.20	82.20	63.00	76.50	45.99	26.25
3.15	19.98	97.65	89.46	86.31	66.15	80.33	48.29	27.56
3.30	20.93	102.30	93.72	90.42	69.30	84.15	50.59	28.88
3.45	21.88	106.95	97.98	94.53	72.45	87.98	52.89	30.19
3.60	22.83	111.60	102.24	98.64	75.60	91.80	55.19	31.50
3.75	23.78	116.25	106.50	102.75	78.75	95.63	57.49	21.81
3.90	24.73	120.90	110.76	106.86	81.90	99.45	59.79	34.13
4.05	25.69	125.55	115.02	110.97	85.05	103.28	62.09	35.44
4.20	26.64	130.20	119.28	115.08	88.20	107.10	64.39	36.75
4.35	27.59	134.85	123.54	119.19	91.35	110.93	66.69	38.06
4.50	28.54	139.50	127.80	123.30	94.50	114.75	68.99	39.38
4.65	29.49	144.15	132.06	127.41	97.65	118.58	71.28	40.69
4.80	30.44	148.80	136.32	131.52	100.80	122.40	73.58	42.00
4.95	31.39	153.46	140.58	135.63	103.95	126.23	75.88	43.31
5.10	32.34	158.10	144.84	139.74	107.10	130.05	78.18	44.63
5.25	33.30	163.75	149.10	143.85	110.25	133.88	80.48	45.94
5.50	34.88	170.50	156.20	150.70	115.50	140.25	84.32	48.13
5.65	35.83	175.15	160.46	154.81	118.65	144.08	86.61	49.44
5.80	36.78	179.80	164.72	158.92	131.80	147.90	88.91	50.75
5.95	37.73	184.45	168.98	163.03	124.95	151.73	91.21	52.06
6.10	38.69	189.10	173.24	167.14	128.10	155.55	93.51	53.38
6.25	39.64	193.75	177.50	171.25	131.25	159.38	95.81	54.69
6.40	40.59	198.40	181.76	175.36	134.40	163.20	98.11	56.00

TABLE 1.1 (Continued)

Comparison of Various Fuels on an "As Received" Basis[1]

Gas $/mm BTU	Oil $/Bbl	TDF $/Ton	TCF $/Ton	Coke $/Ton	Subbitum. $/Ton	Bitum. $/Ton	Lignite $/Ton	Wood $/Ton
6.55	41.54	203.05	186.02	179.47	137.55	167.03	100.41	57.31
6.70	42.49	207.70	190.28	183.58	140.70	170.85	102.71	58.63
6.85	43.44	212.35	194.54	187.69	143.85	174.68	105.01	59.94
7.00	44.39	217.00	198.80	191.80	147.00	178.50	107.31	61.25
7.15	45.36	221.65	203.06	195.91	150.15	182.33	109.61	62.56
7.30	46.30	226.30	207.32	200.02	153.30	186.15	111.91	63.88
7.45	47.25	230.95	211.58	204.13	156.45	189.98	114.21	65.19
7.60	48.20	235.60	215.84	208.24	159.60	193.80	116.51	66.50

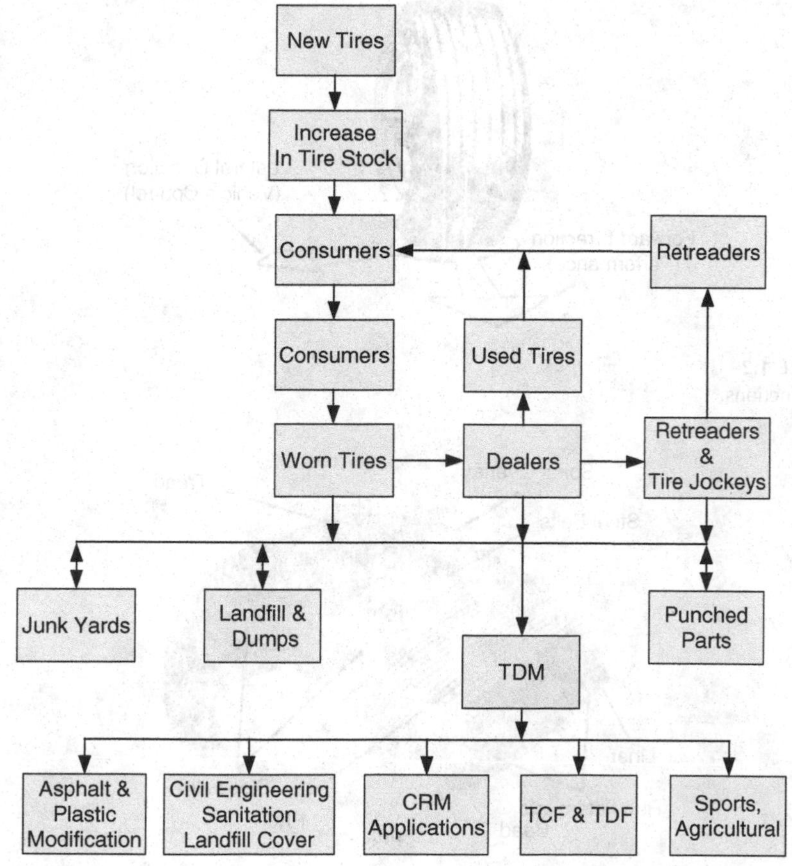

FIGURE 1.1
Eco Life tire flow diagram. (Courtesy of Star Trends, Inc., Sunriver, OR.)[1]

a tire that will perform well when major complex, dynamic forces are involved in forward movement (such as stopping) and that will provide optimal safety in all kinds of weather and pavement conditions. Therefore, due to the many complex polymers and other ingredients being used in the design of tires and compounded into the tire components, the use of recycled tire rubber back into new tire compounds will be very limited.

The tire itself has evolved into a very complex mixture of polymers, textiles, and exotic steel components. This is illustrated in Figure 1.3, which depicts tire construction. This figure reflects the many intricate parts that go into the makeup of today's modern tire. In today's world, the tire manufacturer must design a tire to meet any kind of eventuality. So, tires have to meet all season parameters and conditions to ensure that they meet their intended design and performance requirements.

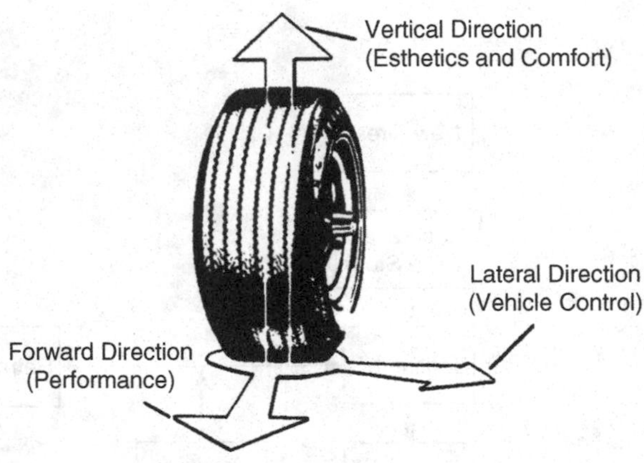

FIGURE 1.2
Tire functions.[5]

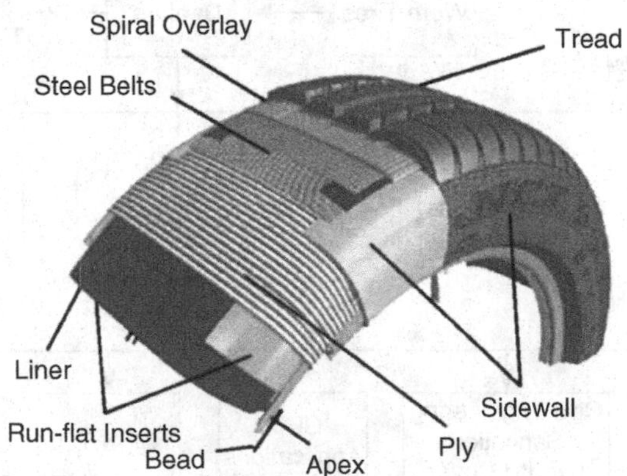

FIGURE 1.3
Tire construction.[5]

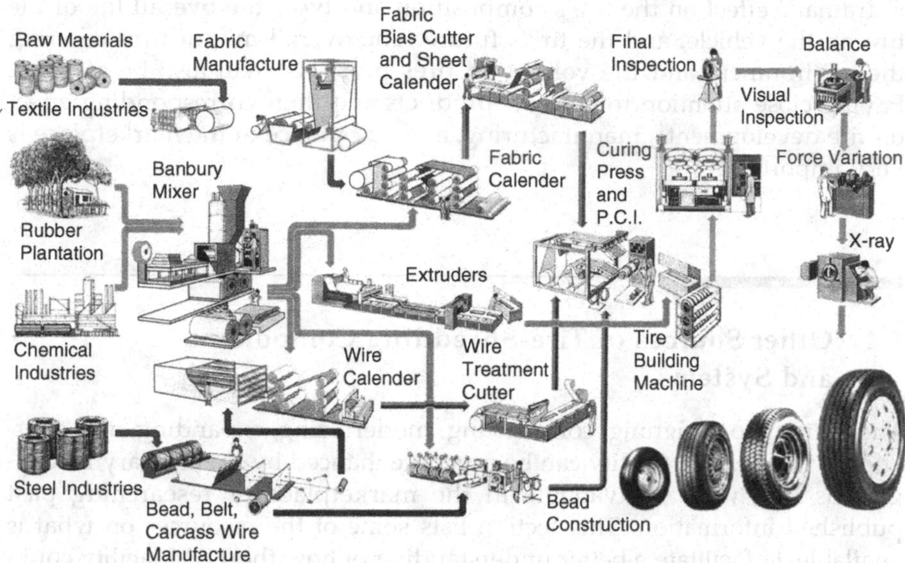

FIGURE 1.4
Cross-section of manufacturing of high-performance tires.[5]

Figure 1.4 represents how tire manufacturing has evolved into a very complex process. Today's scrap tire processor must know the manufacturing process and the general construction of the tire itself to understand the challenge of disassembling a tire back into smaller components, such as CRM, or to understand thermal digestion by pyrolysis, gasification, or some other chemical or thermal-chemical process or newly evolving technology. Since passenger tires represent a significant majority of the tires in units, it is an interesting fact that the total weight of truck tires is nearly equal to all the passenger units in total production or OEM (original equipment manufacturer) weight. This is a very important fact since the rubber composition of passenger tires differs from that of truck tires and off-the-road (OTR) tires. These differences in rubber composition and modes of tire construction can have a dramatic influence on the processing and final application of the intended use of the TDM or CRM in other products or when used back into the manufacture of tires.[5]

It is said that Henry Ford developed today's automobile assembly line by observing what took place in a hog-slaughtering house. He only reversed the process by assembling the necessary components in the correct ratios and set the stage for today's modern car assembly manufacturing line. By observing how a tire is built, its major components, and composition weight ratios in the major broad range of tires, the tire processor can begin to determine what is needed to produce the desired TDM or CRM more effectively. The tire will keep evolving as the automobile responds to new levels of performance. The advent of the electric car and new vehicles with a combination of electrical and mechanical conversions systems will have a dramatic effect on the tire's composition and type, the overall life of the tire on the vehicle, and the tire's future. The overall mix of tires entering the environment and the volume of tires produced will also be affected. Paying close attention to new-car products and their corresponding effect on tire developments, manufacturing, and acceptance in the marketplace is very important.

1.4 Other Sources of Tire-Shredding Components and Systems

The journey to designing, constructing, modernizing, expanding, and operating a tire-derived facility can be greatly enhanced by preliminary investigations into what is available in the marketplace or researching past published information. This section lists some of the resources on what is available, to facilitate a better understanding of how the TDM facility could be improved. This necessary fact-finding will help to avoid obstacles and will provide new insights into innovative ideas. Thomas Edison often noted that his success came from studying the successful practices of other operations and then making adaptations. The many new uses for TDM are continually evolving; this evolution relates directly to machine technology. Some of the many available references concerning starting or improving a TDM operation are described in this section.

TDM manufacturing is a process of manufacturing specific end products, especially on a large scale. It is an industry that employs mechanical power and machinery to manufacture products for one or several of the end uses that are now well established for TDM or CRM. A processor incorporates the manufacturing aspect and performs the many functions central to the TDM plant. The scrap tire professional must view himself as a processor and manufacturer, and act accordingly. The principles and guidelines developed in other manufacturing and processing industries can be very applicable to today's TDM operation.

The following sections are intended to be a source of references for making improvements to existing operations or for developing an investigation of new TDM plants. Use of these resources should help prevent major errors or oversights in future developments.

1.4.1 Publications

Trade journals, books, papers, and advertisements are excellent sources of written information. With today's World Wide Web, surfing the Internet is an excellent method to keep abreast of new developments in the scrap-tire-processing field. *Scrap Tire News*, published by Recycling Research Institute,

Leesburg, Virginia, is an example of a monthly trade journal dedicated exclusively to keeping abreast of all new developments in scrap tire processing, equipment, and processors, and also market developments for whole tire, TDM, and CRM. It includes related updates in governmental regulations. The Scrap Tire Management Council, Washington D.C., is another excellent source of up-to-date information and publications. The Tire Industry Association, located in Reston, Virginia, and Louisville, Kentucky, is an active association with worldwide membership. It is dedicated to recovering worn tires and seeing them back to their highest and best use in a safe and economical manner via recapping or other practices such as bead-to-bead restoration. Members of these organizations have a real interest in finding the highest and best use for tires, or disposing of them in the proper manner. On a broader recycling base, *Resource Recycling*, Portland, Oregon, and *Waste News* publications contain articles that affect tires in the total waste-recycling streams.[6]

Mary Sikora of *Scrap Tire News* and her annual publication of *The Scrap Tire & Rubber Users Directory* set the real standard for professionalism and literary accountability. She has published numerous publications, conducted numerous seminars, conducted investigations, and has traveled to nearly every tire processing facility in North America and Europe. She has been instrumental in setting the standard in the scrap tire processing industry. Through her publications, reliable information concerning this industry has been set and achieved. She has met with leaders in governmental agencies, departments of transportation, educational institutions, tire processors and collectors, and end users of TDM products. Her attention to details and getting the real facts is reflected in her reporting and has helped make the scrap tire industry the profession it has become today.

1.4.2 Patents

Many of the scrap tire processes, equipment, products, and specific uses of individual TDM and CRM materials are discussed and illustrated in listings of the U.S. Patent and Trademark Office. Therefore, the investigator should make a careful search of the patent literature for a specific field of interest before considering the adaptation of any process described.

This resource is a wonderful way to learn about technology and an invaluable reference on know-how. An online search of the U.S. Patent and Trademark Office can produce significant information about what has been developed and which patents are pending or have been issued. By going to the U.S. Patent Office website and looking for databases: patents granted and patent application full-text and full-page images, a wealth of information can be obtained. Quick search, advanced search, and patent number search options are offered. The website also includes operational notices, database content, and a help section.

Expired patents are also listed at the government patent website, such as one issued to Robert L. Thelen and Michael W. Rouse for U. S. Patent Office

No. 4,373,573 entitled "Apparatus for Shredding Rubber Tires and Other Wastes" issued in 1993. It granted exclusive right on the optimal utilization of downsizing the blades in a rotary shearer within the same commutating unit. As the replaceable blades wore down in the center section of the shredder and were sharpened, narrowing their overall width on the sides, they could now be outwardly extended to the periphery of the internal shredder chamber on a slightly narrower mounting base and reused a number of times as they moved to the outside of the shredder's chamber. In the past, this downsizing made the blades useless and resulted in added costs to the processor. This reuse concept was used by Waste Recovery, Inc. for many years; with its recent patent expiration, it is now being fully commercialized by a major tire-shredding manufacturer. In many cases, some patents are issued well before the invented concept becomes commercially accepted or available for others to use. In other cases, it might be just a matter of time before the patent will expire, and the idea will become available for everyone to copy and use without worry of legal infringement. In any case, researching the patent archives is an invaluable resource and a smart way to keep up with developments in scrap tire processing in North America and other locations throughout the world.[7]

1.4.3 Other Industries

By studying other industries, such as the waste plastic recovery, steel reclamation, cable and wire recovery, automotive dismantling, agricultural methods, and electronics industries, new ideas and other methodologies may be discovered that could be applicable to the scrap tire processor. Demands for material recovery, source separation, collection, recycling, handling, and transportation are some of the many facets that are common to all recycling operations. Visits to and participation in these organizations can provide great insight to advantages not yet utilized in the scrap tire process.

1.4.4 Consultants, Brokers, and Equipment Manufacturers

Consultants that have been around the industry for many years are generally well known and can be an excellent source of information and assistance. It is strongly suggested that you investigate consultants' backgrounds and references before hiring them to see if they have designed a TDM facility that has performed or is operating successfully. It is recommended that you visit a consultant's clients to get a firsthand opinion of the person or firm's capability.

For equipment manufacturers, the longer the manufacturer has been in business, the more reliable the quality of the information and the system proposed. Many experienced, original equipment manufacturers (OEMs) who have installed equipment have had hands-on experience with their systems or have available testing facilities at their manufacturing sites. Scrap tire operators and manufacturers welcome visitors who want to see a process or piece of equipment at their facility, if they do not feel threatened to disclose trade secrets. Many scrap tire symposiums sponsored by the Tire Industry Association, *Scrap Tire News*, Scrap Tire Management Council, U.S. Environmental Protection Agency, or regional governments sponsor visits to local scrap tire operations. These meetings also have manufacturers' displays and usually have on hand key company representatives to answer inquiries. These meetings or symposia are generally attended by professionals in the field and are an excellent means to become acquainted with who's who in the industry. Another excellent resource is *The Scrap Tire & Rubber Users Directory*, published annually by Recycling Research Institute, Leesburg, VA.[6]

1.5 Size Reduction Relationships

The energy to shred and break down a tire into particles has not had much theoretical study. Most of the work, however, has been in the mineral-processing field. It is generally felt that the energy consumed in reduction of materials is directly proportional to the new surface area produced, or

$$\frac{\text{volume of particle before}}{\text{volume of particle after}}$$

Rittinger developed a breakage law, as follows:

$$E = K\{1/x_1 - 1/x_2\} \tag{1.1}$$

where x_1 is the initial particle size and x_2 is the final particle size.

On the theory of plastic deformation, the work of Kick, the energy required for shredding a given quantity of materials is constant for a given reduction ratio, irrespective the original particle size. Kick's law may be written as follows:

$$E = K \ln (x_1/x_2) \tag{1.2}$$

where x_1/ x_2 is the reduction ratio.

It has been generally observed that for size reduction processes, the size changes produced are proportional to the energy expanded per unit weight of particle, and that the energy required to bring about the same relative size change is inversely proportional to some function of the initial particle size, x. The relationship between energy and shredding may be expressed as follows:

$$dE/dx = -K/x^n \quad (1.3)$$

Both Kick's and Rittinger's Laws can be derived from this equation into the following equation:

$$E = 2K\{1/(x_2)^{0.5} - 1/(x_1)^{0.5}\} \quad (1.4)$$

which corresponds to case n = 1.5. However, n can vary over a wide range to meet a particular size reduction criterion of the material being processed.[8]

The proportionality constant K is called the work index and is a function of particle size and mode of shredding. Hukki proposed a generalized equation for the energy size reduction relationships that is considered to be a good description of the required energy of particle size reduction:

$$dE/dx = -K/x^{f(x)} \quad (1.5)$$

These equations are applicable to bulk shredding and grinding operations. Figure 1.5, depicting the relationship between energy input and particle size in comminution, gives general indication of the relative magnitudes of the energies required for breaking bulky materials down to smaller sizes.[9]

The overall energy relationship can be significantly affected by heat and pressure induced while shredding and chopping. This, in turn, can have a significant, positive effect on the energy requirements in a tire processing system. The use of thermal fracture, or decrepitation, has not received a lot of attention. However, heat application to highly elastic, resilient materials greatly affects the behavior of the energy required. Heat buildup during operation can have a very pronounced effect on production; in some cases, it improves overall grinding efficiency. Too much heat, however, can result in depolymerization (noticeably fumes) or potential fires. In the case of two-roll mill grinding and extrusion comminution, the viscoelastic behavior of

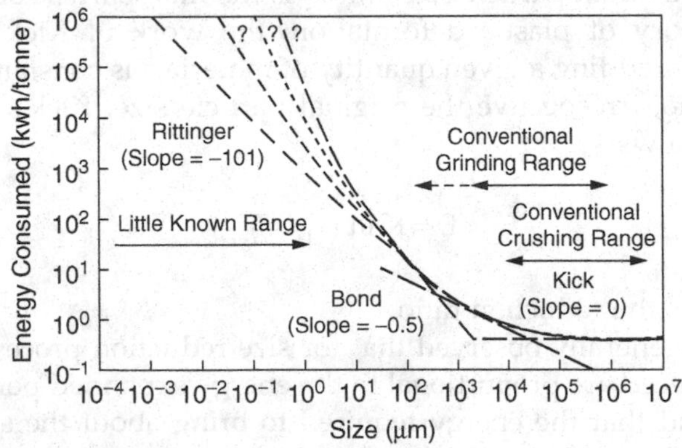

FIGURE 1.5
Typical energy consumption vs. particle reduction.[9]

the tire materials is significantly reduced and particle attraction and abrasion are greatly facilitated as the materials absorb energy in the form of heating up. Temperature monitoring of the mill's performance can be an important aspect to consistent and improved output of these particular units. In the case of granulation, monitoring the grinding chamber temperatures can be very critical in terms of conditions changing within the unit and the unit's performance. Too much heat can result in scorching the CRM, mechanical damage, or even an unexpected combustion of the material being ground.

The introduction of chemicals such as surfactants in the form of a water spray can help particle reduction in much the same way as liquid cutting agent additives are added to water stream while metal is being cut to a specific shape or size by a continuous saw blade.

1.6 Applications

Tires have many properties that can be taken advantage of when the scrap tire is converted into TDM form. Some of the engineering properties of tire rubber are listed in Table 1.2.

In general, the scrap tire processor must have a number of application markets, to survive and be viable. The scrap tire plant can be designed to produce a variety of TDM sizes and varying contents of wire and fiber into final desired products. Many applications do not require overprocessing to satisfy a market need. This means not having to remove all the inherent wire or tire cords if the desired market does not require it. Therefore, overprocessing only adds to the operational costs and minimizes the economic return to the processor. It is suggested that a market approach be the leading theme for any tire-processing scheme. The demise of many crumb rubber facilities has been due to the belief that by simply building a CRM facility, the markets will arise out of nowhere. Be sure to develop and secure the market, and

TABLE 1.2

Engineering Properties of Tire Rubber

Property	Feature
Black	Opaque
Liquid state	Low freezing point
Low density	Specific gravity 1.12 to 1.15
Water resistant	Nonwicking
Low thermal conductivity	Thermal barrier
Low electrical conductivity	Insulator
Absorption	High absorption of most organic liquids
Rheology	Elastic, compliant, and resilient
Enthalpy	High heat of combustion and low ash content
Organic	Nonbiodegradable

FIGURE 1.6
Illustration of a TCF and TDF materials.[1]

follow the U.S. Navy Seals' advice: Identify the target (market), the weapons (equipment), and the response (action steps) to accomplish the mission.[10]

1.6.1 Tire-Derived Materials Applications

Tire-derived fuel can be in the form of tire chips containing the inherent wire. These are called tire-derived chips (TDC); with the removal of 95% of the inherent wire, especially the bead wire, they are called TDF. Figure 1.6 is a picture of a minus 3-inch TDF and TCF produced through a Saturn Granutech Grizzly system. Normally, TCF turns a rust color and does not show a great deal of fiber. As TCF is converted into TDF, the color stays black and more fiber becomes apparent. Applications of scrap tires as a fuel supplement with coal or gas include the production of cement, lime, and steel in their operating kilns. Normally, production of TDF must provide smooth and efficient energy conversion for the purchasing energy user and must have minimal supplemental fuel handling problems, no environmental problems, and no adverse effect on the final material or the operation being produced by the end user.

TDF has been known to reduce sulfur emissions from power plants and improve the overall combustion efficiency. Normally, metering systems are provided to stoichiometrically add the high-value TDF fuel in combination with various coal, wood wastes, or refuse-derived fuels (RDFs) in dedicated combustion furnaces or mass incineration units that convert garbage or household and commercial wastes into energy. Today, facilities that synthesize alcohol from agricultural wastes are considering use of TDF as part of the fuel stream for the distillation process. By researching the U.S. Patent Office archives, you can find U.S. Patent No. 4,806,056, "Fuel Meter-

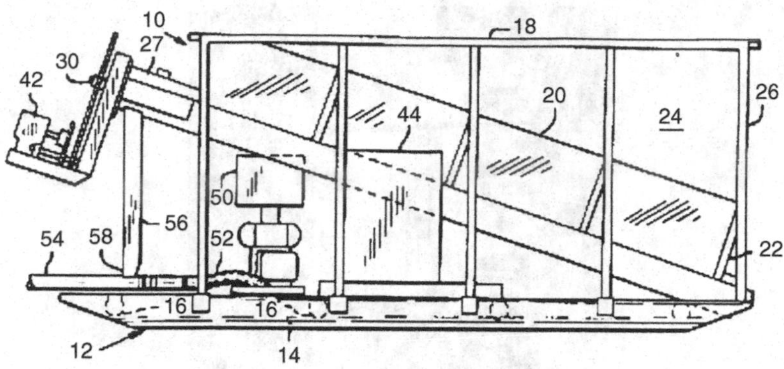

FIGURE 1.7
Modular TDF metering unit.

ing Unit." This patent refers specifically to the design, construction, and operation of a TDF metering unit for the proper addition of tire fuel in combination with coal or wet wood wastes to a combination boiler for process steam generation at a pulp and paper plant. Such a TDF metering unit is illustrated in Figure 1.7, taken from the provided diagrams in the patent literature.

These metering units can also be modified for internal use at the tire processing facility for metering into secondary- or tertiary-stage grinding processes. The construction and use of these components should be investigated for all possible advantages.

A more basic metering system is the energy conversion of whole tires in a cement kiln, such as one developed by TNI-Systems of Dayton, OH (Figure 1.8). Again, the key here is precise control and addition of the tire to the kiln, so proper combustion occurs while producing a high-quality cement product.

TDC has a fuel value equivalent of approximately 14,200 Btu per pound (32,487 kilojoules per kilogram) and about the same heat value for whole tire combustion. A case example is the addition of tires to the cement kiln at Medusa Cements' preheater kiln, located in Clinchfield, GA.

TDF, on the other hand, has an approximate heat value of 15,500 Btu per pound (35,969 kilojoules per kilogram) due to removal of the wire component. The nominal size of TDF is 2 inches (5 centimeters), whereas TDC is usually 4 inches (10 centimeters). The larger the TDF size, the greater the content of wire and the less flowability for handling and metering. The larger sizes take on the configuration of the tire, and the wire reinforcements are harder to deal with. This, in turn, can have a negative effect on energy conversion in the combustion unit (Figure 1.9).

Table 1.3 reflects typical chemical characteristics of a minus 90% wire free TDF compared to an eastern bituminous coal. The comparisons are significant in that TDF has significantly less moisture, lower ash, lower sulfur, and higher volatiles, and less fixed carbon when evaluating the proximate and

FIGURE 1.8
Whole tire metering unit for a cement kiln application.

FIGURE 1.9
Tire fuel control panel with weight indicator at the cement plant.

TABLE 1.3

Comparative Chemical Analyses[11]

Characteristic	Eastern Bituminous Coal	90+% Wire Free TDF
Proximate Analysis (% as received)		
Moisture	7.76	0.62
Ash	11.05	4.78
Volatile	34.05	66.64
Fixed Carbon	47.14	27.96
Total	100.00	100.00
Ultimate Analysis (% as received)		
Carbon	67.69	83.27
Hydrogen	4.59	7.09
Nitrogen	1.13	0.24
Sulfur	2.30	1.83
Ash	11.03	4.78
Moisture	7.76	0.62
Oxygen (By Difference)	5.48	2.17
Total	100.00	100.00

ultimate analysis on an as received fuel basis. These characteristics produce a higher fuel value in favor of TDF.

The ash content of these two fuels is also critical to the use of TDF when used in combination with a coal. Many power plants send their fly and bottom ash to cement plants and other applications. Therefore, this ash by-product can be greatly affected by a change in the final ash composition or the sale of these by-products. Consideration must be given to the ash composition since many users of the ash have stringent restrictions on any change in ash composition or fuel mixture changes. Table 1.4 reflects the elemental metal analysis of an eastern bituminous coal and the same TDF. Also, disposal of ash from a power plant has to meet environmental regulations and the effect of a new alternate or supplemental fuel source must be considered and approved in many cases. Zinc can easily be removed by an electrostatic precipitator or bag house but consideration must be given to its disposal in high concentrations, normally any sanitary landfill. In some cases, the halogens' (such as chlorine and fluorine) effect on the flue gas zones of the combustions unit must be factored in on long-term potential corrosion of the pollution control equipment.[12]

Terry Gray of T.A.G. Resource Recovery of Houston, Texas, contributed Tables 1.3 and 1.4. Mr. Gray is an excellent example of using a knowledgeable professional in making sure that all aspects of tire use are covered before embarking on a venture. He has worked with many institutions and government agencies in defining the environmental and ecological parameters in the handling and proper disposal of tires and tire-derived materials. These

TABLE 1.4

Comparative Metal Analyses[11]

Characteristic	Eastern Bituminous Coal	90+% Wire Free TDF
Aluminum	2.29	<0.01
Barium	—	non detectable
Cadmium	—	0.0006
Calcium	0.36	0.0097
Iron	2.09	0.321
Lead	—	0.0065
Magnesium	0.08	<0.01
Manganese	—	<0.01
Phosphorus	0.07	<0.01
Potassium	0.22	<0.01
Titanium	0.09	<0.01
Silicon	5.30	0.516
Sodium	0.05	<0.01
Strontium	—	<0.01
Zinc	0.01	1.52
Chloride	—	0.149
Fluoride	—	0.001

tables represent general expected average values. However, it is advised that representative samples be obtained from the existing TDM operation and confirmed by a certified testing laboratory.

Whole tires are certainly used in mass incinerators, dedicated combustion units, and cement kilns. However, these applications are limited to certain tire size ranges and generally do not require any special type of preprocessing, only proper metering in combination to fuel requirements.

Recirculating coal fired fluid bed units can be ideal candidates for both TCF and TDF. With TDF, minimal changes are required to the ash handling system. When TCF is used, additional equipment is required to remove the wire, and the wire residue has not found a good market due to contamination of the other combustion by-products. A modern-day fluid bed combustor is illustrated in Figure 1.10.

Typical traveling grate boilers with spreader stockers can be dual fired with supplemental gas or oil in combination with wet hog fuel (bark). The idea of using TDF in combination with the wet wood waste is to reduce the usage of fossil fuels and its associated ever-increasing cost. Traveling grate boilers and their corresponding ash handling systems are sensitive to the inherent bead wires in TDF since they do not oxidize and become entangled in the grate system. Therefore, the TDF supplier must maintain a good quality low percentage of wire content. Figure 1.10 reflects a traveling grate boiler that could also be fired with various grades of stocker coal. Many combinations of industrial and utility and understanding of how the combustion unit operates are critical to successful use of TDF. Pulverized coal boilers (PCB) are very limited as to maximum TDF particle size and amount

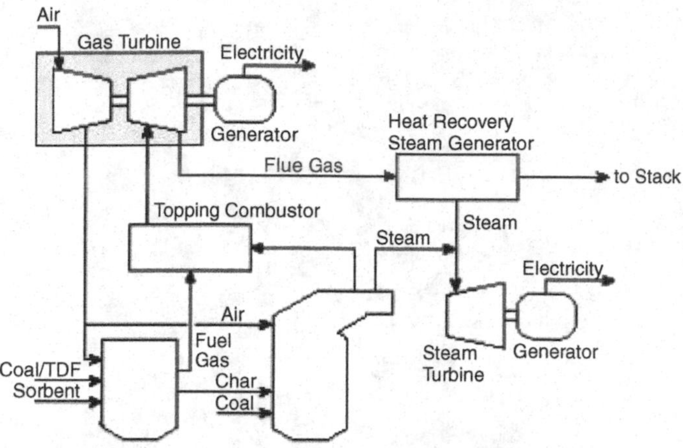

FIGURE 1.10
Modern day fluid bed combustor with TDF or TCF as a supplemental fuel.[1]

of loose fiber in the TDF itself. Generally, 1/2-inch TDF is the maximum size since there are no grates at the bottom of the furnace to burn off the larger TDF chunks. The loose fiber in TDF can carry over into the pollution control system and be another limiting factor as to the quantity used in conjunction with coal. Cyclone boilers can burn slightly larger TDF or minus 3/4-inch and the size is dependent on the size of the boiler cyclones. For traveling grate units, the heat release on the grates is critical. Too high a percentage of minus-2-inch TDF can cause the grates to buckle. Considering good pollution control, 10% by weight is the maximum weight percent for traveling grate and fluid bed boilers and around 3 to 6% by weight addition for suspension type units. Figure 1.11 is a typical industrial traveling grate boiler using multi-fuels.

Fluid bed units can use up to 3-inch nominal TDF or TCF depending on the design and ash handling characteristics of the unit.

TDF or TDC are excellent feedstocks for pyrolysis or gasification units or other forms of thermal mechanical transformation. Generally, reduction in TDMs' nominal size, toward a crumb rubber gradation, is preferable since better thermal and chemical conversion takes place. However, these processes, like other applications, are dependent on the cost of the fuel feedstocks delivered to the site.

1.6.2 Civil Engineering Applications

TDM materials are excellent as lightweight aggregate filler for roadbeds, subgrade thermal insulation, and backfills behind retaining walls. Both TDF and TDC materials can be used in these applications. The volume use of TDM in these applications is enormous. The scrap tire processor can generally find these markets in the community where the operation exists. These are

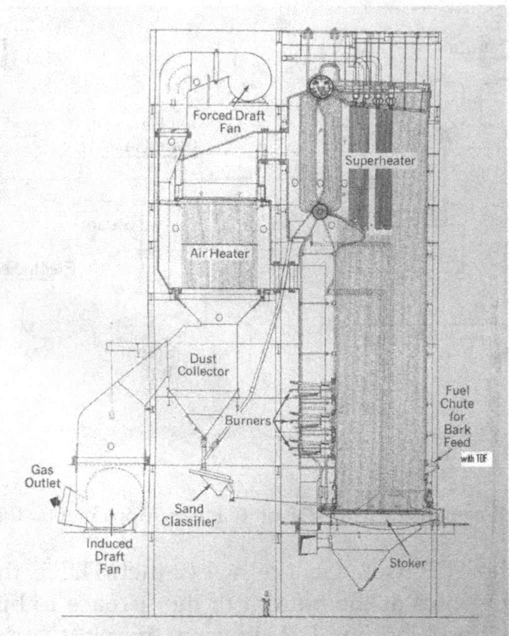

FIGURE 1.11
Typical traveling grate-spreader stoker dual fuel fired hob fuel boiler with TDF blend.[13]

excellent opportunities for new and rehabilitation road and highway projects. A great deal of information exists concerning TDM use for these applications. The processor should contact the local city or county highway/road maintenance department, the state Department of Transportation (DOT), or the regional Federal Highway Administration office. Many state universities work with state DOTs in project developments such as these for enhancing highway and road life cycles. In many cases, there may be grant money for joint development projects of this type in the processor's community.

Figure 1.12 reflects the bundling of truck tires for a crash barrier. Similar bundling can be used for fish reefs.

Figure 1.13 reflects the sizing of tire chips coming across a disk screen coming from a tire shredder for a roadbed, and Figure 1.14 shows an embankment civil engineering project and actual spreading of tire chips for a roadbed application.

1.6.3 Sanitation Applications

There is a growing need to find inexpensive and reliable replacement for wood chips in public sanitation departments. Wood chips are used as bulking agents, but the wood easily decomposes and must be replaced. TDMs can be an excellent substitute because they do not deteriorate and can therefore

FIGURE 1.12
Truck tires used for crash barriers.[1]

FIGURE 1.13
Tire chip production across a disk screen.[1]

FIGURE 1.14
Tire chips for a roadbed application.[1]

improve the life cycle in this application. Also, TDMs can be used for drainage for sanitation fields. Working with local and state governments and home builders can open up markets for TDM end use.

1.6.4 Agricultural, Playground, and Decoration Applications

Refining of TDM for these applications is paramount. The inherent wire, both bead and radial, must be removed for safety reasons. The gradation is also critical. If the TDM is ground too small, it will easily scatter and will blow away, or it may stick to other surfaces such as shoes or clothes. Normally, bias truck, passenger, or OTR tires with the beads removed are excellent feedstock for this application. If radial tires are used, a TDF material must first be produced. Then the TDM chip must be exposed to a wet atmosphere for a monitored time period, to reduce or dissolve the remaining fine radial wires to a powdery form by oxidation. The tire chips for these applications can also be colored for enhanced aesthetics and décor, for equestrian or landscaping applications. Figure 1.15, showing a TDM playground cover, illustrates the beauty and functionality of tire materials for use in gardening or for playgrounds.[6]

TDF has been found to be an excellent soil amendment for certain types of soils. The pH of the soil — whether it tends to be basic, neutral, or acidic — should be checked. Since tires contain compounding agents, accelerators, and other additives, these items can leach out into the soil and over time, can cause adverse effects to vegetation coming in contact with the soil. Generally, long-term studies need to be conducted on the particular soils in question to minimize any product liability risks when using TDF as a soil amendment for food production. In a number of states, the local university or college's agricultural department has conducted soil studies of various

FIGURE 1.15
Playground material produced by Global Tire Recycling. (Courtesy of Global Tire Recycling, Wildwood, FL.[14])

soil amendments. These studies can be an excellent resource. Processors should check with their insurance carriers about product liability coverage in this and other applications where any type of TCM or CRM is being marketed or sold. Working with the local county extension agent can be a way to determine the appropriate soil amendment for the location in mind.

The market for colored TDM is growing for landscaping and playgrounds. This technology is now fairly well understood and established and has excellent market appeal. The prices for both colored and non-coated TDM can be equal to or greater than many of the coarse CRM being produced. These applications do require that the TDM be completely wire free and generally are in the nominal to minus 1/2-inch size and produced from bias truck tires and OTR types such as farm implement tires.

In many cases, you will find tires used for holding down traps in agricultural applications, divides for property lines, or marking entrances to a unique display as shown in Figure 1.16.

1.7 Testing and Quality Assurance

The use of crumb rubber or refined TDC or TDF has the greatest potential for achieving the highest value for tire-derived materials. Generally, CRM maximum rubber size is 6 millimeters, but at that size, it is generally called tire granules and not CRM. Gradations in the 1 to 6 millimeter range are used for running tracks and cushions below the carpet surface. Below 2

FIGURE 1.16
Tire shop decoration in New Zealand.[1]

millimeter or 2000 microns, or 10 mesh in U.S. standard measurements, is the starting point for CRM. Below 80 mesh (180 microns) is considered to be fine CRM powder. The finer the TDMs are ground, the more costly they are, so finely ground TDM requires greater capital resources.

Relative particle size comparisons are illustrated in Figure 1.17, and Table 1.5 is a sieve comparison of various major categories of various TDM materials.

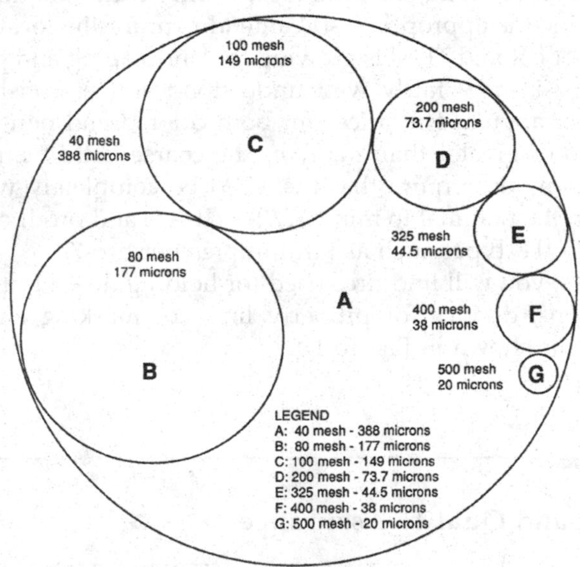

FIGURE 1.17
Relative particle size comparison.[1]

TABLE 1.5

Sieve Comparison Table[1]

Major Grade Size Categories	Designation Size (inch) or Sieve (Mesh)	mm (millimeters)	Microns (1000 × mm)	Inches
Fuel Grade Sizes	2″	50.000	50,000	2.000
	1-1/2″	37.500	37,500	1.500
	1″	25.000	25,000	1.000
	3/4″	19.000	19,000	0.750
	5/8″	16.000	16,000	0.625
	1.2″	12.500	12,500	0.500
	7/16″	10.938	10,938	0.438
	3/8″	9.520	9,520	0.375
	5/16″	7.930	7,930	0.313
	1/4″ or 3	6.350	6,350	0.250
Coarse Grade Sizes	3.5	5.660	5,660	0.223
	4	4.760	4,760	0.187
	5	4.000	4,000	0.157
	6	3.360	3,360	0.132
	7	2.830	2,830	0.111
	8	2.380	2,380	0.094
	10	2.000	2,000	0.079
	11	1.840	1,840	0.072
	12	1.680	1,680	0.066
	14	1.410	1,410	0.056
	16	1.190	1,190	0.047
	18	1.000	1,000	0.039
Fine Grade Sizes	20	0.840	840	0.0331
	24	0.736	736	0.0290
	30	0.590	590	0.0232
	35	0.500	500	0.0197
	40	0.420	420	0.0165
	45	0.350	350	0.0138
	50	0.297	297	0.0117
	60	0.250	250	0.0098
	70	0.210	210	0.0083
	80	0.177	177	0.0070
UltraFine™ Grade Sizes	100	0.149	149	0.0059
	120	0.125	125	0.0049
	140	0.105	105	0.0041
	150	0.099	99	0.0039
	165	0.091	91	0.0036
	170	0.088	88	0.0035
	200	0.074	74	0.0029
	230	0.062	62	0.0024
	250	0.058	58	0.0023
	270	0.053	53	0.0021
	325	0.044	44	0.0017
	400	0.037	37	0.0015

FIGURE 1.18
Testing sieves.[1]

The standard for measuring CRM sizes is the U.S. Standard Tyler screen. Mesh is defined as the number of linear openings per inch (the greater the number of openings across the surface of the screen, the finer the mesh). Figure 1.18 shows a Standard W.T. Taylor & Company Ro-Tap setup, ready for testing. ASTM procedures for testing of CRM materials have been established. The Ro-Tap Method of ASTM D 5603 calls for placing a fixed amount of CRM on the top screen, using 8-inch screens as illustrated in Figure 1.18, and running the shaker shown in Figure 1.19 for a fixed period of time, with the largest opening screen on top progressing to finer screens down to the pan. After shaking, each pan is weighed and recorded. The top screen is called the zero screen and the second from the top is called

FIGURE 1.19
Testing unit.

the designation screen. These two screens presently establish the CRM designation.[14] Figure 1.20 represents an illustrative glossary of millimeter to mesh.

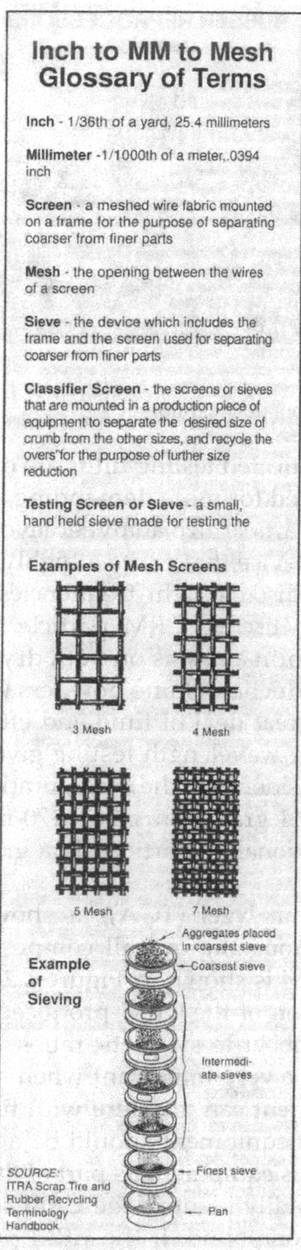

Inch to MM to Mesh
Glossary of Terms
Inch - 1/36th of a yard, 25.4 millimeters
Millimeter -1/1000th of a meter,.0394 inch
Screen - a meshed wire fabric mounted on a frame for the purpose of separating coarser from finer parts
Mesh - the opening between the wires of a screen
Sieve - the device which includes the frame and the screen used for separating coarser from finer parts
Classifier Screen - the screens or sieves that are mounted in a production piece of equipment to separate the desired size of crumb from the other sizes, and recycle the overs"for the purpose of further size reduction
Testing Screen or Sieve - a small, hand held sieve made for testing the
Examples of Mesh Screens
3 Mesh
4 Mesh
5 Mesh
7 Mesh
Aggregates placed in coarsest sieve
Example of Sieving
Coarsest sieve
Intermediate sieves
Finest sieve
Pan
SOURCE: ITRA Scrap Tire and Rubber Recycling Terminology Handbook

FIGURE 1.20
Detailed illustration of mesh definition.[6]

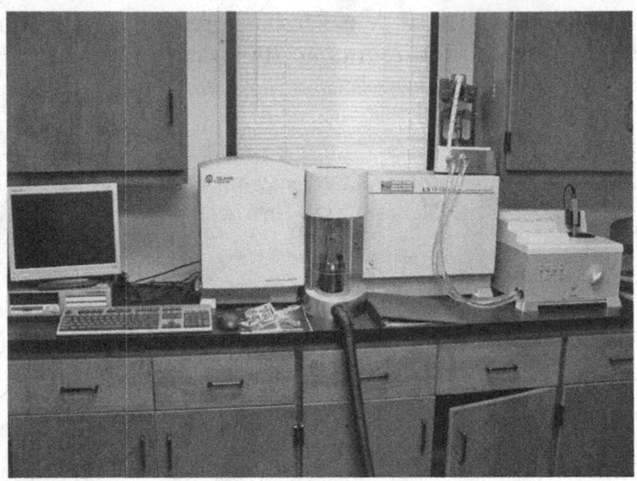

FIGURE 1.21
Laser diffraction size analyzer laboratory set-up.

Today, even more sophisticated testing units are on the market as the finer CRM materials are produced for more demanding applications. Figure 1.21 shows a laser diffraction particle size analyzer layout in Rouse Polymeric's laboratory. The laser is a device that emits a highly amplified and coherent radiation of one or more discrete light frequencies. These frequencies can be used to evaluate each discrete CRM particle. This particular unit by Beckman Coulter can perform analysis on both dry and wet CRM samples. Why wet CRM? In the production of fine powders with wet technology, wet particle analysis saves a great deal of time and effort since the CRM does not have to be dried. Laser wavelength testing gives very accurate particle analysis and takes into consideration the configuration of each CRM particle. Figure 1.22, showing CRM gradation of a –170-mesh CRM produced at Rouse Polymerics International, is particle data graph output from such a testing machine.[1]

The thermogravimetric analyzer (TGA), as shown in Figure 1.23, is an excellent tool for determining the overall composition of the CRM. TGA graph of a –170-mesh CRM is shown in Figure 1.24. The TGA can be used to tell the basic composition of the CRM produced and can reflect the tire type mix for the natural rubber-to-synthetic ratios as well as the overall ash content. Ash content can be very important when selling CRM to tire manufacturers.[15] High ash content can interfere with tire compounding and be a negative influence. TGA equipment should be a standard for most CRM users, who should have this equipment as part of their QA/QC. This equipment can quantify and readily identify the CRM type as being whole tire, truck/bus, or OTR-type compound. If the CRM processor has any doubts about what is being shipped, the TGA is an excellent toll for guaranteeing quality of the produced product. Remember, most tire manufacturers have

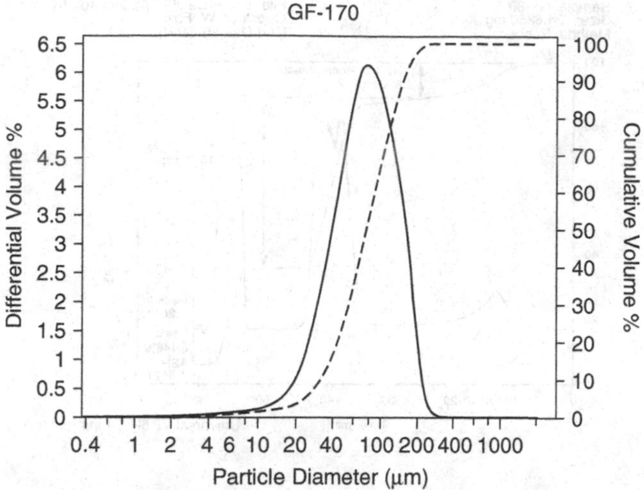

FIGURE 1.22
Laser particle distribution of a −170-mesh CRM.

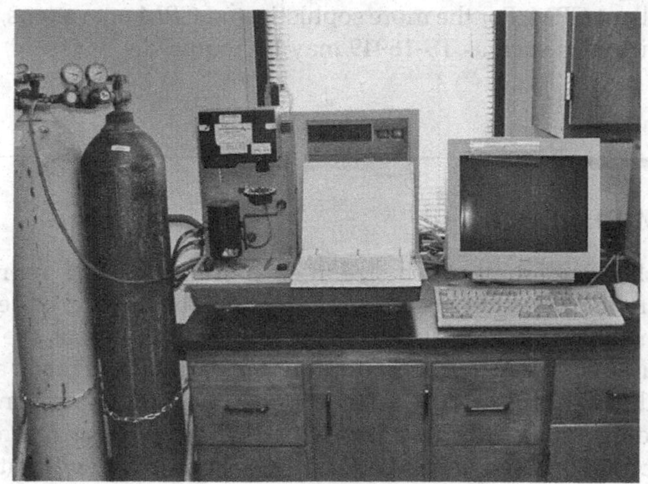

FIGURE 1.23
Thermogravimetric analyzer (TGA) laboratory set-up.[1]

a TGA as part of their laboratory apparatus and can easily quantify the CRM source type.

Moisture testing for all CRM products can be very critical. Standard laboratory ovens and moisture testing machines are available in today's market. There are a number of laboratory supply houses, such as Cole-Parmer, with complete lines of testing supplies from laboratory wares to microscopes.

Testing is vital to the TDM or CRM operation. Testing equipment can also provide a check on the performance of the plant's operation and assure the

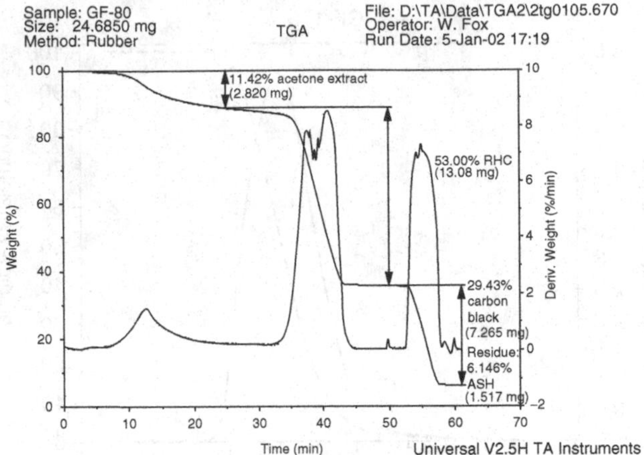

FIGURE 1.24
TGA graph of a –170-mesh CRM.[1]

end user of the CRM or TDM that the plant's personnel has taken all the necessary steps to meet the materials' specification requirements for producing an excellent CRM. For the more sophisticated CRM operations, obtaining other certifications such as TS-16949 may be required.

1.8 CRM Plant Demand

Table 1.6, showing crumb mesh size in each crumb market, represents the market share for different sizes of crumb rubber. The method of determining the crumb rubber range, or classification into four groups, is the subject of much discussion. The ASTM procedures noted earlier have made an attempt to standardize CRM. With the advent of new testing equipment, such as laser technology, standard CRM designations may give way to precise measures and be tailor fitted to the end user's specifications without regard to the ASTM method, but rather, to the exact gradation produced by the CRM producer. The required data may include the aspect ratios of average particle length to diameter of the CRM particles. In some cases, the end user may

TABLE 1.6

Estimated Crumb Rubber Mesh Size and Markets[16]

Group	Size Range	Demand
Large or coarse	(3/8 inch to 1/4 inch)	14%
Mid-range	10- to 30-mesh	52%
Fine	40- to 80-mesh	22%
Superfine	100- to 200-mesh	12%

TABLE 1.7

CRM Designation[17]

Nominal Product Designation	Example ASTM D 5603 Designation	Zero Screen mm	Percent Retained on Zero Screen	Designation Screen mm	Maximum % Retained on Designation Screen
10 Mesh	Class 10-X	2360 (8 Mesh)	0	2000 (10 Mesh)	5
20 Mesh	Class 20-X	1180 (16 Mesh)	0	850 (20 Mesh)	5
30 Mesh	Class 30-X	850 (20 Mesh)	0	600 (30 Mesh)	10
40 Mesh	Class 40-X	600 (30 Mesh)	0	425 (40 Mesh)	10
50 Mesh	Class 50-X	425 (40 Mesh)	0	30 (50 Mesh)	10
60 Mesh	Class 60-X	300 (50 Mesh)	0	250 (60 Mesh)	10
70 Mesh	Class 70-X	259 (60 Mesh)	0	212 (70 Mesh)	10
80 Mesh	Class 80-X	250 (60 Mesh)	0	180 (80 Mesh)	10
100 Mesh	Class 100-X	180 (80 Mesh)	0	150 (100 Mesh)	10
120 Mesh	Class 120-X	150 (100 Mesh)	0	128 (120 Mesh)	15
140 Mesh	Class 140-X	128 (120 Mesh)	0	106 (140 Mesh)	15
170 Mesh	Class 170-X	106 (140 Mesh)	0	90 (170 Mesh)	15
200 Mesh	Class 200-X	90 (170 Mesh)	0	75 (200 Mesh)	15

require the CRM producer to use the same test equipment for identical verification. Table 1.6 represents the generally accepted crumb rubber range and the estimated market distribution. Table 1.7 reflects the CRM designation. A value analysis is discussed in the Economics section.

Applications for CRM in any gradation are expanding, from use as an asphalt modifier or aggregate replacement use as an extender or filler in new tire construction. CRM applications are in a wide range of markets. CRM is an excellent material for rubber mats in commercial use, industrial use, and agricultural use. It is also used as building blocks for enhancing the performance of plastics in under-the-hood automobile components. Certain gradations of CRM are used in paints for texturizing and in latex for providing a frictional surface on boat decks. Blending CRM with asphalts can produce sprayable membranes for pond liners and seal coats for structure protection. These are also called surface-applied membranes (SAMs). An understanding of the potential market will reflect the amount of refinement and control at the scrap tire processor's site. The use of TGA and other analytical instrument data can help establish a high degree of confidence or rapport between the CRM supplier and the end user. One end user could be an asphalt producer who requires very specific and consistent CRM for modified, blended asphaltic product that has long-term life guarantees.[17,18]

1.9 Support Equipment

The support equipment requirements of a scrap tire processing facility really reflect the qualities or traits that make up the mental and emotional resources

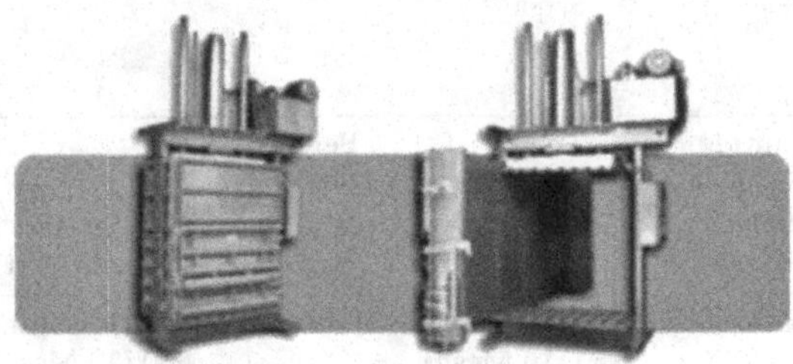

FIGURE 1.25
Typical baler.[18]

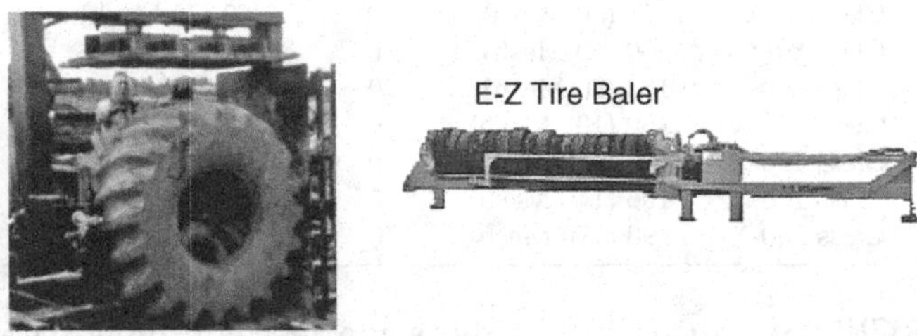

FIGURE 1.26
Tire balers with OTR tire and passenger tires.[19,20]

of the operation. The equipment components discussed in this section may be necessary in part or whole to the success to the entire operation. The equipment components discussed herein may be individual items or multiples such as magnets, conveyors, and shredders or yard equipment for bulk materials handling.

1.9.1 Tire Compaction Units

The use of modern balers can be considered for TDM production for the right applications. If the baled tires are used for crash barriers or for fish habitats, one could make the argument that they are also TDMs. Tire baling can have a significant impact on materials handling in transporting scrap tires to the processing site. In many cases, the worn tires take up a great deal of space and loads end up less than what could actually be transported. Tire compactors can be used to optimize hauling costs and may be justified in freight savings. Special baler-like units are available for compacting used tires for transport in trailers or containers without damaging the casing.

Figure 1.25 and Figure 1.26 show, respectively, a typical baler and a tire baler with an OTR tire. These units can come in all forms and sizes and have self-tying capability with portability.

Figure 1.27 shows a typical bale of truck bead wire. These bead wire materials, when cleanly stripped from the tire carcass and dandified, can have a fairly good scrap iron value. Working with local metal scrap yards

FIGURE 1.27
Typical truck tire bead bale.[1]

in a fairly large geographical region can justify reclaiming this tire material resource.

1.9.2 Packaging

There are many types of packaging available today for the CRM operator. The most common is the bulk bag, illustrated in Figure 1.28. These bags can be thrown away or returned, and have an almost unlimited variety of designs. They come in various sizes, with various mechanical hook-ups to bottom outlets, and are made of various materials. The exact construction of these bags varies according to need.

The plastic bag is normally a low-density polyethylene (LDPE) or a Kraft paper bag type. These two types make up the majority of CRM packaging in today's marketplace. Figure 1.29 shows typical 50-pound (22.5-kg) CRM packaging. Weights can vary from 12.5 to 50 pounds and can also be purchased in different melt ranges, depending on the customer requirements. A packaging specialist should be contacted for a particular requirement.[21]

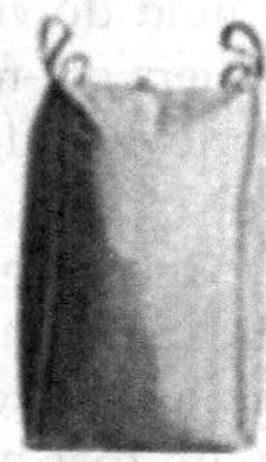

FIGURE 1.28
Super sack.

FIGURE 1.29
LDPE or Kraft bag.

TABLE 1.8

Typical Packing Label Information[1]

Company name and logo
Plant and/or corporate contact information
Health and safety information ratings
Health
Flammability
Reactivity
Personal protection
Notes: Wear dust mask to prevent lung irritation
Patents pending or issue or licensing certifications

There are a number of products of manual and automated packaging machines available in the market. Additional equipment such as a self-sealer is available for the open-top LDPE bags. Valve bags can be designed for specific customer needs. Labeling machines are also available, to stamp the correct product code on the bags. The bag supplier can provide labeling in advance for the CRM company's name, address, and other key information, as noted in Table 1.8.

1.9.3 Transportation

Many deliveries of TDM, such as TDF and TCM or coarse grades of TDM for animal arenas, can be delivered in walking floor trailers. These trailers can have chain floor drives, belts, bottom dump doors, or hydraulically driven walking floor bottoms as in the Hallco Live Floor trailer units. A normal trailer van can be equipped with these floors; a trailer equipped in this manner would be an excellent way to haul whole tires or could be used for metering systems. A typical live bottom-walking floor is shown in Figure 1.30.

Transportation should be an important consideration for TDMs since the efficient handling of whole tires or processed tires is essential to a smooth operation. With escalating diesel fuel prices, care must be taken in assuring the end market user of a fixed long-term delivered price for TDM materials. An understanding and a clause in the purchase orders for the TDM must provide flexibility for the changing fuels. Also maximizing the cubic volume of delivered TDM and lightweight trailers will help hold down costs. The pulp and paper industry found that walking floor trailers offered a unique way to use existing enclosed, open-top trailers to move pulp chips. By installing hydraulically moving slats on the bed floor of a van trailer, the need for special dumps or front-end loaders to remove the pulp chips (or in this case, whole tires or TDM) was eliminated. In some cases, these walking floors are used as metering units for a TDM operation or for adding TDF or TCF (tire chip fuel normally contains the inherent wire that has not been removed) as a supplemental fuel to coal or wet wood wastes. Since TDF has an approximate bulk density of 1,000 pounds per cubic yard (1,683 kg/m^3),

FIGURE 1.30
Illustration of Hallco self-unloading floor system.[22]

the full depth loading of most bulk hauling trailers can be met or the maximum weights achieved for transportation on the roadway to the TDM client.

Figure 1.31 is a common pneumatic bulk trailer for handling a wide range of flowable materials. Generally, CRM less than 10-mesh can be transported by this method. Care needs to be taken to find bulk haulers that have high cube or large volume tanks so maximum weight can be achieved. The use of these trailers requires special receiving containers and, generally due to limited storage capability of the end user for bulk delivered CRM, just-in-

FIGURE 1.31
Pneumatic bulk trailer.[1]

FIGURE 1.32
Passenger tires laced into a standard enclosed trailer.[1]

time scheduling needs to be coordinated. Transportation costs are higher with these units but packaging requirements are minimized.

Overseas containers also offer effective storage for recoverable casings at a tire-processing site. These containers can be stacked and easily moved around the property, and they keep the tires out of sight and protected from the elements. Also TDM itself can be shipped in these containers. Some overseas shippers require a plastic bag lining in the container unit. Normally, maximum weight can be achieved with TDM due to bulk density of the TDM itself.

Whole tire hauling depends heavily on the proper lacing of the tire into the trailer. Figure 1.32 is a trailer of passenger tires laced. The more uniform the tire type the better the lacing and greater the volume of tires that can be loaded into the trailer. Normally, 1,100 to 1,500 passenger tires can be stored in one trailer. Some tire collectors use a tire compression system that greatly increases the volume of tires per trailer.

1.9.4 Magnets

One of the greatest concerns for all CRM products is the removal of inherent steel reinforcing cords, either radial or bead wires, in the tire carcass. Magnets are key to removing this material after initial shredding or processing, depending on the equipment used. When an object is surrounded by a magnetic field and has magnetic properties, either natural or induced, it attracts iron or steel. Nonferrous materials such as aluminum and copper cannot be removed by magnets; therefore, linear induction motors, air tables, or other separation devices need to be used. However, tires do not contain these contaminants unless they have been contaminated at their original source. Magnets can be self-cleaning units, as illustrated in Figure 1.33, and can be permanent or electromagnetic.

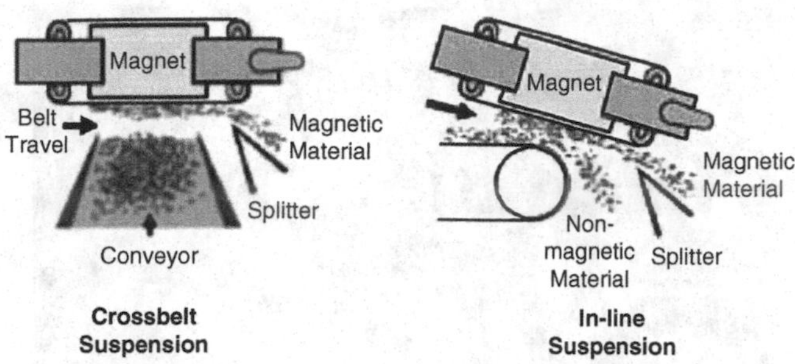

FIGURE 1.33
Permanent and electromagnetic separation setups.[23]

FIGURE 1.34
Action Equipment stainless steel conveyor and magnet system for saleable tire wire recovery.[24]

Figure 1.35 is typical of the wire/rubber by-product after magnets capture it. Some scrap metal processors are willing to take this wire/rubber mix, but in general they want this material to be further refined and a clean wire stream obtained.

The processor may have many variations of magnets to remove the liberated wire after leaving the second or third stage grinder, which depends on the application and location in the process itself. Normally the final rubber wire blend needs to be refined further to recover all the loose metal wires. Action Equipment of Newburg, Oregon, offers a vibrating, abrasive resistant conveyor system with overhead magnets in line to clean up this rubber/wire blend stream. Other manufactures such as Bi-Metals of Ridgefield, Connecticut, and Sweed Machinery Inc., Gold Beach, Oregon, offer alternate systems. The Sweed machine is illustrated in Figure 1.36 for cleaning up a wire/rubber mix with a magnetic separator. Greater recovery of all in-plant TDF plant streams will result in improved economics and min-

FIGURE 1.35
Wire/rubber mix captured by an overhead self-cleaning magnet without wire recovery optimization.[1]

FIGURE 1.36
Sweed Machinery wire rubber separator followed by self-cleaning magnet.[25]

imum disposal from the operation other than normal debris such as rocks, paper, and other contaminants.

Rare Earth magnets (very high magnetic fields) are ideal for removal of fine metal contaminants from very fine CRM powders and are available from a number of major magnet manufacturers.

Table 1.9 represents a typical analysis of tire wire obtained from a TDF operation.[1]

1.9.5 Magnetic Detectors

These units can be very helpful in preventing major damage to a particular shredder, especially if it is a high-speed impact unit, such as a granulator

TABLE 1.9
Typical Chemical Tire Wire Analysis (ASTM E415, ASTM E1024)

Element	% by Weight
Carbon	0.90
Silicon	0.22
Manganese	0.60
Sulfur	0.015
Phosphorus	0.011
Copper	0.079
Nickel	0.032
Chromium	0.053
Titanium	0.01*
Vanadium	<0.01
Lead	<0.01
Niobium	<0.01

or high-impact mill. The sensitivity of these units can be set to bias out the wire in the tire shreds and the belt fasteners. Once the unit detects the particular sensitivity, an alarm can sound and shut off the belt. Generally, personal inspection is required to remove the contaminated object and that causes lost down time. However, the prevention of costly equipment damage may justify the purchase of magnetic detectors and offset the expense of down time.

1.9.6 Pollution Control, Pneumatic Conveying, and Conventional Conveyors

Conveyors are the standard and can be either a belt or screw type. Elevations can vary from horizontal to the vertical position. A typical portable conveyor is illustrated in Figure 1.37.[25] The angle of repose is very important for TDF, TCF, or tire parts since they can tend to roll back down the conveyor. Cleats or some other form of belt conveyor design needs to be considered, along with belt cleaners and wire becoming embedded into the belt itself. For screw conveyors, it is critical to make sure abrasive wear is considered and screws are properly sized for the particular TDM application. However, in most CRM operations, pneumatic conveying of these materials is used; if this is the case, it is important to control any particulate emissions. These air collection units come under environmental regulations and therefore must be considered in the operation of any TDM facility. Figures 1.38 and 1.39 are typical of a bag house and a cyclone operation.[26] Working closely with a particular vendor that supplies these units is important to ensure that proper conveying and capture occur. There are many types of pneumatic conveying systems, and it is considered an art form for a given application. Therefore, proper selection of a conveyor for a particular application is critical, to ensure compliance is met while achieving the process parameters of the plant's

FIGURE 1.37
Typical portable conveyor.[26]

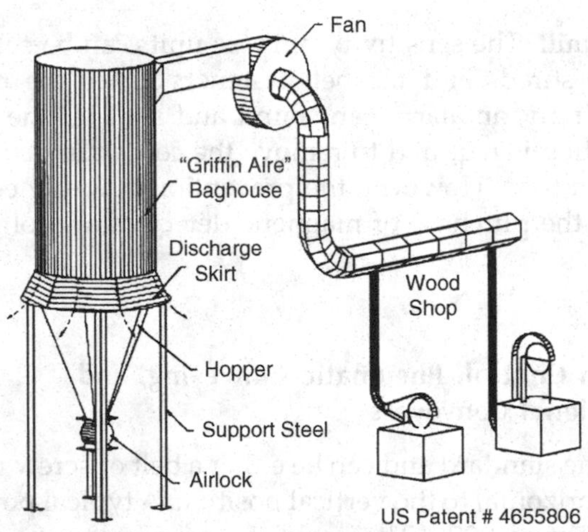

FIGURE 1.38
Typical cyclone operation.

operation. Many of today's high-speed grinders come equipped with some sort of air collection equipment for removing the fibrous and fines generated during grinding. These air systems also remove some of the stringy core fibers liberated during the commutation process.

New types of conveyors are now available that have internal slats on the bottom, which act as a screening device while they move TDM materials to the next destination. These units save space and improve efficiency. Figure 1.40 shows such a conveyor system, manufactured by Action Conveyors.[23] Proper selection of a belt, screw, vibrating, or pneumatic conveyor will depend on the operation of the particular materials handling requirement. Generally, these conveyors are ideal for producing minus 2- to 6-inch TCF. By sizing tire materials to a uniform size before going to a high-speed

FIGURE 1.39
Typical bag house installation.[27]

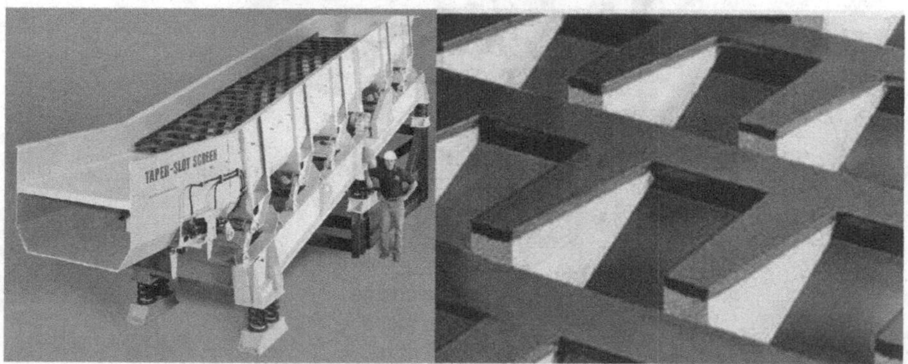

FIGURE 1.40
Vibrating conveyor with slots for screening.[24]

granulator or other type of reduction unit, the more uniform tire chip greatly facilitates the operation of the high- speed machine. In the case of bag houses, cyclones, and air ducts at the facility, it is strongly recommended that sprinkler systems be installed in the internal air-conveying duct and in the collection units.

1.9.7 Gravity Tables and Screening

The use of gravity tables is a unique method to remove the fiber and other inert contents from CRM material. The inerts in CRM can consist of sand

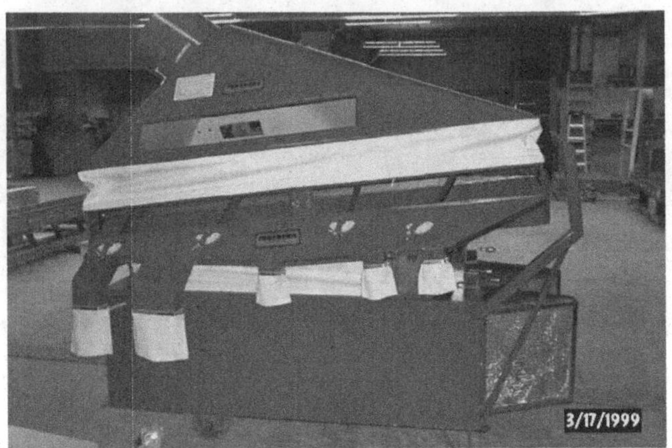

FIGURE 1.41
Actual air table.

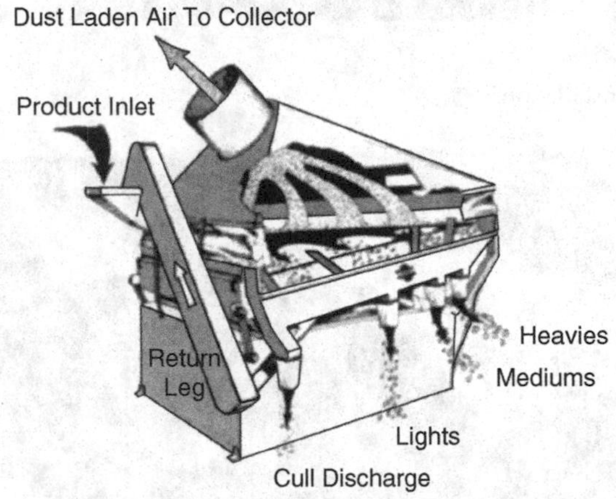

FIGURE 1.42
Cutaway schematic of air table.

and nonferrous materials and can have a negative impact on the final produced CRM. The Forsberg Vacuum Gravity Separators are shown in Figure 1.41, which illustrates the operating unit, and in Figure 1.42, which is a cutaway of the same operating unit. This unit consists of three essential elements, as noted in Table 1.10.[28]

TABLE 1.10
Key Elements for Gravity Table Operation[28]

• Air flow
• Speed vibration
• Deck differential action

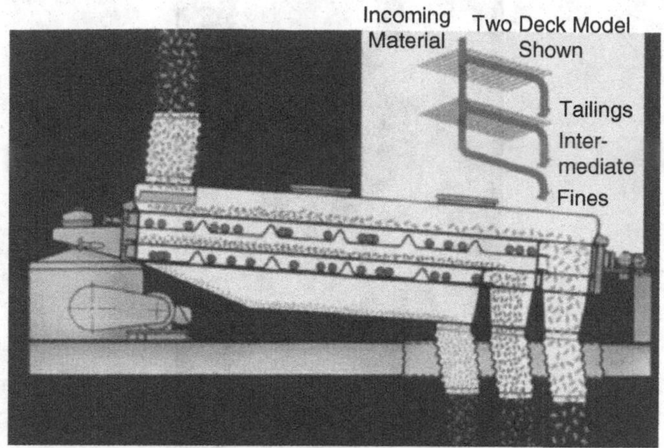

FIGURE 1.43
Operational schematic of an air table.[29]

A vacuum gravity air system assures a complete and accurate product density classification, from the very lightest to the very heaviest. In some applications, such a unit can be used as a screening device. These units are operated in conjunction with bag houses or cyclones.

Figure 1.43 is a cross-sectional view of a Rotex screen. These types of screens are the workhorses of the CRM industry. Multiple screens can be stacked to remove the coarse fraction and also to obtain a finer fraction. Balls can be added below the screen and contained in sections, for keeping the screen clean and greatly facilitating the separation efficiency. Rotary sifters and air classifiers can also be used to produce the various CRM gradations.

Another workhorse of the CRM processing industry is the round screen separator. The separators come in various diameters to meet various production flows; they are operated in the continuous mode, both dry and wet. They are very easy to clean and multiple decks can be used. They can be equipped with balls to be self-cleaning, and these units can be easily moved to various locations. This type of unit is shown in Figure 1.44.

FIGURE 1.44
Sweco round separator.[30]

FIGURE 1.45
Early debeader unit.[31]

1.9.8 Debeaders

Debeaders have played a very historical role in scrap tire recovery. Figure 1.45 illustrates a debeader operating in the 1920s. As in the past, today's beads, especially in truck/bus and OTR tires, possess a great challenge to the TDM operator. Figure 1.46 reflects the progress made in developing sturdy machines to handle the challenge of the many beads that can be found in these tires. Removal of beads greatly improves the shredding of tires, especially truck/bus and OTR tires, regardless of the shredder design. Extracted beads can be used in roadbeds or hangers, and for a number of other applications. Many models are available to pull the bead from the carcass or to cut the bead wire from the tire. Many of the cut beads are used for the bases of traffic diverters.

1.9.9 Fire Suppressant Systems, Safety, and Storm Water Containment

Since many tire fires have occurred, a fire suppressant and prevention system should be a critical part of any TDM or CRM operation. No single operation can have enough of this equipment on hand to ensure the prevention of a fire. Working with vendors, consultants, and local and state fire department agencies is a must. The need for water towers, sprinkler systems, fire extinguishers, conveniently located fire hydrants, and adequate hoses and nozzles cannot be understated. Every operation should have regular fire drills and

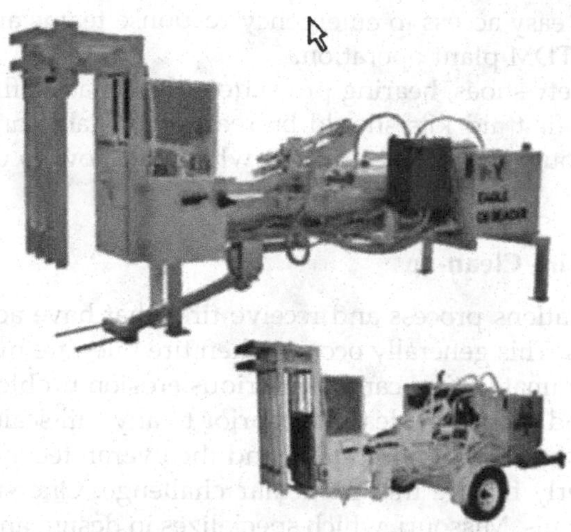

FIGURE 1.46
Modern-day truck and OTR tire debeader.[32]

work closely with the local fire departments in drills at the facility. Fire suppressant and prevention devices should be conveniently displayed and their location should be understood by the operators or other involved parties. In many cases, the fire suppression systems can interlock with the local fire department, so if an incident occurs, the fire department can respond immediately.[33]

Air Products and Chemicals has available an on-site nitrogen gas generator to supply this gas to key components of the CRM process. This nitrogen-rich, gas-generating unit helps prevent static electricity buildup and suppresses sparks or any other type of potential combustion in the screening, conveying, or drying units. Monitors with this system maintain the oxygen content at a minimum level, say 8%, to ensure no negative conditions occur. There are a number of fire suppressant products and systems on the market, and these can be found in *The Scrap Tire and Rubber Users Directory*.[6]

In many cases, discarded tires from site cleanups can introduce a host of other contaminants to the site. Contaminants flowing from these tires during wash down, rains, or some other occurrence can be contained in holding ponds and this water could also be used for fire prevention. With a little imagination, these site holding ponds could be lined with asphalt rubber membranes (SAMs) and provide many years of life over other conventional pond liner systems.

Safety equipment has to be the number one priority. Proper protection and easy access to safety supplies and equipment are very important. Safety class training should be required of all key employees in case of an accident or other incident not related to the plant's operation. An employee may have a respiratory or heart problem. Training in how to respond to an incident of this kind could save a life. Safety training and certification, proper safety equipment, and easy access to emergency response teams are key elements in any CRM or TDM plant operation.

Hard hats, safety shoes, hearing protection, dust masks, first aid training, a stretcher, and first aid kits should be readily available and everyone on the premises should be trained to know when and how to use these items.

1.9.10 Dirty Tire Clean-up

Many TDF operations process and receive tires that have accumulated dirt and other debris. This generally occurs when tire piles are mined or cleaned up. These contaminated tires can pose serious erosion problems to the scrap tire processor and should be dealt with prior to any full-scale production. A number of companies offer equipment and the overall technology and techniques to properly handle this particular challenge. One such firm is Tri-Rinse, Inc., St. Louis, Missouri, which specializes in design and also develops the equipment to shred and rinse tire materials. Incorporating clean-up techniques while processing any type of tire flow is essential to any operation.[6]

1.10 Primary Processing

1.10.1 Overview

There are many types of scrap tire processes. Prior to World War I and especially World War II, the U.S. was very dependent on the import of natural rubber. The development of synthetic rubber, such styrene butadiene rubber (SBR), and other elastomers changed the paradigm of rubber's future. The development of the radial tire and the abundance of available, low-cost crude oil reduced the need to recover tires as a partial rubber hydrocarbon (RHC) replacement for new tire manufacturing. Up to World War II, the vast majority of worn-out tires were reclaimed. All the major tire companies had their own reclaiming operations to process scrap tires back into a reclaim form as a partial replacement for the more costly virgin rubber. Figure 1.47 shows early cost trends of scrap rubber. It reflects how the development of synthetics and availability of energy changed the economics of recycling rubber for tire manufacturing.

Reclaiming was converting the rubber and filler portion of the tire, less the reinforcements, back into a plastic state so the resultant slabbed or baled material could be used in the tire-compounding process or for other types of rubber goods. Reclaiming helped hold costs down for the tire manufacturer, and provided additional advantages such as better processability in the mixing room and in the shaping and release of the final produced tire from the mold. Economics of less costly polymers and the sophistication and full production development of the radial tire drove the tire reclaimers out

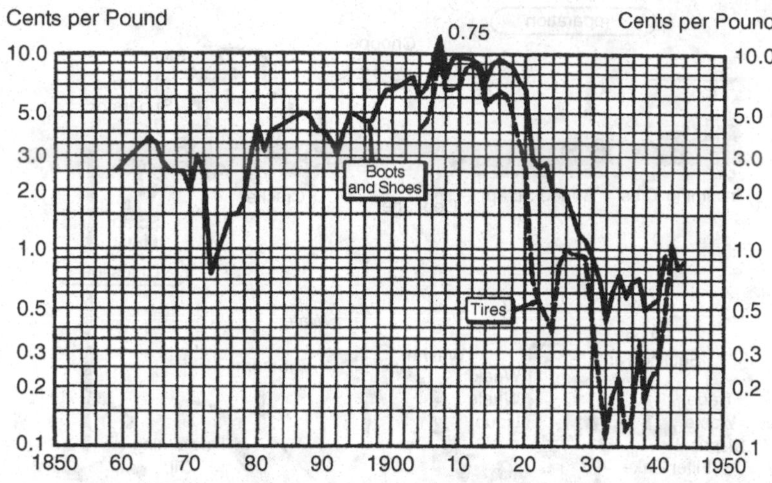

FIGURE 1.47
Early cost trends of scrap rubber.[31]

of business. However, as Figure 1.47 illustrates, reclaim played an important role in holding down costs. It was nearly as valuable as virgin rubber; thus, the economics were in favor of reclaiming the whole tire. As the world price of energy continues to escalate, the value of TDM will increase proportionally. It is very doubtful that reclaim technology will be resurrected.

The development of high-rate tire shredding was a very slow process. Figure 1.48 depicts a flow diagram of the digester process. It illustrates how reclaim was made and continued until the early 1990s. The importance of this diagram is the layout to producing a high-end product based on tire recovery. In essence, it is the model for today's tire processing systems. In today's environment, only butyl and natural rubber are reclaimed in limited quantities. The early reclaiming operations were important because they were the first complete CRM systems fully developed from initial whole tire receiving, processing for wire, fabric, and tire cord removal to converting tire parts into a CRM, normally a –30-mesh into a reclaim, so it could be made into a reclaim and reused back into a tire compound.

In the reclaiming days, the old technologies could be viable compared to the purchase of virgin feedstocks. Today, however, TDM has to compete with relatively low energy costs and the complexity of today's tires. In the past, reclaim could be used back into nearly every portion of the tire. Today, tires are designed to last 40,000 to 100,000 miles and provide exceptional performance in all kinds of weather, climates, and road conditions. A review of past reclaiming practices can be important to understand how tires were handled then and how this knowledge can impact today's shredding operation. For example, it was found that if the beads could be removed, this resulted in improved production and less wear and tear on the cracker mills. The beads now found in tires, not to mention the radial plies, make removal

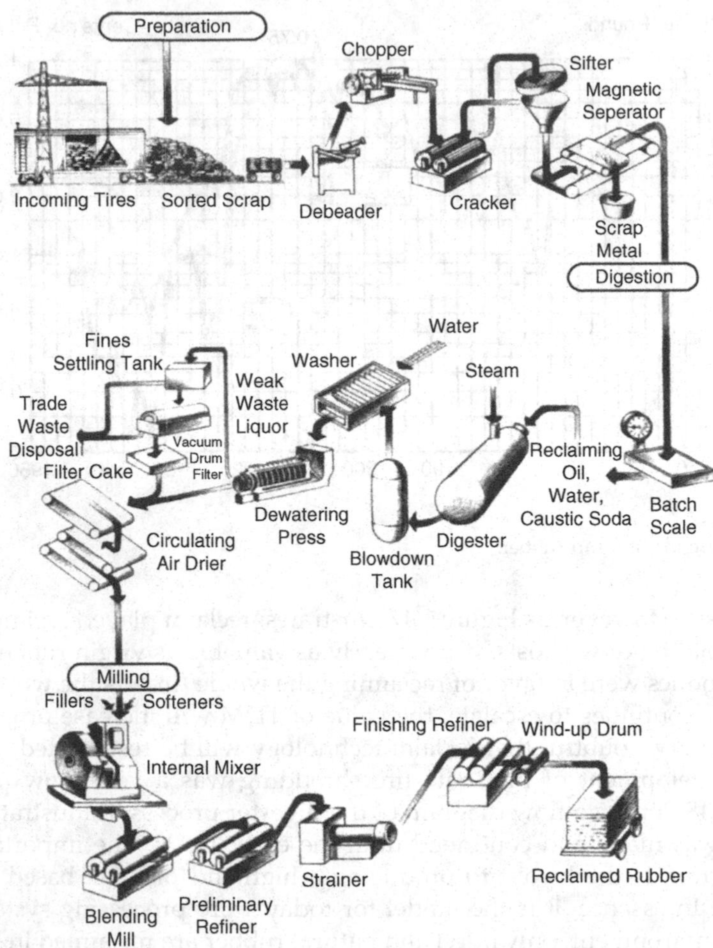

FIGURE 1.48
Flow diagram of digester process.[31]

of the beads an area of investigation and debate as to how best to handle this tire component, especially for truck/bus and OTR. This section discusses some of the many types of processing components that can be used. Tire shredding processes are an art form. The equipment must be rugged and made to last. It also must be safe to operate. This means higher maintenance and costs. In terms of accounting practices, accelerated depreciation is very real on a balance sheet basis. This applies not just to the shredding equipment, but to all the supporting material-handling units — trucks, trailers, and tractors — for whole and finished goods movement.

The reclaiming of tires back into a thermoplastic state, for use back into new tires or tire goods other than retreading was the second highest and best value recovery. The reclaiming process set the standard and the basic flow sheet for tire processing. The first real patent issued on processing whole

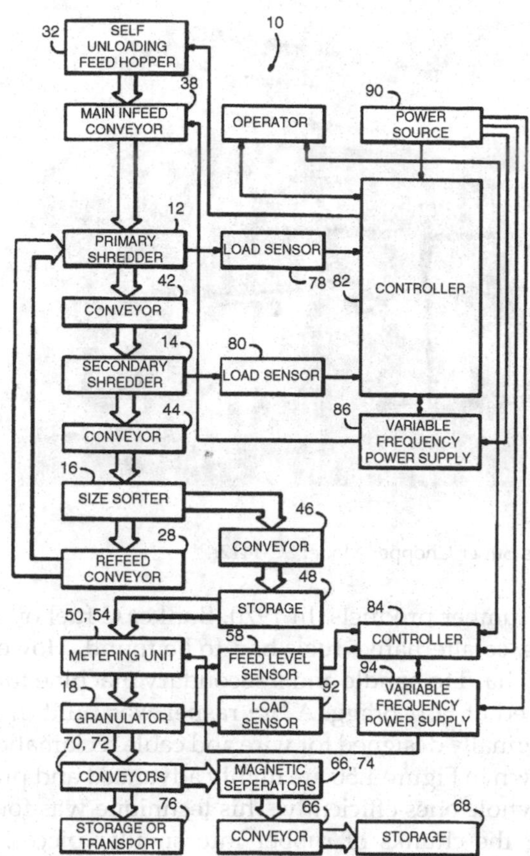

FIGURE 1.49
First patented TDM process.

tires as a total process can be found in the patent files entitled "Tire Processing Apparatus and Mechanics" Patent Number 4,714,201. This patent reflected bulk receiving of tires, process control through shredders to removal of the tire wire for the production of a basically wire-free TDF with minimal labor, and the use of automated equipment to achieve high production capability effectively. The information in Figure 1.49 was developed from the operational experiences in Portland, Oregon, and came about based on the use of high-quality supplemental fuel in northwestern pulp and paper mills, lime kilns, and cement plants beginning in the 1970s.

However, this process of high-rate shredding did not evolve overnight. The markets for a TCF began with Big-O Tires in Oakland, California, using a Saturn shredder (today called the Granutech-Saturn Systems Corporation, Grand Prairie, Texas) that was developed and built in Oregon with the cooperative effort of Georgia Pacific's forest division at Fort Bragg, California. Tires were shredded, baled, and put on truck flatbeds to be used as a supplemental fuel for Fort Bragg's Dutch ovens, used for steam production

FIGURE 1.50
Wendt Corporation's Super Chopper Model SC-1412t.[34]

in the drying of lumber products. In 1970, Bunker C fuel oil was scarce due to oil embargoes, so alternative fuels had to be found. However, coarse tire shreds were very hard to handle and a secondary machine for size reduction had to be installed at Fort Bragg. A tire rasper was built in Denmark. This machine was originally designed for wire and cable reclamation. Today, these machines as shown in Figure 1.50 are highly advanced and process shredded tires as well as whole ones efficiently. This technique was found to be effective in reducing the chunks of rubber into smaller pieces. However, this method was not really effective due to undeveloped tire chip handling methods, and then the high cost of fuel oil rebounded. The machine was moved to Portland, Oregon, and more completely developed by Pacific Energy Recovery Corporation (PERC). Shortly thereafter, Michael Rouse, working with Norman Emanuel (Emanuel Tire Company, Baltimore, Maryland), used a Cumberland granulator, also developed for wire/cable and plastic grinding. A whole new industry was found for Emanuel Tire's machines. Norman Emanuel today stands out as one of the great pioneers in the development of totally integrated scrap tire processing, including the collection and grading of passenger tires, truck tires, and OTR types. Today, these machines in combination are the backbone of the TDM industry. By reviewing the events and the equipment of the past and how it came to be used in the scrap tire industry, one can see that seemingly unrelated equipment used in a different application can have a profound effect on emerging technologies such as tire processing.

Uniquely, the tri-county area around Portland, called the Metropolitan Service District (MSD), had formed a tire monitoring system. This system licensed tire jockeys and processors in the form of a franchise to collect, grade out the good cases, and ensure that the scrap tires were handled in a

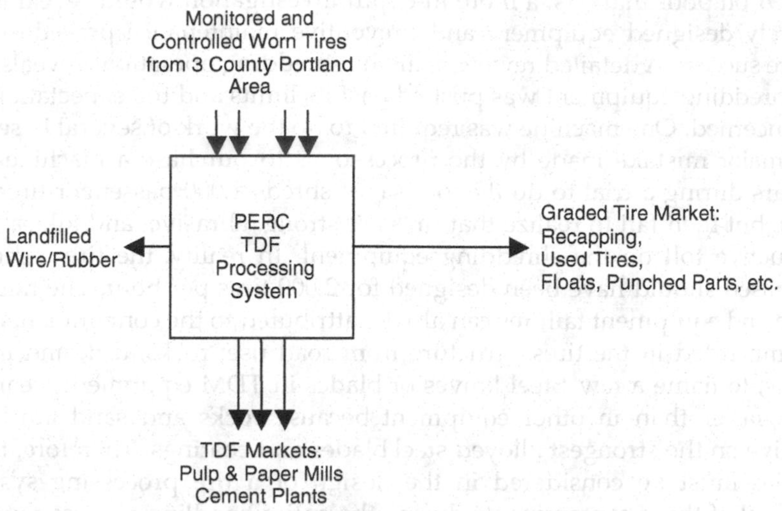

FIGURE 1.51
First TDF processing system with governmental participation.[1]

safe and environmentally sound manner. Refer to Figure 1.51 on how this concept performed.

Michael Rouse (who coined the acronym TDF and who pioneered TDF as a supplemental fuel while working at Georgia Pacific), along with his partner Robert Thelen, were the processors who developed the first high-rate scrap tire to TDF fuel systems. MSD had a supervisor monitor and work with the tire hauler to grade out casing for recapping, reuse, or other applications (such as dock bumpers and punched parts) and to ensure that scrap tires were handled properly. PERC provided a location for grading the tires and their shredding. The truck tires were split prior to shredding and the wire was disposed of in a landfill. The first shredder at this site was a bone crusher from the rendering industry, followed by the Rasper to further refine the tire chunks. Later, the Holman rotary shear, manufactured by Barclay Machine Works in cooperation with Rouse and Thelen, perfected this unit, which was the precursor to today's Columbus McKinnon and the Barclay Roto-Shred. Michael Bungay, from Sacramento, California, operated the actual first use of the rotary shear. He clearly showed that scrap tires could be collected from a geographical area, as a private operator, and efficiently volume-reduced for landfill disposal. The real uniqueness of the MSD project was the cooperation by the government and the private sector (processor, tire haulers, the energy user, and the landfilled operator) for a complete and total system of tire recovery and disposal. MSD had gained worldwide attention for its pioneering efforts to find effective solutions to the scrap tire dilemma. Oregon was also the first state to implement glass recycling.[1]

A review of all the scrap tire activities of the past 20 years shows a very high incidence of scrap tire business failure. Although these failures can be blamed on poor markets, a more in-depth investigation would reveal inadequately designed equipment and preventive maintenance procedures to ensure success. A detailed review of all the shredding operations reveals that tire shredding equipment was pushed past its limits and the expectations of all concerned. One machine was required to do the work of several in series. One major mistake made by the processor is to purchase a machine that appears during a trial to do the job (say it shreds 1,000 passenger tires per hour), but then fail to realize that tires are strong, abrasive, and take a very destructive toll on the shredding equipment. In reality, the tire shredder purchased should have been designed for 2,000 tires per hour. The maintenance and equipment failures can also be attributed to the contaminants that are embedded in the tires' structure from road use: rocks, dirt, and metal objects, to name a few. Steel knives or blades in TDM equipment wear out much faster than in other equipment because rocks and sand are more abrasive on the strongest alloyed steel blades and coatings. Therefore, these realities must be considered in the design of a tire processing system. Removal of these contaminants during the entire handling process must be done and should be considered in the engineered design of the facility. Working with your equipment supplier and a qualified consultant or consultants is imperative to overcome these challenges.

1.10.2 Challenges in Processing Tires

The process of tire shredding or chopping can vary greatly as to the tire type. When a new tire versus a worn-out tire of the same type are shredded, the latter will be much easier to process. The process becomes more complex if the tire is dirty and comes from a stockpile where contamination has occurred. Normally, a contaminated tire is removed from the pile before processing, to be treated separately. This involves additional cost. Some tires still have the rims, which must be removed by use of a derimmer. With the high prices obtainable for steel scrap, this task alone can pay for the labor, operation, maintenance, and amortization of this operation.

Tires should be inspected on the conveyor belt to remove any undesirable contamination such as tire rods, rocks, or anything that hinders or sabotages the tire shredding or chopping process and the quality of the final product. It is absolutely imperative that safety shut-off switches be conveniently located, to stop any part of the process and prevent personal injury.

Safe operation also means making sure the environment is not disturbed in any manner. This could include leakage of fluids or gaseous emissions. All operations involved in processing the tires need to be in accordance with local, state, and federal guidelines and mandates. The Occupational Safety and Health Administration (OSHA) has many guidelines. This information is available from a local OSHA office and can be very important for maintaining a safe operation.

After the scrap tire has begins its journey, the objective is to reduce the tire into uniform pieces. The more uniform the tire pieces are as they pass from one processing unit to the next, the more successful the operation will be. Tire rubber is very abrasive, but the tire wire is a hindrance to shredding. Tire wire is generally a medium carbon steel with a hardness of around a 65 Rockwell and excellent tensile strength. The combination of these features places great stress on the shredder knives or blades. Softer steel blades in the cutting chamber of the shredder will experience greater wear. The harder the toll steel in the shredder, the longer lasting it will be, but the blades are more likely to break. In this case, the shredder is like a pair of specialty steel scissor blades — once dull, it loses its effectiveness. If breakage occurs, removing this steel may be critical, to prevent further damage to the shredder downstream.

Since rubber likes to stretch, sharp cutting or firm shearing is critical for high-rate, efficient production. The configuration of the cutting edges should be closely monitored and recorded. Comparing these measurements to the number of tires processed and the power usage can be an effective preventative maintenance tool and will save time and money in the long run. In other words, when shredding or chopping efficiency slows, amperages will tend to climb and material weight balances during production will fall. This decrease in efficiency will only result in added costs and lost production time. This is where a good preventive maintenance program and operator awareness is essential. Monitoring can also mean the difference in success or failure. By maintaining good metrics (monitoring, measurement, and recording) for all aspects of the TDM operation, preventive maintenance procedures can be implemented accordingly and will deter downtime for the whole operation.[1]

1.10.3 Primary Reduction

There are many ways to reduce a tire. However, the discussions herein are limited to high-rate processes involving equipment components that are very rugged and have relatively low maintenance and labor requirements. The difficulty in this approach, however, is deciding what is the best method. It is the objective here to list and briefly discuss some of the major equipment techniques available in today's marketplace.

Generally, four types of machines are used in the primary reduction of whole tires down to a manageable form. They include the guillotine, the cracker mill, the hammer mill (high impact), and the rotary shearer. The guillotine merely cuts the tire into sections for further processing. Many types of shearers are available in the marketplace. Normally, shearers are self-powered tractors that are used on OTR tire types. Guillotining has not been found to be practical for passenger and truck tires.

1.10.4 The Cracker Mill

The cracker mill has been the workhorse of the scrap tire industry. Smaller versions of this unit are also used for further milling of the tire chips into

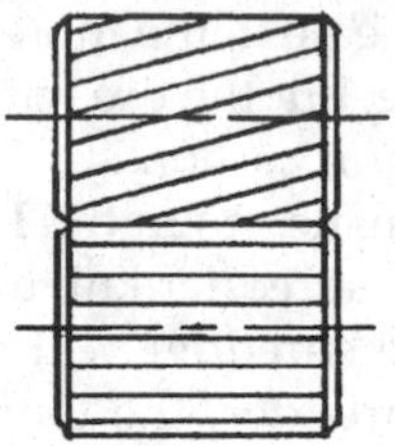

FIGURE 1.52
Typical roll corrugation design.[1]

finer mesh forms. Tires are elastic, and when they are chopped and then ground, they resist grinding even more than they resist cutting. The cracker was developed to address these challenges. It consists mainly of two large-diameter, corrugated rolls that run at differential speeds. The tire chips are caught in the nip of these two rolls and are subjected to a shearing force. The material passing the rolls is screened to a given size, so material over that size is returned to the cracker in a repetitive action. Because this procedure occurs at room temperature, it is called an ambient operation. Figure 1.52 shows a typical corrugated roll design.[47]

1.10.5 The Hammer Mill

Use of hammer mills (horizontal or vertical) is another effective way to reduce tires. These machines generally require significant energy to operate. Tub-like grinders use a hammermill configuration to reduce whole tire effectively, as shown in Figure 1.53. These units are especially effective for high-volume tire pile cleanups where contamination is severe, such as in muddy and/or sandy conditions. Vertical type units are also available and tend to be in the stationary form at a plant site. However, over the years, vertical and horizontal mills have been put on trailers and used as portable units in the field.

FIGURE 1.53
Diamond Z manufacturing tub grinder.[36]

FIGURE 1.54
Open Pulva mill (high-impact unit).[37]

Another type of unit used especially with cryogenic grinding is the high-impact mill such as the Pulva mill, shown as an open unit in Figure 1.54. These units use swinging hammers to break up the frozen tire chips; they are equipped with variable feed screws and internal screen, and are very predictable in their operation. Pulva mills, like the MicroTec, generally operate with cryogenic systems and greatly assist in the fragmentation of the incoming materials.

Existing mills are numerous, and new types of mills are being produced at a growing rate. The use of high-pressure water jets for primary reduction and applications in secondary reduction has been explored and is being tested. Processes or techniques that have been unsuccessful in the past sometimes undergo equipment modifications or new adaptations, to become practical units for processing whole tires or producing CRM. Horsepower requirements vary greatly and noise migration needs to be factored into the layout when installed.

1.10.6 The Rotary Shearer

The rotary shear shredder comes mainly in two types: one has a one-piece mounting of counter-rotating knives (Figure 1.55) and the other has detachable knives (Figure 1.56). The Tyco/Untha is unique in that sizing screens are located beneath the rotating blades and can be easily removed for maintenance or cleaning of the machine. The one-piece units can have multiple hooks and can vary in width, depending on the desired results. Generally, for primary reduction, wide knives from 20 to 6 inches are common, with

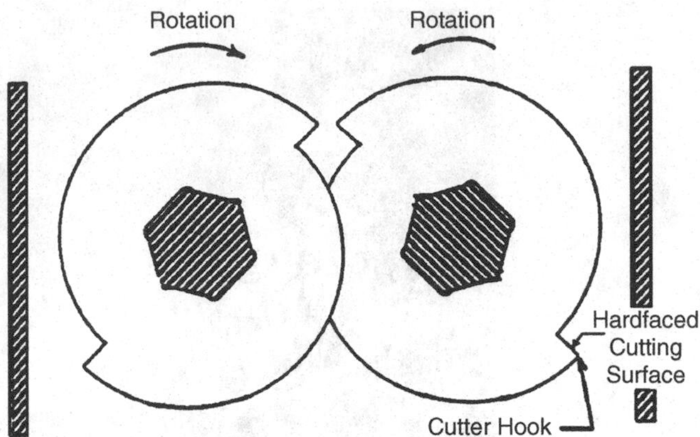

FIGURE 1.55
Rotary shearer with one-piece counter-rotating knives.[35]

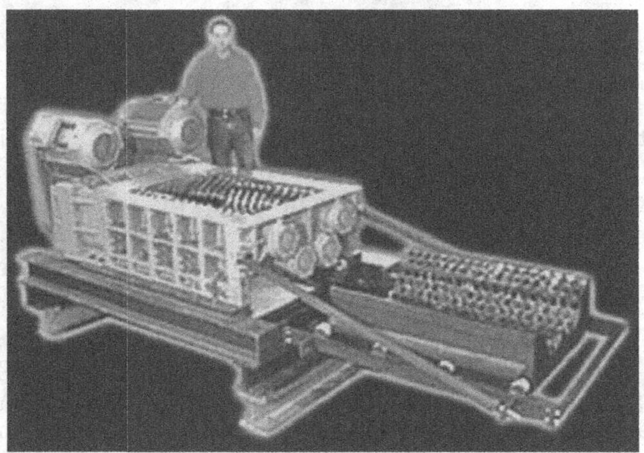

FIGURE 1.56
Tyco/Untha rotary shearer with one-piece counter-rotating knives.[38]

two hooks per knife. Two 2-inch-wide blades can be put together to form a 4-inch-wide blade. Later, as one side of the blades wear, the blades can be switched, exposing the inside blades to the outside for extended knife. These machines are unique because of their relative quietness, slow speed, and lower susceptibility to major damage if an unshredable object enters the machine. Electrical sensitivity can be set for these machines, as is the case with hydraulic-driven units. The energy to operate these units varies greatly depending on the size of the unit. Standard horsepower requirements tend to be around 200. When these machines are operated in series, the second machine generally has multiple hooks to increase the chopping or chipping action more efficiently to produce smaller pieces. Increased hooks increase

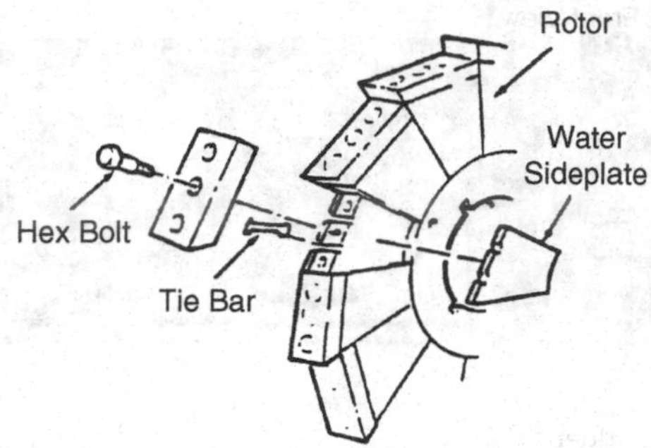

FIGURE 1.57
Working view of detachable shear knives in a Columbus-McKinnon rotary shear tire shredder.[39]

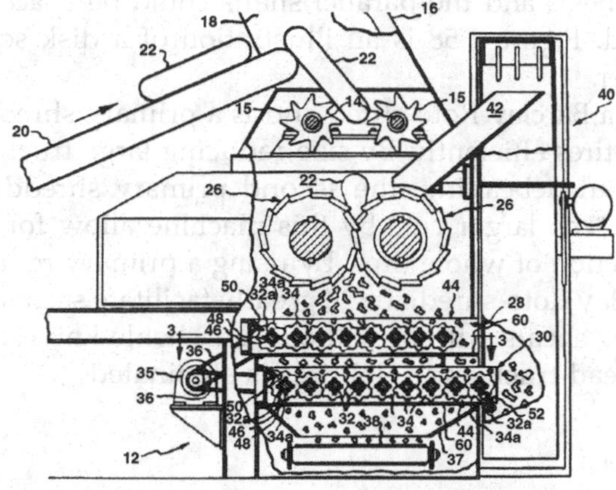

FIGURE 1.58
Cross-sectional knife view of a Columbus-McKinnon rotary shear tire shredding machine with disk screener.[39]

the number of TDM pieces produced but this must be weighed in relation to output and the material desired to be produced.[35]

In the case of the detachable knives, larger knives can be placed in the center of the shredder and the blade widths can be reduced as the knives are placed to the outer perimeter of the shredder. In this way, the knives from the center can be sharpened on the other edges and sized to meet the reduced widths as the knives wear. These designs allow the processor to use more exotic tool steels, for longer life. Figure 1.57 shows a working view of a detachable knife and Figure 1.58 shows a cross-sectional view of a Columbus McKinnon, rotary shear, tire-shredding machine with disk screener, sometimes called a classifier. Michael Rouse and Robert Thelen developed a disk screen from the wood waste industry to handle the stringy, convoluted tire shreds for regrinding back through primary shredders. Later, this action screen became a standard in the industry varying in width and length to

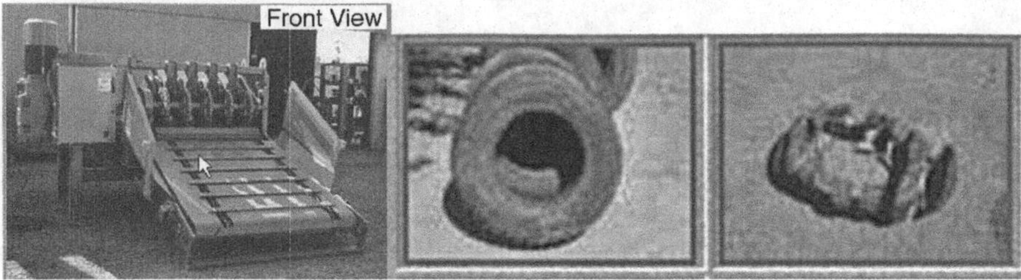

FIGURE 1.59
Primary to a primary shredder.[40]

stacking for a double screening effect. The design of the daisy wheels would vary from star shapes to flower looking appearances. The spacing between the screening wheels and the parallel shafts could be spaced according to the TDC desired. Figure 1.58 is an illustration of a disk screen operating below a shredder.

Figure 1.59 is a Barclay Roto-shred and is a primary shredder developed to handle truck tires efficiently. By size-reducing large truck/bus and OTR type tires without debeading, the second primary shredder can operate more efficiently. The larger cuts by this machine allow for high rate production or reduction of whole tires. By using a primary reduction machine such as the Barclay Roto-shred, it will greatly facilitate secondary shredding with other rotary shears with or without detachable knives. In many cases the need to debead truck and OTR tires is eliminated.

1.11 Secondary Processing

This step can involve many types of machines. However, this area is evolving not only by machine but also by method, that is, ambient or cryogenic. These methods can also include combinations of each other for a desired end material result. Normally, secondary processing is the reduction of coarse tire chips into the granular or crumb rubber range, as discussed earlier. The secondary process can be a single step or multiple steps. Secondary processing usually incorporates the removal of the inherent reinforcements and contaminants found in the scrap tire stream.

1.11.1 The Two-Roll Mill or Grinder

The two-roll mill is similar in operation to the cracker, but it is smaller and requires less horsepower. In earlier days, several of these mills were lined up with one common drive. Until recently, two two-roll mills in tandem had

FIGURE 1.60
24″ × 24″ primary grinder mill.[41]

FIGURE 1.61
The compactness of a Hagglunds hydraulic drive unit on a two-roll mill.[41,42]

a common gearbox and motor or one two-roll mill had a gearbox and motor built on a unitized stand or foundation (see Figure 1.60, showing a 24″ × 24″ primary grinder mill). However, companies such as Rubber Machinery Corporation in Akron, Ohio, have significantly improved these units. Instead of using a unitized motor and gear box arrangement to drive the two rolls, each roll is driven by its own hydraulic motor (see Figure 1.61). This greatly reduces space requirements and improves control and efficiency. This advance in technology allows the TDM operator to hydraulically change the speed of the individual rolls, for increased productivity. Figure 1.62 shows this concept and how the torque can be held constant at a given roll-to-roll speed ratio without sacrificing efficiency. It is reported that greater output can be achieved by increasing the speed of one of the rolls while holding the other constant; a finer gradation is also obtained. These mills are rebuilt and are very rugged. The energy requirements can range from 150 to 1,000 horsepower, depending on the size of the unit or the number of units operated in unison.

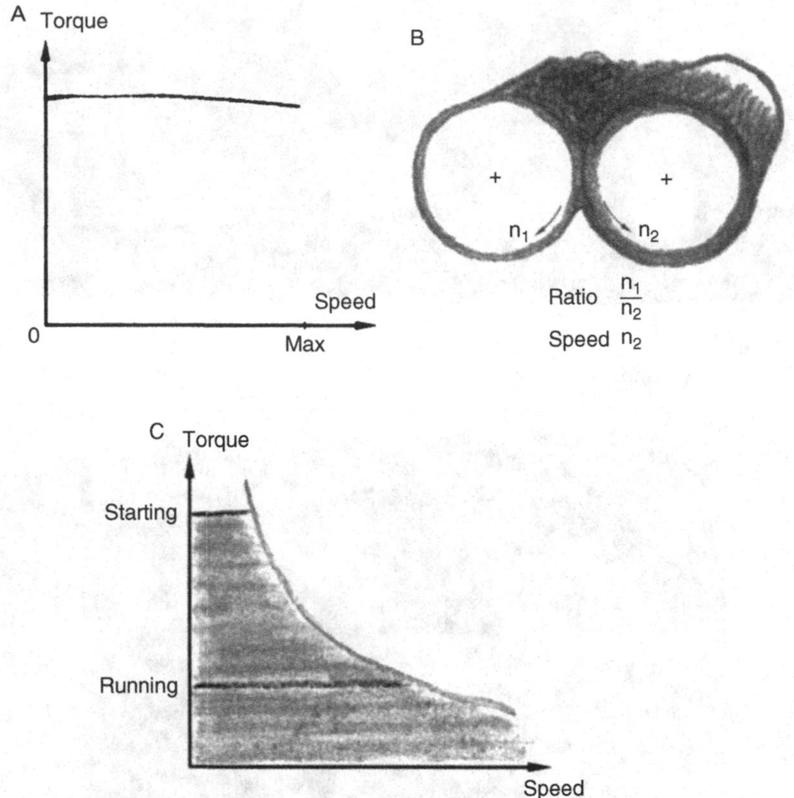

FIGURE 1.62
The dynamics of hydraulics A and B over a gear-driven mill C. (A) Constant torque curve, (B) speed ratios varied between rolls n_1 and n_2, (C) lost torque of a gear-driven unit.[42]

1.11.2 Granulator Type Machines

Granulators are high-speed machines that can reduce tire chips down to less than 10-mesh. From a CRM operations standpoint, the granulator operated better at the π-inch size opening range. These units are constructed as a set of long or staggered knives mounted on a common axle that rotates at high speeds in close clearance to a set of bed knives. The tire chips are fed into the granulator and the chips are chipped between the rotary and fixed-bed knives. The desired granular size continually sifts through the screen opening attached to the unit.

The granulator is built as a rugged machine and can be operated with open or closed rotors. Solid rotors are better suited for chopping large tire chips into 2-inch particle sizes. The open rotor is more desirable when 6- to 2-millimeter particles are desired and the material is removed from the granulator using a pneumatic system. These units have substantial energy

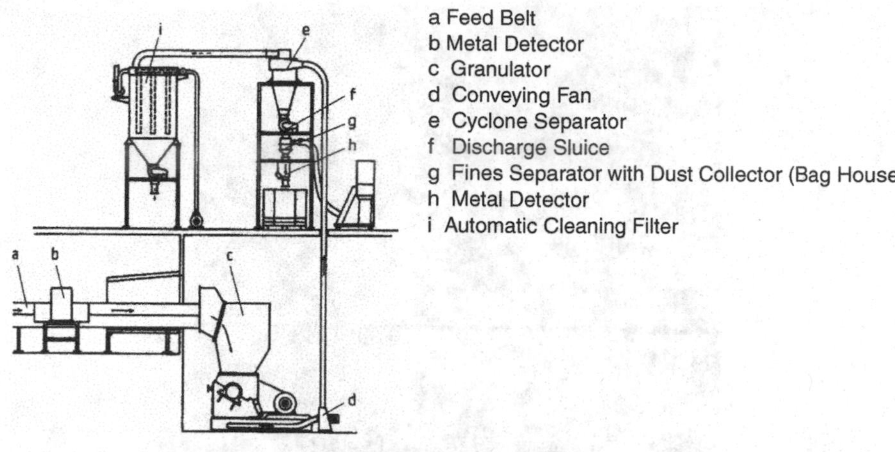

FIGURE 1.63
Typical arrangement of an open rotor air-swept granulator.[43]

requirements ranging from 50 to 500 horsepower. Sound enclosures are recommended due to the high-frequency noise that is generated. For these high-speed units, it is recommended that a metal detector be installed on the feed to the units. The metal detector's sensitivity can be set to bias out the wire fraction in the tire chips. Additionally, open rotors allow for more passage of materials and help cool the unit. Generally, open rotors are used for small screen openings or where the potential for plugging or heat build-up is the case. Solid rotor units under continuous high loading need cooling jackets built within the housing of the granulator. Water mist sprays also assist in grinding and keep the unit cool. Figure 1.63 and Figure 1.64 show a granulator installation and a cross-sectional view of a typical unit. Figure 1.65 shows typical key parts for a standard Cumberland granulator.

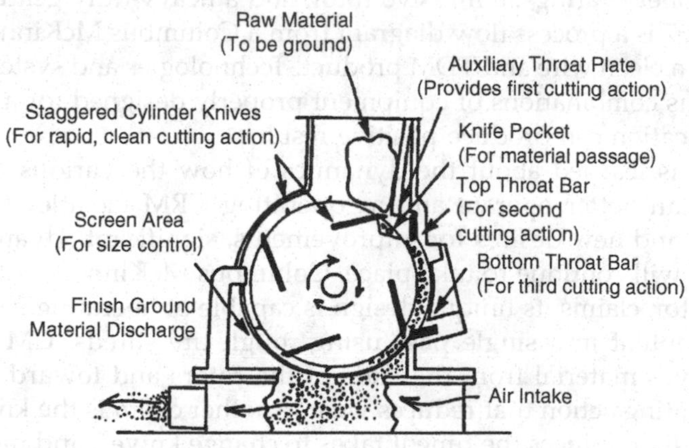

FIGURE 1.64
Cross-sectional view of a typical granulator.[35]

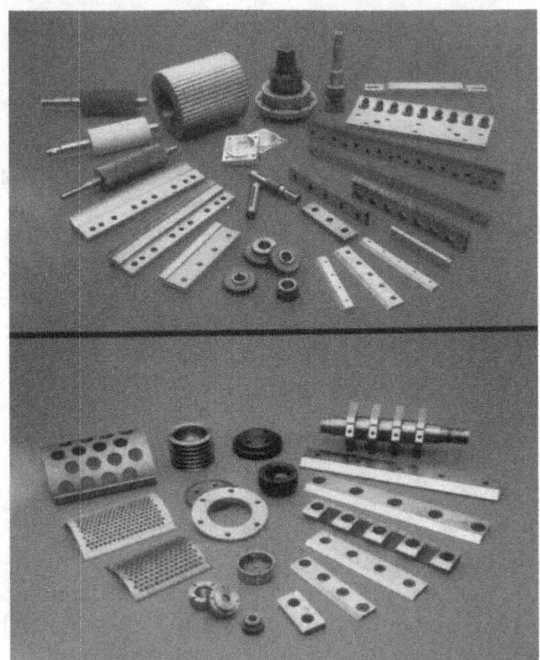

FIGURE 1.65
Typical parts for a standard Cumberland granulator.[44]

The granulator has also met its match in new developments, such as the Grizzly manufactured by Granutech Saturn in Grand Prairie, Texas, as illustrated in Figure 1.66.

This designed machine has had significant internal construction and design improvements to reduce blade wear. Its features include beefed-up spherical roller bearings, a massive rotor, and a heavy-duty gearbox.

Figure 1.67 is a process flow diagram from a Columbus McKinnon system to produce a clean wire and TDM product. Technologies and system designs with various combinations of equipment properly designed for a particular CRM application can produce positive results.

As more is learned about the dynamics of how the various processing machines can better operate and as operating CRM facilities share their experience and new desires for improvements, significant advancement in processing will continue to take place. Columbus McKinnon, with its New CM Liberator, claims its unique design is capable of liberating 97 to 99% of the steel content in a single pass using rough tire shreds. CM claims its design moves material from the ends of the rotors and toward the center of the liberating action that reduces wear. Another claim is the knife clamping system that reduces the time it takes to change knives, and most importantly, the Liberator's improved "flow through" design that allows the machine to purge it of fine pieces of tread wire that can become trapped

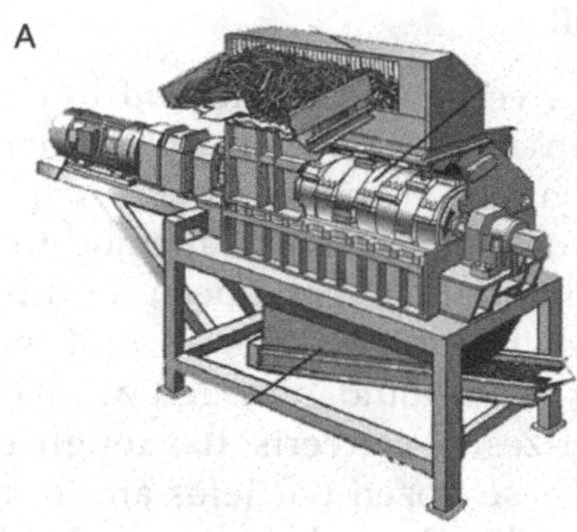

FIGURE 1.66
Views of advanced type granulators. (A) Cutaway view, (B) machine open and screen below, (C) assembled Grizzly.[45]

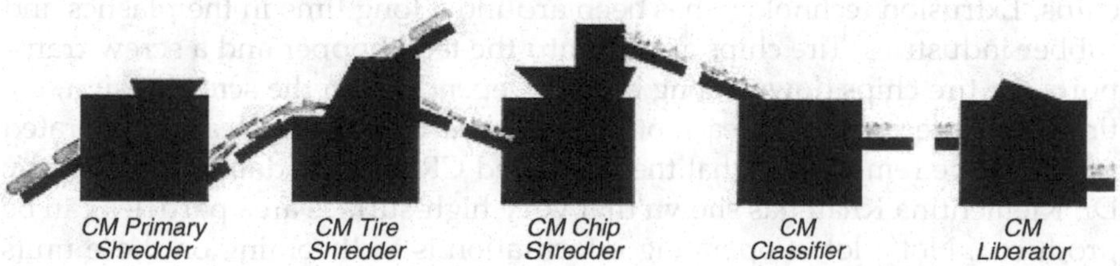

FIGURE 1.67
A turnkey wire and TDM system.[6,39]

between the rotor, rotor housing, and the screen and then cause wear, friction, and heat.[61] Other internal designs incorporated into these machines by OEMs greatly assist in preventing rubber or wire build-up and erosion in the internal workings of the machines. This, in turn, can cause fires. So the flow through design reduces this occurrence. These types of improvements to the granulator allow this type of equipment to be more effective in the commutation of tire materials and should be investigated when designing or enhancing the crumb rubber facility.

1.11.3 The Cryo-Mill

Cryogenics has been a very popular method to reduce tires chips down to the 20- to 40-mesh range. Normally, the finer the mesh, the more intensive the operation, and economics plays negatively on finer grades. However, cryogenics offers an excellent method to separate the fiber and steel from the tire chip. Markets have been found for the fiber and steel by-products. The cryo process accepts 3- to 4-inch tire chips that are spayed in a chamber with liquid nitrogen at –383 °F (–230 °C). As the liquid nitrogen vaporizes, it converts the tough elastic particles into a brittle, glassy state. These frozen particles are then fed to a high-impact mill such as hammermill, where they are easily shattered. Further processing removes the steel and fiber components. The screened overs can be feed back to the cryo-unit. High-impact units such as the Pulva mill (MicroTec) or a hammermill are used for these applications. The design and construction materials of these units are critical to long-lasting performance. Size, hammer configuration, screens, power requirements, etc., all need to be considered.[46]

It should be noted that the finer cryo-particles do not readily reduce efficiently when fed back to these systems. Generally, another processing unit type is required to handle the stream efficiently. Since these are high-speed units, noise abatement has to be considered.

1.11.4 The Extruder

The extruder has been found to be another excellent way to commute rubber chips. Extrusion technology has been around a long time in the plastics and rubber industries. Tire chips are fed into the feed hopper and a screw transports the tire chips down a long barrel. Depending on the screw design, the tire chips intermesh with each other in a shearing action. Heat is generated and must be removed so that the produced CRM is not damaged. Work by Dr. Klementina Khait has shown that very high-surface area particles can be produced. Not a lot of operating information is forthcoming on these units for CRM production.[63]

As scrap tire chunks are fed to the entrance of the Berstorff Maschinenabau extruder, the screw pulverizes the rubber as it passes through the chamber. It is claimed that the rubber is chemically altered or activated. More work needs to be done to verify this change. The extruder has a set of twin screws to pulverize the tire rubber. Figure 1.68 and Figure 1.69 illustrate the extruder concept. These units operate quietly and horsepower is a direct function of output. The uniqueness of these extruder units is that they seem to perform more efficiently with time. As the clearances between the barrel and the screw wear, production and efficiency improve. After a while, however, this does become a diminishing return and the unit needs to be serviced as with all machines.[64]

FIGURE 1.68
Typical extruder set-up.[47]

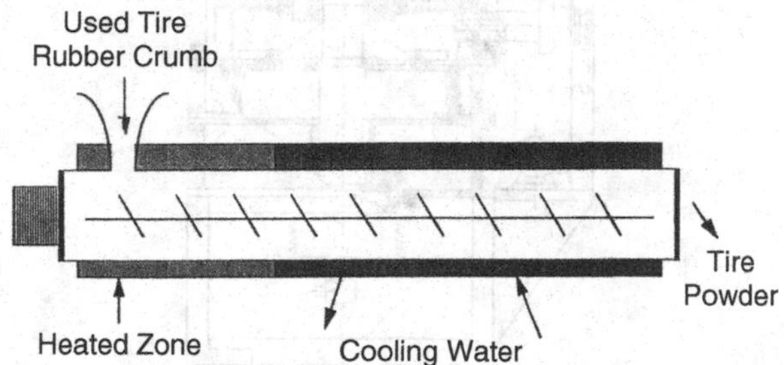

FIGURE 1.69
Typical tire extruder set-up.[48]

1.11.5 High-Impact Air Mill

These units are relatively new to the CRM manufacturing system. According to Crosston Industries literature, "The grinder features a pair of stationary and rotary grinding disks designed to grind rubber material in an ambient and dry environment. The rotary grinding disk can be adjusted to set the desired grinding space between the two disks to achieve the optimum throughput and yield of fine powders." The grinder does produce a very good 40-mesh CRM. If an 80-mesh CRM is desired, the 80-mesh fraction needs to be screened from the stream. Regrinding of the overs, –40-mesh to + 80-mesh fraction is not practical with this system and another method has to be used to convert this fraction into a –80-mesh product. Since this system, and also cryogenics and the two-roll mill systems, produce a wide CRM distribution, they do not economically convert all the CRM efficiently into

FIGURE 1.70
Crosston grinder.[49]

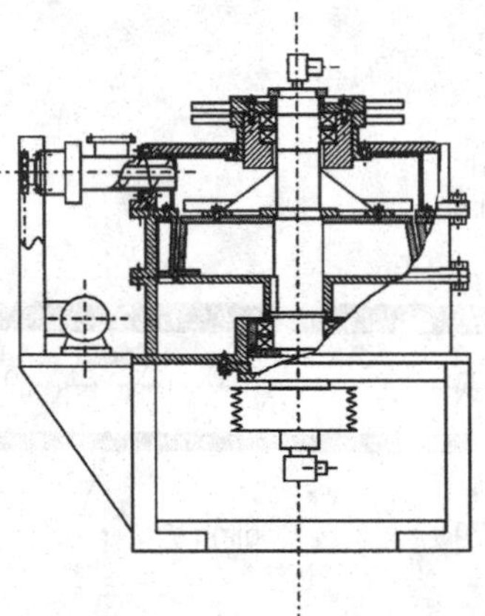

FIGURE 1.71
Structure of Crosston grinder.[49]

fine powders. Fine powders are defined as passing the U.S. Tyler 80-mesh screen. Figure 1.70 shows a picture of a Crosston grinder and Figure 1.71 shows the structure of a Crosston grinder. These units require a total soundproof enclosure, and airflow control is critical to the operation of these units.[49]

1.12 Tertiary Processing

This can be ambient, cryogenic, wet, thermomechanical, chemical, and combinations of all these processes. Tertiary processing is generally the production of the finer grades of crumb rubber, called powders. These are the CRM gradations below 80-mesh (180-microns). All these processes will produce a very different end result, including varying degrees of cleanliness and different surface morphology, particle configurations, and gradation ranges. Therefore, the processor must carefully study and research the process technology best suited for the particular application or market penetration desired. It should be clearly understood that tertiary grinding means 100% conversion to the desired CRM end product and having no residual left-overs in the plus range that are of no value to the desired end market for the CRM material. To date there is only one known true tertiary process for producing fine mesh CRM powders meeting the 100% criteria. This is the wet mill process.

1.12.1 Wet Mill

Wet milling is probably the most efficient method for producing fine powders in a single unit. The overs from this grinder can be fed back to the unit very efficiently and the entire system has a 100% conversion. The only losses are due to any spillage during materials handling. This technology has evolved over last 30 years, beginning with the Gould process. The Goodyear Tire and Rubber Company advanced the Gould work and US Rubber Reclaiming developed a similar process. In its patents, the Goodyear process claimed to make a CRM in the –15 micron range; however, in its operations in Cleveland, Ohio, only a –40-mesh CRM was commercially produced. A very small fraction of minus 15 micron material was produced and was not economically viable to produce in reasonable quantities even after screening. Very sophisticated equipment needs to be developed to handle such a small particle. Later, Baker Rubber Company moved this operation to its Chambersburg, Pennsylvania, facility, where Edge Rubber now operates it. At its Vicksburg facility in 1988, Rouse Polymerics International further refined and commercialized the wet grinding process into a high-rate technology for producing –80-mesh CRM and finer grades of –120, –140, –170, and –200 CRMs at total rates in excess of 10,000 pounds per hour.[1]

At this time, the wet process is proprietary, and detailed discussion as to its technology would jeopardize trade secrets.

1.13 Major CRM Processes

1.13.1 Overview

The production of CRM requires many equipment components in any number of variations and layouts. The unique challenge to the producer of these materials is in the way the equipment components are used and the total process layout. There can be any number of components in series or parallel or they may be completely eliminated as listed in Table 1.11. Experience and

TABLE 1.11

Some of the Key Equipment Components to Consider for CRM Facility[1]

Extruder (single- and twin-feed types)
Packaging (bulk bag (returnable or throwaway), Gaylord boxes, paper and plastic bags, top- or valve-filled)
Heavy-duty stationary and mobile conveyors (belt, steel, vibratory, roto-lifts, bucket, etc.)
Metering systems (hydraulic and mechanical)
Self-unloading trailers
Tug (yard tractor for moving trailers)
Fixed and self-cleaning magnets (permanent or electromagnetic) form overhead to drum type separators
Screens (rotary, flat)
Balers for compaction of whole tires, TDM, tire reinforcements (fabric and steel) with or without automatic tie
Containers of all types from roll-offs, barrels to dumpsters
Gravity separators
Destoners
Aspirators
Fans
Dust collection (bag houses, cyclones, electrostatic)
Wheel crushers stationary or mobile
Yard equipment (front-end loaders, tow motors, etc.) with or without attachments such as grapples, thumbs, forks, buckets, couples, shears, grapple buckets, or tire grapples
Laboratory equipment (Roto-tap, TGA, moisture analyzer)
Electrical power suppler
Metrics (scales, counters, etc.)
Derimmers
Shredders (various)
Granulators (various)
Choppers (various)
Guillotines
Presses
Maintenance shop and store room (spare parts)
Fire protection equipment
Fencing and security monitoring
Communications (telephone, fax machine, computer\office)
Employee facilities
Labeling machine
Packaging equipment
Magnetic detectors
Catch basis
Loading docks
Highway, railroad, or barge access and related material-handling equipment
Continuous lubrication systems

documentation will be the final proof as to the optimal system for a particular plant that seeks certain end results.[1]

The equipment listed in Table 1.11 may have names that are considered to be synonymous with others. Despite the close relationship in names, the intended purpose of each equipment component is considered to be in its own category. As the CRM processing technology rapidly progresses, it is advised that contact be made with the OEM as to new developments or

TABLE 1.12

General Tire Composition[1]

Component	Percent by Weight
Recoverable rubber (RHC)	70%
Steel reinforcements	14%
Fabric	3%
Miscellaneous	13%

improvements to the particular equipment component. Table 1.11 represents some or all the components that may be found in a totally integrated TDM facility, depending on the demands or criteria placed on the facility's design and product output.

CRM processing represents different challenges to today's processor. Table 1.12 shows the scrap tire material contents and details the general composition of a tire on a weight basis. Each facet of the TDM facility must take into account these streams and how they are to be handled. In the past, it was not practical to recover the steel reinforcements due to economics, market logistics, or lack of relative markets. Today, companies such as the Bi-Metal Corporation of Ridgefield, Connecticut (Figure 1.72), Action Equipment of Newberg, Oregon, and Sweed Machinery offer wire recovery systems. Bi-Metal will work with the TDM processor on a contractual basis to locate steel scrap markets for its clients.[69] Today, scrap metal dealers recognize tire wire by-product as a highly valued quality steel by-product, and its value as a commodity is increasing. Consideration to capture the wheel rims is another source of valuable income. Rims can bring a good steel market value and justification for purchasing a tire/truck derimmer can usually be justified by the tire processor. These factors need to be considered in the design. It may or may not be economically favorable to install equipment to recover all these streams, but the remaining streams such as the fabric can be recovered at a later date, as the facility proves to be viable and gains credibility with the financing agency.

FIGURE 1.72
Bi-Metal Corporation's Clean Wire System.[50]

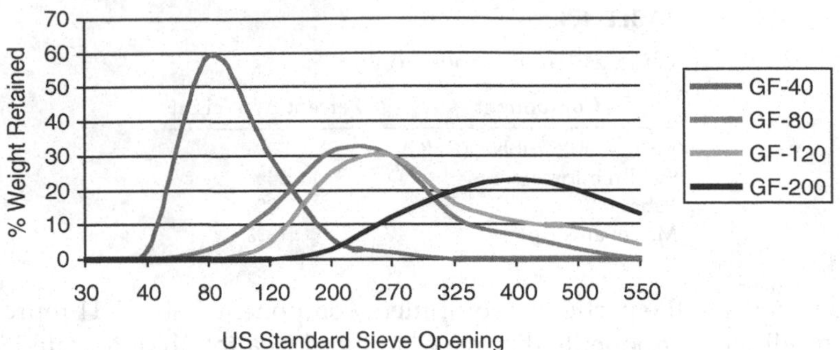

FIGURE 1.73
Typical defined gradation curves of various wet grind CRMs.[71]

Understanding these components completely is critical in the CRM process. What should be done with the nonrecoverable materials? In many cases, the fabric and the steel reinforcements are very hard to clean and convert into usable forms for resale. The yield of most CRM processes is 65 to 70%, which hurts the economics of the operation. Radial truck tires and many OTR types offer even higher yields of clean RHC or crumb rubber, but are more difficult to process due to their construction and size. All these factors need to be taken into account.

The desired gradation of the finished rubber product and what is actually produced are generally quite different. Many systems offered by the OEM promise a desired gradation, high quality, and smooth production output. For coarse systems produced by a granulator, the defined size can be fairly accurate. For example, producing 2- to 6-millimeter granules can be accomplished. The processor can then simply screen this range into more defined ranges, say 1 to 3 millimeters, 4 to 6 millimeters, etc., and sell those products to the sports track, rubber mat, carpet, or similar markets that require those gradation ranges. Production of the coarse CRMs, in the 10- to 40-mesh range, can be fairly predictable also when using conventional ambient (air milling, two-roll mills, and high-impact) and cryogenic processing. Processing the 10- to 40-mesh CRM with these methods can also be accomplished with minimal losses. However, with fine and superfine CRM, the processing and efficiency of achieving a gradation without losses is very hard to achieve. The ambient, cryogenic, and air milling processes require screening of the fine fractions from the coarse. The regrinding of the coarse fractions is expensive and usually not very effective. Only the wet grinding achieves the desired fine gradation in a one-step process. Even though the small percentage of overs are recycled within the wet system, very little efficiency is lost and precise gradations can be achieved whether it is a true –40 mesh or as fine as a 200-mesh gradation. Figure 1.73 illustrates the precise and defined gradations that can be achieved.

Figure 1.74 reflects the general surface effect of the CRM via cryogenics, granulation, and milling of chopped tires. Cryogenics give a very smooth

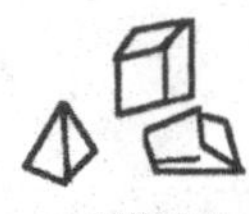

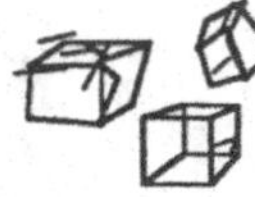

FIGURE 1.74
CRM surface morphology comparisons.[1]

surface area, ideal for the use of urethanes for molding parts or constructing sports tracks. The same can be held true for granulated CRM and more precise control of a coarse gradation can be achieved. Milling via two-roll mills produces a highly fractured CRM better suited for dissolving in asphaltic solutions, tire compounding, or molding into parts by compression and temperature.

1.13.2 Typical Ambient

The two-roll mill system has been the real workhorse of CRM manufacturing since the beginning of reclaiming. Generally, two-roll mills operate at slow speeds and are very forgiving. In reviewing all the systems technologies, it is advisable to go with the tried and proven technologies. Two-roll mill installations may cost more to purchase, but in the long run they generally cost less to operate. A Grade-5 bolt slipping through the system can cause severe havoc in most operations, whereas the two-roll mill will see only slight damage and will flatten the bolt. Figure 1.75 is a typical ambient grinding system designed by Rubber Machinery of Akron, Ohio.[41] Historically, these machines were operated individually with a motor and a gearbox at fixed roll speeds. Today, these units have been greatly modernized with individual hydraulic drives, to improve productivity and the desired product gradation. This is an excellent case where a tried and proven system can be greatly enhanced.

The ambient two-roll mill system illustrated in Figure 1.75 is a two-stage system that can produce 30-mesh very economically and with predictable performance. Fiber separation using typical air tables work in unison with this process. By using hydraulic drives on the two-roll mills, achieving 40- or even 50-mesh CRM can become an economic reality. The ambient system is generally fairly quiet, and additional parallel grinders can be installed to increase productivity if all other process components are sized accordingly.

Another dry grinding system that in many ways resembles a paper mill refiner is the Gabriel reduction mill.

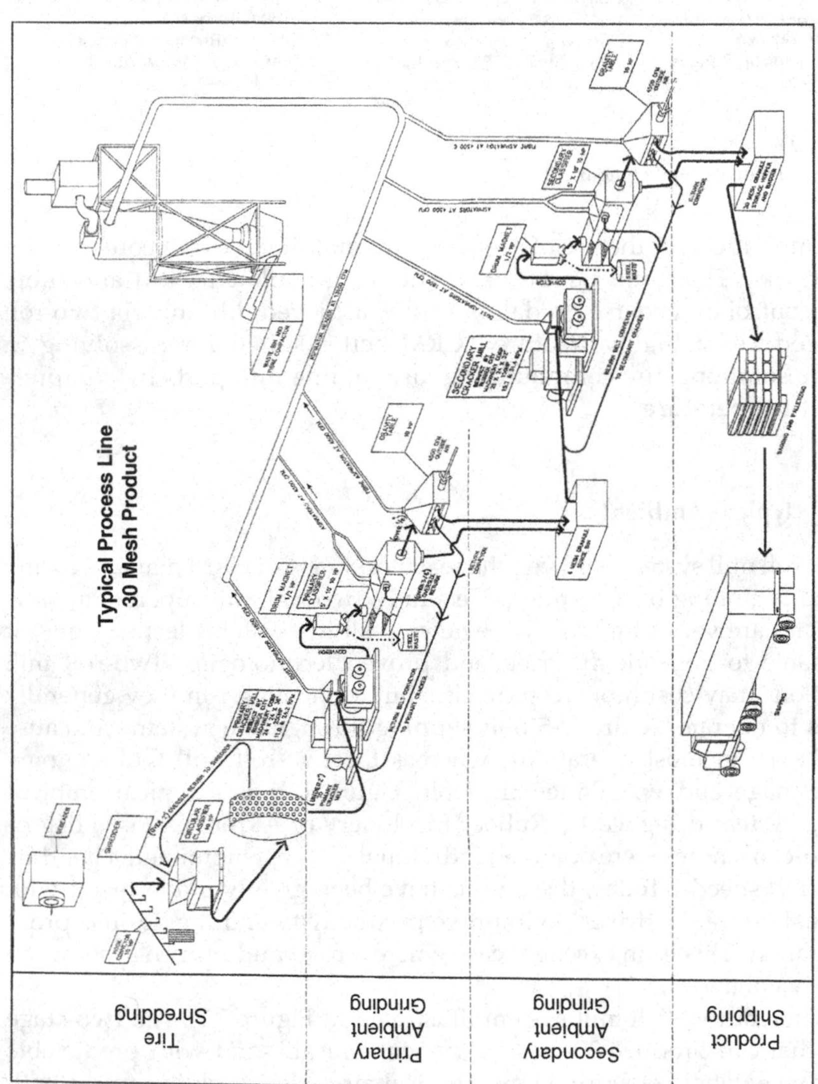

FIGURE 1.75
Typical two-roll mill CRM system.[41]

1.13.3 Typical Cryogenic

The cryogenic CRM systems can be found throughout the world. These systems depend on liquid nitrogen, although cold air systems are available as well as liquid carbon dioxide. Ideally, if these plants are located adjacent to an LN_2-producing facility, costs can be substantially reduced. LN_2 is like other energy-intensive systems that require energy to produce. Special tanks must be brought on site to handle the liquid cryogens, and these liquids evaporate with time. Generally speaking, these systems operate best in steady state and lose efficiency when idle or with interruptions in production.

The general efficiency of a cryogenic facility can vary from 0.4 pounds of LN_2 per pound of CRM produced, to 2.0 pounds of LN_2 per pound of product. The general average of a cryogenic facility is around 0.75 pounds to 1.0 pound of LN_2 per pound produced. The cost of delivered LN_2 varies greatly by location and by the LN_2 supplier and the contract supply arrangements. As with all supply contracts, the contractual agreement should be carefully reviewed and compared to the future expectations of the facility. Normally, there are penalty clauses by the supplier if the minimum quantities are not met, and an additional service charge is usually assessed for the supplier's LN_2 tanks. The liquid level in these tanks can be monitored remotely and the liquid level maintained without being a nuisance to the CRM producer. The LN_2 supplier can also be of great assistance in taking proper safety precautions with these very cold liquids and monitoring any unacceptable nitrogen levels within the facility. These systems come with complete automatic controls for ease of operation.

Other considerations are the type of high-impact mills used to reduce the frozen TDM into smaller sizes, and the liberation of the inherent reinforcing fibers. Contamination such as dirt, sand, and rocks, not caught by the magnets or other cleaning devices, can have a severe impact on the operation of these mills. Care should be taken to enclose the high-impact mills, to reduce noise generated by these units. The fiber separation can be accomplished with standard air separators and other pollution control equipment such as bag houses. Experience has shown that when tire fiber is liberated, it can become airborne and be very explosive and also a health nuisance. Care needs to be taken in the design of the facility to remove this material from the atmosphere. Protective masks should be worn in this work area. Since the final product is very cold, it tends to collect moisture from the atmosphere and must be dried to reduce the moisture content. Vibrating fluid bed driers can effectively remove this surface moisture.

The cryogenic system produces very clean product streams. The steel radial wires and beads tend to be very clean and can be easily sold to a local recycler. The steel recycler can provide the collection service, remove the collected steel product, and send monthly payments based on this service. Some operating facilities install steel balers for compaction and also baling units for the fabric stream. The fabric generally contains residual steel strands (which are sharp like hypodermic needles) that contaminate this product

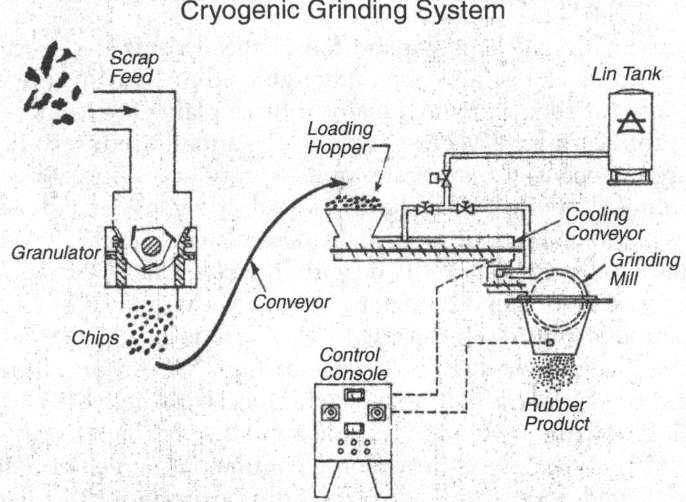

FIGURE 1.76
Basic cryogenic process layout.[1]

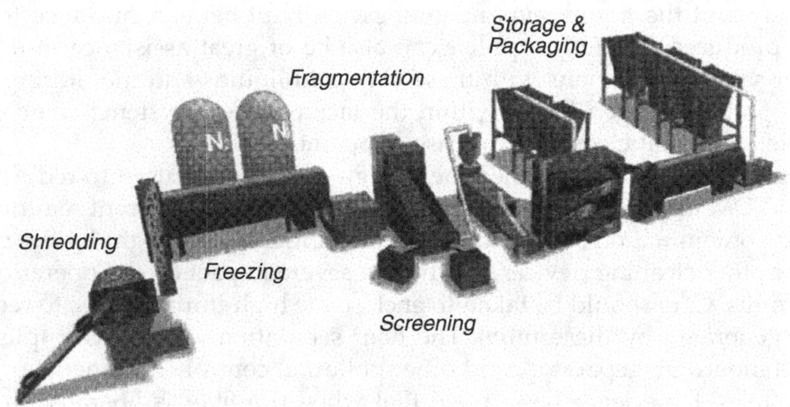

FIGURE 1.77
Conceptual layout of a Recovery Technologies (Canada) cryogenic plant.[57]

stream, as well as other inerts. Further classification of the fabric stream is generally required to make it salable for other uses.[73]

The cryogenic system concept can appear to be very simple as illustrated in Figure 1.76, which shows the basic cryogenic process layout grinding system. A more realistic layout is shown in Figure 1.77.[74]

Figure 1.78 reflects the real sophistication when fine cryogenic grinding is included. All such systems are involved and have many components to achieve the desired product and quality. As the CRM is ground to finer sizes, yields or total conversion of the larger CRM to the final sizes needs to be seriously considered. Recycling the overs to produce a true –80-mesh product

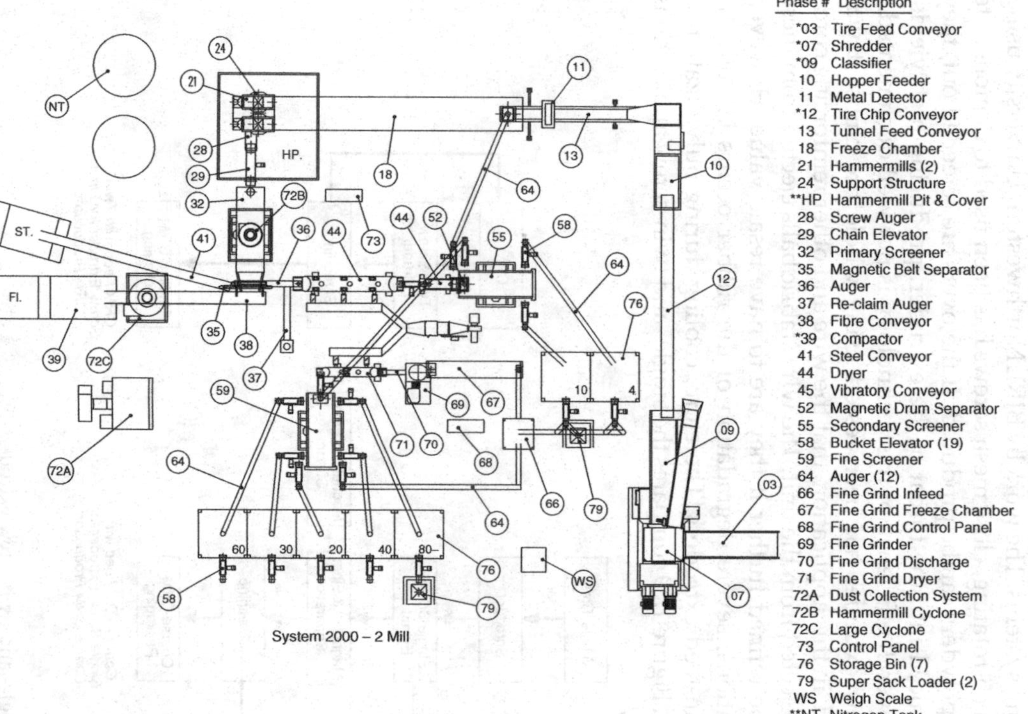
Phase # Description
*03 Tire Feed Conveyor
*07 Shredder
*09 Classifier
10 Hopper Feeder
11 Metal Detector
*12 Tire Chip Conveyor
13 Tunnel Feed Conveyor
18 Freeze Chamber
21 Hammermills (2)
24 Support Structure
**HP Hammermill Pit & Cover
28 Screw Auger
29 Chain Elevator
32 Primary Screener
35 Magnetic Belt Separator
36 Auger
37 Re-claim Auger
38 Fibre Conveyor
*39 Compactor
41 Steel Conveyor
44 Dryer
45 Vibratory Conveyor
52 Magnetic Drum Separator
55 Secondary Screener
58 Bucket Elevator (19)
59 Fine Screener
64 Auger (12)
66 Fine Grind Infeed
67 Fine Grind Freeze Chamber
68 Fine Grind Control Panel
69 Fine Grinder
70 Fine Grind Discharge
71 Fine Grind Dryer
72A Dust Collection System
72B Hammermill Cyclone
72C Large Cyclone
73 Control Panel
76 Storage Bin (7)
79 Super Sack Loader (2)
WS Weigh Scale
**NT Nitrogen Tank
**ST Steel Bin
**FI Fibre Bin
* Options not Included
** Supplied by Others
System 2000 – 2 Mill

FIGURE 1.78
Detailed layout of a Recovery Technologies fine grind cryogenic plant.[51]

may not be very efficient. Trials need to be run with the vendor and performance guarantees need to be in place prior to construction of the facility.

Typically, cryogenic systems produce very smooth or glass-like surfaces with relatively low surface areas compared to the two-roll mill, extruder, or wet grinding systems. This surface morphology can be very ideal for urethane systems, where reduced surface area means fewer urethane requirements. Additionally, cryogenics are ideally suited for elastomers such as butyl and all types of thermoplastics. However, as finer meshes are required, cryogenics begins to have limitations.

1.13.4 Shear Extrusion Pulverization

Shear extrusion process flow sheets look very similar to cryogenics and ambient grinding systems. The work by BIRL Northwestern University using a Berstorff with co-rotating, close mesh screws has been used to demonstrate that fine mesh powders can be produced. If the overs are screened out, they can be reintroduced into the throat of the extruder, producing high yields. However, the overall efficiency of this technology is not well documented.

A single screw extruder has been tested using whole tire chips. It was generally found, in this application, that the wire and other reinforcing cords separate when ejected from the unit. The wire/fabric balls ejected from these units need to be refined further if they are to have resale value. However, in all investigations, severe degradation of the rubber occurs in both the single- and twin-feed extruders. Therefore, cooling during pulverization is very important. Figure 1.79 compares the single- and twin-screw applications

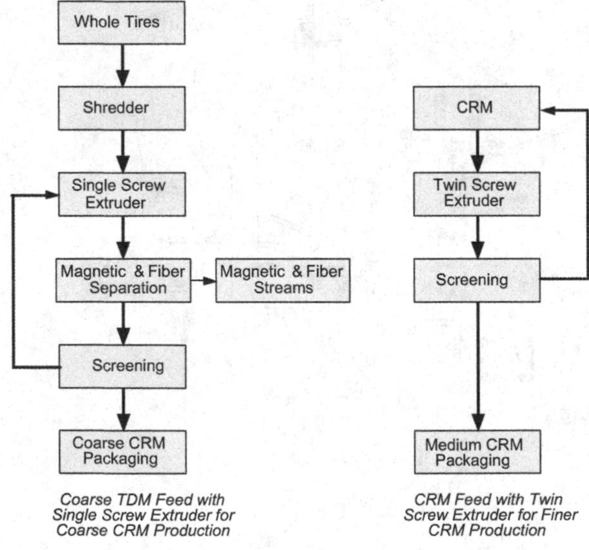

FIGURE 1.79
Comparison of single- and twin-screw extruders.[1,52]

for producing CRM. It is important to consider power requirements in relation to production capability and maintenance costs. Normally, extruders are very expensive to operate and maintain at high horsepower requirements, compared to other technologies. Dr. Klementina Khait has conducted some intensive research into the final particle morphologies of various grinding techniques to this process.[52] The final particle configuration appears to have a cauliflower shape, or the CRM particle appears to have exploded from its interior. The actual life and efficiency of the screw and extruder barrel in relation to maintenance needs more documentation. As noted earlier, improved commutation occurs with a certain amount of wear and then production falls off again. Good metrics by the CRM operator can help optimize the time to retool the extruder.

1.13.5 Ambient Air Milling

The ambient dry grinder system is similar to other systems in that it can be integrated into other TDM equipment components. Figure 1.80 depicts a suggested layout arrangement. Noncontact cooling water is used to maintain the temperature of the rotary air grinders. A wide particle distribution is obtained using this system. Table 1.13 shows the production capability at different operating rates.[79] For the Crosston grinder, the gradation output is similar to the MicroTec, and this gradation is reflected in Table 1.13. When dry milling without LN2, it is very important to ensure explosion suppression systems are installed and working properly. If one is going to produce

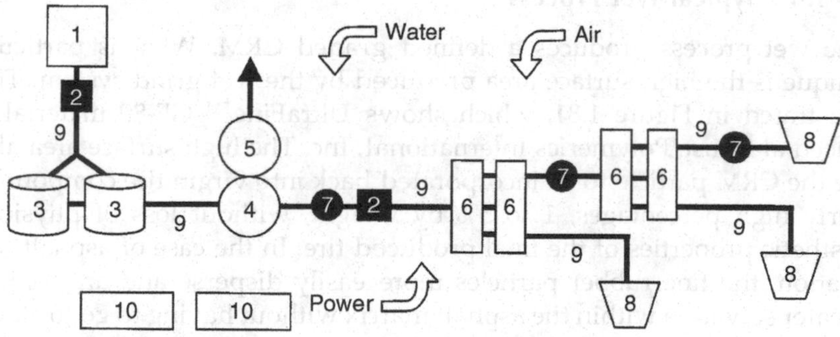

Item	Equipment Description	Qty	Item	Equipment Description	Qty
1	Feedstock Hopper	1	6	FSQ650-180 Air Rotary Screener	4
2	Metal Remover	2	7	Agent Feeder	2–3
3	XJF-1000 Grinder	2	8	Bagging Frames/Silo	3
4	Cyclone	1	9	Conveyor System	1
5	Dust Collector and exhaust	1	10	Control Panel	2

FIGURE 1.80
Process layout and description of an ambient air grinding system.

TABLE 1.13
Anticipated Production Capability and Resulting CRM Percentages[49]

Throughput	Unit	+40 Mesh (%)	40–80 Mesh (%)	–80 Mesh (%)
800–1000	lb/hour	0–10	45–60	40–60
1000–1200	lb/hour	5–10	45–60	35–50
1200–1600	lb/hour	5–20	45–60	30–40

a true –80-mesh product, a maximum yield of only 40 to 60% can be obtained. Recirculating the coarse fraction is not efficient and will not achieve a greater conversion. Therefore, if 100% –80-mesh product is desired, either an alternative method of handling the coarse faction needs to be installed or a coarse market found. Since these machines turn at high speed, they need to be contained in soundproof enclosures and careful air handling must be incorporated into to these units. The feedstock is generally a 5- to 20-mesh product and the more friable the incoming material, the better. Not a lot is known about the operating economics of these units. Additionally, reprocessing the overs developed by these machines is not practical or feasible unless some new commutation mechanism is developed. Machines like Pulva, Crosston, and the MicroTec do process softer rubber and plastics much better. Butyl tubes and butyl inner liners when used in conjunction with cryogenics can produce a very high-quality butyl 40 and even a 60-mesh grade.[49]

1.13.6 Typical Wet Process

The wet process produces a defined graded CRM. What is particularly unique is the high surface area produced by the wet grind system. This is illustrated in Figure 1.81, which shows UltraFine™ GF-80 material produced at Rouse Polymerics International, Inc. The high surface area allows for the CRM particle to be incorporated back into virgin tire compounds at fairly high percentages, 1 to 5% by weight, without loss of physical or aesthetic properties of the final produced tire. In the case of asphalt modification, the fine rubber particles more easily disperse and are subject to greater solvation within the asphalt matrix without having to go to elevated blending temperatures. Higher temperatures result in higher energy costs, degradation of the final polymer modified asphalt (PMA) blend, and reduced shelf life PMA product. Finer grades of wet grind grades are ideal for coatings and epoxies.

Figure 1.82 is a typical schematic trial setup using a fine GF-80 product for producing a PMA. For a full-scale operation, the fine tire powder can be delivered by pneumatic bulk trailer and loaded into a receiving silo. The CRM can then be metered into the blending system. The system came to be known as the *continual process* and was the concept for terminal blending at the asphalt refinery serving many asphalt plants rather than having to have

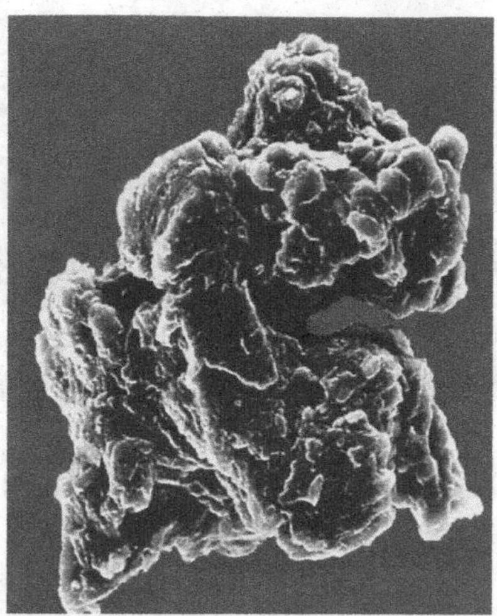

FIGURE 1.81
Surface topography of an UltraFine™ GF-80 CRM.[1]

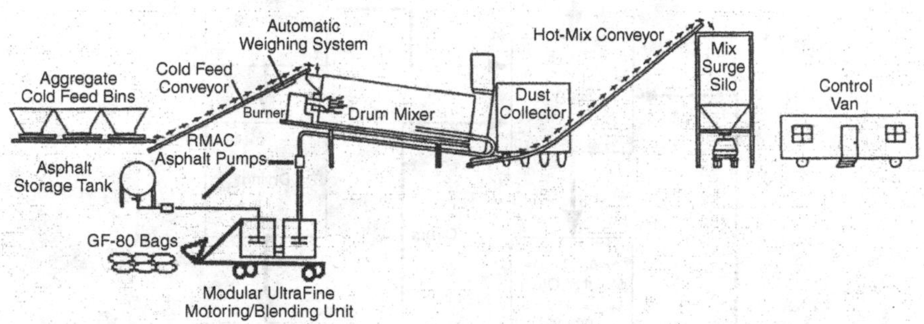

FIGURE 1.82
Schematic of a trial system to produce a PMA from a minus 80-mesh CRM.[1]

a blending system at each asphalt cement plant. An actual bulk delivery of powdered rubber to an asphalt facility is shown in Figure 1.83.

The general process flow sheet for a typical wet grinding process is illustrated in Figure 1.84. The system is unique in that the temperature of the process cannot exceed the boiling point of water. In most grinding processes, high temperatures greatly affect the polymers' integrity, especially in degradation. By maintaining lower processing temperatures such as in wet grinding, the integrity of the CRM is not compromised. This system also allows the easy addition of chemicals, such as surfactants to improve commutation. The surface and internal CRM chemistry can be altered in this process. By using water as a carrier, more efficient grinding can take place and the water can be the carrier to other locations. Specially designed pumps are required for this operation. By employing centrifugal force within the fluid system, light contaminants such as wood can be easily removed by

FIGURE 1.83
An actual asphalt facility using a bulk delivered minus 80-mesh CRM.[1]

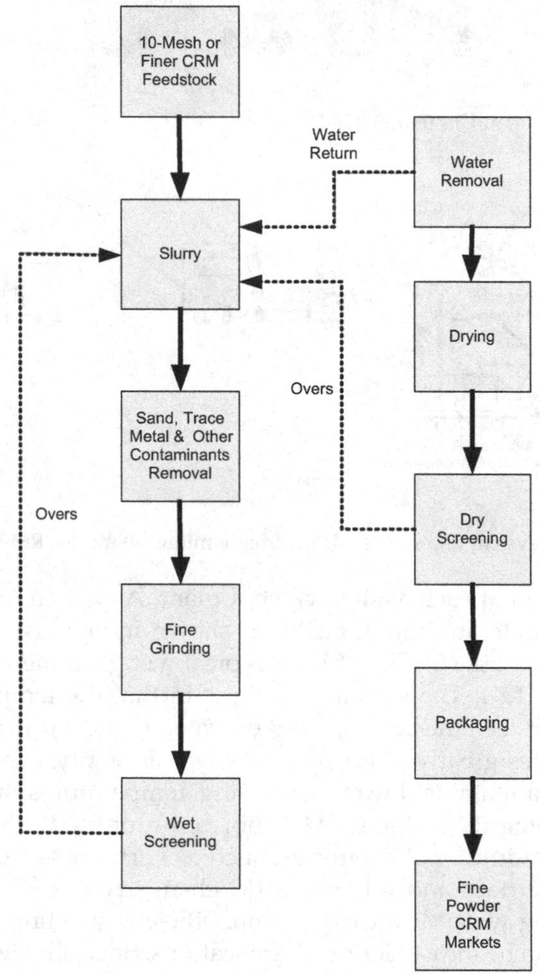

FIGURE 1.84
Wet grinding process.[1]

flotation and the heavier contaminants such as sand and metals can be separated by gravity. Slight improvements in production can be obtained with a finer mesh, such as 16-mesh product. Coarse CRM can be 100% converted to –40- to 200-mesh CRMs. The water in the system is recirculated after the water is removed. The only loss from the system is the 15 to 20% moisture removed by the drier. Care must be exercised in the design, construction, and operation of the drying unit, and to the drier's pollution control, combustion monitoring, and sensing system. By wet water slurry screening, more precise gradation control can be achieved. Final safety screening takes place after drying to assure no overs exist. Very high production rates can be achieved with this process.

1.13.7 Other Processes

Many other systems have been developed to reduce whole tires into chips. These include soaking tires in a solvent such as waste oil or subjecting the whole tire to cryogenics (freezing). High-pressure water spray jets can also be employed to reduce tires into CRM. Abrading mill systems have been developed and will fill in as a process player in the future. As new demands are made for better and better CRM materials, and as new technologies evolve, no one really knows what the future holds. New methods need to be tested over time. It is suggested that if a new concept appears to have merit, it should be operated in a pilot plant mode for an extended period of time to note the potential and/or deficiencies of the machine or the processes, and the necessary adjustments or modifications should be made. Then the new technology should be operated in parallel or in combination with the established system until it has been proven reliable and viable.

Columbus-McKinnon, Granutech-Saturn, Sweed Machinery, and others are developing machines to produce finer TDM in one pass through their machinery. What is needed in this case is screening systems after these machines to separate the rubber particles. For example, the coarse materials, say above 1/2-inch, could be sold as a TDF for crash barriers, while the –1/2 to +1/4 fraction could be sold for playgrounds or landscaping, and the –1/4-inch fraction for sports surfaces or other CRM products.[6]

1.14 Overall Concept

The preceding methods can be used in any combination or arrangement. Normally, a single large operation will do its main receiving, sorting, grading, and primary reduction on the day shift. Today's shredders allow for primary reduction of tires of all types to be done very efficiently. Even the modern truck tire debeader allows the operator to efficiently remove the beads and quickly shred the remaining tire casing, to then be quickly reduced in a primary shredder. Intermediate storages between each operation allow for greater efficiency. For example the primary shredding operation may run very nicely at 2,000 passenger or 100 truck tires per hour (20 tons per hour) and process all the incoming tires in one shift. However, the secondary operation may be able to run at only 1,500 pounds per hour and the tertiary operation at 1,000 per hour. Therefore, an operator will not want to run all the stages to the tertiary plant's rate due to power and labor costs alone. So intermediate storages and personnel considerations are very key elements. Additionally, a TDM operator may want to have an even higher shredding capability in the primary stages, since TCF and TDF markets may be much greater than CRM outlets. Careful analysis of present and future incoming feed tire types to TDM markets and availability needs of those markets is critical to the design and layout of a CRM facility.

It is advised that reinforced concrete flooring be used in handling tires instead of an asphalt pavement or dirt surfaces due to the shear mass and volume of the equipment interplay and debris that is generated. Good drainage and proper run collection systems should be designed. In case of a fire, emergency firefighting equipment needs to have easy access to the stored tire materials. A design that is well laid out will reduce the possibility of contaminants from the tire-shredding operation ending up in the final products.

1.14.1 Plant Design and General Process Considerations

Every CRM operation must take into consideration many elements. Table 1.14 shows plant design elements.[1]

TABLE 1.14

CRM Plant Design Considerations[1]

Adequate land for growth
Accessibility by major road routes (railroad and water preferable for bulk handling)
Utilities (water, power, sewage) for present conditions and future considerations
Operations under cover (preferable metal building with sprinklers)
Loading docks
All surfaces concreted with rebar reinforcement with controlled drainage
Surface runoff containment
Fire hydrants and other fire-prevention controls
Fire department and environmental agencies site map knowledge
All pertinent business and environmental operating permits
Continuous site monitoring, including direct hook-up to local fire departments
Compliance with all OSHA regulations
Safety stations and training and contacts in case of emergency
Metrics for optimization
Aesthetics and neighborhood considerations
Local perceptions of a tire-processing site and its derivatives

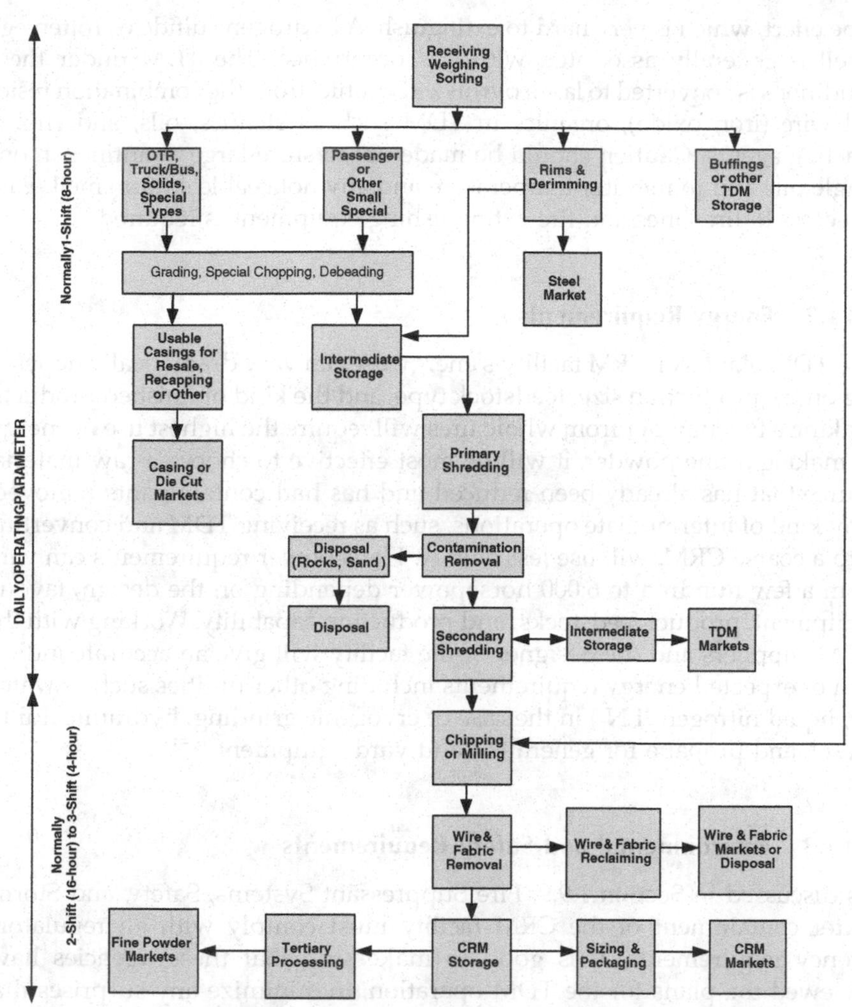

FIGURE 1.85
Suggested key components for a successful CRM facility.[1]

Today's CRM plants have many components that are necessary for a viable operation. Figure 1.85 illustrates some of the key component items and steps that should be considered. A particular process may have several shredders in parallel or series, and other operations may have special cleaning systems to handle landfilled tires. Some operations may be more extensive in some areas and deficient in others, such as tire wire and tire cord reclamation or cleanup for a particular market stream. Figure 1.85 is intended as a guideline or footprint for developing a successful operation and does not reflect a specific plant.[97]

TDM storage needs to be closely monitored, especially in large pile volumes. Spontaneous combustion can occur in the pile and produce a coal-generating-type effect, which is very hard to extinguish. A hydrogen sulfide or rotten egg smell is generally associated with this occurrence. The TDM under these conditions is converted to large pyrolyzed chunks from the combination residual wire (iron oxide), organics in TDM such as stearics, oils, and rubber reacting agents. Caution should be made not to store large quantities in one single pile and to monitor temperature and any noticeable sulfur smell. Ease of access to fire lanes and fire-extinguishing equipment is required.

1.14.2 Energy Requirements

The TDM plant's or CRM facility's energy use can vary dramatically depending on the production size, feedstock type, and the kind of finished products. Making a fine powder from whole tires will require the highest use of energy. In making a fine powder, it will be most effective to choose a raw material source that has already been reduced and has had contaminants removed. Any kind of intermediate operations, such as receiving TDM and converting it to a coarse CRM, will use less energy. Horsepower requirements can vary from a few hundred to 6,000 horsepower depending on the design, layout, equipment, product feedstocks, and production capability. Working with the OEM suppliers and the designer of the facility will give an accurate indication of expected energy requirements including other utilities such as water, air, liquid nitrogen (LN_2) in the case of cryogenic grinding, hydraulic fluids, diesel, and propane for generators and yard equipment.[48,53]

1.14.3 Environmental and Safety Requirements

As discussed in Section 1.9.9: Fire Suppressant Systems, Safety, and Storm, water containment of the CRM facility must comply with all regulatory agency requirements. It is good to make sure that these agencies have reviewed the plans for the TDM operation, to minimize any surprises that might affect the start-up and successful operation of the site.

1.14.4 Preventive Maintenance

Preventive maintenance is very critical to the reliability of a plant's success. Installation of continuous lubrication systems instead of doing manual greasing can usually be justified in labor savings and ensures continuous operation. Use of metrics and following a prescribed and documented routine can also be a significant cost savings. There are a number of companies that provide such services. They will come in and document each piece of equipment; evaluate the manufacturer's recommendations for maintenance; and write up schedules, the prescribed routine, and other critical items such as the type of oil recommended or a replacement schedule.

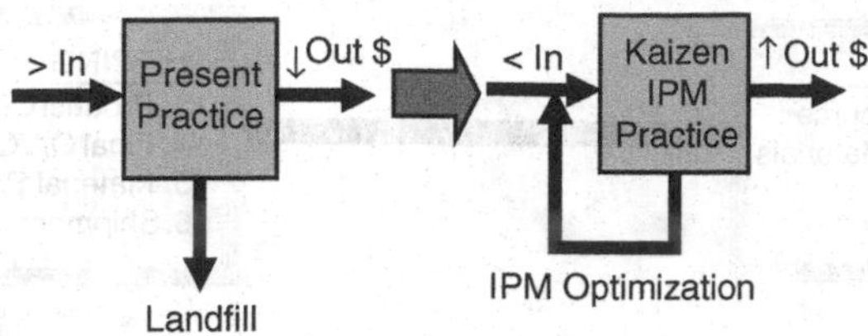

FIGURE 1.86
Custom grinding opportunity for a potential client.[1]

1.14.5 Custom Grinding

Although the CRM plant is focused on processing TDM materials, if the operation can be performed and maintained in an orderly and clean manner, processing of many other types of rubber materials can be done at the CRM site. This might include processing industrial scrap by tread rubber manufacturers, who may want their plant scrap tolled. By tolling, the scrap at the rubber manufacturer's plant is collected, transported to the CRM facility, and processed to generally a –40 mesh or finer. With few exceptions, nearly all tolling customers in these applications want a true –80-mesh tolled CRM. It is critical that these tolled materials be kept separate and the CRM to be clean of any contaminants prior to processing. Remember that the user of tolled materials closely monitors QA/QC. Closely monitored samplings, using analytical testing equipment, need to be inline and continuous.

Figure 1.86 details the methodology of how custom grinding can provide significant savings to the custom grind client. Previously designated for the landfill or some lower end use such as a supplemental fuel, the rubber polymer can now be converted into a useful raw material feedstock for the

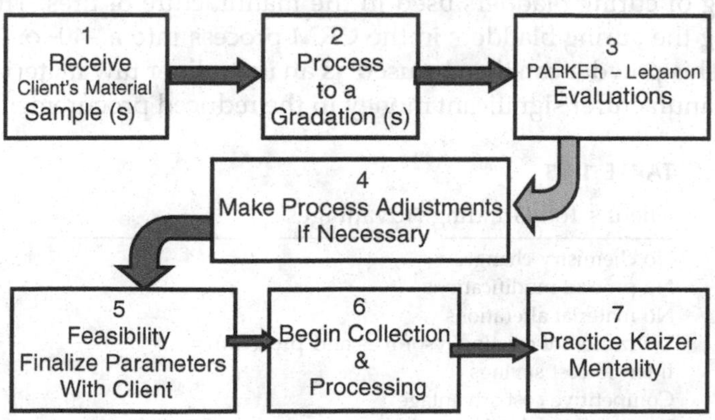

FIGURE 1.87
Developing the IPM or tolling opportunity.[1]

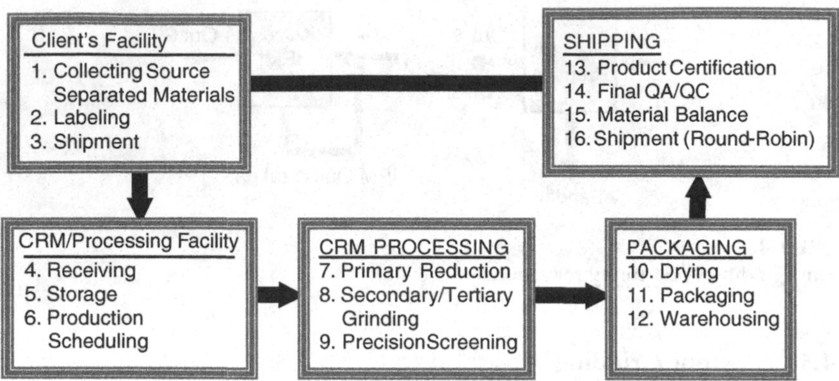

FIGURE 1.88
Custom tolling loop.[1]

client. IPM is the client in plant materials and the total operation should reflect constant improvement (Kaizen). This concept is depicted in Figure 1.87.

Once the potential client has shown an interest in having their rubber scrap processed, the IPM process can commence. Figure 1.88 illustrates a typical process to take the prospective client to the next level. This means receiving the plant's rubber material to be ground and evaluated by the client. Several repetitions may be necessary to achieve the optimum rubber gradation. Sometimes the entire process of evaluating, resubmitting, reevaluating, and optimizing the toll grinds, plant trials, and long-term tests may take up to several years. In this IPM process, the specifications, quantities, packaging, and pricing are quantified and agreed upon.

Table 1.15 shows some of the benefits of IPM or toll processing. Toll processing for the tire industry can be processing reject tread materials used in the manufacture of truck, bus, and OTR treads. It can also include the reclaiming of curing bladders used in the manufacture of tires. This means processing the curing bladders in the CRM process into a –40- or –60-mesh powder. This powder can then be used as an inner liner raw material, saving the tire manufacturer significant money in the reduced procurement of butyl

TABLE 1.15

Client's Toll-Grinding Advantages[1]

No chemistry changes
No process modifications
No material alterations
Utilization of existing resources and procedures
Internal cost savings
Competitive cost advantage
Improved production performance
Internal business operation (trade secret)
Landfill cost savings and future environmental cost avoidance

feedstocks. If butyl is to be run in the CRM, care must be taken to have a thorough clean-out, so no other tire contamination is present.

1.15 A Market-Driven Approach

A market-driven approach to developing a successful CRM facility with TDM markets can best be illustrated in the work performed by Chuck Smith, a professional engineer with a unique background in developing processes from tire manufacturing to TDM/CRM facility operations. In Figure 1.89, Smith first identifies the target (the Market), and then evaluates the weapons (the Equipment). These steps are shown in a flow sheet illustrated in Figure 1.90. The final step is the strategy (the Method) that is part of the process flow sheet. These two illustrations go a long way in determining the steps to begin the path toward a successful CRM operation. A CRM facility plan may also include the necessary steps over an extended time period; milestones to achieve on this path can act as measurements for what will be accomplished next. This may mean reevaluating new pieces of equipment that may have come on the market, a potential acquisition, a new market, a lost customer, refinancing, possible relocation, or a merger. TDM will be dynamic and continually evolving; the CRM professional today must be ever watchful of the changes and opportunities arising out of these events.[54]

1.16 Economics

Figure 1.91 reflects the generally accepted value assessed for the various CRM gradations. Location of the facility will have an impact on what the end user is willing to pay. The closer a CRM operation is to the purchaser's facility, the lower the cost of transportation and the greater the advantage to the CRM producer. Maintaining handling costs at a minimum, maximizing on economy of scale, and getting the best prices will have a pronounced effect on the overall success of a TDM facility.

The economics of a tire-derived-material operation are varied. Each step of the operation needs to be understood and measured. Metrics offers such a method but has yet to be totally employed in any operation. Metrics is a system of documented measurements of the process, and covers all phases of the operation. This includes the number of tires received and the number processed, power consumption by individual equipment components, the flow into and out of a particular piece of equipment or process section, and the overall dynamics of the entire operation. It is absolutely essential to understand all the operating parameters from a cost perspective and to determine what it takes to break even. In all the failed CRM operations

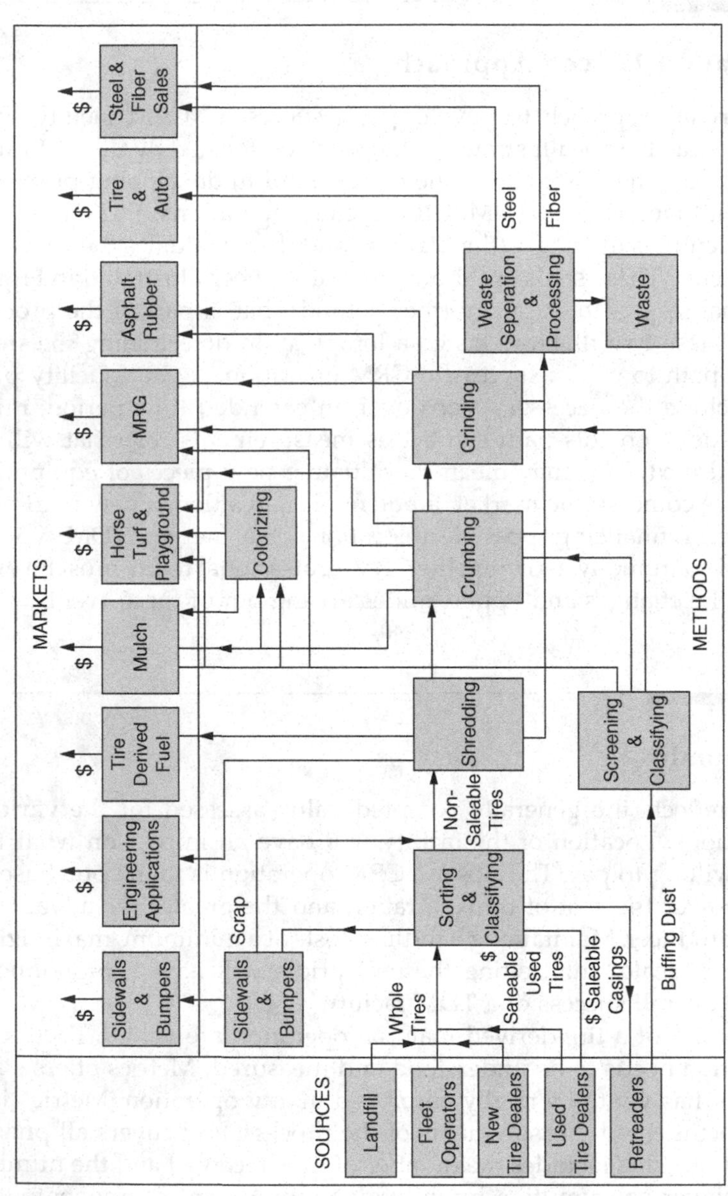

FIGURE 1.89
A market approach.[54]

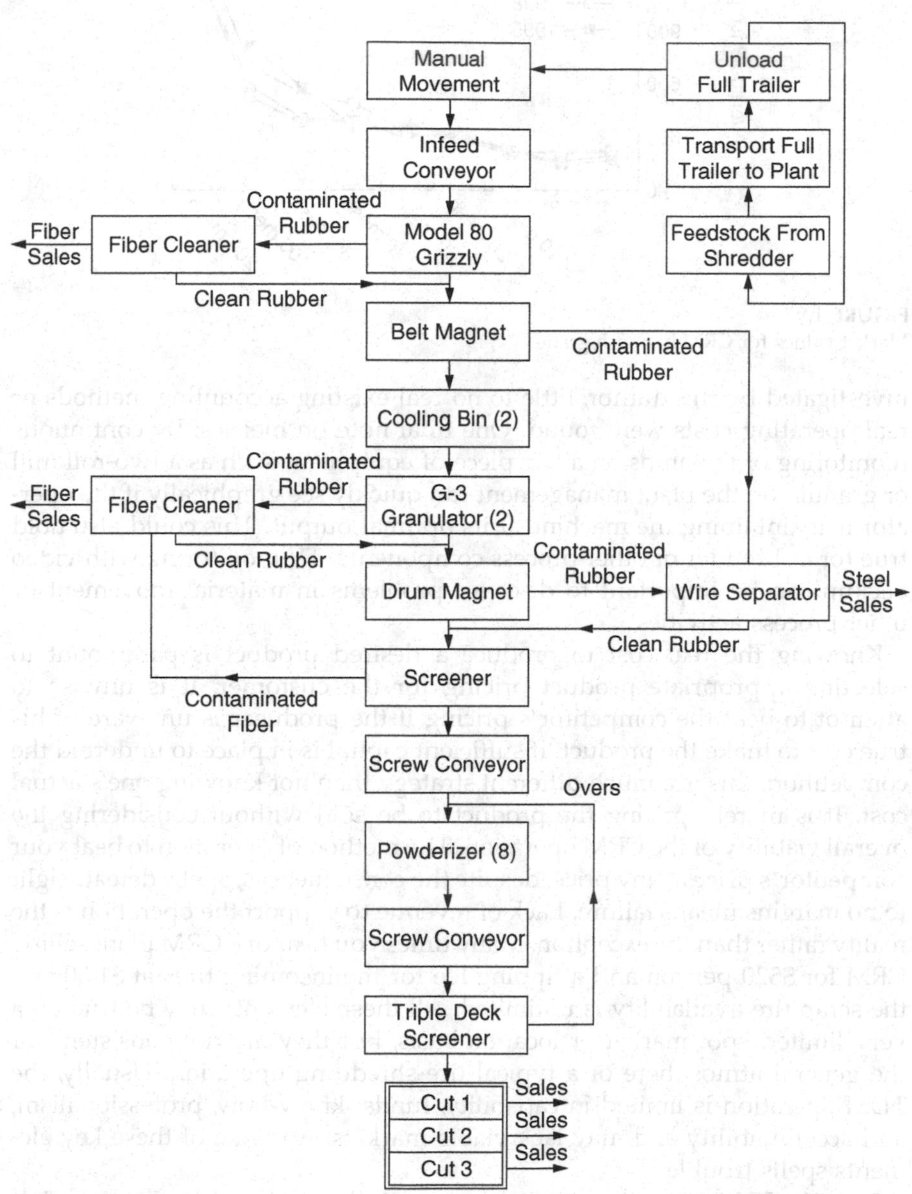

FIGURE 1.90
The flow sheet.[54]

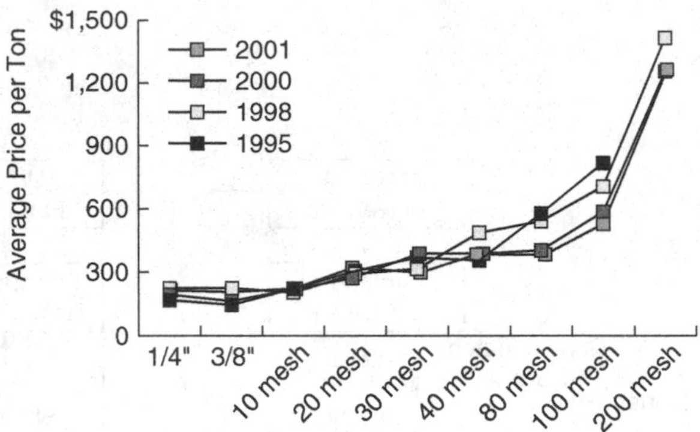

FIGURE 1.91
Market values for CRM by mesh grade.[55]

investigated by this author, little to no real existing accounting methods or real operating costs were found. One final note on metrics: By continuous monitoring of the amps on a key piece of equipment such as a two-roll mill or granulator, the plant management can quickly see graphically if the operator is maintaining the machine at its optimal output. This could also hold true for a shredder or other process components. TV monitoring with video backup can be important to discover problems in material movement or other process activities.

Knowing the real cost to produce a desired product is paramount to selecting appropriate product pricing for the customer. It is unwise to attempt to beat the competitor's pricing if the producer is unaware of his true cost to make the product. If sufficient capital is in place to underbid the competition, this is a much different strategy than not knowing one's actual cost. It is merely pricing the product to be sold without considering the overall viability of the CRM operation. The method of operation to beat your competitor's price at any price, despite the consequences, spells defeat. Tight to no margins means failure. Lack of revenue to support the operation is the reality rather than the exception. Many times you hear of a CRM plant selling CRM for $320 per ton and a tipping fee for the incoming tires at $1.00 and the scrap tire availability is unlimited. All these elements may be true on a very limited spot market or location basis, but they are not consistent for the general atmosphere of a typical tire-shredding operation. Usually, the TDM operation is limited in capability, funds, know-how, professionalism, and accountability and may lack viable markets. Any one of these key elements spells trouble.

Today's TDM operation must have multiple markets for the materials produced at the plant site, a detailed plan of action, and alternate plans if thing go in a different direction. Good metrics and of base selling prices must be known. Any operation should be held accountable and the funds wisely spent. The TDM equipment should be oversized, despite what initial trials at an OEM or reseller suggest. Spare parts and excellent mechanics with a preventive maintenance program need to be integral parts of the plant's operating parameters.

The grading of recoverable casings cannot be underestimated. Fabric tires from passenger, truck, and OTR are excellent feedstocks for producing playground and equestrian TDMs. The recovery of truck/bus and truck duals to be repaired and recapped can help bring in excellent income to a shredding and tire collection system. Finding reputable and reliable casing dealers may take time and experience, but the effort is well worth it. One word of caution, as liability insurance cases increase, this will have a predominant future effect on recovered casings to be used back on transportation vehicles.

Good relationships with governmental entities and customers must be established. Local laws and ordinances must be known. In some locations, there are restrictions on the storage of whole or partially shredded materials. The fire marshal in a local area might have certain demands that need to be met for a company to operate. Keeping updated in all these areas and exploring these conditions requires full-time personnel and is a cost of doing business. All of these considerations need to be built into the final selling price of the produced TDM or CRM. To date, no operating entity has clearly demonstrated that it has made a reasonable profit or a profit at all on any consistent basis. It is hoped that by employing some of the ideas presented herein, the TDM industry will develop and document proven and viable economic results.

The Scrap Tire Management Council, 2002, and the Recycling Research Institute, 1998 have broken the general CRM market down by size and distribution. This information is reflected in Table 1.16, showing CRM market distribution. These percentages will change as the CRM market matures expands with time. This table does demonstrate the growth in CRM and TDM.[2]

Due to the high cost and maintenance of CRM systems, and taking this into account against the current prices paid for these materials, it appears

TABLE 1.16

CRM Market Distribution[2]

Market	Weight in Millions of Pounds	Percent	Mesh Range
Molded products	307	31	3/8-inch–40-mesh
Asphalt modification	292	29	4–200-mesh
Sport surfacing	141	14	1/4-inch–40-mesh
Tire and automotive products	112	11	10–40 mesh and 80-mesh for tires
Plastic blends	38	4	10–120-mesh
Animal bedding	37	4	1/4–3/8-inch
Surface modifications	36	3	10–40-mesh
Construction	28	3	10–40-mesh
Other	5	1	1/4-inch–200-mesh

only large, efficient operations can hope to remain sustainable and make a potential profit in this industry. Consideration must be given to the rise in whole tires being consumed by cement plants and in some cases dedicated boilers. Tire availability can be greatly affected, along with disposal fees being asked in the general vicinity of a major operating cement facility. These plants can burn passenger tires and many are equipped to handle truck and bus types also. Therefore, care must be taken in planning a shredding or tire collection facility and the impact of these facilities.[110]

1.17 Other Considerations

1.17.1 Energy Hierarchy of TDM

There are many processing alternatives available. In general, the value of developing a specific TDM can be placed on a hierarchy scale that was developed by a Department of Energy study in 1983. This study made a comparison of the replacement value of TDM to its replacement in the marketplace. It took into consideration the amount of energy required from the time the hydrocarbon or crude oil was pumped from the ground, refined, and converted into the tire itself. This table illustrates that asphalt applications for roads or shingles can have the potential for the greatest savings. Obviously, the use of CRM back into tires is the highest use, but is limited in the total reuse of scrap tire stream. The new rubber-thermoplastic combinations and the use of CRM as an asphalt modifier for highway pavements, sealants, and roofing applications can and will consume the greatest portion of worn-out tires in a safe and environmentally sound manner. The greater the initial savings, the greater the potential profit margin for the scrap tire processor and the greater opportunity to have a viable market for the long term. The results of this study are summarized in Table 1.17, which shows selected processing alternatives for scrap tires.

TDM systems are now depending on each other and are becoming interconnected. This means a primary tire processor producing TCM or TDF can be supplying a CRM operation. The CRM operation could be supplying a tertiary operation to produce fine powders well before the CRM is used in a final product. If there could be a synergistic interdependence between the entire scrap tire processing system and processors, this could result in each TCM entity operating at maximum efficiency. Vertical integration is fine, but has not become a reality yet. The capital and general resources required make the task very challenging.

1.17.2 The Organization

The real essence of any organization is the management and/or the leader, that is, the person or persons who control or direct a business. The leader is

TABLE 1.17

Selected Processing Alternative for Scrap Tires[56]

Process Type	Initial Savings	Later Combustion	Plant Comments	Scale	Potential
Whole tires	15,000		Requires special furnace	Large	Could consume all tires
TCF, TDF	15,000		Can be used in many types of combustion units	Medium	Could consume all tires
Pyrolysis (fuel products only)	11,000–14,000		Could replace up to 12,000 Btu/lb	Large	Could consume all tires
Pyrolysis (recovered fuel products and carbon black	19,000–23,000	4,000	Quality of any given by-product in question	Large	Could consume all tires
Reclaim	27,000	15,000	Considered inferior product	Small	Limited
Biological degradation	30,000	15,000	Experimental	Small	Could consume all tires
Microwave devulcanization	33,000	15,000	In development	Small	Could consume portion of tires
Asphalt modifier	90,000		Great potential	Medium	Could consume all tires
Plastic modifier	90,000		Great potential	Medium	Could consume all tires

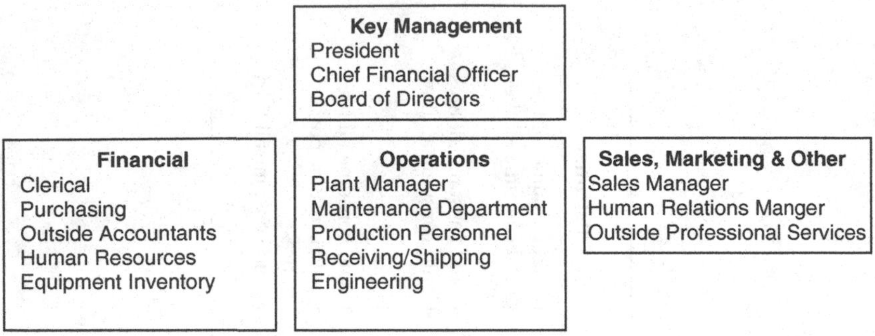

FIGURE 1.92
Organizational considerations.[1]

the one who is in charge, leads the others, and is responsible for the overall direction, credibility, and spirit of the company. The president normally fills this function and reports directly to the board of directors if the organization is not a sole proprietorship. The leader and his management team must have the skills to operate the many facets of a CRM facility. Careful selection in this process is very important. Many companies in other industries find that a blend of talents within and outside the industry is essential in maintaining proper focus and flexibility to meet a wide variety of challenges that every business must face.

Assuming the CRM facility has all the equipment, financing, permits, markets, and the optimal location, no progress will be fruitful without the key annual intervals and staff in place to make success a reality. Figure 1.92 shows just some of the key personnel to consider for a major TDM or CRM facility. Personnel are listed within grouping areas to consider, rather than in order of hierarchy. Sales and marketing can work with outside advertising agencies and with a broker who may have markets for TDF. Engineering may be under the operations manager or report directly to the president or a particular department. R&D may be put in a special category within the organization or have an outside consulting role, reporting to someone within the organization. A number of these individuals can be added to the staff as revenues and needs arise.

The professional operation today and in the future will require all of the above areas to be handled responsibly. For any initial start-up, some of the key management may have to wear several hats or perform a number of these functions. In the longer scheme of things, if the operation is to function smoothly and be flexible in meeting future opportunities, all of the basic managerial functions need to be addressed and performed. In some cases, outside professional agencies and consultants can be of great assistance, and a clerical person on the staff can collect the required information to be reviewed by the retained professional agency — information such as accounting, environmental permitting, or key market development segments. A good local accounting firm can help ensure that the books are in proper order for financial packages, IRS audits, etc. An outside consultant can greatly assist with plant layout and future plant expansions, or by working with local agencies to ensure all bases have been covered for OSHA requirements and air and water permits. Local employment agencies can be of great assistance in locating personnel on either a full-time or temporary basis, and can act as the company human relations go-between for an interim period. Flexibility, open attitude, planning, and developing good relationships are vital human components that will make a CRM facility sustain itself well into the future.

1.18 Summary

Tire-derived materials (Figure 1.93) are finding their use in a wide variety of applications, such as for supplement fuel, fuel by-products, playground applications, sanitation leach fields, composting materials, equestrian uses, and civil engineering applications. Whole tire parts will still be used for a number of applications (Figure 1.94). Sophisticated engineered polymers such as thermoplastic vulcanizates are being used successfully in new tires. The process used is widely varied, depending on the processor, and in many ways it is shrouded in trade secrets. By participating in trade organizations, researching patents, and reviewing publications, a good perspective of existing tire-shredding entities and the persons involved can be gained.

Tires are designed to last. They have many unique, intrinsic qualities that make them valuable after their normal use as a tire. Taking advantage of the inherent chemical and rheological properties of tires, to make or modify many new materials, is not only the wave of the future but also an excellent use of one of our national resources. Tires are an excellent modifier for asphalt. America's roads are deteriorating at an alarming rate as we grow as a nation and depend on our road infrastructure for our growth and prosperity. A working knowledge and understanding of the many different

FIGURE 1.93
TDM products.

FIGURE 1.94
Whole tire parts.

ways to build, maintain, and operate a tire-shredding system are key elements for success.

Competing markets for whole tires, such as cement kiln, need to be recognized as a concern to existing and planned tire processing facilities. A rising consciousness of global warming and enforcement of the Clean Air Act of 1990 will play a vital role in permitting of facilities to continue to burn or who plan to use TDF. How all these different arenas play out will only be known in time. Having and developing multiple markets for whole tires will be essential to any tire operation.

Tire shredding is very capital intensive and requires a professional approach for its success. Tire-derived materials such as CRM must realize their maximum economic value as needed resources and products in America's marketplace. Undervaluing TDM will only spell disaster for the industry. Tire wire markets will continue to grow due to the need to recover our precious resources found in scrap tires. Proper resources and adequate capital, market positioning, dedicated professional employees, optimal plant layout, and operational savvy will spell success for the scrap tire processors of the future.

References

1. Personal notes of Michael W. Rouse 1977–2005.
2. Conversation and information from Michael H. Blumenthal with Scrap Tire Management Council Washington, DC, May 2002.
3. Conversation with Tire Retread Information Bureau, Harvey Brodsky, Pacific Grove, CA, May 2002.
4. Scrap Tires: A Resource and Technology Evaluation of Tire Pyrolysis and Other Selected Alternative Technologies, Idaho National Engineering Laboratory, November 1983, Prepared for the U.S. Department of Energy.
5. *The Vanderbilt Rubber Handbook,* R. T. Vanderbilt Company Inc., New York, NY. 1968.
6. *The Scrap Tires & Rubber Users Directory 2005,* 14th ed., Rubber Research Institute, Suffield, CT.
7. *McGraw-Hill Concise Encyclopedia of Science & Technology,* 5th ed., McGraw-Hill Book Company, 1984.
8. Particle Attrition, Trans Tech Publications, Federal Republic of Germany, 1987, pp 15–19.
9. *Principles of Powder Technology,* John Wiley & Sons, New York, 1990, pp 234–235.
10. Scrap Tires Disposal and Reuse, Robert H. Snyder, Society of Automotive Engineers, Inc., Warrendale, PA, 1998.
11. Tire Derived Fuel: Environmental Characteristics and Performance, prepared by Terry Gray, President, T.A.G. Resource Recovery, 2005.
12. *Random House Webster's Unabridged Dictionary,* Random House, New York, Second Edition, 2000.

13. *Steam: Its Generation and Use,* Babcock & Wilcox, New York, NY, 1972, p. 22-6.
14. Information courtesy of Global Tire Recycling, Wildwood, FL, www.gtrcrumb rubber.com, 2004.
15. *Science and Technology of Rubber,* 2nd ed., Academic Press, San Diego, 1994.
16. Scrap Tire Management Council, Scrap Tire Use/Disposal Study, Final Report, September 11, 1990.
17. Glossary of Terms Relating to Rubber and Rubber Technology, Special Technical Publication 184A, (04-184001-20), American Society of Testing Materials, Philadelphia, PA, 1972.
18. Information courtesy of International Baler Corporation, www.intl-baler.com, Jacksonville, FL, 2004
19. Information courtesy of Encore Systems, Inc., www.tirebaler.com, Cohasset, MN, 2004.
20. Information courtesy of R. M. Johnson Company, www.ezcrusher.com, Annodale, MN, 2004.
21. Information courtesy of Carpo International LLC, www.carpo.net/bulk bgs.php, Los Angeles, CA, 2004.
22. Information courtesy of HALLCO Live Floors, www.hallcomfg.com/live floors/index.html, 2000 Hallco Mfg. Co. Inc., Tillamook, OR, 2004.
23. Information courtesy of Dings Company Magnetic Group, www.dingsco.com/overhead_magnetic_separators.htm, Milwaukee, WI, 2002.
24. Information courtesy of Action Equipment Company, www.actionconveyors .com, Newberg, OR, 2005.
25. Information courtesy of Sweed Machinery Inc., www.sweed.com, Gold Hill, OR, 2004.
26. Information courtesy of Multilift, Inc., www.multilift.thomasregister.com, Denver, CO, 2004.
27. Information courtesy of Griffin Environmental Company, www.griffinenviro .com, Syracuse, NY, 2004.
28. Information courtesy of Forsberg, Inc., Three Rivers Falls, MN, www.forsbergs.com, 2004.
29. Information courtesy of Rotex, Inc., Cincinnati, OH, www.rotex.com, 2004.
30. Information courtesy of Sweco, A business unit of M-1 SWACO, Chicago, IL, 2004.
31. Reclaimed Rubber, The Story of an American Raw Materials, John M. Ball, Rubber Reclaimers Association, Inc. New York, 1947.
32. Information courtesy of Eagle International, LLC, www.eagle-equipment.com/html.com, Dakota, NE, 2005.
33. *Materials Handling Handbook,* 2nd ed., John Wiley & Sons, New York, 1985.
34. Information courtesy of Wendt Corporation, Tonawanda, NY, www.wendtcorp .com, 2005.
35. Scrap Tires Disposal and Reuse, Robert H. Snyder, Society of Automotive Engineers, Inc., Warrendale, PA, 1998.
36. Information courtesy of Diamond Z Manufacturing, Caldwell, ID, www. diamondz.com, 2002.
37. Information courtesy of Pulva Corporation, www.pulva.com, Saxonburg, PA, 2003.
38. Information courtesy of Tryco/Untha International, Decatur, IL, 2004.
39. Information courtesy of Columbus McKinnon Corporation, Sarasota, FL, 2003.

40. Information courtesy of Barclay Roto-Shred Inc, Stockton, CA 95213, www.tireshredders.com, 2004.
41. Information courtesy of Rubber Machinery Corporation, Akron, Ohio 44309, 2004.
42. Hugglunds Drives AB Brochure, EN 350-2A, Sweden
43. Information courtesy of Tire Recycling Consultants (TRC Inc.), www.recyclingtires.com/index.htm, Fort Lauderdale, FL, 2004.
44. Information courtesy of Cumberland Engineering (Division of John Brown, Inc.), S. Attleboro, MA, Web site: www.cumberland-plastics.com, 2005.
45. Information courtesy of Granutech-Saturn Systems Corporation, www.granutech.com, Grand Prairie, TX, 2004.
46. *Perry's Chemical Engineers Handbook,* 6th ed., McGraw Hill, New York, 1992.
47. New Solid State Shear Extrusion Pulverization Process for Used Tire Rubber Recovery, Paper No. 13, Dr. Klementina Khait, Rubber Recycling and Technology, June 3-5, 1998, Rubber Division, ACS, Akron, Ohio.
48. *Rubber Technology Handbook,* Hanser Publications, Munich, 1989.
49. Information courtesy of Crosston Industries, Drummondville (Quebec), J2C 8M6, Brochure Package, 2003.
50. Information courtesy of Bi-Metal Corporation, www.bimetalrecycling.com, Ridgefield, CT, 2005.
51. Information courtesy of Recovery Technologies (Canada) Inc., Cambridge, Ontario N1R 7E5, 2002.
52. New Solid State Shear Extrusion Pulverization Process for Used Tire Rubber Recovery; Paper No. 13, Dr. Klementina Kait, Rubber Recycling an Technology, June 3-5, 1998, Rubber Division, ACS, Akron, Ohio.
53. *Polymer Recycling Science, Technology and Applications,* John Wiley & Sons, Chichester, 1998.
54. Conversations and flow diagrams with and by Chuck R. Smith, PE, Brownsville, Texas, on his experiences and examples of a market driven approach to producing TDM and CRM, May, 2003.
55. *Resource Recycling,* April, 2003 Vol. XX, No. 4, pp. 11–15.
56. Scrap Tires: A Resource and Technology Evaluation of Tire Pyrolysis and Other Selected Alternative Technologies, Idaho National Engineering Laboratory, November 1983, Prepared for the U.S. Department of Energy.
57. Information courtesy of TMI Systems, www.tmi-systems.com, Dayton, OH, 2004.

2

Quality Performance Factors for Tire-Derived Materials

Michael W. Rouse

CONTENTS

2.1 Abstract

The use of crumb rubber materials (CRMs) from scrap tires has met with a great deal of resistance by industrial users, mainly due to the perception that quality cannot be met on a consistent basis. These hurdles and mindsets can be overcome if these tire-derived materials can be consistently demonstrated to be a viable raw material component in the final compound for high-end applications, mainly new tire manufacture. This chapter discusses the strategy, practices, and philosophy necessary to provide high-quality raw materials derived from scrap tires, that is, CRMs. Areas covered include strategic management, manufacturing, quality control, and quality assurance practices to ensure that CRM will be a valuable resource for the compounder.

2.2 Introduction

During World War I and World War II, shortages of essential materials led to collection drives for silk, rubber, and other commodities. In recent years, the environmental benefits of recycling have become a major component of waste management programs. However, success in using tires, a valued resource, takes on a different paradigm — how to effectively convert a waste material into a useful commodity that can be used in a wide variety of applications with assurance. By evaluating and instituting various quality control practices, the processor today can employ procedures to assure the operation meets or exceeds desired, consistent quality of a specified material. Most importantly, the client or final end user can be confident that procedures and practices have been put in place to assure that the whole tire-derived materials (TDMs) meet the highest standard for quality and consistency in a normal mode of operation.

The recycling process is the recovery and reuse of waste products. These may include household products, postmanufacturing material, agricultural waste, and for very select uses, scrap tires. The reuse of tires is generally perceived as reducing the burden on the environment. With increased understanding, waste tires can be seen as a valuable resource.

Quality performance is generally not associated with recycling. Success in supplying high-quality recycled material replacements must come under the same scrutiny that is demanded of virgin materials. Quality, quality control (QC), and quality assurance (QA) go hand in hand. Quality is defined as "an inherent or distinguishing characteristic; a property." Quality control is "a system for ensuring the maintenance of proper standards in manufactured goods, especially by periodic random inspection of the product." Quality assurance is "a program for the systematic monitoring and evaluation of the various aspects of a project, service, or facility to ensure that standards of quality are being met."

2.2.1 The Average, Standard, and Bound of Measurements

In this chapter, statistics most used to characterize observed data and look at lower bounds on the percentage of measures can be defined as,

The average:

$$X = \sum_{i=1}^{n} x_i / n \tag{2.1}$$

Standard deviation:

$$\sigma = \sum_{i=1}^{n} (x_i - X)^2 / (n-1)]^{1/2} \tag{2.2}$$

where $x_1, x_2, \ldots, x_n$ are the individual observations in a random sample from the TDM process. The standard deviation gives a measure of how the "n" observations vary about the average. The square of the standard deviation is called the variance. The lower bound on the percent of measures is at a given interval X – 2 to X + 2 for 75%, X – 3 to X + 3 for 89% of the observations or the confidence level desired, such as X – 6 to X+ 6 for Six Sigma.

2.2.2 The TDM Challenge

Before the tire processor can convert scrap tires into a useful product, he must have an understanding of the complexity of the tire, including its various types and compositions. Different types include passenger (such as the new sophisticated run flat type illustrated in Figure 2.1), SUV, truck and bus, agricultural, and off-the-road (OTR). The general composition of tires is noted in Table 2.1.

Concerning a tire's composition, the rubber components can vary widely. Passenger types tend to be high in styrene butadiene rubber (SBR), a general purpose rubber. Truck tires tend to be a 50:50 blend of SBR and natural rubber (NR). OTR types are composed mainly of NR, which has excellent fatigue resistance and high strength; these tires are mainly designed for cut and chip ruggedness. Knowing the tire source — the raw material — and its relation to the intended application can play a major role in the success or failure of the processed TDM. Consider a situation where crumb rubber material (CRM) is designated for an asphalt application. If the initial test

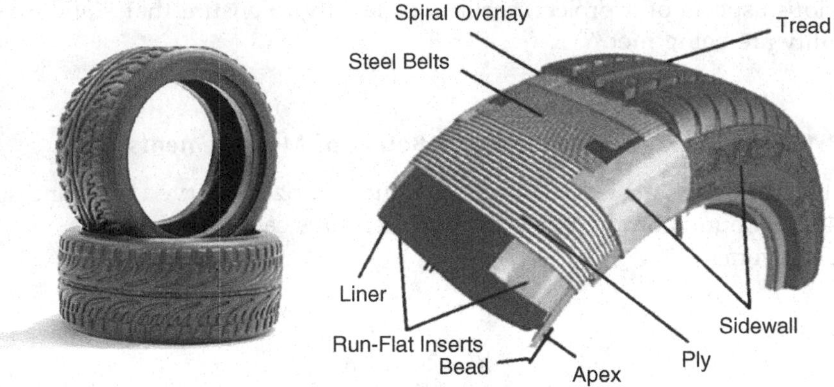

FIGURE 2.1
Typical passenger tire and components of a run flat tire.

TABLE 2.1
General Composition of a Tire

The Materials of Construction	Approximate Weight Percent
Rubber hydrocarbon composition (RHC)	48
Carbon black and silica	22
Metal reinforcements	15
Oil, antidegradants, wax, stearic acid, etc.	8
Fabric	5
Zinc oxide (ZnO)	1
Curing agents	1
Total	**100**

work was with OTR rubber but the tire processor shipped passenger type CRM instead, the result could be disastrous at the paving site.

Let's analyze why such a mix-up could be a disaster. CRM from OTR tires or NR will dissolve more readily in a given asphalt cement (AC) at lower temperatures and will provide low-temperature performance for a given climate, such as very cold weather. The passenger tire CRM, on the other hand, will require higher blending temperatures and be more suitable for hot climates. This dichotomy could produce very different results. Today, the Federal Highway Administration (FHWA) Superpave™ program has developed very specific performance tests to meet highway pavement criteria. Switching CRM source types from initial investigations could seriously hamper a highway pavement construction project. If this happened, significant penalties could be assessed and the processor's credibility and future projects could be jeopardized. It is therefore critical to know the source and composition of the CRM.

Another example is the use of TDM in playgrounds. Bias tires that have the beads removed would be an ideal raw material for this application. However, with all tire types now containing steel cords, the liability issue becomes very critical. The wire contaminants, including the bead wires, must be completely removed and rendered as a non-issue when it comes to this application.

Trace amounts of wire powders left in powdered CRM for tire compounding applications can interfere with the cure mechanism, resulting in adverse effects to the performance of the final produced tires. It is the responsibility of the CRM processor to find or develop suitable tests for iron content. These test procedures should become part of the certificate of analysis (COA) and be agreed to by the end user, so no surprises occur.

A general understanding of the chemical components in a tire is essential even when considering TDM materials as a soil amendment. Zinc oxide (ZnO, noted in Table 2.1) is a significant metal in TDM, along with other ingredients. These components can have the adverse effect of leaching into the soil or plants over time, under certain conditions of the various soil types. Preparation for the application, in accordance with the defined guidelines, needs to precede the time of CRM procurement and continue through manufacturing (processing) to delivery at the end user's site.

2.2.3 Identifying the TDM

The American Society for Testing and Materials (ASTM) has attempted to designate some of the many parameters for TDMs, but much work is still needed. Accurate identification of the various processing steps is essential, and the form and function of scrap tire in its various forms of reduction and clarification need to be understood. Obviously, the whole tire is a single unit or entity. But as it is reduced in size, various nomenclatures can be assigned to the tire parts as they are reduced to smaller components. Table 2.2 is intended as a guideline, to distinguish the various forms of tire parts as some type of mechanical, chemical, or thermomechanical means reduces the tire.

ASTM D 5603 has made an attempt to designate the particle gradation. However, what is considered a –80 mesh by one process might be classified very differently in another process. What is retained on the lower-level

TABLE 2.2
Tire-Derived Materials (TDM)

Tire-Derived Fraction	Approximate Size Ranges
Cuts	1/2- to 1/8-inch of the whole tire
Shreds	12- to 2-inch (300- to 50-mm)
Tire Chips (with Inherent Wire)	2- to 1/4-inch (50- to 10-mm)
Tire Chips (95% Wire Free)	2- to 1/4-inch (50- to 10-mm)
Tire Granules	10–1 mm
Typical Crumb Rubber (CRM) Coarse	2000–420 microns (10–40 U.S. Mesh)
Typical Crumb Rubber (CRM) Fine	420–180 microns (40–80 Mesh)
CRM Powder	180–75 microns (80–200 Mesh)
Super Fine CRM	< 75 microns (200 Mesh)

screens tells a different story in each process. Therefore, using a number of different testing techniques and comparing the various CRM suppliers is recommended for quality assurance. In other words, a –80-mesh rubber produced by cryogenics is very different from that produced by a wet grind. The cryo-grind will result in less-defined gradations that tend to be very random, whereas the wet grind process will produce a CRM product with much finer particles, over a defined gradation curve.

Normally, tire chips have a tire-derived fuel (TDF) grade designation. This came about in the early 1970s, when utilities and other industry users were strapped to find alternative energy materials. Boilers equipped with traveling grates to suspension-fired units needed a wire-free tire chip, since the bead wires caused problems with fuel-handling systems. On the other hand, cement kilns and mass-fired incinerators did not necessarily need to have the inherent wire contaminants removed. However, some standards were set for these TDMs when these fuels began to be utilized as a viable supplemental energy source by the energy procurer. In cement, the residual wire in TDF is an asset, since it can replace the more expensive iron oxide or slag purchased by the kiln operation as a key additive in the manufacturing of cement.

2.2.4 Tests for the Various CRM and TDM Materials

As the CRM technology has evolved from the typical coarse grades of 10 to 40 mesh (2000 to 420 microns) to the finer (180 to 75 microns), the need to find alternative or improved test procedures and definitions has grown. The dynamics of the CRM particles change as the particles are ground finer and finer. In the case of a wet grind, the particles resemble the surface of carbon blacks. This surface morphology tends to make the particles want to interlock, and they appear to agglomerate. When slight pressure was applied to this mass, the particles easily crumbled. However, when tested on a shaker screen, these agglomerates did not readily break apart; thus, reproducibility of the test procedure was unreliable. Therefore, a new set of testing parameters had to be developed for fine and superfine CRM materials. Now, small balls are placed on each sieve screen to facilitate deagglomeration. There are also a number of elegant instruments on the market that use laser technology to measure not only the particle distribution but also the topography and volume of CRM particles. These instruments provide very accurate analysis of the CRM produced.

2.2.5 Contamination

For the tire recycler to meet the demands of the compounder or the final end user, the recycled tire materials must meet the same standards of excellence required for virgin materials. A higher level of quality assurance must be in place when factory scrap is utilized. Even the factory or what the factory calls "in process materials" are subject to contamination. To overcome these obstacles, certain practices need to be in place, not only after the tire scrap is collected and shipped to the recycler, but also at the recycler's processing site where the scrap is converted into materials of higher value. Therefore, understanding the process layout and operations at the factory where the scrap is generated is necessary. There must also be close cooperation with individuals at the plant to assure that material is source-separated for refinement at the recycler's site.

If a plant's polymer type is allowed to become contaminated or commingled with other polymeric materials, no amount of source separation techniques are practical, due to the logistics and economics involved. Therefore, practices need to be in place at all levels (from the factory floor to the final shipping to the recycler) to assure that a high-quality material can be produced that will meet everyone's expectations. As required at the manufacturing facility, standard operating procedures (SOPs) must be set in stone and be implemented, and a total commitment to these SOPs must be the guiding light.

This chapter discusses the key elements for making scrap materials a value-added material for compounding. The strategies for the production and utilization of high-quality material produced from scrap have been greatly influenced by the teachings of W. Edwards Deming. He has probably done more for the modern industrial revolution than any other person in the twentieth century. His principles reflect the desire for achieving and practicing a high level of excellence in all endeavors. His teachings should be the foundation of recycling activities.

The principles that Deming formulated, and the concepts he proposed, such as Total Quality Management and Continuous Improvement, are not practiced by every company, but they are there for the taking. (If we would be wise enough to do so.) Key to Deming's philosophy about management to enhance manufacturing is "doing it right the first time." This slogan suggested that a task should not have to be done over and over and that taking back rejected goods from a complaining customer should not have to happen. If the "doing it right the first time" slogan is ignored, the customer confidence is adversely affected and remains extremely difficult to regain. In that spirit, doing the job correctly the first time cannot be stressed enough.

If mistakes are made, and they will be, make absolutely sure to evaluate them. Make checklists to ensure that jobs are done correctly and are reviewed. This practice can be a tedious path; however, the rewards will be reflected in strong customer relationships and a strong reputation that reflects pride in the recycling industry.

2.3 The Tire Processor's Commitment

Adding value is the recycling processor's commitment. This can be broken down into basically five major areas, as noted in Table 2.3.

TABLE 2.3

Guiding Principles

- The Customer
- Quality
- Empowerment
- Teamwork
- Speed

TABLE 2.4

Quality

- Quality is everything we do.
- Being a low-cost producer/supplier provides focus.
- Six Sigma mentality in servicing our customers.

The customer is the "king," but this notion is lost many times in daily operating practices. In many companies the employees, the investor, or other entities seem to come first. But without the customer, the recycler has no income to keep the operation running. Thus, if the customer is made the "king" and the recycler is his partner, this mindset will go a long way to ensure that every job is done correctly, the first time. The recycler will then realize the same joys as being a "king." Growing a recycling business is like growing a mighty tree; it requires a good environment for the roots to grow and nourish the planted tree.

The quality of the product will reflect the recycler's commitment to maintain the customer in developing future business. The "speed" or reaction time in which the recycler can handle an order, through successful delivery to the customer, reflects the vitality of the recycler's organization.

In today's environment, just in time (JIT) is the standard for all industries. "Speed" is critical to achieve success. Six Sigma is a method for building a long-term, viable business. Table 2.4 reflects the essence of what quality is all about, in conjunction with Six Sigma. Six Sigma in many organizations simply means a measure of quality that strives for near perfection. It is a disciplined, data-driven approach and methodology for eliminating defects. The statistical representation of Six Sigma (Figure 2.2) describes quantitatively how a process is performing. To achieve Six Sigma, a process must not produce more than 3.4 defects per million opportunities, or a 99.0066% quality level. A Six Sigma defect is defined as anything outside of a customer's specifications. A Six Sigma opportunity is, then, the total quality of chances for a defect.

"Empowerment" applies to all levels of recycling. The customer empowers the recycler to collect, process, and deliver the goods. The recycling business must expect its people to do their jobs at all levels, from the salesperson taking the order, to the person receiving materials at the plant site, to the production staff making sure the right quantities are processed

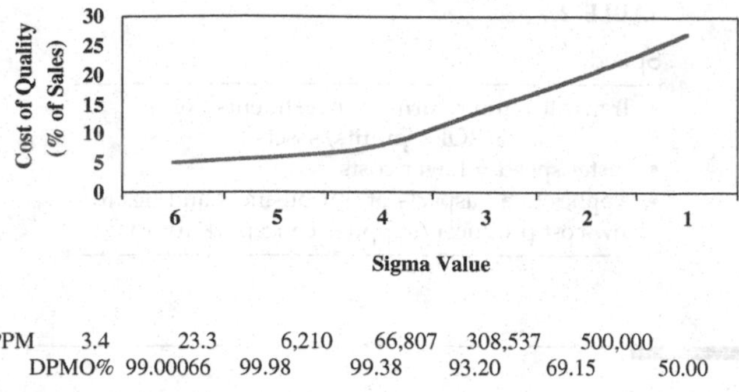

PPM	3.4	23.3	6,210	66,807	308,537	500,000
DPMO%	99.00066	99.98	99.38	93.20	69.15	50.00

%DPMO = Defect percentage per million opportunities

FIGURE 2.2
Cost Reduction Opportunity Six Sigma performances.

TABLE 2.5
Empowerment

- Means fully utilizing people closest to the work — empowering them to make decisions.
- Requires the proper environment — allowing people to understand and grasp the concept.

in a timely manner. Quality control has to ensure that the product meets all specifications required by the customer. Combined, the concepts discussed above represent quality assurance. Table 2.5 reflects the concept of "empowerment."

The entire network of parties involved in recycling make up a team. Having "value added" to the process will require "teamwork." This goes beyond the organization itself, to the customer, the shipper, and the performance of the recycled products application at the customer's manufacturing site. The key principles of "teamwork" are listed in Table 2.6.

TABLE 2.6
Teamwork

- Empowerment is
  - Customer focused
  - Problem/issue focused
  - Cross-functional (business and operations)
- Only way to achieve targeted levels for customer is
  - Service
  - Quality
  - Speed
- Well beyond the capabilities of any one person is
  - Teamwork

TABLE 2.7

Speed

• Translates into return on investments (ROI) ROI = profits/assets
• Faster speed = lower costs
• Applies to all aspects of our business and ties to low-cost producer/supplier concept (Kaizen)

2.4 Management's Direction

Management must emphasize the need to serve the customers from every aspect of the business. In so many instances, the fatal flaw by recycling operations is assuming "all they have to do is build a facility, the customers will send in orders, and profitability will be like a cosmic experience" The recycler in many ways has to be an internal customer; employees must recognize how they would want to be served, how they would want to be contacted, and the fact that they demand high quality. When we buy a car or a camera, we demand the highest quality of these items. The recycler must always strive to provide the necessary tools and information to the customer. This means supplying the highest-quality product for the application, in accordance with what has been promised.

The receiving, processing, and delivery of a recycled materials in a timely fashion will result in increased profits, improved efficiency, and overall lowering of costs of the recycled materials. This, in turn, allows the recycled product to be more competitive with virgin materials. Table 2.7 illustrates the concept of "speed" in the manufacturing process. The Kaizen concept is discussed in a later section of this chapter.

2.5 Lean Manufacturing

One of the most important aspects in making recycling a viable entity is the practice of *lean manufacturing* within the recycler's operation. Lean manufacturing is an organized strategy that focuses on the elimination of waste. Waste is defined as a non-value adder. A non-value adder is defined as any cost that the customer is not willing to pay for. Like the manufacturer, the recycler is trying to serve by providing high-quality end products. The recycler, who processes polymers back into useful materials, must also be cognizant of carrying out certain practices to ensure that costs are kept in line. Examples of non-value-added actions, or malpractices, that can occur and drive up operating costs are illustrated in Table 2.8.

TABLE 2.8
Areas of Waste

Areas of Waste	Example
Motion	Materials overhandled
Overprocessing	Recirculation
Transportation	Forklifts/loaders
Overproduction	Poor planning
Waiting	Grinding (commutation)
Inventory	Work in progress
Scrap	Trash

TABLE 2.9
Quality Control Practices (Four Methods)

ISO-9000	Certification
QS-9000	Certification
Kaizen	Actively involved
Six Sigma	An in-process, ongoing endeavor

The examples of waste shown in Table 2.8 can be expounded upon at length, individually. In many instances, a major cost can incur when the processor does not realize these areas of waste are occurring. It is suggested that an outside consultant, recognized for manufacturing and materials-handling expertise, be brought in to discover, advise, validate, and assist production in converting inappropriate areas of waste activities (see Table 2.9) into positive outcomes.

Many factors need to be considered when processing scrap tires without reliable markets. Understanding of the total synergy must occur with all aspects of the recycler's operation. The main aspects to be considered are disposal of the fiber, reliable markets for the steel/rubber stream, the know-how to identify potential resalable casings, understanding and budgeting for preventive maintenance, and factoring in low disposal fees to attract scrap tires. Other aspects to manage are well-trained, sophisticated professionals handling the sales of TDM and the procurement of the raw materials. A critical component for success is a management team that understands the dynamics of the tire business, including the funding to develop the business. A strong management team is required, and committed markets and a sound technology need to be in place before the business can become a lasting reality.

In reviewing the experiences of the past, the tire recycling business is built on perceptions that do not meet reality. Normally, no technical persons or persons with hands-on experience are involved, so a very poor concept of what the scrap tire operation can really achieve is the norm. The acumen needed is a balance of all these factors of financing, market position, technology, a strong management team, and an understanding that developing the markets will take time and patience. The tire recycling operation is definitely an entrepreneurial business and must be treated as such. In many cases, tire recycling is underfunded, with many entrepreneurs lacking the knowledge of all the aspects of the business. Without the funding to purchase sufficient and required equipment and lacking the ability to understand the evolving markets, the prospect for a successful venture is limited. The tire recycling business is becoming a dynamic and professional business, requiring individuals to meet the demands of technology in this industry.

Improved process control can also result in reductions in direct labor costs, reductions in damage to the final product, and an increase in "work in progress." The more times a product is handled, the greater the chances are that it can become contaminated or damaged. The appearance and quality of the final product is paramount in selling recycled materials. Education, enthusiasm, and endurance are keys to making lean manufacturing a success.

2.6 Quality Assurance and Quality Control (Continual Improvement)

Lean manufacturing practices can develop by applying a number of control practices and manufacturing methods. Many articles have been written on this subject; the purpose of this chapter is to familiarize the recycler with these practices and, perhaps, influence companies to become certified in quality assurance standards. The automotive industry is demanding the supplier to be at least ISO certified, and now ISO is being combined with the QS certifications to go to the new TS-16949 standard. Initially, there was much resistance to this international standard; those manufacturers who have embraced the ISO, QS, and the other standards, however, have found many other benefits, from improved monitoring and tracking of their own internal activities to in-process improvements. In many cases, it has brought a high degree of consistency to their everyday work practices and has helped their business operations to be more cross-functional and consistent.

A definition of quality was given in the Introduction. However, the practice of "quality" becomes "Quality Control." Quality control is a system for ensuring the maintenance of proper standards in manufactured recycled goods, especially through random and frequent inspections. This takes time and personnel; however, returned material is costly, and lack of customer confidence is very hard to regain. In some instances, "lost customer confidence" is never regained. What is the final financial loss to the processor if this occurs? This section discusses the four methods of quality control, which are also listed in Table 2.10. [Note: ISO is not being replaced. It has been revised. ISO certification can still be held independent of TS. QS will be replaced by the TS standard.]

TABLE 2.10

ISO-9000 Certification

- Internationally adopted and recognized system.
- Emphasis on documentation of process and procedure. Includes document and control and calibration of gauges, tools, and other measuring devices.
- Provides vehicle for training and subsequent consistency.
- Application of every function of business. ISO-9000 is the standard for manufacturing operations

TABLE 2.11

QS-9000 Certification

- Originally developed by U.S. automotive manufacturers.
- Mirrors ISO-9000 in many respects and modules.
- Customer requirements drive the details.
- Supply chain management is a key element.
- Supplier requirements vary by manufacturer (Ford, GM, Daimler Chrysler).
- Incorporates concept of continuous improvement.

TABLE 2.12

TS-16949 Certification

- Internationally adopted and recognized system.
- Emphasizes process flow, customer requirements, and documentation. Includes document control and calibration of gauges, tools, etc.
- Places responsibility for the system with the top levels of management.
- Provides vehicle for training and subsequent consistency.
- Application of every function of business. ISO-9000 is the standard for manufacturing operations.

Until recently, many recycling operations were not required to have ISO-9000 certification. However, as demands are placed on all areas of the manufacturing industry, suppliers, including the recycler, are being required to become certified in order to do business. To service the automotive industry as a tier one supplier, TS certification is required. The TS standard includes all of the requirements of ISO-9000; therefore, ISO-9000 certification is the first step toward TS certification. So the requirements of being in the supply chain as a supplier, especially to the automotive industry, have been raised to a new, higher level of commitment. Recognizing these standards is very important for the growth of the recycling polymer industry — not just in automobile manufacturing of parts and assembly, but in many other nonautomotive manufacturing and service industries as well. Certifications and the maintenance of these certifications are expensive; maintenance is a very intensive, ongoing process, with yearly audits to ensure that compliance is being met and to keep the certification. Table 2.10 and Table 2.11 list the definitions for ISO-9000 and QS-9000 certifications. Table 2.12 is the framework for the new TS-16949 definition.

TABLE 2.13

Kaizen Definition

- Definition — Literally means to break down and put back together by continuous improvement.
- Key Focus — Reduces waste and identifies non-value-added activity. Stresses workplace organization, cleanliness, and general housekeeping. "Poke Yoke" is the process that utilizes involvement of the employees, measurements, and the reduction of risk and errors. Fast change is a concept that means "just do it mentally."

TABLE 2.14

Six Sigma Practice

- A heavy emphasis on customer satisfaction.
- Mathematically means 3.4 defects per million.
- GE, Motorola, and Cooper Standard are some big proponents.
- Focuses on quality improvement through waste reduction and employee involvement.
- Lower costs are an end product of the process.

Unlike in the past, the new standards directly involve top management, including the president's interaction with the entire certification and maintenance process. Commitment and cross-functional activities at all levels of the company are now involved. This is no longer a delegated responsibility. It will take teamwork to keep an active status in TS.

To fully implement the certifications, to improve the operation within a recycling company's facility, and to minimize variances from quality, Kaizen and Six Sigma practices are the tools for achieving success. These practices have been demonstrated as proven paths for achieving quality and performance. The definitions of Kaizen and Six Sigma are listed in Table 2.13 and Table 2.14. And, again, Figure 2.2 shows Six Sigma performance in ppm (parts per million) or %DPMO (Percent Defect Performance per Million Opportunities); Six Sigma performance can be directly related to cost reductions, which means long-term viability for a polymer recycling facility. The concept of Six Sigma for a 3 process results in 6.6% defects when the customer specification is for no defects. By reducing variability, we can make a robust product or process. The before-to-after effect of reducing variations is shown in Figure 2.3. Good operating practices mean staying in business for the long term; again, examples of these practices are shown in Table 2.13 and Table 2.14.

Kaizen practices include greater employee involvement and making visual measurements. "Poke Yoke" pursues standardization and change over reduction. Better lighting, housekeeping, and general adherence to safety rules, along with daily meetings, result in changes to improve the process and reduce bottlenecks. These practices, followed by tracking performance to goals, will help to develop a first-class recycling operation and to create a climate of excellence.

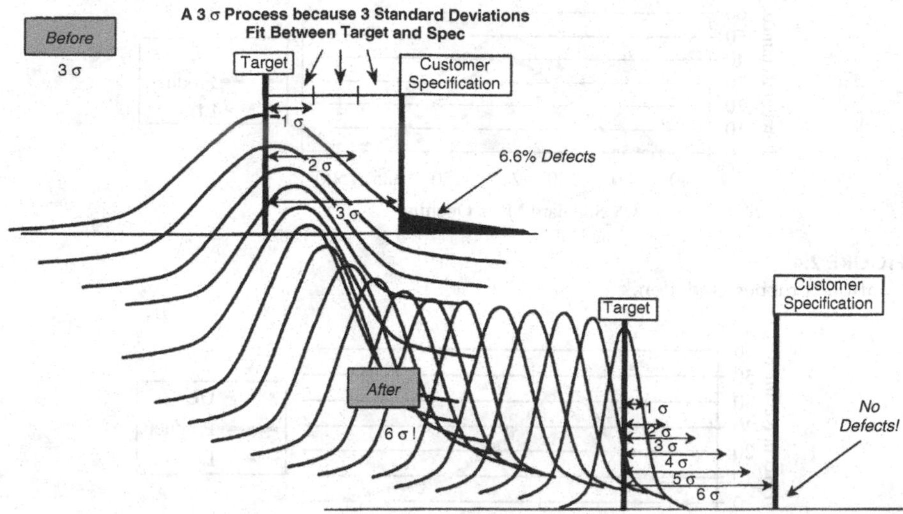

FIGURE 2.3
A 3-Sigma example.

Six Sigma examples are a reduction of product manufacturing cycles and a reduction of non-value-added time, labor, processes, parts, and products. Six Sigma incorporates the concept of continuous improvement. As the recycler begins the path of statistical control to an initial target of a one-sigma, value, operational results, and the customer will be realized. Six Sigma is the ultimate target or goal.

Achieving sigma values of greater than one is no small feat. A returned truckload can mean over 44,000 units of defective parts alone if the tracking system is based on one pound as a unit. Other incidents, such as throwing away 50,000 pounds of good rubber that may have been lost due to mishandling or poor processing, also add to the defect list. These two incidences alone bring the total to over 90,000 defective units, putting the recycling operation between two- and three-sigma. This does not take into account other incidents that may lead to lost opportunity and increased costs. It is very difficult to achieve a one-sigma in any organization but applying the principals discussed will open up new vitality for an operating business.

2.7 Analyzing the Outcomes

The tire recycler must analyze the customer requirements. Figure 2.4 depicts what the customer wants and Figure 2.5 shows what the customer does not want. These figures illustrate an example of supplying precise rubber grada-

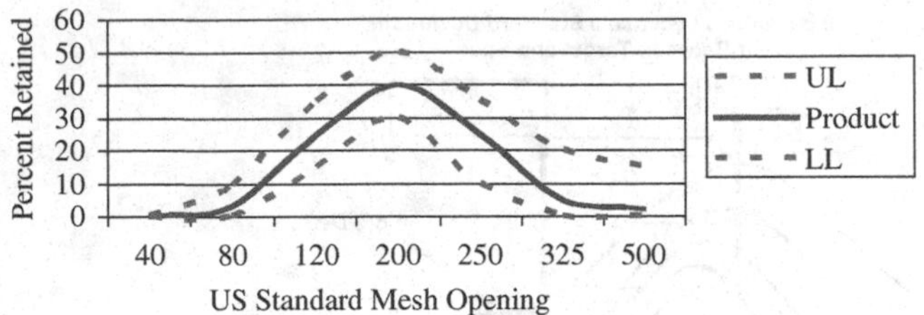

FIGURE 2.4
Controlled rubber gradation.

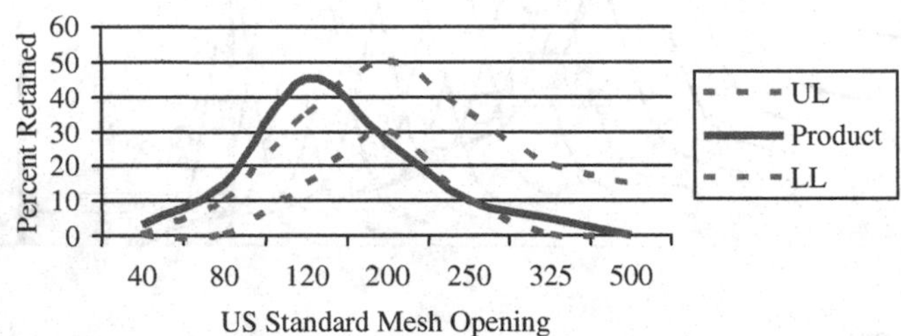

FIGURE 2.5
Uncontrolled rubber gradation.

tion by staying in the boundaries of the upper control limit (UL) and the lower control limit (LL). In Figure 2.5, the product curve is outside these limits and, therefore, not acceptable. With today's sophisticated analytical testing equipment, the customer has on hand the necessary equipment to measure and analyze many of the parameters set forth in their expectation. Examples of commonplace testing equipment include particle analyzers and chemical analysis instruments. When a simple matter of controlling the ash content in a CRM is not met, poor tensile properties in the final compound can result. Figure 2.6 and Figure 2.7 show a comparison between not controlling the ash content in an 80-mesh whole tire crumb rubber vs. tight control of the partitioning agent. The customer's product is seriously jeopardized by insufficient attention to one simple processing practice that needs to be standardized. Attention to reducing errors by monitoring and calibration of the partitioning agent keeps the overall ash content in line with the customer's expectations.

Figure 2.8 summarizes good process control, being off target, and excessive variation. The goal is to reduce variation in the scrap tire manufacturing process and to reduce the overall number of defects.

Many times, the tire recycling industry looks only to improvements in production and not to see if the TDM is transported and delivered in a timely and professional manner. However, to the customer, on-time delivery is also very critical. Late deliveries can mean lost production and incurred costs on the part of the supplier and the end user. Perception can play a big role. An example is how a producer thought an improvement had been made to a product, but the customer had the opposite view. Figure 2.9 illustrates this example and

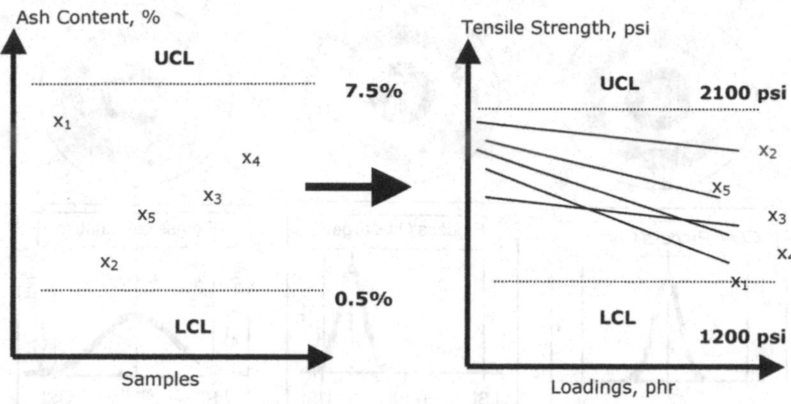

FIGURE 2.6
Poor control.

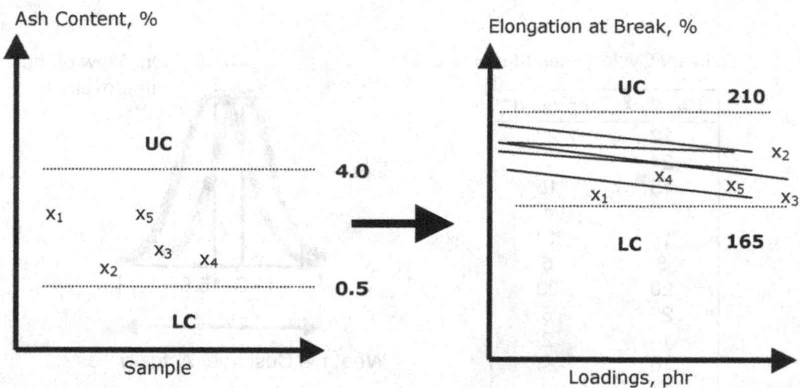

FIGURE 2.7
3-Sigma control.

shows that the mean time for delivery cycles did improve from 15.8 to 11.2 days, but the standard deviation did not, 7.0 to 9.0 days. These results have a negative perception by the purchaser of the processor's material, whereby only the combined outer limits of both mean variations are viewed in combination. The processor must be careful to analyze the net effect of improvements and the scope of the improvement on the impact to his customer.

2.8 An Overall Strategy for Success and Perception

For starters, attempting to improve things by repeating the processes of the past simply doesn't work. Innovation, imagination, and persistence will be

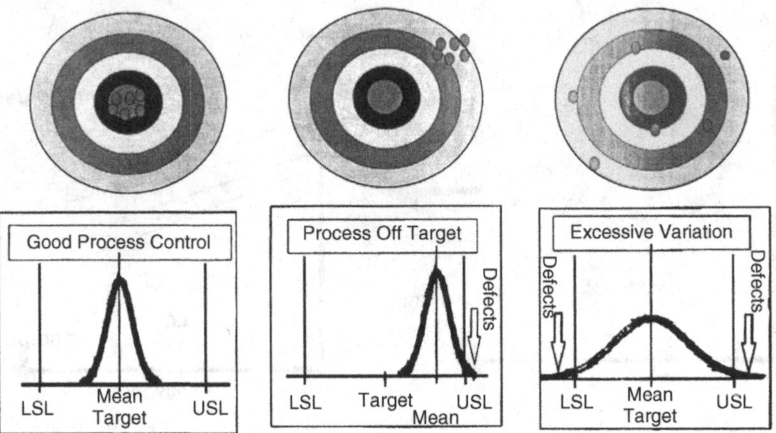

FIGURE 2.8
Process control.

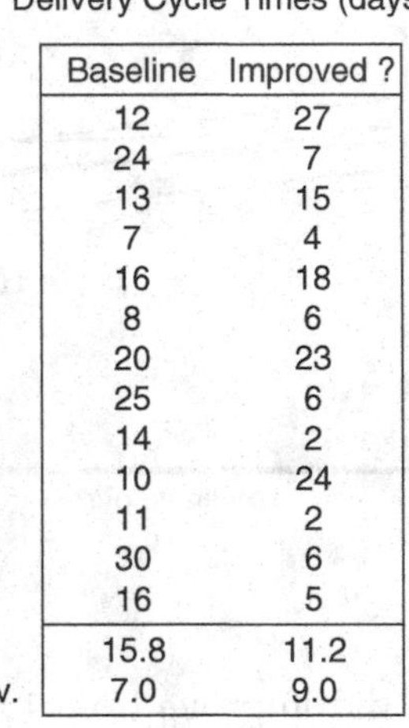

Delivery Cycle Times (days)

	Baseline	Improved ?
	12	27
	24	7
	13	15
	7	4
	16	18
	8	6
	20	23
	25	6
	14	2
	10	24
	11	2
	30	6
	16	5
Mean	15.8	11.2
Std. Dev.	7.0	9.0

Our View of the Improvement

11.2 15.8

What the Customer Actually Feels

- Using the traditional mean-based thinking approach, we would think we had an average performance improvement of 29%
- But, our customers only feel the variance and would not see an improved process

FIGURE 2.9
Example of data perception.

the foundation for tomorrow's survival. If the position of mediocrity is taken, the result will be the inability to promote it, to fire it, or to win with it. Remember that quality is an inherent distinguishing characteristic, a personal trait, and superiority of kind.

2.8.1 Chemical Characterization Methods

Running a lean manufacturing facility and applying the practices of Kaizen and Six Sigma philosophies are the foundations for developing quality assur-

TABLE 2.15

Chemical Characterization Methods

TGA (Thermal Gravimetric Analysis)
How much of each piece is present?
FTIR (Fourier Transform Infrared Spectroscopy)
What chemical pieces are present?
DSC (Differential Scanning Calorimetry)
How do the pieces change with temperature?
SEM (Scanning Electron Microscopy)
What does the material look like?
XRF (X-ray Fluorescence) Spectroscopy
How much of each element is present?

ance for recycled polymers. What is needed next is to have in place a system for conducting polymer characterization. Most companies use physical testing such as tensile, tear, elongation, modulus, etc. A few companies use chemical characterization. Characterization tells the technician what is there, how much, and what it is like. It can also show how the chemical is put together and what it can do. We must familiarize ourselves with the characterization techniques for polymers and select the appropriate equipment for the operation and customer needs.

Table 2.15 lists the chemical characterization of the methods now available to the recycler. Probably the most valuable tool for characterization in quality assurance is thermal gravimetric analysis (TGA). The TGA provides information on the quantity of each piece or ingredient present. The information is quantitative and is an exact percentage of each piece. The formulation of the compound can be reconstructed from the data. However, the TGA should be used as a QA/QC tool and not as instrument to inspect your client's materials. The TGA should be used to make sure that there are no other contaminants, such as other polymer types, in the customer's product. Thermal stabilities can also be determined, and the various gaseous environments can be used. A typical TGA graph is shown in Figure 2.10 of a whole tire –80-mesh CRM produced from Rouse Polymerics' wet process.

Fourier transform infrared spectroscopy (FTIR) provides information about the chemical functional groups in the material. A spectrum is a chemical fingerprint of the compound and can be shared via libraries. Additionally, spectral software and library matching techniques make this tool fast and efficient. The attenuated total reflector (ATR) unit is an instrument with the ability to quote instant "information" on most polymer compounds. The microscope attachments to the FTIR allow up to 10 micron views and surface maps. For high-end compounds, the FTIR gives the assurance that the processed materials are the exact materials the customer has shipped. These data also provide the backup confidence that this is the material the customer has requested. It is like receiving blood; you want a chemical match.

The DSC provides information on chemical transitions that compounds go through with temperature changes. It gives the glass transition temper-

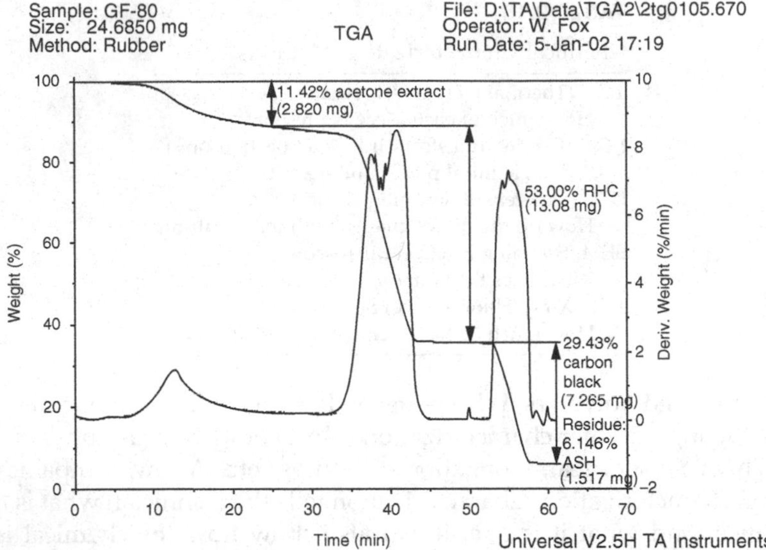

FIGURE 2.10
Typical TGA of whole tire –80-mesh CRM.

atures for rubber-based materials. It can also give the melting points of TPEs. Additionally, the DSC gives recrystallization points of TPEs and purity information on rubber chemicals and additives.

The SEM provides pictures of the surface of the compound from 10 times to 150,000 times. It also provides elemental information on spots via energies vs. x-ray analysis and can give a chemical map of the surface. This can be very useful in fracture analysis, dispersion analysis, and contaminant identification.

XRF provides elemental information from a few percent to ppm of all elements from Na to U. It can evaluate solid, liquid, or powder forms. It is also useful in identifying trace ingredients and for positive identification of materials.

Many of these instruments cost under $100,000 each. The instruments in Table 2.16 are in order of priority. From these instruments, information can be obtained about what is happening from a quality assurance standpoint in a recycling operation, especially in handling and processing very critical polymers used in seal applications. The TGA is the workhorse in many plants and is a must if the recycler is going to provide high-quality materials to a manufacturer. Tests need to be run often and randomly. These results can be summarized in statistical report form. This detailed information is critical to having a successful operation. In the case of the DSC, FTIR, or the SEM, testing can be done by many reputable laboratories to obtain the necessary information, on an as-needed basis.

For an operating recycling business, a collection of representative samples can be sent from the production site to a laboratory overnight, and the information can be e-mailed back to the facility prior to shipment. Charac-

TABLE 2.16
Particle Equipment Comparisons

Sieve Analysis	Laser Particle Size Analysis
Limited to 6 data points	Hundreds of data points
Limited by screening efficiency to 125 micron	Gives accurate measurements 0.2 microns below
Smallest particle diameter measurement	Largest particle diameter measurement
Large sample size	Small sample size less than 5 grams
Materials must be dry	Can be either dry or wet materials

terization methods are necessary in very high-quality, specialty polymers such as fluoroelastomers and the silicone families. The Web is very useful in locating analytical laboratories that can perform the characterizations that may be required in a certificate of analysis.

2.8.2 Other Methods

For most rubber recycling processors, size measuring equipment is the most necessary and most valued piece of equipment. In addition to particle gradations, consideration should be given to surface morphologies, particle configuration, total surface area, specific gravity, and bulk density. All of these are important values concerning the quality and type of grind the polymer was subjected to. They can also indicate the quality of resultant polymer processed when the polymer is compared against a known similar polymer gradation as a standard. Such data can be important for the compounder to know when evaluating polymers, especially on a volume cost basis.

In most compounding applications, the rule, "the finer the particle is ground, the better it can be compounded into the mix," applies. The finer the particle is commutated or ground, the greater the surface areas; this can be an important key in ensuring that the material can be uniformly dispersed in a compound. When compounds are ground to finer sizes from a general mass of the same compound, it is easier to remove contaminants and thus improve the quality of the final product. However, cost must be taken into consideration, since grinding to fine size, normally less than 80 mesh (125 micron) or less than 200 mesh (75 micron), increases the cost.

The pan screening sieve analysis has been the standard. This technique uses pans stacked on each other, with decreasing screen openings as the powders or granules move down through consecutively smaller screens by gravity and the help of small rubber balls. However, as the commutation or grinding technologies have advanced, the need to find better ways to measure particle size has become important. Today, laser instruments can be used to measure the exact dimension of particle size over a wide range of 1

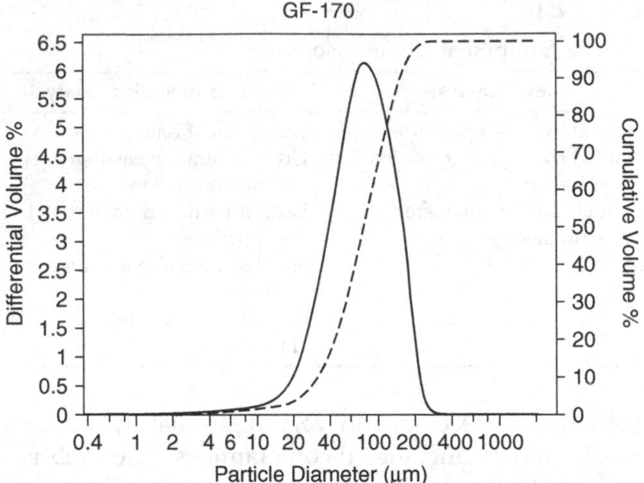

FIGURE 2.11
Typical printout of a laser particle analysis of a –80-mesh CRM.

millimeter to 1 micron very accurately and efficiently. For coarser materials, the standard shaker screen test is sufficient. Figure 2.11 represents a state-of-the-art particle analysis of a wet processed –80-mesh CRM. The software in many of these modern laser analyzers can configure the data into many different formats, from the one shown in Figure 2.11 to S-curves.

Table 2.16 lists the comparison of the standard sieve analysis to the laser technology.

Quality assurance is only as good as the instruments that are available. The technologist trained to operate these instruments and evaluate the resultant data plays a key role in maintaining the standards of the final shipped product. More test data enables a more appropriate response to maintain continuous improvement and avoidance of defects.

2.9 Implementation

The practice of processing polymers recovered from the factory or the consumer stream requires attention to details, as outlined in Table 2.17, Quality Control Flow Diagram. These simple steps are often overlooked or bypassed. Omissions will lead to errors, and errors are costly. The processor seeks to establish procedures that are satisfactory to the end user, thereby enabling consistency in a quality product.

Figure 2.12 illustrates the essence of a balanced approach to all of the activities mentioned in this paper. On one hand, there has to be "focus in all our activities," which includes the business and profitability section with the

TABLE 2.17
Quality Control Flow Diagram

Customer order received
↓
Order entered into system
↓
Production schedules the run
↓
Raw materials are received and inspected by QC department
↓
Material is manufactured and tested by QC department
↓
QC inspects the final packaging to ensure proper packaging and marking
↓
Material is shipped to customer

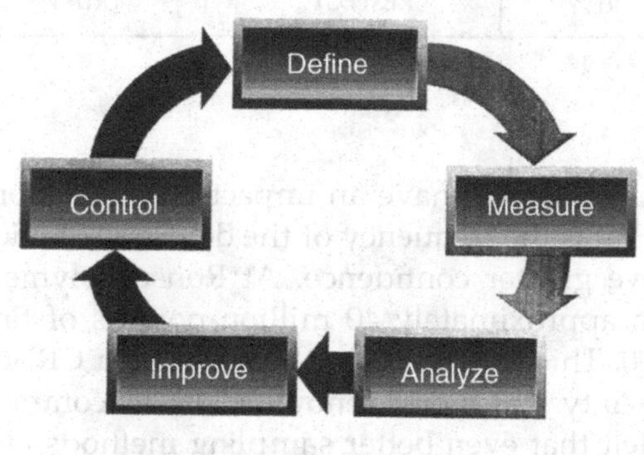

FIGURE 2.12
A total systems approach.

cultural transformation. On the other side are the "principles" that give the direction and alignment, the competitive factor, the performance, and most important, the foundation comprising the ethics and integrity, mutual respect, and the openness and trust within the organization and with the customer.

2.9.1 A Suggested Approach

General Electric, in its approach toward being objective about quality assurance, uses Figure 2.12 to illustrate its focus on constant improvement. The circular movement in a systematic, scientific, and fact-based process allows for continual improvement to reduce defects and improve long-term profitability and a healthy viability.

The Define is "What are the Customer's expectations of the process?" As discussed earlier, the type of process, say a mechanical grind vs. a cryogenic

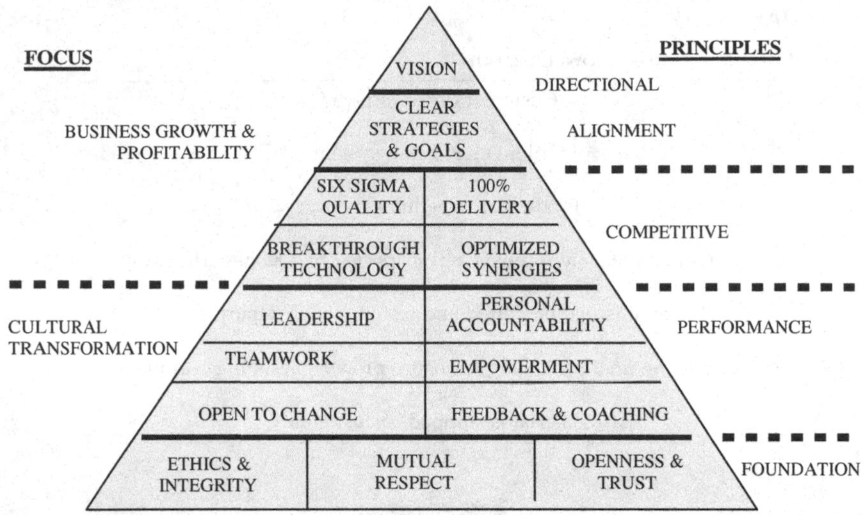

FIGURE 2.13
Balanced approach.

impact to a wet method, will have an impact on the customer. By Measure we are asking "What is the frequency of the defects?" Obviously, with fewer defects, we achieve greater confidence. At Rouse Polymerics, over 90,000 TGA were run on approximately 40 million pounds of fine CRM powder production in 2000. This was totally unheard of by a CRM producer. However, the Rouse facility was world renowned for its commitment to quality. It was generally felt that even better sampling methods and more samples needed to be run to ensure that the quality was not compromised in any way and the process was operating at its optimum level. What you can measure, you can analyze. By this we mean why, when, and where do defects occur. This leads to improvement. To improve, we can now ask the question, "How do we fix the process?" And finally, we can implement a process for control so we can ask the question, "How can we make the process stay fixed?"

As noted earlier, mediocrity cannot be promoted, it cannot be fired, and you cannot win with it. Success comes to those who pay attention to details, apply QA/QC as the cornerstone, and are the very best in all endeavors.

Generally, all tire processing operations are businesses that can dramatically improve performance. Applying your best resources will produce the best results. Customer satisfaction always seems to improve as quality improves. A dedicated client means continued revenues. Employee job satisfaction generally increases as a result of quality improvements. The bottom line will typically become lower, resulting in improved profitability if we define, measure, analyze, improve, and control our facilities operations. Figure 2.13 represents the final essence of balance between the processor's focus and principles. Focus is the business growth and profitability, based on a firm cultural transformation as the base. On the other side of the equation are the principles that guide the operations. These include directional alignment, competitiveness, performance, and the foundation of the company. The Balanced Approach pyramid of Figure 2.13 contains the key elements for achieving true balance within an operation.

2.10 Summary and Conclusion

Various ideas have been discussed. Many of the practices presented can be grouped together, such as lean manufacturing and Six Sigma to become a "lean Six Sigma." Certain analytical tools can be employed to monitor quality and a strong mindset by management. The operating personnel have to be in place to achieve a first-class quality assurance system. By adhering to the practices presented in this paper, the use of recycled polymers as a value-added material can be a reality. This reality, in turn, can spell success for the processor of valuable polymers to be used as a strategic raw material back into existing or new compounds.

References

1. *The American Heritage® Dictionary of the English Language,* 4th edition, Houghton Mifflin Company, 2000.
2. *The Rubber Roller Group Handbook,* 1968, page 106.
3. Barnawal, K.C. ASTM Standards & Testing of Recycled Rubber, Akron Rubber Development Laboratory, Rubber Division, ACS Meeting, San Francisco, CA, April 29, 2003.
4. *Marks Standard Handbook for Mechanical Engineers,* McGraw-Hill, 1978, pages 17-19—17-20.
5. 9th *Annual Global Plastics Environmental Conference (GPEC), Proceeding Book,* February 26th & 27th, 2003, Detroit, MI, pages 127–135.
6. Personal notes of Michael Rouse (1988 to 2003).
7. Rouse, Michael, Quality Assurance as a Value Added from Recycled Polymers, American Chemical Society, 163rd Technical Meeting, San Francisco, CA, April 23–30, 2003.
8. International Silicone Conference, April 22–23, 2003.

3

Untreated and Treated Rubber Powder

Wilma Dierkes

CONTENTS

3.1 Introduction

A lot of effort is put into the improvement of the quality of recycled rubber, reflecting on the one hand the high interest in rubber recycling, and on the other hand indicating the difficulties of the recycling process. Rubber recycling, in general, is a multistep process that begins with grinding the rubber material, separating contaminants and reinforcing material, and possibly upgrading the rubber powder by surface treatment or reclaiming. Some processes recently developed for this purpose are more academic in nature, mainly due to the high costs of the treatment. Others have a good potential to be introduced in the rubber recycling industry for the production of a high-quality recycled rubber product that can compete with virgin material in terms of consistency, purity, and costs. This chapter provides an overview of all processes for rubber grinding and surface treatment of rubber powder.

3.2 Grinding Technologies

The processes used for grinding of rubber are based on cutting, shearing, or impact, depending on the equipment (knife, shredder, granulator, extruder, disk grinder, or impact mill) and the grinding conditions (ambient, wet, or cryogenic grinding). The choice of the process is based on the requirements for the final product, such as particle size and particle size distribution, morphology of the particles, and purity of the rubber powder.

3.2.1 Cutting and Shearing

3.2.1.1 Cutting

Cutting, in general, is only a precomminution step for rubber parts, such as tires, in which rubber products can be reduced to a particle of size a few centimeters. In a subsequent step, the coarse material can be ground to a finer rubber granulate or powder by using a different grinding process. The equipment used for size reduction of large rubber pieces by cutting are hydraulic or mechanical knives. Processes used on a larger scale are based on a combination of cutting and shearing, due to the high wear of the cutting equipment. Once the cutting blades get blunt, the material is no longer simply cut but also sheared, and the final process is a mixture of cutting and shearing.

The next level in comminution is shredding of the material. A shredder consists of two sets of counter-rotating cutting blocks with a speed of 20 to 40 rpm (revolutions per minute). The final size of the shredded material is a few centimeters (average 5 × 5 cm), depending on the geometry of the cutting blocks.[1–5] Granulators consist of a rotating shaft equipped with knives. They rotate in a stationary barrel also equipped with knives. The rotation speed is 100 to 1200 rpm. The finest particle size which can be produced on this type of equipment is approximately 2 mm (millimeter), and the granulometry can be adjusted by a sieving unit separating the fine fraction from the coarse fraction, which is transported back to the mill.[1,6] The same principle is used for rotational cutting. For this process, the mill is equipped with stationary and rotating knives, but the knives are not necessarily parallel with the axis of the rotor. Depending on the angle between the knives and the rotor axis, the gap between the knives varies in shape and size.[7]

3.2.1.2 Milling

Specially adjusted mills, called crackermills, are used for rubber grinding. At least one of the rolls has a corrugated surface, and the rolls counter-rotate at a speed of 30 to 50 rpm with a very narrow gap and large friction. The final particle size is determined by the profile of the rolls, the gap and the friction between the mill rolls, as well as the cooling efficiency.[1,6] Alternatively, the corrugated roll can be placed in a stationary barrel, and the material passes the clearance between roll and barrel wall. The roll can be centered in the barrel but can also be mounted excentrically.[8,9]

A flow sheet of a typical grinding installation is given in Figure 3.1.[6] The material is pre-ground in a shredder, and fine grinding is done on a crackermill. A fabric separator is included to remove reinforcing material, for example when passenger car tires are ground. The particle size is controlled by sieve screens. The flow sheet also includes a reclaiming unit.

3.2.1.3 Extrusion

Twin screw extruders can be used for grinding of rubber. A grinding extruder, as described in literature, is a twin-screw extruder equipped with two independent co-rotating screws with a set of grinding elements.[10] Starting with a material having a granulometry of some millimeters, the final powder can have an average particle size of a few hundred micrometers.

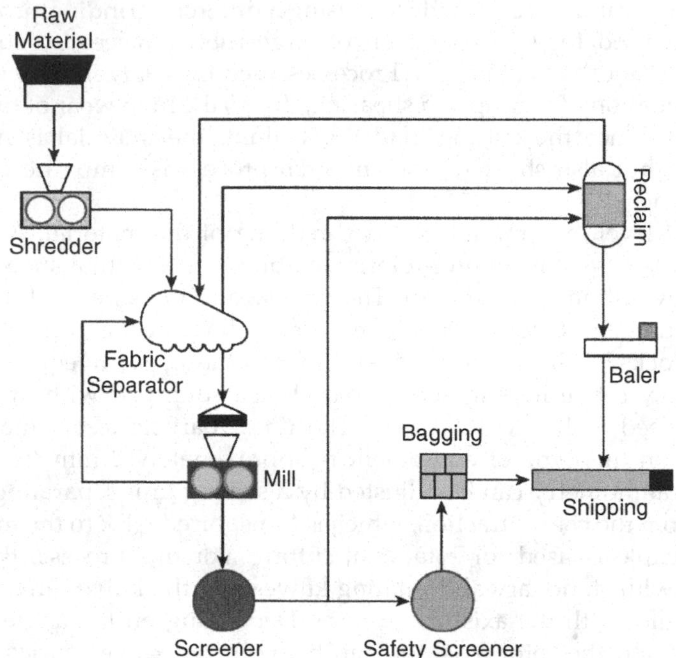

FIGURE 3.1
A typical ambient grinding process.

During this process the rubber powder is thermally degraded on the surface, resulting in a coarse and tacky surface. The extruder needs to have a very high cooling efficiency for reducing the degrading effect. This process is particularly suitable for grinding of soft materials.[11]

Solid-state shear extrusion (SSSE) is a mechanochemical grinding process. The twin-screw extruder used in this technology applies high shear forces and pressures to the material, which disintegrates under these conditions and at the same time partly devulcanizes by rupture of sulfur cross-links.[12–14]

A comparable process is the elastic-deformation grinding (EDG) technique. This technique is based on the fracture of the rubber material in multiple-stressed state under all-around shear and compression at elevated temperatures.[15]

In the high-temperature, shear-induced grinding process, a specially adjusted screw is used: the middle part of the screw consists of a cylinder. In this area, the material is heated up very fast, resulting in a superfacial degradation and formation of microdefects. At a certain level of temperature, pressure, and shearing forces, these conditions cause disintegration of the material, resulting in a fine powder. The material can be ground down to a particle size of less than 100 μm (micrometer) and is characterized by a high relative surface area (3.0 m^2/g).[16]

3.2.1.4 Disk Grinding

In this process the material is ground between two grinding wheels that are rotating with a high friction. It is possible to use direct water cooling during the grinding process to increase the efficiency (wet grinding).[6] The final particle size that can be achieved in this equipment is 500 μm to 1 mm, but the energy consumption for the production of the fine fractions is reported to be high. When water is used as the cooling medium, the final particle size can be 0.25 mm and smaller.[17,18] A variation of this equipment is a mill containing a single, vertically placed grinding wheel, where pre-ground material enters the grinding chamber through a plunger.[19,20]

3.2.1.5 Grinding in a Pellet Press

In this process, the coarse rubber chips (50 mm) are consecutively passing several pellet presses until the final particle size (approximately 2 mm) is reached. The disadvantage of this process is an insufficient separation of the reinforcing material.[21]

3.2.2 Grinding by Impact

Impact grinding is based on collisions between rubber particles or between rubber particles and parts of the grinding equipment. The particles need to have a high kinetic energy to disintegrate during a collision; this is achieved by acceleration to a very high speed inside the mill. Different types of processes have been developed for this purpose.

3.2.2.1 Grinding in the Rubbery State (Ambient and Wet Grinding)

The concept of jet mills is an acceleration of the pre-ground rubber granulate by the air jet. Once the kinetic energy of the particles is high enough, collisions with other particles or different parts of the mill cause disintegration of the particles.[22,23] Disintegration of the particles can also be achieved by turbulences in the mill. A strong upside-down current of air is carrying the rubber granulate through the mill. Inside the mill the current is turbulent, and in this area particles collide and break. A high amount of energy is stored in the rubber material, making efficient cooling necessary. Cooling is preferably done by liquids, for example water or liquid nitrogen.

Water can be the impact-exerting medium and at the same time remove the thermal energy from the rubber particles. An ultra-high-pressure water jet allows the disintegration of a coarse granulate down to a rubber powder with a particle size of 200 μm. The strong water jet induces kinetic energy into the particles and causes disintegration by impact. Disadvantages of this process are the insufficient separation of rubber and reinforcing material, as well as a high energy consumption caused by the drying step.[21,24]

In the resonance disintegration process, a very fine powder with a particle size of less than 100 μm can be generated, with a good separation of rubber and reinforcing material. The rotors of this machine are horizontally placed disks in vertically placed consecutive grinding chambers. The disks are equipped with guiding rails, and the rubber granulate is transported over the disks and through a centered opening of the separating plates. The grinding effect is based on sudden direction changes of the particles (shearing, impact), as well as resonance forces.[25]

3.2.2.2 Grinding in the Glassy State (Cryogenic Grinding)

The rubber material can be cooled down to a temperature in the glass transition area by liquid gases (for example, nitrogen or air).[26] In the glassy state, rubber can be efficiently ground by impact.

The most common cooling medium is liquid nitrogen. Cooling can be done either prior to the grinding process or during grinding. The optimal grinding temperature is just above the glass transition point.[27]

Cooling of the rubber granulate before entering the mill can be done directly on a conveyor belt, in an extruder, in a tunnel-style chamber, or by immersing it in a nitrogen bath. Indirect cooling is done in a heat exchanger. The liquid nitrogen can also be injected into the mill before and during the grinding process. The type of mill used is an impact mill or a hammer mill, and the disintegration of the rubber particles is caused by impacts between the particles or with the different parts of the grinding equipment. The final particle size that can be achieved is less than 0.6 mm.[6,28,29]

The main difference between cryogenically ground rubber powder and material ground in the rubbery state is the surface structure: the cryogenically ground material has a very smooth surface and a low specific surface area (Figure 3.2),[21] making it less active for revulcanization than ambiently ground material.

When cryogenically ground powder is blended with a rubber compound, it fails to physically bind to the polymer. The ambiently ground material has

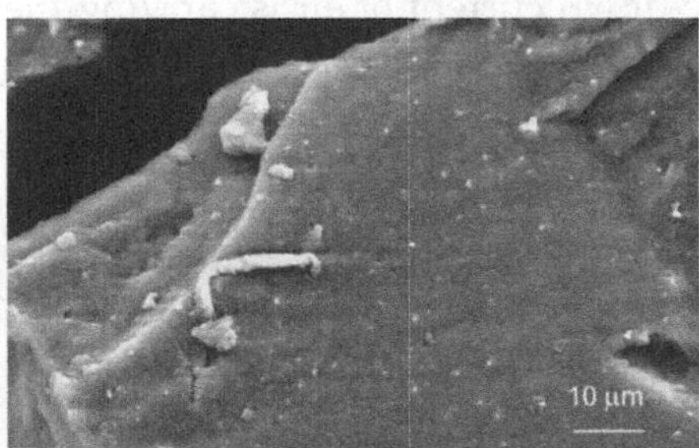

FIGURE 3.2
The surface structure of cryogenically ground powder (left) and ambiently ground powder (right).

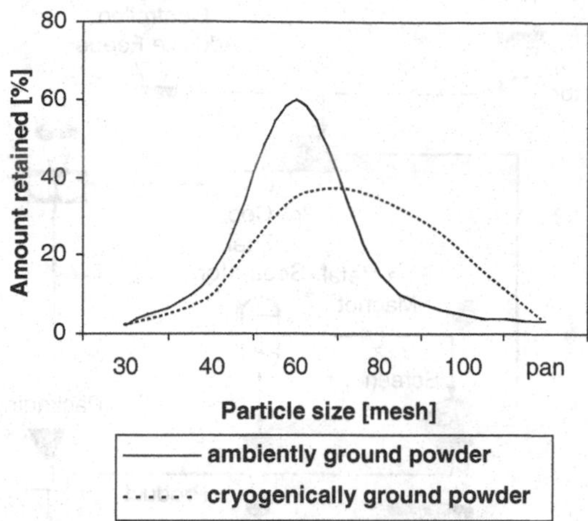

FIGURE 3.3
The particle size distribution of ambiently vs. cryogenically ground rubber powder.

cavities on its surface where the polymer chains can penetrate, resulting in a physical binding between polymer and rubber particle.

The nitrogen acts as an inert medium, resulting in a lower degree of surface oxidation for the cryogenically ground rubber compared to the ambiently ground material. Due to the exclusion of oxygen, the risk of a dust explosion is also reduced.

The cryogenic grinding process, as such, is consuming less energy than the ambient grinding process, but this comparison does not take into account the energy consumption for pre-grinding and drying of the powder. The wear of the equipment is lower for cryogenic grinding. This process has a higher efficiency than the ambient grinding process, and powders with a lower particle size can be produced. Cryogenically ground material differs from ambiently ground powder in particle size distribution; cryogenically ground powder has a broader particle size distribution compared to ambiently ground material (Figure 3.3).[5] The separation of rubber and reinforcing material (steel and fibers) is more effective. Ambiently ground rubber powder has a fiber content of 0.5% and a steel content of 0.1%. In cryogenically ground powder, the contamination level is below the detection limit.[29,30] The rubber feedstock does not necessarily have to be fully vulcanized; unvulcanized material can be ground as well, but handling of uncured ground material is more difficult because the granulate tends to agglomerate after grinding due to its tackiness.

The values for the nitrogen consumption differ depending on the equipment used. As an indication, the amount of approximately 0.5 kg (kilogram) liquid nitrogen is reported to be necessary for grinding 1 kg of a 50 mm rubber granulate down to a 2 mm rubber granulate.[31] When the rubber material is cooled prior to milling, the efficiency is found to be twice as high

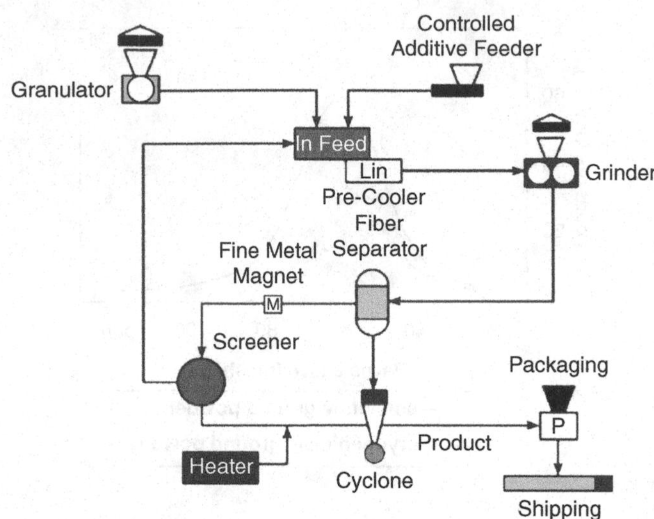

FIGURE 3.4
A typical cryogenic grinding process.

compared to cooling of the mill together with the rubber during grinding. The cryogenic grinding process, making use of liquid air for cooling of the rubber granulate, has a higher cost-efficiency compared to the liquid nitrogen process.[24,26,32]

Figure 3.4[6] shows a typical cryogenic grinding process, in which the granulated material is cooled prior to grinding. Comminution takes place on a grinder, followed by separation of fibers and metal.

3.2.3 Grinding by Physical or Chemical Processes

A coarse granulate can be ground by swelling in a solvent and reducing the particle size of the swollen powder by pressing it through a sieve or putting it on a mill or on a grinder. The resulting fine powder is physically and mechanically treated to remove the swelling agent, and then dried. Particle sizes of 100 μm can be achieved.[33,34]

Fatty acids can be used for the same purpose. The material is milled together with the acids and afterwards dispersed in water. After separation from the watery phase the crumb is reported to have a particle size of 2 to 20 μm.[35]

In the ozone cutting process, the rubber articles, for example tires, are exposed to high ozone concentrations. The material degrades by ozone cracking, and in a subsequent mechanical grinding step the degraded material is ground to a fine powder. The ozone treatment affects the material, and after final grinding the resulting powder has a low surface activity due to the oxidation by ozone.[36,37]

3.3 Applications of Untreated Particulated Rubber

3.3.1 Rubber Products

3.3.1.1 Products from Pure Rubber Crumb

Shredded tires are used as filler material in highway construction because they increase the abrasion resistance and enhance the resilience.[38–40] Concentrations of tire rubber used for these purposes are 10% to 20%. They are also used as a soil improver because they increase soil aeration and as backfill material for noise barriers. No negative effect of leaching of, for example, zinc oxide was found.[41] Rubber chips are used in foundations of roads, and finer rubber crumb is used in blends with asphalt. Other applications are absorbents for oil or wastewater.[42]

Rubber crumb can be cured by compression molding, but the quality of such a product is rather poor. Typical values for a vulcanizate of ambiently ground rubber are tensile strengths of 2 to 3 MPa (megapascal) and elongations at break of less than 100%, with tear strength and abrasion resistance also being low. An adjustment of the curing conditions alone can improve the properties of a product containing untreated rubber powder.

It has been found that a blend of a virgin compound with 70% rubber crumb gave the same property profile as the virgin compound, when curing pressure was increased to 150 bar.[43] A higher mechanical stability can be achieved when the high-pressure, high-temperature sintering (HPHTS) process is used. The curing pressure influences the properties of the vulcanizate; in the pressure range between 20 and 80 bar, the tensile strength of the vulcanized products increases. The higher the curing temperature, the higher the tensile strength and the lower the modulus. For natural rubber (NR), the optimal curing temperature is 200°C, and the curing time is in the range of several minutes. It is stated that under the above-mentioned conditions, free radicals are formed on the surface of the rubber particles. These radicals combine to form bonds between the two phases, making the particle-matrix interface indistinguishable from the matrix. NR and styrene-butadiene rubber (SBR) powder provide a vulcanizate that has a lower strength compared to the original material, while polysulfide rubber has the same stability after revulcanization. Vulcanizates from NR powder can be further improved by the addition of dienophiles (phthalimide, maleic anhydride, or maleic acid) or dipolar organic compounds prior to sintering. The addition of cross-linking agents and anti-reversion agents reduces the loss in properties. The reason for such a loss is assumed to be degradation of the polymer backbone followed by a formation of conjugated dienes, resulting in reversion effects. This effect can be reduced by the addition of sulfur and anti-reversion agents.[43–46]

The HPHTS treatment influences NR and SBR powder differently: for NR, the balance between cross-link scission and the formation of new cross-links is on the side of the cross-link scission, finally resulting in a superfacially de-cross-linked material. For SBR, the balance is on the side of recombination, resulting in a higher cross-link density and a higher modulus. Polysulfide rubber does not undergo a net change of cross-link density.[48]

3.3.1.2 Products from Untreated Rubber Granulate and a Binder

Rubber granulate is widely used in the athletic and recreational areas (playground surfacing, parking lots, equestrian footing, sports floorings, and running tracks), in molded products (isolating sheets, solid tires, truck bed liners, flooring tiles, railroad crossings, and traffic control products such as crack-sensitive tiles), and in coatings.[49–53]

Depending on the application, different particle sizes are required. Brake linings, for example, can contain a powder with a maximum particle size of approximately 0.8 mm. In carpet backing, a rubber granulate with a particle size between 0.8 and 1.6 mm can be used, whereas in a carpet underlay the particle size is between 0.6 and 2.0 mm. Molded products can be made using a fine powder with a particle size of 0.5 mm, but can also contain a granulate with particles sizes of 5 mm. Playground tiles, sports fields, and running tracks can be produced using a granulate with a particle size of 1.6 to 2.5 mm. Shoe soles can contain fine rubber powder in the range of 0.4 to 1.6 mm. For road paving material, a powder with a maximum particle size of 0.8 mm is used, and railroad pads can be made from rubber powder with a particle size of 0.4 to 1.6 mm.[54] In most of these products, a high percentage of rubber granulate is blended with a few percents of a binder. The most commonly used binding system is polyurethane. The advantage of a polyurethane binder is a low curing temperature and pressure, allowing the production of bigger preforms out of which the final products are cut or stamped.[55] The disadvantage of this system is the sensitivity of the binder to humidity, making a drying step of the rubber granulate necessary. Due to the polyurethane coating, the final product is hydrophilic, enabling water to drain through a slightly porous sheet.[56]

Other binding systems like latex, together with a curing system, are also used. One example for the application of such a product is stable mats. They have the advantage of being resistant to hydrolysis under use conditions. A further application is small, massive tires used for waste containers, for example.[57] Further, a coarse whole tire granulate bound with latex can be used for products such as protection walls in highway building.[58,59] A disadvantage of this technology is the more expensive curing process compared to the polyurethane systems, because high pressures and temperatures have to be applied to vulcanize these products. Therefore, they are less commonly used than the polyurethane binding systems.

Solid ethylene propylene diene (EPDM) rubber granulate is also used as a binding system. It is blended with the rubber granulate and other additives, such as curatives and chalk, and cured into sheets that are used as floorings.[60]

Rubber sheets and mats are widely used for noise and vibration damping. No difference between waste material and virgin material has been found in terms of damping efficiency. For railway applications, a noise reduction of approximately 40% and a vibration reduction of approximately 30% can be achieved.[54]

For special applications, other additives can be added, such as aluminum hydroxide trihydrate for material requiring a low inflammability.[61] Rubber powder is used for making products with a special surface finishing, such as a slightly rough surface.

3.3.1.3 Use of Untreated Rubber Powder in Rubber Compounds

Rubber powder is used in tire compounding, but the concentration used in these compounds is limited to approximately 5%. This is due to the fact that blending of rubber powder with a virgin compound influences the processing behavior as well as the final properties of the product. The average loss in mechanical properties by the addition of crumb is 1% for each percent of powder added to the compound.[62] This value has been found for a rubber powder with a particle size in the range of 0.5 mm. Advantages of the addition of rubber powder to a compound are shorter mixing cycles, improved green strength and form stability of uncured products, reduced curing times, improved air-venting during curing, and improved mold release. Rubber powder can be used in several parts of the tire, such as in the innerliner, ply coat, and treadcap of a radial tire.[63]

Other applications are hoses, belts, friction materials (brakes), trailer bumpers, mud flaps, truck bed liners, brake pedal covers, door step pads, air deflectors, and splash guards.[48]

3.3.1.3.1 Processing Properties

The addition of rubber powder to a virgin compound results in an increase of the compound viscosity, an effect generally observed with the addition of spherical particles, such as fillers, to viscous substrates. In the case of rubber powder, the increase in viscosity can be described by the Einstein-equation, describing the relation between the final viscosity η of the blend, the initial viscosity η_0, and the volume fraction of the filler ϕ.[64]

$$\eta = \eta_0(1 + 2.5\phi)$$

This equation allows a rough estimate of the viscosity increase as a result of the addition of rubber powder. It does not take into consideration the morphology and chemistry of the powder surface, which are further influencing factors for the viscosity. Ambient ground rubber powder leads to a higher viscosity compared to cryogenically ground material, due to the rougher surface. The virgin polymer is trapped in the cavities on the surface of the rubber powder, resulting in an increase of the effective volume fraction of the rubber powder in the matrix.[64]

Scorch time and curing time of sulfur-cured compounds are influenced by the addition of rubber powder; scorch time is decreased whereas curing time is increased The final cross-link density is reduced by the addition of untreated rubber powder. An explanation for this effect is the migration of curing agents between the two phases; migration of accelerators from the rubber powder into the matrix of virgin rubber results in an increased accelerator-concentration in the matrix and, therefore, a shorter scorch time. Migration of sulfur from the matrix into the rubber powder particles causes a lower cross-link density.[65]

For resin-curing systems, the curing kinetics and final cross-link density are not influenced by the addition of rubber powder. In contrast to sulfur-curing systems, no migration is taking place.[43]

Other compounding ingredients, such as pepticides and anti-aging additives, also migrate from the matrix into the rubber powder, influencing the performance of the final product.[43]

3.3.1.3.2 Mechanical Properties

The influence of the addition of rubber powder on the final mechanical properties of a compound depends on the following factors:

- History of the rubber waste
- Composition of the rubber powder
- Grinding process
- Purity of the powder
- Particle size and particle size distribution
- Concentration of rubber powder
- Type of compound it is blended into

The history of the rubber powder refers mainly to influencing the aging resistance of the final product. Ozone resistance of a virgin compound is reduced by the addition of ground rubber powder, and the final ozone resistance decreases with aging of the feedstock used for the rubber powder. This effect is caused by a very low concentration of antiozonants in the rubber powder, causing a migration of the protective agents from the virgin compound into the rubber particle.[66]

Ambiently ground rubber powder has less influence on the properties of a rubber compound compared to cryogenically ground material. The polymer of the matrix can penetrate into the cavities of the rough particle surface, resulting in a physical-mechanical anchorage of the rubber particle to the matrix. The surface finishing of a product containing ambiently ground powder is smoother compared to an article containing the same amount of cryogenically ground material.[67]

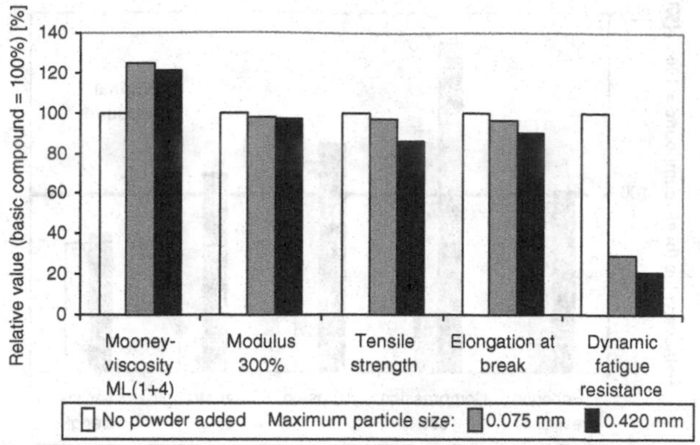

FIGURE 3.5
Influence of particle size on the properties of a NR-based compound (truck tire tread powder, concentration: 20%).

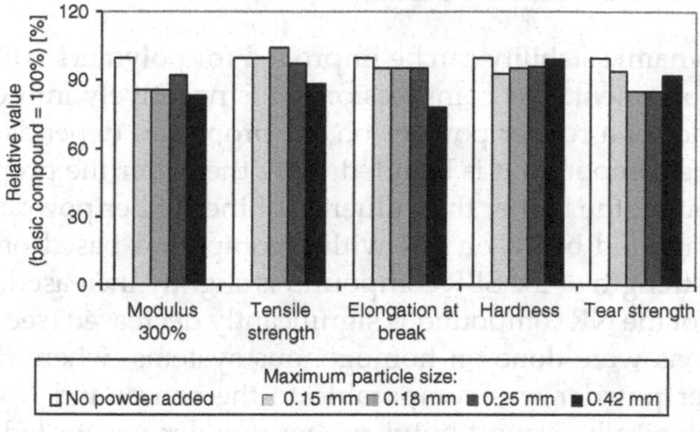

FIGURE 3.6
Influence of particle size on the properties of an EPDM compound (EPDM rubber powder, concentration: 20%).

Examples of the influence of powder particle size on the properties of a compound are shown in Figure 3.5 and Figure 3.6.[68,71]

The compound viscosity increases by the addition of rubber powder, but this increase is independent of the particle size. A decreasing particle size results in less influence on the mechanical and dynamical properties of the compound. The influence of a NR-based rubber powder with an average particle size of 100 micrometers has been tested in a NR compound. Concentrations up to 50 phr (parts per hundred rubber) have been used, and only a slight influence on the compound properties was found.[69]

The tensile strength of a compound containing rubber powder is reduced due to the difference in modulus between the rubber particles and the matrix, resulting in an increased tension at the interface. Tear strength can be improved by the addition of rubber powder; the tear stops once it reaches a rubber particle, because the strain concentrated at the crack tip is dissi-

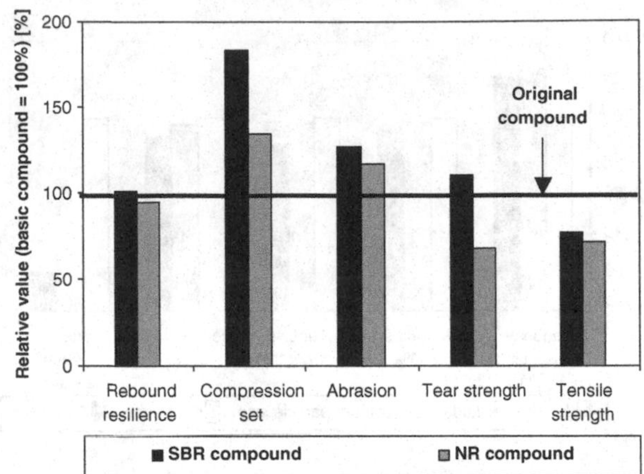

FIGURE 3.7
The influence of 20% rubber powder on the properties of SBR and a NR compound (truck tire tread powder, maximum particle size 0.42 mm).

pated. The dynamic stability can be improved for polymers without crystallization reinforcement. The compression set is negatively influenced.[70]

The influence of a rubber powder on the properties depends on the composition of the compound it is blended with; the better the properties of the initial compound, the higher the influence of the rubber powder. A comparison of a compound based on NR with a compound based on SBR shows that the tear strength of the SBR compound is slightly increased, whereas the tear strength of the NR compound is significantly decreased (see Figure 3.7).[68]

Investigations were done on homogenous systems, where the feedstock for the rubber powder was comparable to the compound it was blended with. A cryogenically ground butyl rubber powder was tested in an innerliner compound with a concentration of 15%. Blending of the rubber powder with the virgin compound resulted in a decrease in tensile strength of approximately 10%, a decrease in modulus of roughly 15%, and a reduction of the air permeability by 12%.[71]

EPDM powder has been tested in an EPDM compound in concentrations up to 100 phr. The overproportional increase in viscosity of this blend was explained with the rough surface structure of the rubber powder. Scorch time was decreased when the rubber powder was added, caused by migration of curing additives between the EPDM matrix and the EPDM powder. The torque values as well as the cross-link density increased up to loadings of 50 phr rubber powder; at higher loadings they decreased again. The increase of the torque at lower concentrations was explained to be a filler effect. At higher concentrations, the migration of curing additives is the dominating effect, resulting in lower cross-link densities.[72]

Polychloroprene (CR) powder with a maximum particle size of 0.32 mm was used in a CR hose compound. The scorch time of this compound was reduced by the addition of the powder. Tensile strength, elongation at break, and dynamic properties deteriorated, whereas modulus, hardness, compression set, aging properties, and low temperature behavior were not influenced.[73]

3.3.1.4 Blends with Thermoplastic Materials (Polyblends)

Thermoplastic elastomers commonly contain a thermoplastic and a rubbery elastic component. One way to produce thermoplastic elastomers (also called polyblends) is to mix the two components; in this case, the elastic material can be a ground rubber powder. The polyblend, consisting of an elastic and a plastic phase, can be cured during processing, resulting in a thermoplastic vulcanizate. This material is used for shoe soles, mats, sheets, and containers.

The rubber powder can be added to the plastic material prior to polymerization. This process results in better properties for a system containing, for example, a whole tire powder and polymethyl methacrylate, compared to blending of the rubber powder and the ready polymer. Further improvements were achieved when the rubber powder was treated with oxygen plasma.[74] Improvements of the properties of a blend made from high-density polyethylene (HDPE), low-density polyethylene (LDPE), or linear low-density polyethylene (LLDPE) and rubber powder can be achieved when the particle size of the rubber material is reduced during the blending process. A double-screw extruder can be used for blending and size reduction.[75] Blends of different whole tire rubber powders with LLDPE have been investigated concerning the influence of particle size, particle size distribution, morphology of the particle, and grinding process. For these systems, wet-ground rubber powder results in a high impact resistance due to their higher degree of oxidation. The properties of the blend are improving with decreasing particle size of the rubber powder.[76] The impact strength of polypropylene (PP) is increased by the addition of rubber powder from tires. For a blend of this powder with polyamide-6 (PA-6), the impact strength is reduced. Swelling of the rubber powder in paraffinic plasticizing oil to alter its surface properties does not improve the properties.[77]

Virgin EPDM in a blend of 70% EPDM and 30% PP blend can be replaced by ground vulcanized EPDM; the maximal amount that can be replaced is 45%. The limitation in concentration is caused by processing difficulties at higher concentrations. The mechanical properties of this blend depend on the morphology, the state of cure, and the degree of packing of the rubber powder. It was found that the mechanical properties of the blend are initially negatively influenced with increasing concentrations of the rubber powder, but for higher concentrations an improvement of the properties was found. The explanation given is that the degree of cross-linking is reduced with increasing powder loadings, but at higher loadings this effect is counterbalanced by a filler effect, resulting in an increase of the mechanical strength.[78] The melt flow rate of the plastic component is another key factor for the properties of a blend with rubber powder. This conclusion was the result of a study with ambiently and cryogenically ground rubber powders (0.177

mm and 0.088 mm) based on SBR and EPDM blended with PP. Particle size and blending time played a minor role.[79]

An important application area for thermoplastic elastomers is the shoe industry. During the production of sports shoe soles, buffings are generated that can be reused in soles. Blends from polyvinyl chloride (PVC) with 20% rubber powder based on ethylene vinyl acetate (EVA) show higher impact strength compared to the standard system (PVC blended with 5% of acrylonitrile-butadiene-styrene-copolymer). The influence of the buffings on the properties of the PVC was dependent on type and concentration of the rubber powder, but independent of particle size. The addition of compatibilizers gave further improvements in toughness. Tested were NR, SBR, BR (butadiene rubber), liquid natural rubber, ethylene acrylic copolymer elastomer, styrene-butadiene-styrene block-copolymer, EVA, epoxidized natural rubber, and polyolefins as compatibilizers.[80] Other compatibilizers are grafted or block-copolymers, and their components are adjusted to the type of rubber powder and the polymer of the matrix.[81,82] Another method to improve the properties of a blend of a plastic material (polyethylene) with rubber powder is a chemical reaction with ethylene and an acrylic acid copolymer. This treatment enhances the impact strength of the blend.[83]

The properties of a blend of EPDM rubber crumb with PP are improved by using a radical initiator (t-butyl peroxide) as a compatibilizer. For a blend of SBR rubber powder with PP, a compatibilization can be achieved with a phenolic resin and a Lewis-acid catalyst (tin chloride).[84]

The polyblends can be vulcanized during blending. The curing systems used in this process are both sulfur- and peroxide-based. The mechanically stabilized blend can still be processed like a thermoplastic material, but the properties of this blend are improved by the vulcanization. A morphological investigation has shown that the rubber particles are chemically bound to the matrix.[85,86]

3.4 Surface Treatment of Rubber Powder

3.4.1 Chemical Activation

3.4.1.1 Addition of a Polymer and/or a Curing System

Several surface treatment processes are based on the addition of a polymer alone or in combination with a curing system. When the treated powder is blended with a virgin rubber compound, it forms chemical bonds between the rubber particle and the virgin polymer. This chemical bonding results in an improvement of the overall properties of the product containing the recycled material.

The addition of a cross-linkable polymer and a sulfur-based curing system can be done in a single-step process. The polymer added to the rubber powder is tuned to the polymer of the rubber waste, and the curing system is adjusted to the curing system used in the final product. This method can be used for NR, SBR, NBR (nitrile rubber), CR, and EPDM rubber. An example is given for a truck tire tread rubber powder containing SBR and NR and having a particle size of less than 0.42 mm. When this material is cured without any further additives by compression molding, the resulting product has a tensile strength of 8 to 12 MPa and an elongation at break of more than 200%.[87,88] A variation of this process is the addition of a liquid polymer and a curing system. The polymers used in this process have molecular weights of 1000 to 50,000 gram per mol; therefore, their viscosity is low. Water-free copolymers containing styrene and butadiene or isoprene and homopolymers from isoprene or butadiene are used. These liquid polymers are premixed with the curing system and added to the rubber powder. A powder based on truck tires with a particle size of 0.6 mm, treated according to this method and compression molded without any further additives, resulted in a product with a tensile strength of 7.5 MPa.[89] The addition of only a curing system is also sufficient to improve the properties of a powder vulcanizate. For a tire-based rubber powder the addition of 10 phr sulfur and 2 phr tetramethylthiuram disulfide (TMTD) gives the best properties after vulcanization at 150°C for 15 minutes.[90]

Another process is based on the addition of a combination of the following additives: an activator or accelerator, a retarder, a peptizer, a peroxide, and an antioxidant. This process can be used for sulfur-cured rubber materials. The rubber powder is mechanically devulcanized on the surface in an impact mill at a temperature between 105°C and 135°C during a period of 1 to 4 minutes. The additives react with the generated active sites on the surface of the rubber powder. The accelerators and activators react with the active chain ends of the broken cross-links, making them available for a new vulcanization reaction. The peroxide increases the cross-link density at the interface between the rubber powder and the matrix. The final cross-link density of the sulfur bridges in this area tends to be lower because curing agents migrate into the rubber particle, reducing the concentration of curatives at the interface. The antioxidants prevent oxidative degradation of the polymer during the breakdown reaction.[91] A chemical activation can also be achieved by the addition of an amine, an oxide, and an alcohol. A dispersion of equal amounts of a tertiary amine and an aliphatic or aromatic oxide in 1 to 10 parts of alcohol is added to the powder.[94]

Another process for coating a rubber powder with a polymer is to blend the polymer with an oil. Ten percent to 25% polyoctenamer can be dissolved in paraffinic oil at a temperature between 80°C and 100°C. This mixture is blended with the rubber powder in a fluid mixer or a mixing extruder; the concentration of the polymer-oil blend is 3 to 35%. A curing system and other additives can be added simultaneously. This material can be vulcanized without any further treatment, and the product has an improved tear resistance compared to the untreated powder. The process can be used for NR, SBR, EPDM, butyl rubber (IIR), BR, CR, and NBR rubber. Examples are given for EPDM powder having a particle size between 0.5 to 1.5 mm and resulting in a product with a tensile strength of 2.7 MPa and an elongation at break of approximately 200%. The results for the NR-based material are 3.3 MPa and 271% for a particle size of 0.2 to 0.5 mm, and for SBR-based material, 4.0 MPa and 207% for a particle size of 1 mm.[92]

3.4.1.2 Grafting

Silanization was tested as a surface treatment for rubber powder. First, the rubber particles were treated with maleic anhydride, followed by an epoxidation or hydroxylation, to create active groups on the surface of the powder. These groups are then able to react with the alkoxy groups of a silane coupling agent. The coupling agent is also able to react with the polymer during the curing reaction due to the polysulfide-moiety. This treatment results in an improvement of the properties.[93]

Acrylamide is another additive, and can be grafted onto rubber. Grafting of a NR powder (particle size less than 0.25 mm) with acrylamide and blended with a rubber compound resulted in an improvement of tear strength, compared to a blend with untreated powder. Grafting with p-nitroso diphenyl amine improved the chemical and thermal stability and the aging resistance of this system. The dynamic stability, especially for polymers with crystallization, was found to be improved as well.[43] Polystyrene can be grafted onto the surface of a rubber powder following an oxidative treatment. Different acids and peroxides as oxidizing agents have been compared in terms of the final performance of the treated rubber powder in a polystyrene matrix. The best results were found with a benzoic peroxide oxidation.[35]

A grafting reaction can also be achieved by swelling the rubber particle in an epoxidized or carboxylated oligomer. Grafting of a rubber powder with an epoxidized oligomer (5 to 13% oligomer) resulted in an improvement of the mechanical and dynamical stability after blending with a virgin polymer, compared to the properties of the blend with untreated powder.[43]

3.4.1.3 Swelling

Rubber powder can be swollen in an oligomer, and oligomer chains then migrate into the surface cavities of the rubber powder. When the treated rubber powder is subsequently blended with a virgin polymer, the oligomers can be chemically bound to the matrix, as well as to the rubber particle, under the formation of a stable chemical link between the two phases. The final products perform better in terms of mechanical and dynamical properties.[43]

3.4.1.4 Halogenation

Halogenation is a method to increase the polarity of the rubber powder. The compatibility with polar polymers such as polyurethane is increased, and the surface of the material becomes more hydrophilic. It is assumed that chlorination dehydrogenates the polymer and forms additional double bonds, resulting in an increased curing activity. A positive effect of this treatment was found for a blend of a tire-based rubber powder with polyurethane; the adhesion of the powder to the polyurethane was increased by a factor of 5, compared to the untreated rubber powder.[94] A comparative study was done with untreated and chlorinated rubber powder based on a blend of NR, SBR, and BR. Chlorination of the powder increased the hardness of the blends. In cases where a polar polymer was part of the matrix (NR/NBR and SBR/NBR), the mechanical strength of the blend was improved by the addition of chlorinated powder, compared to the blend with untreated powder.[95]

3.4.2 Physical Activation

Rubber granulate can be activated in a low-temperature plasma field. Depending on the medium in which the electrical discharge is taking place, and the presence of reactive gases, the chemical composition of the surface of the rubber powder is changed. The surface of the rubber particles becomes more polar and hydrophilic. This improves the compatibility of the rubber powder with polar binders or polymers. Media for the discharge are noble gases, air, oxygen, or nitrogen. Reactive gases used for a chemical reaction during the discharge are water, carbon dioxide, alcohols, and esters.[96]

3.4.3 Microbial Treatment

Different microorganisms are able to oxidize or reduce the sulfur in crosslinks. A complete oxidation of the sulfur leads to sulfate, whereas the reduction results in hydrogen disulfide.

Thiobacillus bacteria are able to oxidize the sulfur in polysulfidic bonds to sulfate. The reaction is limited to a surface layer with a thickness of less than 1 µm, and the total degree of desulfurization for a rubber powder with a particle size in the range of 0.1 to 0.2 mm is 4 to 5%. The reaction medium is a watery dispersion with other nutrients for the bacteria, and the process takes several weeks.[97]

The thiophilic bacteria *Sulfolobus acidocaldarius* is able to split carbon-sulfur bonds. The reaction is a stepwise oxidation of the carbon-bound sulfur into a sulfoxide, a sulfone, and finally to a sulfate. No improvement was found when the material was oxidized to the stage of the sulfate. It is assumed that the sulfate dissolves in the watery phase and the sulfur is, therefore, no longer available for revulcanization. Only when the reaction is interrupted at an intermediate stage, the sulfoxide or the sulfone stage, the sulfur remains bound to the rubber and is still available for vulcanization.[98–100]

The bacteria *Pyrococcus furiosus* is able to reduce sulfur-sulfur bonds. It is an anaerobic desulfurization resulting in cross-link-scission and reduction of the sulfur to hydrogen disulfide.[101]

3.4.4 Superfacial Devulcanization

A superfacial devulcanization of a rubber powder can be achieved by a mechanical breakdown of the rubber powder, together with the addition of a reclaiming agent and other additives, such as curing accelerators (for example, 2-mercaptobenzothiazole, tetramethyl thiuram disulfide), or activators (for example, aliphatic or aromatic amines). Different rubber powders having particle sizes between 0.5 and 5 mm have been activated on a mill using these additives. When blended with a virgin SBR compound, the activated powder results in better aging properties and a higher abrasion resistance, compared to untreated powder.[102]

The mechanical activation process is based on bond scission (breaking of cross-links as well as of polymer backbones) on the surface of the rubber particle. The rubber powder is passed through a mill with a friction of 1:1.3. The shearing forces are increased stepwise by reducing the gap between the two rolls down to 100 μm. The material heats up during this process, but the temperature should not exceed 100°C. The preferable temperature for this process is around 20°C to 30°C. The whole process takes 7 minutes, and the material passes the mill approximately 100 times during this period. The activated powder can be compression molded without any further additives, and the vulcanization of a tire granulate with a particle size of 1.0 mm results in a product with a tensile strength of approximately 5 MPa. This is a significant increase compared to the untreated material, which results in a product with a tensile strength of less than 1 MPa. The properties of the final product are independent of the particle size of the rubber powder. The following polymers can be activated according to this process: NR, isoprene rubber (IR), IIR, SBR, BR, NBR, silicone rubber, and polyurethane rubber.[103]

An internal mixer can also be used for mechanical activation. This process is shorter, compared to the activation on a mill (approximately 20 seconds), and the processing temperatures are higher. Less oxygen is present in the internal mixer during the activation process. Tests were done with a rubber granulate having a particle size of 2 mm, and the breakdown of the material has been analyzed by measuring the sol fraction and the cross-link density. The sol fraction increased during the process, and the cross-link density was reduced to 30% of the original value. It is assumed that mainly the weaker carbon-sulfur and sulfur-sulfur bonds are broken during this process.[104,105]

A treatment with caustic soda (NaOH) is reported to devulcanize the surface of a rubber particle by removing the sulfur. This method has been compared with an oxidative treatment with hydrogen peroxide in caustic soda, and a plasma treatment in oxygen, hydrogen, and ammonia. The material was blended with polymethyl methacrylate, and for this system the oxygen-plasma treatment gave the best results.[106]

3.5 Applications of Surface-Treated Rubber Powder in Rubber Products

Rubber powder can be cured without any further additives, but the resulting product has a low mechanical stability. Some surface treatments consist of the addition of a cross-linkable polymer and a curing system, providing the basis for a good vulcanization of the powder as such. Curing can be done by compression molding and also in continuously working presses, to produce mats and sheets. The minimum pressure necessary for a good vulcanization is 12 bar, and sheets with a thickness of approximately 3 mm can be produced in a few minutes at a temperature of 150°C. The exact curing time is depending on the curatives added to the rubber powder.

Surface-treated rubber powder has the same influence on the processing properties of a blend as untreated powder does. The surface treatment normally aims at improvement of the properties rather than improvement of the processing behavior. The main advantage in processing is an improved ventilation during curing, especially in the case of tire treads. Other areas of application are sieves in the mining industry, roofings, profiles, conveyor belts, bladders, and carpet backings.[107]

The influence of a rubber powder based on truck tire treads with a maximum particle size of 0.42 mm has been studied in a NR and a SBR compound. Figure 3.8[68] shows the influence of a surface treatment based on the addition of a polymer and a curing system, in comparison to untreated powder. Tensile strength is reduced for both polymers, but the decrease is less when surface-activated powder is used. In this case, the rubber particle is bound to the matrix by the vulcanization reaction, resulting in a higher stability of the material at the interface between particle and matrix. The effect of an untreated powder is more comparable to the effect of an inert filler. The same trend is found for compression set and rebound resilience; treated powder results in less influence compared to untreated material. Abrasion resistance is increased when treated powder is used. Tear strength is increased when the treated rubber material is blended with SBR.[68]

A similar investigation was done with increasing concentrations of treated IIR powder in a butyl compound. The powder was based on resin-cured bladder material, and was blended with a sulfur-curing butyl compound. The surface treatment had been chosen to match the curing system of the matrix. Despite the adjustment of the curing system, it was found that scorch time was shorter and curing time longer. Tensile strength and moduli were reduced. The mechanical properties of the material were improved by using

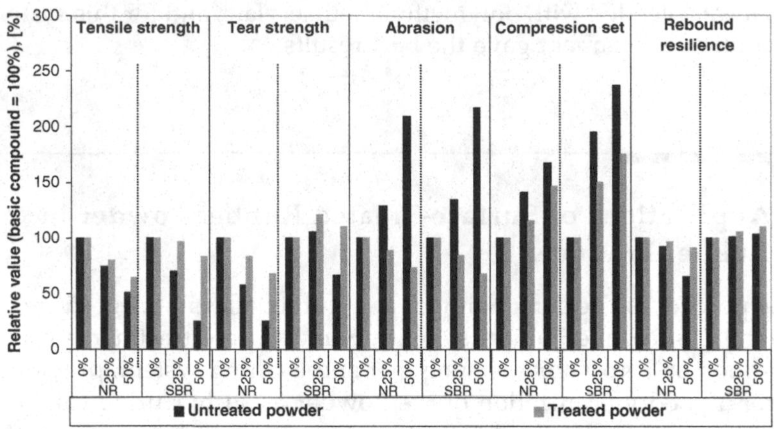

FIGURE 3.8
The influence of surface-treated and untreated powder on the properties of an NR and an SBR compound (truck tire tread powder, ambiently ground, maximum particle size 0.42 mm, surface treatment: addition of polymer and curing system; percentage in the graph: concentration of the rubber powder).

a finer powder, down to a particle size of 0.6 mm. The application of even finer powders did not result in further improvements.[108]

Nitrile-based powder (buffings from shoe sole production) was chemically activated and used in the same type of compound. When activated powder was blended with the same amount of a virgin nitrile compound, the result was a decrease in tensile strength and elongation at break, and an increase of the modulus at 300% elongation.[89,109]

In heterogeneous systems where the matrix is more polar than the rubber powder, halogenation can improve the properties. A chlorinated whole tire powder was blended with polyurethane, and a positive influence was found on the coefficient of friction, the thermal stability, and the humidity absorption. Hardness and dynamic properties of the final blend were not influenced by the addition of the treated rubber powder, but tensile strength and tear strength were negatively influenced. Typical applications for this type of blend are automotive parts, massive tires, and containers.[110,111]

3.6 Summary

Generally, the first step in the recycling of rubbery materials is grinding. Depending on the further treatment of the material (direct application, surface treatment, or reclaiming), definite particle sizes are required. Factors influencing the quality of a rubber powder are particle and particle size distribution, purity, surface morphology, and chemistry as well as the feedstock used. Most of these factors are determined by the grinding process, and the predominant differences between the processes are the temperature of the material during grinding (around the glass transition temperature or at ambient or higher temperatures) and the grinding mechanism (cutting, shearing, impact, or chemical processes).

Quality improvements of the rubber powder can be achieved by a surface treatment, resulting in an increased reactivity toward chemical or physical bond formation with a polymer matrix in which it is embedded. Chemical, physical, mechanical, and microbiological processes are used for this purpose.

References

1. Astafan, C.G. and Kohler, R., Comparison of scrap tire processing technologies — Ambient processing vs. cryogenic processing, presented at the 148th ACS Rubber Division Meeting, Cleveland, OH, USA, October 1995.
2. US 5 695 131, Shredder separator, December 1997.
3. US 5 299 744, Granulating separating and classifying rubber tire materials, April 1994.
4. Jones, K. and Lawson, K., Environmental challenges in the rubber world, presented at the Rubbercon, Gothenburg, Sweden, May 1995.
5. http://home.snafu.de/kurtr/str/en_sr.html, 06.02.2003.
6. Scrap Tire News Online, www.scraptirenews.com/areas/crumb/process.html, 14.10.2002.
7. Konieczka, R., Kaluzny, W., Sykutera, D., Feinzerkleinerung von Gummi durch Drehschneiden, Kautschuk Gummi Kunststoffe, 50, 9, 1997.
8. EP 0 567 759, Vorrichtung zum Pulverisieren von Gummibrocken, March 1993.
9. EP 0 567 760, Verfahren und Vorrichtung zum Zerkleinern von Gummibrocken, March 1993.
10. http://www.geocities.com/desislava20/rubber_extruders.html, 06.02.2003.
11. Röthemeyer, F., Aufbereitung von Altgummi für den Einsatz in Kautschukmischungen, Kautschuk Gummi Kunststoffe, 46, 5, 356, 1993.
12. Novelli, L.R., Recycling tires back into tires, Scrap Processing and Recycling, 75, November–December 1995.
13. Khait, K., Carr, S.H., and Mack, M.H., Solid-state shear pulverization: a new polymer processing and powder technology, Lancaster: Technomic, 2001.
14. Khait, K., Solid-state shear extrusion pulverization process, *Rubber World,* 216, 2, February 1997.
15. Kuznetsova, O.P., Tchepel, L.M., and Prut, E.V., Physical and mechanical structure of powders after rubber grinding, http://dicpm.unipa.it/modest/html/symp9/P_9_Th_02.pdf
16. Arinstein, A.E. and Balyberdin, V.N., High-temperature shear-induced multiple cracking and grinding of polymeric materials, presented at the 154th ACS Rubber Division Meeting, Nashville, TN, USA, October 1998.
17. LaGrone, R.D. and Lynch, J., New process makes small-particle recycled rubber, *Rubb. Plast. News,* 16, 15, 14, February 1987.

18. Sommer, F., Stoffliches Recycling von Altgummiabfaellen, Kunststofftechnik 'Elastomere und Umwelt', VDI Gesellschaft Kunststofftechnik, Duesseldorf, Germany, 1991.
19. US 6 425 540, Method and apparatus for grinding rubber, February 2000.
20. http://www.wegrind.com/services.html
21. Umsicht (Fraunhofer Institut Umwelt-, Sicherheits-, Energietechnik), Einsatz neuer Technologien beim Altgummi-Recycling), Workshop, Oberhausen, Germany, 26.09.2002.
22. Technical information, Turborotor, Mahltechnik Görgens GmbH.
23. Technical information, Ultrafein Turbo Mühle, Microtec
24. US 5 482 215, Method of reclaiming rubber from vehicle tires, January 1996.
25. Technical information, Resonance disintegration, Velox GmbH.
26. http://www.hn-oed.com/english/scrap%20tyre%20recycling%20tech/, 06.02.2003.
27. Orsen, M., Cryogenics processing and recycling opportunities, *Rub. Tech. Int.*, 253, 1996.
28. US 4 813 614, Method and apparatus for treating waste products to recover the components thereof, March 1989.
29. CWC Best Practices in Scrap Tires & Rubber Recycling: Ambient versus cryogenic grinding, http://www.cwc.org/tire_bp/t_bp_pdf/2-03-04.pdf, 1998.
30. Kohler, R., Cryogenic processing of scrap tires, presented at the 148th ACS Rubber Division Meeting, Cleveland, OH, USA, October 1995.
31. Kohler, R., Advances in cryogenic fine grinding, presented at the 152nd ACS Rubber Division Meeting, Cleveland, OH, USA, October 1997.
32. Hausmann, J.M., Frozen claim, *Tire Techn. Int.*, 69, June 1998.
33. EP 0 608 157, Fast process to reclaim cured or semi-cured rubber, January 1994.
34. Leca, M. and Runcan, I.F., Reclaiming of rubber from waste tyres by physical methods, *Kautschuk Gummi Kunststoffe*, 49, 12, 855, 1996.
35. Yamashita, S., Reclaimed rubber from rubber scrap (2), *Int. Polym. Sc. Techn.*, 8, 12, T/77, 1981.
36. US 5 492 657, Method for processing rubber products, 1996.
37. Trofimova, G.M. et al., Effect of the method of tire grinding on the rubber crumb structure, *Polym. Sci.*, A, 42, 7, 825, 2000.
38. Cecich, V. et al., Use of shredded tires as a lightweight backfill material for retaining structures, *Waste Man. Res.*, 14, 5, 433, October 1996.
39. Rogers, J.N., Vanini, T., and Crum, J.R., Simulated traffic on turfgrass topdressed with crumb rubber, *Agron. J.*, 90, 215, March – April 1998.
40. Adding value to crumb rubber, *Biocycle*, 39, 3, 48, March 1998.
41. Groenevelt, P.H. and Grunthal, P.E., Utilisation of crumb rubber as a soil amendment for sports turf, *Soil and Tillage Res.*, 47, 1-2, 169, January/February 1998.
42. *Scrap Tire Users Directory*, Recycling Research Institute, Suffield, CT, USA, 1994.
43. Makarov, V.M. and Drozdovski, V.F., Reprocessing of tyres and rubber wastes, Ellis Horwood, Great Britain, 1991.
44. Williams, D.E. et al., Redefining thermosets: A method of molding vulcanized rubber into new products, *Rubber World*, 40, June 2002.
45. Tripathy, A.R. et al., Rubber recycling: understanding the mechanism of high-temperature high-pressure vulcanized rubber sintering, presented at the ACS Rubber Division Meeting, Cleveland, OH, USA, October 2001.

46. Williams, D.E. et al., High-temperature high-pressure sintering: a method of recycling waste rubber, presented at the 160th ACS Rubber Division Meeting, Cleveland, OH, USA, October 2001.
47. Farris, R.J. et al., Powder processing techniques to recycle rubber tires into new parts from 100% reclaimed rubber powder/crumb, Chelsea Center for Recycling and Economical Development, technical report no. 40, December 2001.
48. MacKnight, W.J. and Tobolsky, A.V., Polymeric sulfur and related polymers, Interscience Publishers, New York, USA, 1965.
49. European Tire Recycling Association (ETRA), ETRA News, 6, 3, 15, fall 2002.
50. Scrap Tire News Online, www.scraptirenews.com/areas/crumb/prod_app.html, 17.01.2003.
51. Klingensmith, B. and Baranwal, K., Review of rubber recycling and reuse in the USA, *Rub. Techn. Int.*, 216, 1997.
52. http://www.scraptirenews.com/99jan03.html, http://www.scraptirenews.com/98nov1.html, http://www.scraptirenews.com/98dec2.html, http://www. scraptirenews.com/99mar3.html, http://www.scraptirenews.com/99nov5.html
53. Zhu, H., Thomg-On, N., and Zhang, X., Adding crumb rubber into exterior wall materials, *Waste Man. Res.*, 20, 5, 407, 2002.
54. European Tire Recycling Association (ETRA), *Introduction to Tyre Recycling*, Shulman, V. L. (Ed.), 1998.
55. Robinson, S., Rosehill binds up scrap rubber, *Plast. Rub. Wkly.*, 1568, 8, 13.01.1995.
56. McNeill, J.D., Advanced technology tire recycling system for improved, cost effective polyurethane parts, presented at Polyurethanes '95, Chicago, USA, September 1995.
57. White, L., A wheelie good nice, *Europ. Rub. J.*, 31, November 1996.
58. US 5 316 708, Method for making products made of recycled vehicle tyres, May 1994.
59. DE 195 08 166, Verfahren zur Herstellung von gummihaltigen Formteilen, gummihaltiges Formteil und Verwendung eines solchen, September 1996.
60. US Patent 5 468 539, Precast surface paving overlay comprising rubber crumbs and clay particles, 1995.
61. Iwainsk, H., Wiederverwendung von Gummiabfällen, Entsorgungspraxis, 657, November 1990.
62. Eastman, A.L., presented at the 140. ACS Rubber Division Meeting, Orlando, FL, USA, October 1991.
63. Smith, F.G., The benefits from a partnership between the recycling specialist and tire maker, *Rubber World*, 16, September 1992.
64. Gibala, D. and Laohapisitpanich K., Cure and mechanical behavior of compounds containing ground vulcanizates. Part II — Mooney viscosity, presented at the 148th ACS Rubber Division Meeting, Cleveland, OH, USA, October 1995.
65. Gibala, D. and Hamed, G.R., Cure and mechanical behavior of compounds containing ground vulcanizates. Part I — Cure behavior, presented at the 144th ACS Rubber Division Meeting, Orlando, FL, USA, October 1993.
66. Boruta, J. and Behal, M., The effect of degradation degree of the dispersed phase on the ozone aging of rubbers filled with crushed rubber powder, *Die Angewandte Makromolekulare Chemie* 176/177, 173, 1990.
67. Hausmann, J.M., Crumb rubber manufacturing: cryogenic vs. ambient, *Rub. Tech. Int.* '97, 223, 1997.

68. Technical information Rubber Resources, The Netherlands, 1999.
69. Jacob, C., De, P.P., and De, S.K., Utilisation of powdered natural rubber vulcanizate as a filler, *Rubber World*, 43, November 2002.
70. Klüppel, M. and Kuhrke A., Wiederverwendung von Altgummi in technischen Elastomerartikeln, Kautschuk Gummi Kunststoffe, 5, 373, 1997.
71. Klingensmith, W. and Baranwal, K., Recycling of rubber: an overview, *Rubber World*, 41, June 1998.
72. Jacob, C., De, P. P., Bhowmick, A. K., and De, S. K., Recycling of EPDM waste. I. Effect of ground EPDM vulcanizate on properties of EPDM rubber., *J. Appl. Polym Sc.*, 82, 3293, October 2001.
73. Baranwal, K. and Rogers, J.W., Catalytic regeneration of rubber: Part 1 — Polychloroprene, presented at the 154th ACS Rubber Division Meeting, Nashville, TN, USA, October 1998.
74. Chidambaram, A. and Min, K., Reactive blending of chemically and physically treated waste rubber with polymethyl methacrylate, presented at the ANTEC '94, San Francisco, CA, USA, 1994.
75. Khait, K., New solid-state shear extrusion pulverization process for used tire rubber recovery, *Rubber World*, 38, May 1997.
76. Rajalingam, P. and Sharpe, J., Ground rubber tyre (GRT) / thermoplastic composites: effect of different ground rubber tyres, *Rub. Chem. Technol.*, 66, 4, 664, 1993.
77. Fuhrmann, I.and Karger-Kocsis, J., Einfluss von Altgummipartikeln auf das mechanische Eigenschaftsprofil von Thermoplast-Blends, *Kautschuk Gummi Kunststoffe*, 52, 12, 836, December 1999.
78. Jacob, C. et al., Recycling of EPDM waste. II. Replacement of virgin rubber by ground EPDM vulcanizate in EPDM/PP thermoplastic elastomeric composition, *J. Appl. Polym. Sci.*, 82, 3304, October 2001.
79. Liu, H.S. et al., Development of novel applications for using recycled rubber in thermoplastics, Chelsea Center for Recycling and Economical Development, March 2000.
80. Phinyocheep, P. and Axtell, F.H., Development of low cost toughened plastics utilizing recycled rubber, presented at the 144th ACS Rubber Division Meeting, Orlando, FL, USA, October 1993.
81. Rajalingam, P. and Baker, W.E., Compatibilization of ground rubber tires in recycled polyethylene/polypropylene waste plastics, presented at the ANTEC, Detroit, MI, 1992.
82. US 5 514 721: Thermoplastic compositions having altered reclaimed rubber therein and method of making same, May 1996.
83. Rajalingam, P. and Baker, W.E., Role of functional polymers in ground rubber tyre-polyethylene composites, *Rub. Chem. Technol.*, 65, 5, 908, 1992.
84. Liu, H.S., Mead, J.L., and Stacer, R.G., Process development of scrap rubber/ thermoplastic blends, Chelsea Center for Recycling and Economical Development, March 2001.
85. Michael, H. and Scholz, H., Blends from recycled rubber and thermoplastics, presented at the National Seminar on Recycling and Waste Management, CIPET, Madras, India, 1997.
86. Scholz, H. et al., Morphology and mechanical properties of elastomeric alloys from rubber crumb and thermoplastics, *Kautschuk Gummi Kunststoffe*, 55, 11, 584, November 2002.

87. US 5 425 904, Process for activating vulcanized waste rubber particles and a process for producing a rubber-like article using said activated waste rubber particles, December 1991.
88. Dierkes, W., Surface activated crumb rubber: a new development in rubber recycling, presented at the 17. IRMRA Rubber Conference, Madras, India, December 1995.
89. US 4 481 335, Rubber compositions and method, November 1984.
90. Kim, J.K. and Lee, S.H., New technology of crumb rubber compounding for recycled waste tires, *J. Appl. Polym. Sci.*, 78, 1573, September 2000.
91. DE 100 52 450, Devulkanisat aus Gummiabfällen, Devulkanisatcompound, Verfahren zu deren Herstellung und Verwendung zum Wiedereinsatz in Frischmischungen und zur Herstellung von Spritzgussformteilen, October 2000.
92. DE 41 11 158, Vernetzbare Mischungen aus Recycling-Gummigranulat und Kautschuk, 1992.
93. Amash, A., Giese, U., and Schuster, R.H., Interphase grafting of reclaimed rubber powder, *Kautschk Gummi Kunststoffe*, 55, 5, 218, 2002.
94. Smith, F.G. and Daniels, E.J., Testing and evaluating commercial applications of new surface-treated rubber technology utilizing waste tires, *Res. Cons. Rec.*, 15, 133, 1995.
95. Kim, J.K., Utilization of recycled crumb rubber as a compounding tool, *Int. Polym. Proc. XIII*, 358, 1998.
96. EP 0 788 870, Process for surface treatment for vulcanized rubber and process for production of rubber-based composite materials, August 1977.
97. Löffler, M. and Neumann, W., Mikrobielle Entschwefelung von Gummigranulat — ein Beitrag zur stofflichen Wiederverwertung von Altgummi, *Kautschuk Gummi Kunststoffe*, 48, 6, 454, 1995.
98. Romine, R.A. and Romine M.F., Microbial processing of waste tire rubber, presented at the 148th ACS Rubber Division Meeting, Cleveland, OH, USA, October 1995.
99. Tire-eating bacteria under study, *Rub. Plast. News II*, October 7,1995.
100. Holst, O., Stenberg, B., and Christiansson, M., Biotechnological possibilities for waste tire rubber treatment, *Biodegradation*, 9, 3–4, 301, September 1998.
101. Bredberg, K. et al., Anaerobic desulfurization of ground rubber with the thermophilic archaeon *Pyrococcus furiosus* — a new method for rubber recycling, *Appl. Microbiol. Biotechnol.*, 55, 43, 2001.
102. Ovtscharov, V.I., presented at the Technomer, Chemnitz, Germany, 1995.
103. WO 96/02372, Method of reactivating used rubber particles or shredded rubber waste, February 1996.
104. Kolinski, A. and Barnes T., Thermo-mechanical re-activation of tire-rubber crumb, presented at the 154th ACS Rubber Division Meeting, Cleveland, OH, USA, October 1998.
105. Kostanski, L.K. and McGregor, J.F., Physico-chemical changes during re-activation ("devulcanization") of tire-rubber crumb, presented at the 154th ACS Rubber Division Meeting, Cleveland, OH, USA, October 1998.
106. Chidambaram, A. and Min, K., Reactive blending of chemically and physically treated waste rubber with polymethyl methacrylate, presented at the ANTEC, San Francisco, CA, USA, 1994.
107. Stark, F.J., Surface treated recycled crumb rubber, presented at the Australian Plastics & Rubber Institute's 10th National Conference, Leura, Australia, October 1997.

108. ADVAC Elastomers of Wisconsin, technical information on Tirecycle®, October 2002.
109. Wagner, D. and Stark, F., Case histories of utilizing scrap rubber, Rubber Recycling & Technology Seminar, Cuyahoga Falls, OH, USA, May 1997.
110. Smith, F.G. and Daniels, E.J., Testing and evaluating commercial applications of new surface-treated rubber technology utilizing waste tires, *Res. Cons. Rec.*, 15, 133, 1995.
111. Baumann, B.D., PU incorporating surface-modified particles and fibres, *Rubber World*, 208, 1, 16, 1993.

4

Tire Rubber Recycling by Mechanochemical Processing

Klementina Khait

CONTENTS

4.1 Introduction

Utilization and disposal of used tires represent an enormous challenge to industrialized countries worldwide. At the present time, there are over three billion scrap tires in the U.S., and approximately 270 million are added annually.[1] Incineration of tires would be attractive, but there is concern about atmospheric contamination by particulate emissions. Stockpiles of tires cause environmental hazards such as mosquitoes breeding and spontaneous fires that are difficult to extinguish. Fire fumes emit smoke containing volatile organics, metals, and other potentially harmful compounds. When worn tires are landfilled, they leach some additives that may kill useful bacteria in the soil.

According to Blumenthal,[2] tires pose a paradox. Although an average new tire made today has the longest wear, about 45,000 miles and some up to 80,000 miles, the number of scrap tires generated annually continues to rise due to both an increased number of cars on the road and mileage driven every year. The three major markets for scrap tires are tire-derived fuel, civil engineering, and ground rubber applications. The latter use is of strong interest with respect to an increased amount of recycled tire rubber in new tire compounds. Several tire producers that use up to 5% of recycled tire rubber conducted successful field tests. For playground surfaces, tire rubber is applied in two different forms. One is a mat, sometimes covered by recycled rubber, and the other is loose 1/4- to 3/8-inch pieces free of steel and cord. This loose fill material has strong cushioning effect, lasts for a long time, and is moisture resistant.

Although some quantities of scrap tires are used as a fuel due to their high calorific value, the most desirable approaches to their reutilization are recycling and reclaiming of raw materials. Tire rubber has been recycled and reused for a long time, through the application of various techniques involving size reduction by grinding, at either ambient or cryogenic temperature. Each of these grinding techniques produces crumb rubber of different particle size and particle-size distribution and shape. Both characteristics (size and shape) vary widely and affect the applications of the tire rubber. Rubber crumb is utilized in many products: as a toughening agent for thermoplastics, in rubberized asphalt, in athletic surfaces, and in boat docks, among others.

4.2 Brief Review of Existing Tire Rubber Recycling Processes

At present, the major effort in tire rubber recycling is to reuse it as a finely ground crumb, produced by mechanical grinding. This crumb is added as a partial replacement of virgin rubber in various compounds. Unfortunately, the amount of ground rubber used in these compounds is limited, due to some loss in its physical properties. The search for better technologies that will allow larger quantities of used tire rubber to be incorporated into new products continues, and several new approaches have been successful. These approaches include surface modification, ultrasonic devulcanization, mechanochemical pulverization, new reclaiming techniques, and others. Current recycling technologies used to convert scrap tire rubber into crumb include size reduction and chemical processes to recover oil and energy. Manuel and Dierkes,[3] and Myhre and MacKillop[4] categorized existing technologies as follows:

- **Recycling:** All processes where scrap tire rubber is converted into a reusable material to make new products.

- **Reclaiming:** Processes that break down vulcanized rubber by heat in the presence of chemicals or high-pressure steam. The breakdown of scrap rubber can be accomplished by devulcanization (breaking sulfur-sulfur bonds) or by depolymerization (breaking polymer chains).
- **Pyrolysis:** In the absence of oxygen, tire rubber generates oil, gases, and chemicals. Depending on temperature, various products are produced. At temperatures below 400° C the major product is carbon char, and at higher temperatures more liquids, oil, and gases are made.[4]
- **Tire-derived fuel (TDF):** The process by which shredded tires are burned for their calorific value of 13,000 to 15,500 Btu/lb, which is close to that of coal (13,000 Btu/lb) according to Myhre and MacKillop.[4]

Another method of reuse of tire rubber is retreading of worn tires, which helps to divert old tires from landfills. This method, however, only postpones recycling or reclaiming because eventually, worn tires must be taken out of service.

Most equipment used to grind tire rubber produces materials with very limited reactivity. According to Stark,[5] the main emphasis has been to reduce the particle size to the level of carbon black, but that was unattainable by any existing grinding processes. When concentration of tire rubber powder (used as a filler) in a new tire formulation exceeded 5 wt%, the physical properties of the compound declined.

Vulcanized rubber scrap is used as a filler in polyethylene composites, to be extruded into open pour pipes for aeration and soil irrigation.[6] The addition of tire rubber as a partial replacement of virgin rubber in various compounds offers some processing advantages,[7] such as shorter mixing time, enhanced release during molding, and reduced blemishes due to air entrapment.

New technologies continue to emerge in the U. S. and Europe involving devulcanization and surface modification of rubber. The latter is beneficial since it increases the adhesion between rubber and plastics in rubber/plastic composites.

Bauman[8] developed a new surface-modification process for scrap tire rubber utilizing a reactive gas atmosphere, which causes a permanent chemical change in rubber. It was indicated that surface-activated used tire rubber had improved interfacial adhesion in the rubber/plastics composites.

Another method of reprocessing rubber through surface-activated crumb Surcrum® (Vredestein Rubber Recycling, Maastricht, Netherlands) has been reported by Dierkes.[9] Waste rubber stream included natural rubber, nitrile rubber, styrene-butadiene rubber, and ethylene-propylene diene monomer (EPDM) rubber. These materials were ambiently ground to a maximum particle size of 40 mesh (420 μm) and treated via a thermomechanical method that made the surface of crumbs very reactive. The activated rubber has high specific area of 1 to 2 m^2/g. Dierkes stated that during the activation step, a cross-linkable surface layer consisting of a polymer with curing agent was added to the crumb. Significant increase in tensile strength of compounds containing Surcrum® was reported.

Isayev, Yushanov, and Chen[10] carried out extensive studies to develop a novel, continuous ultrasonic devulcanization process. They showed that ultrasonic waves under pressure and heat rapidly break up the three-dimensional network in vulcanized rubber. The resultant product can be reprocessed, shaped, and revulcanized similar to virgin rubber.

Watson[11] reported a new process for recycling used tire rubber by mechanochemical means, by which rubber is softened and sheared during mastication.

In a typical cryogenic grinding system,[12] tires are first ambiently shredded into 1- to 3-inch pieces, chilled with liquid nitrogen to a temperature ranging from –70° to –130° C, followed by grinding on a mill, separation of fiber and metal, and classification of the crumb rubber. In this process, up to 80% of the resultant crumb is smaller than 10 mesh (2,000 μm) size. If the system uses multiple ambient temperature granulators and cryogenic grinding of precooled chunks, the produced crumb is smaller than 30 mesh (590 μm) size. The advantages of cryogenic grinding include high throughput and ease of fiber and metal separation. Disadvantages include the cost of liquid nitrogen and the production of flat powder particles, with sharp corners like crushed stone.

Ambient grinding is practiced as either a dry or wet process, and it produces powder with higher surface area than that of cryogenically ground tire rubber. It is well known that powder with high surface area has better adhesion to virgin rubber or to thermoplastics, which leads to improved physical properties of the compounds. In addition, powder with higher surface area requires a shorter mixing cycle during compounding and results in products with better appearance.

4.3 Principles of Size Reduction

As previously noted, the most commonly used method to obtain crumb rubber is size reduction or comminution. Comminution processes are energy intensive and costly to operate; frequently, their costs are greater than 25% of the total raw material processing.[10,13] The choice of size reduction equipment is dependent upon the nature of the rubber feedstock and the particle size needed for specific application. A coarse size reduction (from large pieces to 1 cm) is achieved by crushing or shredding. Fine size reduction (from 1 cm to 10 μm) is achieved by grinding in tumbling mills, while ultrafine size (below 10 μm) is accomplished in pulverizing mills. Numerous grinders and mills produce crumb rubber with different particle morphology, depending on the equipment design. As mentioned earlier, the energy consumption[10] to produce crumb rubber is very high; most energy is spent on heat dissipation (85%), moving material within the equipment (13%), and only about 1% is spent on size reduction.

The finely ground tire rubber is needed to increase its reuse as a partial substitute for virgin rubber in order to reduce cost. The size of ground rubber used in compounding varies from 10 mesh to 200 mesh (2000 to 74 μm). Mechanical grinding is conducted at ambient or cryogenic temperatures. In both cases, in order to produce fracture, a solid material must be stressed.[14,15]

If a solid is subjected to an applied force, the resultant stress induces a deformation (strain) within the solid,[16] which will respond by either a change in shape and dimensions, or in fracture. The former response is often reversible, while fracture is irreversible and may take place with little deformation (brittle fracture) or with considerable strain (ductility).For many materials, stress is proportional to strain, at least at low strains (Hooke's law); polymers have a more complex relationship between stress and strain, which is represented by a stiffness parameter that includes shear modulus and tensile modulus.

In general, solids are divided into the following categories: homogeneous or heterogeneous, brittle or ductile, and crystalline or amorphous. Most of the commercial materials are heterogeneous and brittle and will break when stressed, producing brittle fracture. Behavior of these materials rendering stress, fracture, and disintegration is a complex phenomenon. It is important to realize that during grinding, there is a concern not only with the operating conditions at which fracture occurs, but also with the particle size distribution (PSD) resulting from the fracture.

During grinding, a material is subjected to various forces sufficient to break pieces into smaller particles. There are four basic methods of force application[17]: impact (particle concussion by a single force), compression (particle disintegration between two rigid forces), shear (wrenching apart of a particle by rigid force), and attrition (the scraping of a particle against a rigid face or particle against particle). Particle size and PSD are affected by changes that occur as grinding proceeds. According to Prasher,[17] primary breakage of solids results in a bimodal distribution of particles, followed by further breakage toward a monomodal distribution. During breakage, pieces progressively become subdivided into ever-smaller particles. The terms "breakage" and "rupture" used in the technical literature are synonymous with fracture.

Kausch and Williams[18] developed a model to explain polymer fracture as follows: fracture is defined as stress-based material disintegration through a formation of new surfaces within a solid. The physical principle of fracture mechanics is based on the introduction of the energy necessary to propagate a crack. Griffith[19] made two assumptions: first, that all solids contain flaws or inhomogeneities that propagate to cause failure, and second, that fracture strength of a material is characterized as the energy per unit area required to produce new surfaces. Williams[20] described a relationship between the stress at fracture and the size of the flaw, assuming that when a flaw grew in the stressed solid, there was a decrease in the energy of the solid, so energy was released to form the new surfaces of the growing flaw. The stress-induced breakdown occurs by the cleavage of either primary (covalent) or secondary (van der Waals or hydrogen) bonds or both.[21]

It is known that rubbers consist of long, flexible, chain-like molecules that are interconnected by cross-links to form a network.[21] Rubbers are able to undergo large deformations, and when the stress is released, they return to their original shape. The initiation sites of microvoids and cracks may arise from accidental scratches, dust or dirt, or mechanical damage. Other sites for the initiation and growth of internal voids are filler particles. A high degree of cross-linking reduces the flexibility of long molecules (which can coil, bend, and straighten), so these molecules become weak to shear stresses.

4.4 Mechanochemical Processes for Rubber Recycling

4.4.1 Elastic-Deformation Grinding (EDG)

In the early 1980s, a group of researchers at the Institute of Chemical Physics, Academy of Sciences in Moscow, Russia, under the direction of the academician, Enikolopyan, studied the behavior of solids under high pressure and shear.[22,23] Their work can be traced back to the well-known research of compressibility of solids by Bridgman.[24] The unusual effects of the application of high pressure and shear deformations of solids were first described by Bridgman, who tested 300 materials (including rubber) in an anvil-type apparatus, later called a Bridgman anvil. Bridgman noted that compressibility of rubbers differs greatly from that of other polymers. Tsirule showed that the presence of sulfur as the vulcanizing agent changes compressibility dependence on pressure since sulfur transforms the soft, elastic material into a rigid, low-compressible rubber over a wide temperature range.[25]

The Russian researchers observed a formation of fine powder in a high-shear Banbury mixer when a low-density polyethylene (LDPE) was subjected to high pressure and shear accompanied by cooling (instead of heating). Enikolopyan proposed that under these conditions, elastic energy was "pumped" into the polymer under shear deformation until stored energy was released with a formation of new surface. This discovery resulted in the development of a new grinding process that was initially termed elastic deformation (ED) or elastic deformation grinding (EDG).

By varying the processing parameters (temperature, pressure, feed rate, etc.), it was possible to select the conditions by which the deformability of the polymer was minimal and, therefore, substantially decrease the energy of fracture.[26] The ED method was characterized by rather low energy consumption, which was two to three times lower than that of conventional grinding methods.[27]

Enikolopyan described the fracture mechanism of the EDG as an explosive, lava-type process, based on two observations: first, a destruction of the solid takes place over a narrow "grinding" zone (between 3 and 8 mm in length with a total rotor-grinder length of 100 mm) and second, destruction takes place suddenly and simultaneously across the volume of the grinding zone[27], and the particle size remains unchanged. This process was very different from conventional grinding, where sequential size reduction caused particles to decrease in size as the grinding time increased.

4.4.1.1 Mechanism of EDG

The extensive studies by the Russian researchers have provided a basic understanding of the pulverization of polymers under high shear and compression. The theory of pulverization, however, is not yet fully developed. Enikolopyan described this process as having a branched chain character, where the energy released for the creation of a new surface was, in turn, spent on the formation of the next surface (powder). The initiation of a single crack and propagation of a crack that led to fracture were part of the explanation of the proposed mechanism of EDG. Enikolopyan applied Griffith's theory of fracture to explain the EDG phenomenon. Griffith[19] proposed that a fracture starts from a microcrack initiated by stress concentrations in the vicinity of a preexisting flaw. The microcrack propagates from the flaw, and at some point, becomes a crack that eventually results in fracture. Enikolopyan and co-workers discovered that in addition to size reduction, the process is capable of intensive mixing and chemical modification of polymers via so-called "mechanochemistry." This term was initially introduced in the 1960s, in the former Soviet Union, by Baramboim,[28] who investigated chemical reactions in polymers induced by mechanical energy.

When polymers are stressed, chain scission occurs, with the formation of free radicals at the ruptured end. Because these radicals are formed as a result of the application of mechanical load, they are called mechanoradicals.[28] Baramboim proposed that the rupture of the molecular chains in polymers under the application of mechanical forces takes place when the stress exceeds the critical value for the covalent bonds between the atoms. This rupture can change the properties of the polymeric solid, introducing modification of the chemical and physical structures, which in turn results in a change in the melt rheology.

Baramboim differentiated between comminution (milling for reducing particle size) and grinding (milling for change in surface properties, although it also reduces particle size). According to Baramboim, comminution of polymers consists of the following stages: 1) change in particle size and particle shape, and cleavage of inter- and intramolecular bonds, 2) reduction of the molecular weight, 3) release of macro-radicals on the new surfaces, and 4) change of physical properties. These processes may occur simultaneously and depend both on the type of mechanical action and the polymer structure. Baramboim and a large group of Russian researchers studied the mechanochemistry of polymers in the solid state, including the formation and nature of free radicals during milling. Electron spin resonance (ESR) spectroscopy was used to study free radicals formed during application of mechanical load to various polymers. Baramboim concluded that the fracture took place along weak bonds and that free-radical concentration decreased exponentially with distance from the fracture surface (about 85% of all radicals formed within 2 μm of the surface). Chain rupture is related to the reduction and distribution of particles achieved by the mechanical action (shear intensity). During milling, the rupture is caused by localized stresses that exceed the strength of bonds.[29]

The above-mentioned mechanochemical reactions during EDG showed a strong potential for *in-situ* compatibilization of ordinarily immiscible polymers, due to a formation of block copolymers or graft copolymers. Several researchers[28,30,31] have successfully used mechanochemistry to create copolymers through free radicals, using various polymer substrates.

The applied stress can cause movement of the chains relative to each other (disentanglement) or scission of the chains (degradation). Both processes are temperature dependent, and the dominating process is determined by the chain length and a critical strain rate.[28]

The Romanian researchers Simionescu and Oprea[32] studied mechanochemistry of polymers during an application of intense mechanical forces concentrated on surfaces, causing a sudden formation of submicroscopic cracks followed by a breakage of chemical bonds. An occurrence of uneven strain at the site of the rupture of stressed polymers leads to several complex phenomena, such as submicroscopic cracks, change of relative position of cracks at different temperatures, the formation of new surfaces, and the cleavage of bonds along the chain (intermolecular bonds). An increase in reactivity of the polymers during rupture due to a presence of free radicals was explained by the formation of new functional groups, detected by infrared spectroscopy (IR).

Knowledge of the effect of shear, temperature, time, polymer structure, and composition is essential in understanding solid-state mechanochemistry. Stress-induced reactions were observed with both rubbers and plastics. The conversion of the mechanical energy applied to a polymer shifts from thermal dissipation energy to storage for higher molecular weight polymers. According to Porter and Casale,[33] that energy storage causes several changes such as bond rotation, chain orientation and elongation, bond bending, and finally, the rupture of chemical bonds. Free radicals released during chain breakage affect molecular weight and its distribution, branching, and cross-linking. Most mechanochemical processes reduce molecular weight, with the exception of the effect of shear for rubbers and polyolefins (under oxygen-starved conditions), where cross-linking and branching reactions result in an increase in molecular weight. Weak bonds along the chain can occur from inhomogeneity, adjacent double bonds, or catalysts. Porter and Casale suggested that shear, as one of the primary variables in mechanochemistry, can cause a change in molecular weight distribution (MWD). That change has practical implication of processability and properties; at rupture of a monodispersed polymer, MWD broadens. Another important variable is temperature, which affects the physical state of the polymer (glassy, crystalline, elastic, or fluid) and the bond rupture. It is known that the specific temperatures for each physical state depend on polymer composition and molecular weight. When temperature is rising, it softens the polymer, lowers stress, and reduces chain rupture. At lower temperature, the range of molecular motion is decreased and the rate and extent of mechanochemical reactions are generally higher.[33]

Sohma,[34] in Japan, defined mechanochemistry as "a branch of chemistry in which chemical phenomena, such as chemical reactions and changes in crystalline structures induced by mechanical actions like fracture and large deformation, are studied." Researchers working on rubber chemistry were the first investigators that observed chemical changes, such as a considerable reduction of molecular weight of rubber, after roll milling. One of the reasons, however, for not labeling roll milling as a mechanochemical process was the fact that strong mechanical action was accompanied by friction that produced heat. Therefore, the observed chemical change might be postulated as caused by heat. Sohma stated that the formation of free radicals at low temperature (21° C) under strong mechanical action has been reported based on ESR data.[34] Since the free radicals were generated at low temperature, the possibility of their production from heat was ruled out. These free radicals served as a proof that mechanical action induced chemical change, such as the scission of chemical bonds.

Information obtained by ESR spectroscopy includes positive, direct evidence for the existence of unpaired electrons, either in free radicals or at broken bonds, and concentration of free radicals from spectrum intensity. In addition, radical species can be identified. The ESR intensity is proportional to the quantity of broken bonds. Presence of free radicals can be also determined by end-group analysis using IR, change in molecular weight by gel permeation chromatography (GPC), and change in rheological characteristics by measuring melt viscosity.

The micro-cracks in a solid polymer can be detected by small-angle x-ray scattering. According to Sohma,[34] micro-cracks formed under mechanical action were found to be as high as 10^{12} to 10^{17} cm^{-3}. During fracture, especially at a temperature as low as 21°C, many cleavages occurred and new surfaces were formed. Mechanoradicals are produced simultaneously by fracture.

Enikolopyan and his group observed that when polymers are subjected to a simultaneous action of high pressure and shear, they undergo a drastic change in both physical and chemical properties. The rate of fracture was strongly dependent on a polymer structure, temperature rate of deformation, and other factors.

4.4.1.2 Equipment for EDG

Over the years, ED grinding was also termed elastic-strain powdering (ESP). When modified single- or twin-screw extruders were used in place of a Banbury mixer, ED process was also referred to as extrusion grinding (EG), solid-state shear extrusion (SSSE), and solid-state shear pulverization (SSSP or S^3P). In the more recent past, Russian researchers used the term intensive stress action compression and shear (ISAC&S).

Macroscopic fracture of solids is usually preceded by the accumulation of numerous micro-cracks, in a somewhat explosive manner. Enikolopyan found that, during the fracture, micro-cracks formed at all the sites of the largest stress concentration.[35] [He stated that the elastic energy stored in stretched macromolecules at their fracture transforms into heat, causing the temperature to rise. It is also spent on the fragmentation of free ends of broken chains, with formation of free radicals and breaking of several adjacent polymer chains at once.[35] Pavlovskii and co-workers[36] observed some peculiarities during grinding of used tires by the ED method, where fracture occurred at elevated temperature under complex deformation of compression and shear. They used a modified co-rotating twin-screw extruder ZSK-53 by Werner-Pfleiderer, Germany, consisting of five zones capable of maintaining independent temperature of each zone during processing. In the first zone, tire rubber was subjected to compression, followed by shearing in the zones where shearing (kneading) elements were located. Used tire rubber in this study contained about 15 wt% of cord; the finest powder has been obtained at screw rotation up to 150 revolutions per minute (RPM). Pavlovskii noted that a presence of fine fraction in a resultant rubber powder caused agglomeration, especially at 150 RPM. The authors proposed that at very high shear deformation, rubber is subjected to mechanical destruction, which is typically more pronounced for small particles (less than 0.5 mm in diameter). It was also observed that during EG, two competing processes took place: grinding and agglomeration. Both processes are dependent upon the speed of screw rotation; the higher the speed, the more agglomeration occurred, but it lowered the energy consumption.[36,37]

Chaikun and colleagues[38] emphasized another feature of the ED process as it relates to combining grinding and mixing, which is especially important for the development of new composite materials. They used both single- and twin-screw extruders and obtained powder of various sizes as a function of temperature and clearance between screws and the barrel. Analysis of rubber powder showed that finer fractions were agglomerated irrespective of the type of extruder used. GPC analysis revealed a presence of oligomers with a molecular mass from 300 to 2,000, which were formed as oxidation products near the surface of the powders. These oligomers acted as lubricants and negatively affected the grinding process (in some cases, grinding even stopped) due to the change in a mode of friction (dry vs. lubricated friction).The authors also investigated the influence of sulfur as a vulcanizing agent during ED grinding on properties of recycled vulcanizates. They found that a presence of sulfur-sulfur bonds resulted in a formation of finer rubber powder. By varying the amount and the type of vulcanizing agents during ED, the authors obtained rubber with higher tensile strength than that of the used tire rubber that was ground without the vulcanizing agents.

In another study, Polyakov and co-workers[39] showed properties of recycled vulcanizates obtained by ED technique and by conventional milling. They stated that during ED, rubber is subjected to multiple shear deformations that allow production of much finer powder than that produced by using roll mills. The authors indicated that agglomeration of fine powder during ED was a function of the particle size of resultant tire rubber powder (the smaller the particle size, the more agglomerates were observed). Physical properties of ED-made rubber were higher for a powder containing finer particle size powder.

Pavlovskii and co-workers[40] studied properties of mixtures of tread rubber with up to 60% of used rubber obtained by EDG with a single-screw extruder-grinder designed and built at the Institute of Chemical Physics (ICP), Russian Academy of Sciences (RAS). Ground tread rubber (consisting of three different rubbers) was converted into powder at 140° C with bimodal particle size distribution from 0.315 to 0.8 μm. Intensity of mechanochemical processes during EDG was studied by analysis of sol- and gel-fractions. The authors found that gel-fraction was decreased for finer particle-size powder, while sol-fraction localized on the entire surface of particles; large particles with less dense networks were capable of absorbing more of the sol-fraction and had more of toluene extract.

An increase in mechanical properties of materials based on mixtures of tread rubber with used tire rubber was explained by an interaction of dispersed phase and a matrix.[40] Recycled rubber obtained by EDG had the larger surface area, by a factor of two, as compared with that made by conventional grinding. The authors reported that tear strength and flexural modulus of mixtures containing EDG-made used rubber powder were higher than those of virgin tread rubber compound, while a small decrease in plasticity was observed. When sol-fraction was removed from used rubber prior to mixing with virgin rubber, both flexural modulus and elongation at break were increased.

Their work confirmed results obtained by Pavlovskii's earlier work[36,37] and studies by other researchers that EDG process is not simply mechanical grinding, since several chemical processes such as oxidation take place. It was concluded that the EDG process is a mechanochemical process by which an application of mechanical energy causes several chemical changes to occur.

When EDG-made tire rubber powder was added to a virgin compound, some loss in elasticity was noticed. It did not, however, affect the processability of the compound. Some calculations of energy consumption during the EDG process were published in a study by Enikolopyan and co-workers[41] as 0.8 to 0.9 kwt/ton of product using a laboratory extruder. They claimed that these values were two to three times lower than the energy consumption of conventional grinding (particle size of powder was ranging from 10 to 50 μm, with the mean size of 20 μm). When the Banbury mixer was used for the EDG process, the energy consumption was reported as 40 to 45 kwt/ton for LDPE, with particle size of powder in the range of 150 to 300 μm. Enikolopyan and co-workers[42] also studied surface characteristics of the polymeric powders made by the EDG method. Scanning electron microscopy (SEM) revealed a well-developed surface area and elongated shape of the particles. They described particles having microchannels and a microporous structure, which are advantageous in the development of new composite materials with either virgin rubber or thermoplastics.

Fridman, Petrosyan, and Kazaryan[43] referred to the EDG process as a "rheological explosion" with sudden fracture of solid materials under compression, shear, and torsion. They noted that while some polymers like high-density polyethylene (HDPE) cannot be ground using EDG, in the presence of another polymer such as polypropylene (PP) it was possible (so-called co-grinding). The authors obtained composites containing used tire rubber with wood flour, several thermoplastics (using polyolefins, polystyrene, and polyvinyl chloride), and other ingredients.

Danielyan and co-workers[44] showed that during mixing of tread rubber and LDPE at a 70:30 ratio by EDG process at 165 to 230° C, chemical bonds are formed; the structure of these composites was different than that of analogous composites made by conventional melt-mixing.

Knunyants and co-workers[45] further investigated the EDG process for composites based on tire rubber and polyethylene. They found that processing parameters during mixing and grinding influence toughness of composites. Melt flow rate (MFR) measured at 190° C and 5 kg load was 0.8 g/10 min, which allow for processing of that composite material by injection molding at conditions typical for polyethylene. Analysis of gel- and sol-fractions suggested that significant grafting of rubber on LDPE took place during mixing and grinding. No chemical structural change of LDPE was observed (melting point and glass transition temperatures of LDPE were unchanged after ED processing). At higher temperature during ED processing, the amount of gel-fraction decreased (the amount of sol-fraction increased). Physicomechanical properties of tread rubber with LDPE composites were mostly affected by the amount of thermoplastic component (the higher the LDPE content, the lower the properties).

Pavlovskii and co-workers studied the mechanism of grinding cross-linked elastomers by EDG using extruders.[46] Since during EDG two process take place simultaneously, namely fracture and agglomeration, it was important to determine which of these processes is dominant. Pavlovskii and co-workers established that it could be determined by the particle-size analysis of the resultant powders. The authors also noted that initial particle size of crumb rubber, type of rubber, and the design of equipment influenced the mechanism of grinding. Pavlovskii proposed two mechanisms that govern the EDG process for cross-linked rubber: The first mechanism leads to a formation of small particles and related to viscoelastic properties of rubber and was insensitive to a change in deformation rate. The authors suggested that a large quantity of small particles was caused by friction between rubber particles and the extruder barrel. The second mechanism was sensitive to a deformation rate and followed established physics of fracture by forming micro-cracks at the stressed sites, accumulation of elastic energy, and its release with a creation of new surface (i.e., powder).

Polyakov and co-workers[47] studied the ESP process for tire rubber, using a modified twin-screw co-rotating Berstorff extruder (40 mm in diameter, length to diameter ratio of 28) at 70 to 120° C temperature and 110 to 120 RPM, and using rubber as a feedstock (particle size 1.5 mm in diameter), with and without vulcanizing agents. Particle size of resultant powder corresponded to the clearance between shearing (kneading) blocks of the screws and the barrel. The authors stated that grinding temperature did not affect PSD of the rubber powder, while the presence of vulcanizing agents resulted in an increase in 0.2 to 0.4 mm fraction. Throughput of the extruder was substantially higher when the grinding temperature was increased in the presence of a vulcanizing agent. Physical properties of ESP-made rubber produced with a twin-screw extruder (measured on compression-molded samples) were higher for materials ground with vulcanizing agents. The authors concluded that these properties warranted reuse of ESP-ground tire rubber in various new products.

4.4.2 Solid-State Shear Extrusion (SSSE)

Schöcke, Arastoopour, and Bernstein[48] studied noncryogenic processing of vulcanized natural rubber obtained under high compression and shear using a modified Bridgman anvil apparatus. They determined the effects of temperature, shear forces, and residence time on particle size of the powder with minimum agglomeration. The rubber particles had an irregular shape, and a wide distribution of particle sizes was observed (from 40 to 1500 μm).

Bilgili, Arastoopour, and Bernstein[49,50] investigated solid-state shear extrusion of rubber, using a small-scale modified single-screw Brabender conical extruder.[51–53] They studied particle size and physical properties of natural rubber as a function of extrusion parameters, and the effect of several passes through the extruder. Due to an excessive agglomeration of small particles during SSSE, a high-speed laboratory blender was used to break the agglomerates into discrete particles. The agglomeration increased with the multiple passes through the extruder, possibly due to a surface oxidation of the powder at 135° C. The investigators found that specific surface area of SSSE-made rubber powder was three times higher than that of cryogenically ground rubber. The cross-link density of produced rubber was significantly lower than that of feedstock, which was indicative of the breakage of sulfur-sulfur bonds in vulcanized rubber, due to a partial devulcanization under excessive mechanical stresses. The gel-fractions of SSSE-made rubber powder were also lower than that of the feedstock, suggesting main chains scission due to a degradation of natural rubber. The authors proposed that bonds breakage occurred with sulfur-sulfur bonds rather than carbon-carbon bonds, at cross-links.

In another study of pulverization of natural rubber by the same authors[52] a number of independent variables during the SSSE process were examined. These independent variables included degree of compression, number of extruder passes, and the effects of the processing parameters, including material and screw temperature, RPM, and the feed rate. The effect of these conditions on the dependent variables, such as particle size and PSD, and power consumption, were investigated. The authors concluded that nonlinear stress-strain behavior of rubber under compression and shear, and the storage of strain energy were essential to cause rubber to fracture into powder. Their data suggested a narrowing of PSD with multiple passes through the extruder. As in their earlier work,[49,50] significant, undesirable agglomeration of the small particles less than 250 μm in diameter was observed. The investigators noted that some particles formed clusters, which may have resulted from oxidation and surface tackiness of the softened rubber. The extent of the agglomeration increased with the formation of fine powder (causing more compaction) and at a higher temperature in the pulverization zone of the extruder. The authors concluded that there was a minimum particle size to be obtained during the SSSE process, which was defined by the specific combination of processing parameters and the screw design, especially using the screw with a high compression ratio.

The authors further hypothesized that the degree of compression could be the most important variable of the process, since it affected the storage of high-strain energy and prohibited slip at the rubber-metal interface. These two effects caused shear strain of the rubber granulate, and the combination of high compression and shear resulted in more strain energy accumulated by the rubber. At some critical energy level, cracks were initiated and propagated, with the dissipation of the stored energy.

Bilgili, Arastoopour. and Bernstein[54] reported the characterization of rubber powder made by the SSSE process. The observations by SEM revealed that the particles had irregular shape and rough surface; the larger particles were mostly clusters of small particles. These authors determined that the extend of thermo-oxidation of rubber was a function of the surface area of the powder. Their data suggested that the extent of devulcanization and degradation of the rubber particles ranging from + 40 to – 30 mesh (420 to 590 μm) size increased with the multiple reprocessing of rubber through the agglomeration of particles. The authors found that the agglomerates had lower cross-link density and gel-fraction than those of the single particles of similar size. They postulated that the extent of devulcanization was higher than that of the degradation, since the sulfur-sulfur bonds at the cross-links were weaker than the carbon-carbon bonds.

A different method for pulverization of thermoplastics and thermosets has been proposed by Patel and Shutov,[55] called pressure shear pulverization (PSP). They described PSP as a nonextrusion, low-cost process that is realized with a specially designed pulverization device. The authors claimed that PSP is different from both SSSE and cryogenic grinding and that PSP has high throughput and low energy consumption. They described their device as a pulverization head between very smooth "mirror-like surfaces"; the design is based on a Bridgman anvil and utilized a combination of pressure and shear to achieve pulverization. The authors stated that there was neither change in chemical structure of the polymer chains nor degradation after the PSP process. Microscopy revealed "lettuce leaf–like" shape and smooth surface texture of particles. Particles smaller than 200 μm had elongated shape and aspect ratio of 5:1. The authors concluded that PSP process is strictly a mechanical disintegration process that preserves all physical properties of the feedstock.

4.4.3 Developmental Tire Rubber/Plastic Composites

Wolfson and Nikol'skii[56,57] published review papers summarizing research results of the Russian scientists dedicated to EDG from the early 1980s through 1997. In spite of the authors' considerable time and effort spent at the several leading scientific institutions on the theory of the EDG of polymeric materials, this unique process is not yet fully understood. The authors emphasized advantages of the EDG process that include, but are not limited to, continuous production of polymeric powder of controlled particle size, large surface area of powders, development of new composites based on tire rubber and thermoplastics, recycling of large waste streams including used tires, etc. Since EDG discovery in the early 1980s, numerous devices to practice the process have been developed in Russia, Germany, and the U.S. The modified Banbury mixer, single- and twin-screw extruders, the extruder-grinder, the rotor-dispergator, and other machines were built for the scale-up effort. Wolfson and Nikol'skii indicated that large-size, twin-screw extruders with screw diameters of 90 mm and 120 mm were efficient in producing rubber powder from used tires and other rubber wastes. Both units were capable of producing 500 to 1000 kg/hr of rubber powders, with an average particle size of 300 to 800 μm. The specific energy consumption was 120 to 160 kwt/ton, which is five to six times lower than that of conventional milling. In the 1990s, a few companies in Russia built several versions of so-called rotor-dispergators of different sizes and throughputs.[58] In Germany, modified co-rotating twin extruders were designed and built by Berstorff for powder production on a laboratory scale (25 mm in diameter) and on a production scale (60 mm and 90 mm in diameter).

Arinstein and co-workers[59] investigated physicochemical properties of single polymers and composites subjected to high pressure and shear-induced strain. Their studies have revealed some previously unknown phenomena of theoretical and practical importance. One of the most significant observations was that the multiple cracking of polymers at high temperature under compression and shear resulted in a formation of polymeric powders with a sharp increase of the surface area. This process, according to the authors, intensifies with an increase in temperature of the process. They proposed that each polymer has its own temperature range, where the shear-induced multiple-cracking process is most efficient at the minimal mechanical energy applied. A family of devices called rotor-dispergators were built in Russia for internal use and for export to several countries in Europe, the U.S., and Asia. It has been reported[59] that the size and the shape of powder depends on the type of polymer, type, and amount of additives, cross-link density (in case of thermosets), and processing parameters during pulverization. For many powders with particle size of 10 to 200 μm, the authors reported the specific surface area ranging from 0.5 to 10 m^2/g. They observed that the average particle size of powder decreased with an increase in temperature, although the energy consumption was lower. Based on these observations, they concluded that ISAC&S grinding could not be considered as a consecutive size reduction, but rather, should be viewed as an avalanche-like process. At present, there is no traditional theory explaining the above-mentioned peculiar behavior of polymers' fracture under intense shear and compression, although some models have been proposed. One of them by Arinstein[60] is based on the assumption that a polymer subjected to simultaneous compression and shear at the elevated temperature accumulates micro-defects, and when its concentration reaches a critical level, the new surface (powder) is formed.

During conventional mechanical grinding, the micro-defects appeared at the most stressed areas of the material. In contrast, Arinstein stated that during EDG, the micro-cracks are the main stress concentrators; he assumed that, due to these micro-defects, the local regions of the stressed solid relax toward the unstressed state, dissipating the accumulated energy as heat (i.e., induced warming).

At 80 to 230° C various vulcanized rubbers have been successfully ground; ethylene-propylene rubber, however, could only be converted into a powder at 200 to 230° C. The specific surface area of these powders was high (0.3 to 3 m^2/g). The rubber particles produced by the ISAC&S method had "sponge" morphology, with a high concentration of open and closed pores.[59] Several composites comprising virgin isoprene rubber, butadiene-styrene rubber, and nitrile rubber and 10 to 15 wt% of waste of the respective type of rubber, exhibited physical properties similar to those of virgin rubber. These data suggested an opportunity to recycle and reuse tire rubber with some reduction in cost, as compared to that of using virgin rubber.

Goncharuk and co-workers[61] studied an effect of the surface area and the shape of ground tire rubber on the mechanical properties of rubber-filled thermoplastics. They found that rubber particle morphology (which depends on a grinding method) significantly influenced stress-strain properties of the resultant composites with LDPE. Both the tensile strength and elongation at break were increased for the powders with higher specific surface area (for similar particle shape of the powders). The authors investigated rubber/LDPE composites using tire rubber made by EDG and by cryogenic grinding for comparison. The composites with 60 wt% of rubber were compounded with a single-screw extruder at 120 to 150° C. Physical properties of the rubber/LDPE composites were tested on compression-molded specimens. Goncharuk stated that the stress-strain properties of the composites made with cryogenically ground used tire rubber were not affected by a particle size of rubber powder. When EDG-made rubber powder was used in the composites, the stress-strain properties showed dependence on a particle size (tensile strength and elongation were lower for a larger particle size powder). Overall, composites made with cryogenically ground used tire rubber showed substantially lower tensile strength and elongation than those measured for composites containing EDG-made powder.[61]

SEM of tire rubber powder obtained by different grinding methods revealed that specific surface area of powders increased with a decrease in particle size, as expected. The surface area was measured by gas absorption using the Brumauer-Emmett-Teller method as 0.3 m^2/g for EDG-made rubber. The particle shape varied from smooth for cryogenically ground rubber to "snow-like" shape for EDG-ground powder, and it was concluded that the particle morphology affected the specific surface area of the powder. Goncharuk also stated that the shape of particles was partly responsible for the areas with high stress concentration and reduced interactions between the matrix and the rubber interphase. The authors proposed that these highly stressed sites cause critical defects during tension, facilitating earlier rupture. It was concluded that the method of preparation of rubber powder had a significant effect on the stress-strain properties of rubber/LDPE composites.

Erikolopyan and Fridman[62] indicated that Bridgman's discovery of plastic flow of metals using Bridgman anvils resulted in further advances in the chemistry of solids. They proposed that in solids subjected to high pressure coupled with intense shear, the molecules and atoms become unusually mobile and activated. This mechanism was supported by measuring molecular mass of sieved fractions of powder by GPC. During brittle fracture that takes place in a device like a ball mill, particles are getting smaller over time and molecular mass is decreased. Unlike with a ball mill, during EG, molecular mass of the powder does not depend on the particle size. The authors postulated that all particles are formed simultaneously and not consecutively. Energy consumption and yield were calculated as 0.8 to 1.0 kwt/ton of polyethylene powder of 20 μm size; these values were a basis for designing equipment to be run at increased yield and energy savings, compared with current grinding practices. More recent investigations of EDG-made powders by microscopy showed that EDG of polymer blends is accompanied by homogenization of those mixtures, which could not be achieved during conventional melt-mixing.

Knunyants and co-workers[63] discussed high-temperature EDG under shear deformation of polymers at a temperature near transition (crystallization). They explained high-temperature fracture of polymers based on heterogeneity of materials subjected to grinding (in the case of polymer mixtures, heterogeneity is caused by the presence of multiple phases; for single polymers, it is caused by the presence of crystalline and amorphous regions). The authors believe that high-temperature grinding is a process where micro-cracks are accumulated in the polymer, and at some critical level powder is formed. It is also possible that the coefficient of friction is influencing the deformability of a polymer during grinding.

Pershin and co-workers[64] studied high-temperature, multiple fracture of various plastics and rubbers. They found that dispersity of powder increased with the lower elongation at break value in a given polymer, that is, more "brittle" vulcanizates typically produced finer powder. These authors proposed a model for the EDG process based on the physical aspects of the process; they believe that during grinding a gradual accumulation of defects leads to size reduction of a powder (as opposed to earlier models stating "explosive" character of the EG process).

In the most recent technical literature, researchers continue to develop new models to explain the relationship between the polymers' structure, their mechanical properties, and friction during high-temperature grinding under compression and shear. Pertsov and co-workers[65] proposed new approaches for a better understanding of EDG, following a principle, "strength through fracture," established by the academician Rebinder. In the EDG process, the comminution occurs as a result of shear deformation of polymers in the compressed state. According to previous experimental data and theoretical considerations,[27,28,52] the EDG process was conducted at a temperature close to the melting point of a crystalline polymer, and the specific surface area of the resultant powder was correlated with a coefficient of friction of that polymer.[63]

Pertsov[65] indicated that stored elastic energy may cause fracture only under very limited conditions. The fracture and the formation of new surfaces, leading to an increase of surface energy, must be accompanied by a decrease of stored elastic energy. The latter is easily realized during fracture of brittle materials with micro-cracks under stretching, according to the Griffith theory.[19] Contrary to that, Pertsov stated that compression forces applied to polymer cannot cause its fracture, so a possibility to realize stored elastic energy during EDG required new considerations. Pertsov believed that a structure's nonuniformity in the case of polymer mixture is responsible for conversion of elastic energy into the work of grinding. More likely, under complex stresses, numerous micro-cracks initially occurred, eventually leading to material disintegration, and the dispersity of powder was related somewhat to friction between polymers (in the case of polymer mixtures).

Prut[66] published a review concerning EDG of thermoplastics and elastomers. According to Prut, in elastomers, viscoelastic deformation is prevailing over plastic deformation, and the defects in a shape of spiral cracks appear at some critical value of accumulated viscoelastic deformation. At increased deformation, these spiral cracks are well pronounced, and the fracture proceeds as more small spiral cracks occur. During complex loading at certain processing conditions, the deformation is lower than its critical value. This deformation is reversible and no fracture occurs. As deformation exceeds critical value, micro-defects form and accumulate, and partial release of elastic energy takes place. The remaining energy strains the material, followed by a sharp increase of surface fracture. Multiple fracture continues because there is no other way to release the elastic energy other than to form a new surface (as the deformation increases, the surface area also increases).

Prut stated that formation of defects varies for different types of rubber and it is dependent on temperature during EDG as well as on the degree of vulcanization. During EDG at least two competing processes take place: formation of particles with a size smaller than that of a feedstock, and agglomeration. Which process prevails depends on the size of a feed material and on the design of the grinding zone of the equipment. Prut suggested that for rubber powders with a wide particle size distribution, small powder particles act as antifriction additives, thus decreasing the efficiency of large particles to fracture. It was noted that small particles are formed primarily due to a friction between particles and friction of particles against the surface of the grinding equipment. The amount of fine powder corresponded to the removal of small particles from the surface of ground rubber crumb. Prut stated that the EDG method is not just a conventional size reduction process. Mechanical forces during EDG initiated several chemical processes, including but not limited to oxidation, destruction, secondary structure formation, and others.

Chemical analysis of powders from used tread rubber showed that during EDG, oxydated oligomers with the molecular mass between 300 to 2000 were formed. Their presence influences the mechanism of grinding due to a change in the character of friction. Prut[66] believed that polydispersity of powders during EDG depended on a type of processed rubber waste. EDG-made powder had particles with an asymmetric shape, with high surface area. By adding this rubber powder to a virgin compound for shoe soles, the resistance to multiple bending increased tenfold, while other physico-mechanical properties were unchanged as compared to a compound made with virgin rubber.

During EDG, mechanical destruction takes place near the surface. Since grinding is accompanied by intensive mixing, sol-fraction is distributed over the whole surface of the powders. After grinding, sol-fraction is redistributed in accordance with the adsorption of the particles, so the large particles with looser networks are capable of better absorption of sol-fraction than the smaller particles. Removal of sol-fraction allowed larger amounts of rubber powder to be incorporated into shoe sole compounds.

Goncharuk and co-workers[67] studied mechanical properties of modified thermoplastic elastomers based on used tire rubber powder and low-density polyethylene. When EDG-made rubber powder was made, some surface oxidation of the powder resulted in a poor adhesion with polyolefin matrix, which in turn led to low mechanical properties of the rubber/thermoplastic composites. The investigators used compatibilizing agents such as ethylene vinyl acetate (EVA) and a cross-linking agent such as dicumyl peroxide to improve the interfacial adhesion between EDG-made rubber (both treated rubber and whole tire product) and LDPE. The composites containing 60 wt% rubber and 40 wt% LDPE were made by first melt-mixing rubber with additives followed, by melt-mixing with LDPE. A single-screw extruder was used for melt-mixing. Both melt-mixed materials were then co-ground via EDG, utilizing a Berstorff twin-screw extruder.

Goncharuk and co-workers[67] reported that elastic modulus and elongation at break of 60/40 tire rubber/LDPE composites containing different amounts of EVA were improved due to the reaction of polar vinyl acetate groups and reactive groups on the surface of the rubber. Tire rubber/LDPE composites made with dicumyl peroxide resulted in an increase of both tensile strength and elongation at break, without sacrifice in elastic modulus. The authors believed that the improvement in mechanical properties was due to chemical bonding between polymer matrix and rubber at the interface and because of cross-linking with LDPE. The use of another additive such as maleimide C changed the elasticity of rubber component in the rubber/LDPE composites (elastic modulus increased).

Polyakov and co-workers[68] studied the influence of modification of extrusion-ground rubber on dispersity and physicomechanical properties of recycled vulcanizates. EDG was carried out with a single-screw extruder at different temperatures, ranging from 75 to 150° C. Small amounts of sulfur (2%) were added to used rubber during EDG. The authors found that at 75 to 90° C finer powder was made; chemical analysis showed that vulcanization of rubber had taken place. Tensile strength and elongation of EDG-made material were lower in the presence of vulcanizing agents, which was explained by the increased density of powder during compression molding and reduced contact between particles, due to an increased rigidity of the vulcanized powder.

Pavlovskii and co-workers[69] studied the influence of additives on EDG of used tire rubber with a twin-screw extruder ZSK-53 by Werner-Pfleiderer. Three sieved fractions of tread rubber (1.0, 3.5, and 20 mm) were ground in two passes. After one pass through the extruder, the amount of large particles decreased, as expected. Optical microscopy revealed that the second extrusion pass increased the tendency for agglomeration. Several additives, such as calcium stearate, silica, and others, were used to reduce agglomeration and to chemically modify the surface of EDG-made rubber powder. The authors found that these additives increased the amount of particles below 0.5 mm up to 70%; throughput was higher with an increase of the amount of a small fraction of the powder.

Abalichina and co-workers[70] studied five types of used rubber, including porous and oil- and wear-resistant shoe sole rubber obtained by EDG on a ZE-53 twin-screw extruder by Berstorff. They stated that powders made by extrusion-grinding were finer than those obtained by conventional milling using a Unimax mill (84.8% vs. 8.4% of particles, respectively, were below 500 μm). These investigators concluded that at higher torque and increased shear deformation during extrusion-grinding, the resultant powders had larger surface area, especially for powder with particle size smaller than 250 μm. Best physical properties were achieved when 35% used shoe sole rubber was added to a virgin compound; resistance to multiple flexing increased tenfold as compared to virgin material.

Kaplan, Chekunaev, and Nikol'skii[71] have developed a mechanochemical model of multiple crack rupture of stressed polymers. They addressed the phenomenon of crushing polymers under simultaneous application of shear stress and pressure at elevated temperature, conducted over many years at the ICP in RAS. Kaplan's model described the polymer's rupture as a two-stage process. The first (physical) stage involved the critical size crack propagation and generation of active centers. The second (chemical) stage consisted of the recombination of active centers or an interaction with chemically active molecules in the polymer, leading to the formation of micro-defects that can combine with existing cracks of precritical size to form a critical crack. Their model considered multiple cracking of mechanically stressed polymers as an "avalanche-like" fracture of the initial micro-cracks, in accordance with the Griffith theory.[19]

The Kaplan model emphasizes the role of micro-defects available in solids to be transformed into self-propagating cracks of critical size during fracture. The micro-defects started as a result of interaction of chemically active impurities in polymers with macro-radicals, or as a result of recombination of free radicals formed in the areas of concentration of elastic energy near tips of micro-cracks. Contrary to known theories that it is possible to fracture stressed solid with one critical crack that would propagate and break a polymer into pieces, Kaplan's model proposed the possibility of stopping the critical crack propagation, and therefore eliminating it from the fracture. The model is based on multiple-crack formation leading to fracture. In the EDG where a polymer or a mixture of polymers are subjected to high compression and shear, cracks are formed far from the critical crack, leading to rupture of material. The authors observed this type of rupture of polymers by multiple cracking at elevated temperature of 106 to 126° C. Kaplan's model contributed to the new theory of polymer fracture by extrusion-grinding at elevated temperature, under combined action of compression and shear.

Trofimova and co-workers[72] analyzed used tire rubber crumb ground by two methods: so-called "ozone-knife" and EDG, and investigated the effect of grinding method on the mechanical properties of resultant materials. The relatively new "ozone-knife" method invented by Danschikov and co-workers[73] involves treatment of tire rubber with about 0.5% ozone under pressure of at least 0.5 kg/cm^2 for less than 15 minutes. During this process, both metal and plastic cord are separated from rubber, with minimal energy consumption. Rubber obtained by the "ozone-knife" method was additionally ground by EDG in a rotor-dispergator. The intensity of chemical processes during grinding tested by sol- and gel-fractions analysis was found to be the same for all powders, with the particle size not exceeding 0.315 mm, irrespective of the grinding method. For smaller particles, however, intensity of chemical reactions was higher for the "ozone-knife" method than that for EDG. The authors interpreted these results to be based on difference in particle morphology, specifically, having more dense core and less dense shell layer. Since mechanodestruction takes place near the particle surface, solvents are penetrating the shell more easily than the more dense core. In addition, due to intense mixing during EDG, sol-fraction localized on the particle shell is absorbed by all particles formed during EDG. When the "ozone-knife" method is used, sol-fraction dissolved in acetone was lower than that for EDG-formed particles. At the same time, due to ozone diffusion during the "ozone-knife" method, the extent of rubber destruction was higher than that during EDG; therefore, the amount of fraction dissolved in toluene (fragments of vulcanized network, products of mechanodestruction, and non-bound rubber molecules) from particle core was greater than that for EDG. The authors concluded that sol-fraction localized on the particle shell during EDG and on the particle core during the "ozone-knife" process. The surface area of EDG-made rubber powder was higher than that for "ozone-knife"-made rubber powder.

Materials obtained by compression molding showed that tensile strength, elongation at break, and elastic modulus were higher for finer powder.[72] With the addition of vulcanizing agents, such as sulfur, tensile strength and elongation of rubber increased, but elastic modulus decreased. Mechanical properties of composites consisting of used tire rubber made by EDG and "ozone-knife" methods and LDPE (11 to 23 wt%) showed that elongation at yield for composites based on the EDG method was greater than that for analogous composites based on the "ozone-knife" method; these data were explained by a more active surface for EDG-made tire rubber powder.

4.4.4 Solid-State Shear Pulverization (S^{3}P)

Khait[74] developed the solid-state shear pulverization process (SSSP or S^{3}P) for polymeric powder production, polymer blends, dispersion of additives in the solid state, and plastics recycling. This mechanochemical process is also suitable for recycling of used tire rubber from whole tire product (WTP) or tread rubber (TR) for composites with thermoplastics. A laboratory-size (25 mm in diameter) pulverizer, a pilot-scale pulverizer (40 mm in diameter), and production-size pulverizers (60 mm and 90 mm in diameter) by Berstorff were used. This work was a continuation of a previous joint development effort between the Academy of Sciences, the former Soviet Union, and Berstorff in the 1980s.[75,76]

The Berstorff PT-25 co-rotating, intermeshing twin-screw pulverizer (screw diameter 25 mm, length to diameter ratio of 26.5) utilizes a patented barrel design for cooling, which allows it to maintain a processing temperature below the melting point (for semicrystalline polymers) or below the glass transition temperature (for amorphous polymers). The screws, consisting of a series of kneading (shearing) elements, are designed to compress, shear, and mix the polymer or polymer mixtures as the feedstock moves along the barrel by the use of conveying elements. Due to a modular screw design, the combination of elements is tailored to achieve various degrees of shear and different residence times. In addition, screw speed, feed rate,

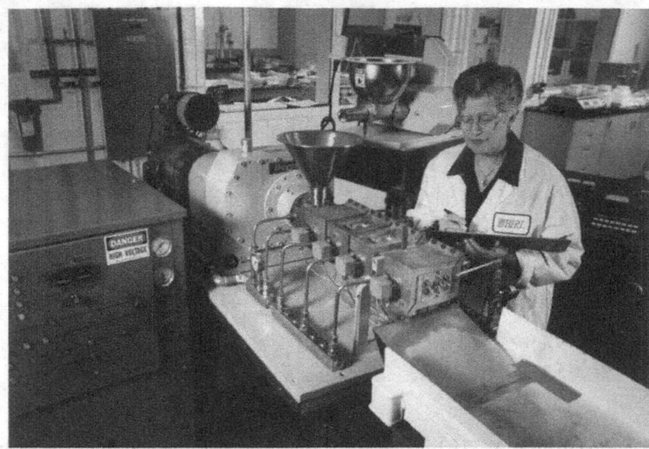

FIGURE 4.1
Dr. Klementina Khait with laboratory-scale PT-25 Berstorff pulverizer, volumetric feeder, chiller (shown on the left), and vibratory cooling conveyor. (From Khait, K. et al., *Solid State Shear Pulverization: A New Polymer Processing and Powder Technology*, Technomic Publishing Co., Lancaster, PA, 2000. With permission.)

and temperature are varied for a specific material(s), to control the degree of shear and the particle size of the resultant powder. It has been shown that, under certain processing conditions, immediately after pulverization the polymeric powders have a significant number of free radicals detected by ESR, which is indicative of chain scission during S^{3}P and typical of mechanochemical processing.[77]

Rubber powders were made from used tires (WTP containing textile cord and TR), using both the laboratory-scale PT-25 pulverizer and production-scale PT-60 pulverizer at The Polymer Technology Center at Northwestern University, Evanston, Illinois, and the commercial-size PT-90 pulverizer with a patented Berstorff cooling system.[78] Powders were made from shredded and coarsely ground 1/4-inch tire rubber by subjecting this feedstock to the simultaneous actions of compression and shear in the co-rotating twin-screw pulverizers. A laboratory-size PT-25 pulverizer and a production-scale Ultra Torque® PT-60 pulverizer are shown in Figure 4.1 and Figure 4.2, respectively. The Ultra Torque® pulverizers have higher torque transferred by spline shafts and increased speed up to 1200 RPM, depending on the process and the machine size.

Particle size of S^{3}P-made tread rubber powder produced with a commercial-size Berstorff pulverizer (90 mm in diameter) is presented in Table 4.1. Particle size was determined with a mechanically operated Ro-Tap sieve shaker and U.S. Standard sieves.

As can be seen from the data in Table 4.1, the majority of powder (38 % by weight) was of 60 mesh (250 μm) size followed by 80 mesh (178 μm) size and small amounts of even finer powder. Some coarser powder of 35 mesh (500 μm) was present as well (7.7% by weight). It is important to note that

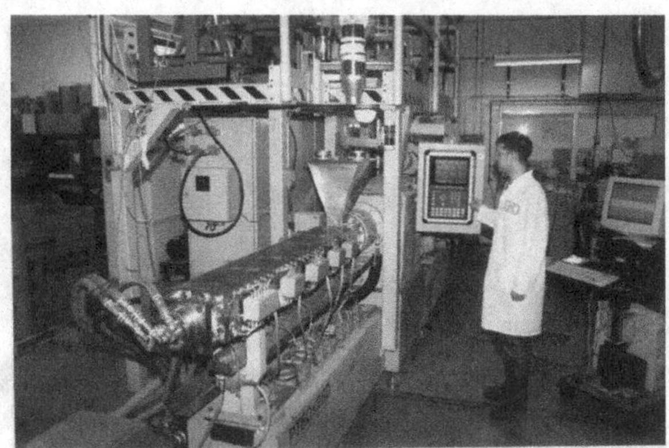

FIGURE 4.2
Research engineer Matthew A. Darling with production-scale PT-60 Berstorff pulverizer.

TABLE 4.1
Sieve Analysis for S³P-made Used Tread Rubber Powder

Mesh Size	Micron (μm)	Weight	Percent Retained
35	500	7.7	7.5
60	250	38.0	36.5
80	178	20.0	19.6
100	150	6.5	6.3
140	104	11.5	11.2
200	74	8.7	8.5
230	63	2.1	2.0
325	45	8.1	7.9

the particle size is controlled by changes in the modular screw design and by adjustments of the processing parameters.

Due to high shearing forces during pulverization, the resultant powder has a well-developed surface, which is advantageous for further use in rubber processing as a partial substitute for virgin rubber in many applications, such as a filler or as an impact modifier for plastics. A degree of shear was controlled by the equipment design. Co-rotating twin-screw pulverizers provide the flexibility to tailor screw design to a polymer or polymer mixture to be pulverized, and to be mixed as well. In the co-rotating, fully intermeshing twin-screw pulverizers, an intense mixing (both distributive and dispersive mixing) occurs depending on mechanical energy input and screw geometry. Of course, feedstock properties, such as interfacial surface tension, elasticity, and coefficient of friction, among others, play an important role during S³P processing.

In addition to having a well-developed surface, particles have a unique shape, described as "cauliflower." In contrast, conventionally ground tire rubber powder has particles with a flat, angular shape and a relatively small surface area.

TABLE 4.2

Key Physical Properties of Used Tire Rubber/Virgin LDPE Composites

Composition/Ratio	Tensile Ultimate Mpa (PSI)	% Elongation	Properties Tear Strength KN/m (lb/in)	Hardness Shore D
PTR/LDPE$_V$ 60/40	3.7 (540)	39	25.5 (146)	33
GTR-30/LDPE$_V$ 60/40	3.0 (440)	27	21.7 (124)	32
PTR/LDPE$_V$ 60/40	5.3 (770)	40	43.2 (247)	38
GTR-30/LDPE$_V$ 40/60	4.9 (720)	33	39.9 (228)	38

Note: GTR-30 = ground tire rubber, LDPE$_V$ = virgin low-density polyethylene, and PTR = pulverized tire rubber.

During S^{3}P, some chemical changes in rubber have occurred. Since these changes are a result of an application of mechanical energy, the S^{3}P is categorized as a mechanochemical process, which causes bond rupture, leading to partial devulcanization of rubber. Khait and co-workers[79] developed several composites based on S^{3}P-made used tire rubber and thermoplastics.

Some of the scrap tire rubber mixes contained virgin plastic, to understand what improvement in properties would be obtained compared to those achieved with the use of recycled resin. Table 4.2 compares properties of compression-molded scrap tire/virgin LDPE composites at two ratios: 60/40 and 40/60. The co-pulverized rubber/plastic blends showed increases in properties over composites made with the conventionally ground tire rubber. The pulverized tire rubber (PTR)/LDPE blends had higher tensile properties and tear strength than those for the ground tire rubber (GTR)/LDPE blends. These increases were due to the larger surface area of the PTR, which resulted in better bonding between the rubber and the plastic.

Khait[79,80] investigated physical properties of developmental tire rubber/postconsumer plastics composites made by co-pulverization of tread rubber and various polyolefins with ethylene-vinyl acetate (EVA UE 645) at 28 wt% as a binder. Postconsumer polyolefins used in those composites were linear low-density polyethylene (LLDPE), HDPE, and LDPE. Key physical properties of these developmental materials are shown in Table 4.3. All pulver-

TABLE 4.3

Physical Properties of Developmental Used Tread Tire Rubber/Postconsumer Polyolefins with EVA as a Binder at 50/40/10 Ratio

Components	Tensile Ultimate Mpa (PSI)	% Elongation	Strength Tear Strength KN/m (lb/in)	Hardness Shore D
TR/LLDPE/EVA	6.1 (890)	34	42.5 (243)	40
TR/HDPE/EVA	9.0 (1,310)	28	67.4 (385)	50
TR/LDPE/EVA	6.7 (980)	168	53.7 (307)	35

Note: TR = tread rubber.

ized materials were melt-mixed at 180° C at 80 RPM for 5 minutes using a small Brabender mixer with cam blades followed by compression molding at 220° C.

As can be seen from Table 4.3, the highest values for elongation were achieved utilizing postconsumer LDPE, while highest tear strength was exhibited by composite containing HDPE. These results suggest improved compatibility between rubber and plastics due, in part, to intimate mixing that accompanied the pulverization process.

Large prototype parts have been made from TR/polyolefin composites using production-scale fabrication equipment. Composites containing 60 wt% of pulverized tire rubber, postconsumer LLDPE, and EVA have been compression molded into heavy-duty mats measuring 18 by 18 inches, which used 8 pounds of the rubber mix. The mats were textured on both sides. One side had large knobs measuring 1/4 inch in length and 3/8 inch in diameter; these knobs were sharp and firm, which was indicative of good flow of the developmental compound. Overall stiffness and surface appearance of the heavy-duty mats made from S^{3}P-made tire powder, LLDPE, and EVA were exceptionally good.

Deitering[81] studied several techniques to characterize rubber powder, and developed a procedure combining SEM and digital-image analysis to physically characterize the size, shape, and surface texture of S^{3}P-made rubber powders and cryogenically ground rubber. A Hitachi S-510 microscope was used to view the rubber particles; the microscope was equipped with a system to capture the SEM images and to acquire digitized gray-level images. Gray-scale analysis was used to detect the size and the number of protuberances on the surface of the particle. Deitering used Fourier analysis to characterize the size, shape, and surface texture of the particles by two-dimensional projection.

The SEM image shown in Figure 4.3 is typical for rubber particles made by S^{3}P; particles are produced by the combination of pressure and high shear by a modified tearing mechanism, generating prominent folds and nobules.The SEM image shown in Figure 4.4 is typical for rubber particles made by cryogenic grinding; particles are produced by brittle fracture, resulting in sharp edges and ridges on the relatively featureless particle surface.

S^{3}P-made rubber powders show a highly developed surface texture. This property is very beneficial in several applications such as impact modification of plastics, use as a filler, and in rubber compounding. Small particle size (less than 500 μm), well-developed surface, and narrow PSD are advantageous for an increased adhesion at the interface during the above-mentioned applications. When S^{3}P-made powder is used as a filler, physical properties of the final product are improved. Due to small particle size and large surface area, faster incorporation of the filler into a compound is achieved. It has been established by the rubber processors that the smaller the particles, the more rubber may be introduced into a formulation with other components while maintaining the key mechanical properties at a required level. In addition, finer rubber particles have resulted in superior

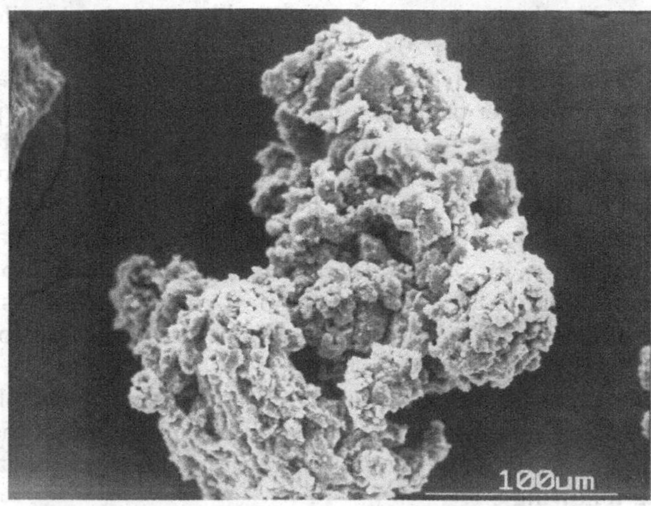

FIGURE 4.3
SEM image of ground tire rubber produced by solid-state shear pulverization (300X magnification, 4.7 cm = 100 μm). (From Khait, K., et al., *Solid State Shear Pulverization: A New Polymer Processing and Powder Technology*, Technomic Publishing Co., Lancaster, PA, 2000. With permission.)

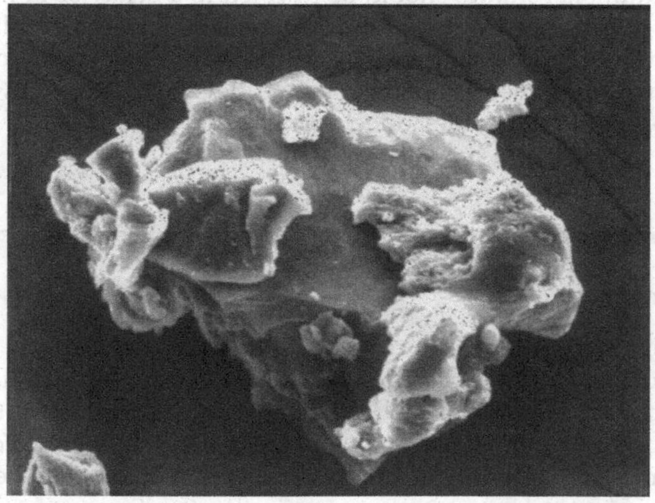

FIGURE 4.4
SEM image of ground tire rubber produced by cryogenic grinding (300X magnification, 4.7 cm = 100 μm). (From Khait, K., et al., *Solid State Shear Pulverization: A New Polymer Processing and Powder Technology*, Technomic Publishing Co., Lancaster, PA, 2000. With permission.)

surface appearance of the rubber products, as compared with those containing coarser rubber powder.

Khait reported[79] that S^3P-made tire rubber powders made from both WTP and TR were used at 5 wt% in soft-tread rubber formulations. The physical properties, such as modulus, hardness, and tear strength of the compounds were unaffected by the presence of used tire rubber powder; tensile strength, however, decreased by about 20%.

Lipp,[82] at Berstorff, reported the specific energy consumption data on S³P-made tire rubber powder using Berstorff PT-90 commercial-size pulverizer in Germany. The specific energy consumption ranged from 0.3 to 0.9 kw/kg on the product, depending on the particle size of the resultant powder. Two particle sizes were made, ranging between 0.05 and 0.20 mm or 0.4 and 0.6 mm, respectively. According to Lipp, the best results were obtained with 57% of the powder with particle size from 0.05 to 0.20 mm. To eliminate agglomeration of fine powder, some chemical additives were used during pulverization. The throughput of the PT-90 pulverizer was up to 1600 kg/hr, depending on the feedstock.

Khait and co-workers demonstrated that properties of rubber/plastic composites could be tailored by varying the rubber/plastic ratio of the composites and changing processing parameters during both co-pulverization and melt-mixing following S³P.[83]

In spite of many years of research geared toward better understanding of physics and chemistry of mechanochemical processing, the S³P process is not yet fully understood. Future research should include more work on the theory of fracture of various polymeric materials exposed to complex stresses. Additional studies should be devoted to a behavior of mixtures of various polymers, filled systems, and polymers with reactive additives during S³P. These studies would enable scientists and engineers to create new materials with designed properties and to develop better knowledge for tailoring properties for value-added applications.

4.5 Summary

Tire rubber recycling is one of the most important directions in utilization of waste rubber, which is inevitably accumulated during the initial production, fabrication, and use of tires. Ironically, many advantages of tire rubber composition such as toughness and wear resistance, which are accomplished by using many types of rubber (up to seven types) in a tire, make recycling very challenging and not always economically attractive.

Further research and development efforts to increase recycling rates for used tire rubber are needed. In spite of continuing increase in utilization of used tires, there are huge quantities that are still being landfilled. Environmental concerns affect new routes to waste management in the twenty-first century. Stockpiling of used tires causes a threat to the environment that requires new "cradle-to-grave" approaches. These tires are a valuable material that could be recycled and reused in a variety of applications as a partial replacement of virgin rubber as well as a filler, impact modifier for plastics, and as TDF. It is apparent that powdered rubber and composites based on used tire rubber products have a significant role to play in future recycling efforts.

Fine grinding of tires consumes a lot of energy and causes high abrasive wear to the grinder. The conversion of electrical energy through mechanical action to surface energy of fracture is very inefficient. Development of more energy-efficient processes represents a challenge for scientists and engineers to be addressed in a new millennium.

Investigations that started in the early 1920s of various substances under high pressure and shear need to be continued. Controlled mechanochemistry is a useful tool for tailoring properties of polymers for specific applications. The pioneering research by P. Bridgman, Nobel laureate, stimulated systematic investigations on the effects of high pressure on properties of solids. The discovery of new mechanochemical processing technology such as EDG greatly contributed to higher utilization of used tires.

Although fracture of polymers has been studied for a long time, chemical phenomena during fracture were largely ignored and a purely physical approach was adopted. There is, however, a direct relationship between application of mechanical forces and chemical reactions in polymers during fracture that occur during mechanochemical processing.

Saving raw materials through the use of secondary feedstocks gains more acceptance due to a concern for a "green" environment. The continuous S^3P process for obtaining fine tire rubber powder offers many possibilities for an increased reuse of that powder in many applications, especially in composites with recycled thermoplastics, thus utilizing both waste streams. The S^3P technology will make a valuable contribution to eco-efficiency of post-consumer polymeric waste recovery worldwide. The recycling of used tires and plastic wastes will continue due to the increased cost of fossil fuels and because of public and government concerns for protecting our environment. In the future, disposal of waste tires in landfills will decrease due to government regulations. In some European countries, such as Germany, regulations already require tire manufacturers to take used tires back and to convert them into new products.

Further advances in emerging processing technologies will increase reuse of tires in a wide variety of existing and new applications and markets.

References

1. Morin, J.E. and Farris, R.J., Recycling of 100% cross-linked rubber powder by high-temperature high-pressure sintering, in *Proc. Annual Techn. Conf. Soc. Plast. Eng, ANTEC 2001,* Dallas, May 6-10, 2001, 2758.
2. Blumenthal, M., What's new with ground rubber, *Biocycle,* 39, March 1998.
3. Manuel, H.J. and Dierkes, W., Recycling of rubber, *Report 99, RAPRA Review Reports,* 9, 3, 1, Shawbury, Shrewsbury, Shropshire RAPRA Technology, Ltd., Great Britain, 1997.

4. Myhre, M. and MacKillop, D.A., Rubber recycling, *Rubber Chemistry and Technology,* 75, 3, 424, 2002.
5. Stark, F.J., Introducing new rubber and plastic compounding techniques with a new raw material utilizing surface treated vulcanized particulate, *Rubber India,* 43, 17, 1991.
6. Kowalska, E. and Wielgosz, Z., Scrap rubber re-used – a new process produces porous pipes from worn-out tires, *Polymer Recycling,* 2, 3, 213, 1996.
7. Reschner, K., The economics of ground tire rubber, *Scrap Tire News,* 10, 12, 1996.
8. Bauman, B.D., Surface-modified rubber particles in polyurethane, *Rubber Plastic News,* 23, 35, July 4, 1994.
9. Dierkes, W., Solutions to the rubber waste problem incorporating the use of recycled rubber, *Rubber World,* 25, May 1996.
10. Isayev, A.J., Yushanov, S.P., and Chen, J., Ultrasonic devulcanization of rubber vulcanizates: process model, *J. of Appl. Pol. Sci.,* 59, 809, 1996.
11. Watson, W.F., Novel process for recycling used tyre rubber, *Rubber Asia,* 12, 4, 81, July-Aug. 1998.
12. Kohler, R., Cryogenic processing of scrap tires, *Polymer Recycling,* 2, 2, 83, 1996.
13. NRC Committee on Comminution and Energy Consumption, *Comminution and Energy Consumption,* Report NMAB-364, National Academy Press, Washington, D.C., 1981.
14. Austin, L.G., Size reduction of solids: Crushing and grinding equipment, in *Handbook of Powder Science and Technology,* Fayed, M.E. and Olten L., eds., Van Nostrand Reinhold, New York, 1990, 562.
15. Falkner, B.P. and Rimmer H.W., Size reduction in *Encyclopedia of Chemical Technology,* Vol 21, 3rd edition, John Wiley & Sons, New York, 1983, 132.
16. Birley, A.W., Haworth, B., and Batchelor, J., *Physics of Plastics,* Hanser Publishers, New York, 1991, chap. 6.
17. Prasher, C.L., *Crushing and Grinding Process Handbook,* John Wiley & Sons, Ltd., Chichester, Great Britain, 1987, chap. 2.
18. Kausch, H.M. and Williams, J.G., Fracture and fatigue in *Encyclopedia of Polymer Science and Engineering,* 2nd edition, John Wiley & Sons, New York, 1987, 328.
19. Griffith, A.A., The phenomenon of rupture and flow in solids, *Phil. Trans. Royal Soc. A,* 221, 163, 1920.
20. Williams, J.G., *Fracture Mechanics of Polymers,* John Wiley & Sons, New York, 1994, chap. 6.
21. Kinloch, A.J. and Young, R.J., *Fracture Behavior of Polymers,* Elsevier Science Publishing, New York, 1983, chap. 3 and chap. 10.
22. Enikolopyan, N.S., Akopyan, E.L., and Nikol'skii, V.G., Some problems of strength and fracture of polymer materials, *Macromolecular Chemie Supplement,* 6, 316, 1984.
23. Enikolopyan, N.S., Physico-chemical aspects of plastic flow, *Macromol. Chem.,* Suppl. 8, 109, 1984.
24. Bridgman, P.W., *The Physics of High Pressure,* McMillan Co., New York, 1931, chap. 1.
25. Tsirule, K.I. and Tyunina, E.L., *Compressibility of Polymers in High-Pressure Chemistry and Physics of Polymers,* Kovarkii, A.L., ed. CRC Press, Boca Raton, 1993, chap. 1.
26. Akopyan, E.L. et al., Elastic deformation grinding of thermoplastics, *Dokl. Akoal. Nauk SSSR,* 291, 1, 133, 1986 (in Russian).

27. Enikolopyan, N.S., Some aspects of chemistry and physics of plastic flow, *Pure and Applied Chemistry,* 57, 11, 1707, 1985.
28. Baramboim, N.K., *Mechanochemistry of Polymers, Rubber and Plastic Research Association of Great Britain,* McLaren & Sons Ltd. 1964, chap. 3.
29. Heegn, H., Stressing of solids by fine grinding and mechanical activation, in *Proc. First Intl. Confer. on Mechanochemistry,* InCome '93, Kovsice, Slovak Republic, March 23-26, 1993, 127.
30. Sakaguchi, M.H. et al., Electron spin resonance study on molecular motion of chain end radicals of polyethylene molecules anchored on fresh surfaces of polyethylene and polytetrafluoroethylene, *Macromolecules,* 26, 2612, 1993.
31. Guerrero, R., Beinert, G., and Herz, J.E., Synthesis of block copolymers through free radicals macroinitiators, *J. Appl. Polym. Sci.,* Applied Polymer Symposium, 49, 43, 1991.
32. Simionescu, C. and Oprea, C.V., *Mechanochemistry of Polymers,* Mir Publishers, Moscow, 1970 (in Russian).
33. Porter, R.S. and Casale, A., Mechanochemical reactions, in *Encyclopedia of Polymer Science and Engineering,* 2nd edition, John Wiley & Sons, New York, 1987, 467.
34. Sohma, J., Mechanochemistry of polymers, *Progress in Polymer Sci.,* 14, 451, 1989.
35. Enikolopyan, N.S., Zarkhin, I. S., and Prut, E.V., Primary molecular products of mechanical fracture of polymers, *J. of Appl. Polym. Sci.,* 30, 2291, 1985.
36. Pavlovskii, L.L. et al., Some trends during grinding of used tires by elastic-deformation grinding, *Production and Use of Elastomers,* 4, 23, 1990 (in Russian).
37. Pavlovskii, L.L. et al., About mechanism of grinding of vulcanized elastomers, *Vysokomol. Soed.,* 32, 10, 784, 1991 (in Russian).
38. Chaikun, A.M. et al., Extrusion-grinding of crumb rubber from tires, *Scientific Research Institute of Tire Industry,* 3, 681, 1994 (in Russian).
39. Polyakov, O.G. et al., Strength of used vulcanizates made by extrusion-grinding of tread tire rubber, *Production and Use of Elastomers,* 12, 17, 1992 (in Russian).
40. Pavlovskii, L.L. et al., Investigation of properties of tread rubber with used vulcanizates, *Production and Use of Elastomers,* 8, 18, 1992 (in Russian).
41. Enikolopyan, N.S. and Fridman, M.L., About a mechanism of elastic-deformation grinding of polymers, *Dok. Akad. Nauk SSSR,* 290, 2, 379, 1986.
42. Enikolopyan, N.S. et al., Structure and morphology of powder polymer materials obtained by elastic-deformational grinding method, *Vysokomol. Soed.,* Series A, 30, 11, 2397, 1988 (in Russian).
43. Fridman, M.L., Petrosyan, A.Z., and Kazaryan, T.A., Rheological properties and new processes to recycle polymeric waste, *Plasticheskie Massy,* 6, 16, 1986 (in Russian).
44. Danielyan, V.G. et al., Mechanochemical transformation in combined extrusion grinding of low density polyethylene with butadiene-styrene rubber, translated from *Dok. Akad. Nauk SSSR,* 293, 2, 386, 1987.
45. Knunyants, M.I. et al., An influence of conditions on properties of composites based on polyethylene and vulcanized elastomers, *Mekhanika Kompozicionych Materialov,* 5, 927, 1988 (in Russian).
46. Pavlovskii, L.L. et al., About grinding mechanism of vulcanized elastomers, *Vysokomol. Soed.,* Series B, 33, 10, 784, 1991 (in Russian).
47. Polyakov, O.B. et al., Grinding of crumb rubber in twin-screw extruder and properties of recycled vulcanizates, *Production and Use of Elastomers,* 6, 28, 1993 (in Russian).

48. Schocke, D., Arastoopour, H., and Bernstein, B., Pulverization of rubber under high compression and shear, *Powder Technology,* 102, 207, 1999.
49. Bilgili, E., Arastoopour, H., and Bernstein, B., Analysis of rubber particles produced by the solid-state shear extrusion pulverization process, *Rubber Chemistry and Technology,* 73, 2, 340, mid-June 2000.
50. Bilgili, E., Arastoopour, H., and Bernstein, B., Solid-state shear extrusion pulverization of rubber granules, in *Proc. Symp. American Institute Chemical Engineers,* 95, 321, 83, 1999.
51. Shutov, F., Ivanov, G., and Arastoopour, H., U.S. Patent 5,415,354, 1995.
52. Arastoopour, H., U.S. Patent 5,704,555, 1998.
53. Bilgili, E., Arastoopour, H. and Bernstein, B., Pulverization of rubber granulates using the solid-state shear extrusion (SSSE) process: Part I, Process concepts and characteristics, *Powder Technology,* 115, 265, 2001.
54. Bilgili, E., Arastoopour, H., and Bernstein, B., Pulverization of rubber granulates using the solid-state shear extrusion process: Part II, Powder characterization, *Powder Technology,* 115, 277, 2001.
55. Patel, T. and Shutov, F., Pressure shear pulverization (PSP) process for thermoplastic and thermoset waste, in *Proc. Annual Techn. Conf. Soc. Plast. Eng.,* ANTEC '99, New York, May 2-6, 1999, 1351.
56. Wolfson, S.A. and Nikol'skii, Strain-assisted fracture and grinding of solid polymeric materials: Powder technologies, *Polym. Sci.,* Series B, 36, 6, 1294, 1997.
57. Wolfson, S.A. and Nikol'skii, V.G., Powder extrusion: Fundamentals and different applications, *Polym. Eng. & Sci.,* 37, 8, 1294, 1997.
58. Kelly, K.F. et al., Improved method for re-utilizing rubber materials from factory scrap and their consequent remolding characteristics, in *Proc. 153rd Meeting of the Rubber Division, Am. Chem. Soc.,* Cleveland, OH, Oct. 21-24, 1997, paper No. 98.
59. Arinstein, A.E. et al., High-temperature shear-induced multiple cracking and grinding of polymeric materials, in *Proc. 154th meeting the Rubber Division, Am. Chem. Soc.,* Nashville, TN, Sept. 29-Oct. 2, 1998, paper No. 41.
60. Arinstein, A.E., Mechanism of anomalously fast diffusion in a solid-phase polymer matrix under strong shearing-pressure-type action, *Physical Chemistry,* 355, 4-6, 237, 1997.
61. Goncharuk, G.P. et al., Effect of the specific surface area and the shape of rubber crumb on the mechanical properties of rubber-filled plastics, *Polym. Sci.,* Series B, 40, 5-6, 166, 1998.
62. Enikopolyan, N.S. and Fridman, M.L., Chemistry of solids — new science and technology, *Nauka I Chelovechestvo,* 261, 1987 (in Russian).
63. Knunyants, M.I. et al., An influence of low molecular fractions on high-temperature grinding of high-density polyethylene, *Polym. Sci.,* Series B, 39, 5-6, 197, 1997.
64. Pershin, S.A. et al., Simulation of high-temperature multiple destruction of polymers, *Vysokomol. Soed.,* Series B, 39, 3. 533, 1997 (in Russian).
65. Pertsov, A.V. et al., Physico-chemical mechanics of high-temperature fracture of thermoplastics, *Colloid Journal,* 60, 5, 573, 1998.
66. Prut, E.V., Plastic flow instability and multiple fracture (grinding) of polymers: A review, *Polym. Sci.,* 36, 4, 493, 1994.
67. Goncharuk, G.P. et al., Mechanical properties of modified thermoplastic elastomers based on low-density polyethylene and rubber powder, *Polymer Recycling,* 5, 5, 161, 2000.

68. Polyakov, O.G. et al., Modification of ground rubber during extrusion-grinding and properties of recycled vulcanizates, *Production and Use of Elastomers,* 5, 22, 1993 (in Russian).
69. Pavlovskii, L.L. et al., Investigation of modifying additives on elastic-deformation grinding of used tire rubber, *Production and Use of Elastomers,* 3, 20, 1990 (in Russian).
70. Abalichina, T.M. et al., Properties of finely ground rubber vulcanizates, *Leather and Shoe Industry,* 6, 40, 1989 (in Russian).
71. Kaplan, A.M., Chekunaev, N.I., and Nikol'skii, V.G., Mechano-chemical model of multiple crack destruction of polymers under stress, *Russian Polymer News,* 4, 3, 16, 1999.
72. Trofimova, G.M. et al., Effect of the method of grinding on the rubber crumb structure, *Vysokomol. Soed.,* Series A, 42, 7, 1238, 2000 (in Russian).
73. Danshikov, E.V. et al., U.S. Patent 5,492,657, 1996.
74. Khait, K., New used-tire rubber recovery process for value-added products, *Proc. Rubber Div. Meeting of ACS,* Chicago, April 1994, Paper No. 24.
75. Enikolopov, N.S. et al., U.S. Patent 4,607,796, 1986.
76. Enikolopov, N.S. et al., U.S. Patent 4,607,797, 1986.
77. Ahn, D., Khait, K., and Petrich, M.A., Microstructural changes in homopolymers and polymer blends induced by elastic strain pulverization, *J. Appl. Polym. Sci.,* 55.1431. 1995.
78. Mayer, D. and Freist, B., U.S. Patent 5,273,419, 1993.
79. Khait, K., S. H. Carr and M. Mack, *Solid-State Shear Pulverization: A New Polymer Processing and Powder Technology,* Technomic Publishing, Lancaster, 2001, chap. 8.
80. Khait, K., New solid-state shear extrusion process for used-tire rubber recovery, in *Proc. Rubber Division Meeting ACS,* Louisville, KY, Oct. 8-11, 1996, Paper No. 49.
81. Deitering, M.L., Particle characterization of recycled polymeric powders, *M.S. thesis,* Northwestern University, Evanston, 1994.
82. Lipp, R., Pulverization technology for recycling rubber and plastic scrap on extrusion lines, in *Proc. Fourth Annual Intern. Forum and Exposition, Recycle '91,* Davos, April 3-5, 1991.
83. Khait, K., Application development for used-tire rubber recovered by a novel solid-state shear extrusion process, in *Proc. Rubber Division Meeting ACS,* Cleveland, October 1995.

5

Recycling Cross-Linked Networks via High-Pressure, High-Temperature Sintering

Richard J. Farris, Drew E. Williams, and Amiya R. Tripathy

CONTENTS

5.1 Introduction

The first mention of thermosetting materials is found in the scientific literature between 1837 and 1844 in patents by Charles Goodyear, Nathaniel Hayward, and Thomas Hancock wherein they describe the vulcanization of rubber.[1-5] In the 1844 patent detailing his gum elastic composition, Goodyear stated the following:

- No degree of heat, without blaze, can melt it.
- It resists the most powerful chemical reagents: Aquafortis [nitric acid], sulfuric acid, essential and common oils, turpentine, and other solvents....

With these patents and much of the work that followed, Goodyear and others created one of the most useful materials of the modern era, and helped spawn the industrial revolution. Unfortunately, they also created one of the most difficult recycling problems ever encountered.

Due to the enormous demand for it, and its inability to be recycled in an economic manner, rubber (especially rubber tires) has become a significant waste problem over the last century. Studies estimate that there are approximately 300 million scrap tires currently in U.S. landfills, with over 280 million additional tires reaching the waste stream each year.[6] This equates to approximately 5.7 million tons of waste per year, or an astonishing 360 pounds of rubber reaching the waste stream per second! Of these 280 million tires, 115 million are burned for fuel, 95 million are incorporated into a variety of uses (civil engineering ~ 40 million, ground rubber ~ 35 million, exported ~ 15 million, etc.), while the remaining ones end up in landfills.[6] Once in landfills, tires are enormous fire hazards and, due to the stagnant water that collects inside of them, they are also breeding grounds for disease-carrying pests such as mosquitoes and rodents.

Currently, waste rubber materials are recycled/reused in only a few low-technology applications. Some specific applications of waste rubbers include the following: [7,8]

- Oil spill–absorbing media (rubber powder)
- Synthetic turf for athletic uses (rubber powder with a binder)
- Asphalt/concrete and virgin rubber goods additive (rubber powder as a filler)
- Retread industry (tire carcass reused)
- Artificial ocean reefs/highway crash barriers/erosion control (whole tires bundled together)

In the 150 years since the first thermosetting patents, numerous ideas have been formulated to combat the problem of recycling rubber, all with varying degrees of success. Interestingly, Goodyear was one of the first to recognize the need for methods of reuse of waste rubber articles, and in 1853 he patented the process of adding ground waste rubber to virgin rubber.[9] A few years later in 1855, Charles Morey patented a process he described as, "Forming or molding scrapings, filings, dust, powder, or sheets of hard vulcanized India-rubber into a compact solid mass by means of a high degree of heat and pressure…."[10] This concept was revisited in 1981, when Accetta and coworkers described molding ground rubber tires at high temperatures and pressures while incorporating additional vulcanizing agents, such as sulfur.[11–13] Phadke and De discuss vulcanization of cryoground (CGR) reclaimed rubber, and blends of CGR with virgin rubber.[14,15] Law talks about a proprietary method of molding polyurethane powders into films.[16] James also mentions employing compression molding at high temperatures to recycle polyurethanes.[17] Corbett and Wadie discuss a similar technique to make solid sheets from polyurethane foam as a mechanical testing technique, but fail to realize its potential application for recycling.[18] Arastoopour and co-workers received a patent for "processing recycled rubber," and mention a sintering technique, although the majority of the patent focuses on rubber grinding.[19] Adhikari and fellow investigators offer an excellent review of the past and present techniques for rubber recycling, and list numerous references for further study.[20] The previous references highlight work that complements and yields insight into the development of high-pressure, high-temperature sintering (HPHTS).

5.2 High-Pressure, High-Temperature Sintering

5.2.1 Brief History of Polymer Sintering

The technology of sintering metals and ceramics dates back thousands of years to the first formations of bricks and pottery. Sintering of polymer powders is a limited science, with some specific uses such as powder-coated metals and porous filters for fluids. Poly(tetrafluoroethlyene) (PTFE) and ultra high molecular weight poly(ethylene) (UHMWPE) are two common polymers for which powder processing is preferred over traditional processing methods.

PTFE and UHMWPE, like natural rubber, have extremely high molecular weights and do not readily flow in their melt state. The high melt viscosities of these polymers make them very difficult to process by conventional techniques such as extrusion and injection molding.[21] Instead, PTFE and UHMWPE are processed through various sintering techniques such as compaction molding, rotational molding, powder coating, and ram extrusion. The pro-

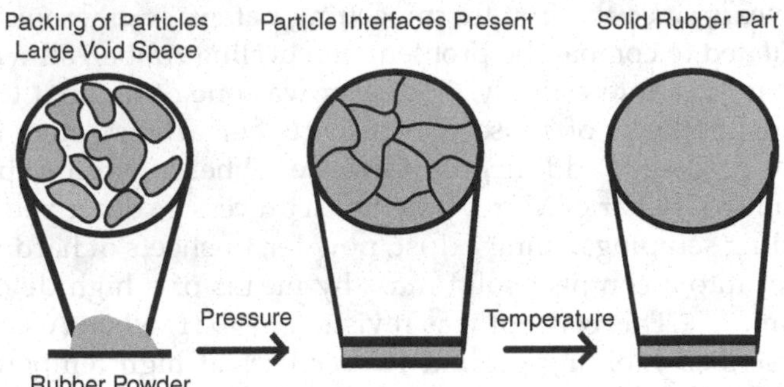

FIGURE 5.1
Schematic of the high-pressure, high-temperature sintering (HPHTS) process.

cess of sintering PTFE and UHMWPE is similar in concept to the high-pressure, high-temperature sintering (HPHTS) of rubber powders. Sintering of PTFE and UHMWPE results from surface crystal melting, diffusion, and recrystallization. The rubber powders in HPHTS are believed to "sinter" due to rearrangement of chemical bonds across the rubber particle interfaces at elevated temperatures.

5.2.2 Schematic of Sintering Process

Until recently, there have been few techniques mentioned in the literature that produce new products from 100% waste rubber. This is the main advantage of high-pressure, high-temperature sintering. The technique works for most rubbers and even blends of different rubber powders have been produced.[22] Figure 5.1 depicts the "reknitting" of the particles by HPHTS. Bond breakage and reformation within and between individual particles fuses the interfaces between the particles to form a new network. Afterward, the particle interfaces are indistinguishable from the bulk material. Pressure is needed for intimate contact of the particles to occur. Temperature adds the necessary energy to break the existing cross-links and backbone bonds of the rubber, allowing them to rearrange and reform throughout the material. It is believed that the mechanism of particle adhesion is these chemical exchange reactions that occur in most rubbers at elevated temperatures. Chemical interchange reactions, especially in rubbers, have been known for many years and date back to work done by A.V. Tobolsky in the 1940s and 1950s.[23–29]

5.2.3 Influence of Molding Parameters on Mechanical Properties

The following work was undertaken in an effort to understand how different molding variables influence the mechanical properties of sintered rubber parts. A sulfur-cured, carbon black-filled natural rubber was vulcanized to its optimum t_{90} rheometer value. Vulcanized sheets were then cryoground

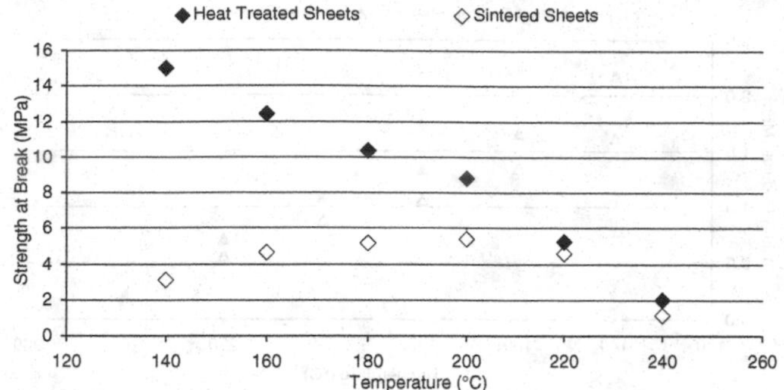

FIGURE 5.2
Strength at break vs. temperature for heat-treated and sintered sheets of natural rubber (NR).

into rubber powder, and the rubber powder was used to produce the sintered sheets. Heat-treated sheets (HTS) were produced by taking the vulcanized unground rubber sheets and exposing them to the sintering conditions.

5.2.3.1 Effect of Molding Temperature on Mechanical Properties

Figure 5.2, Figure 5.3, and Figure 5.4 illustrate the influence of molding temperature on the mechanical properties of tensile strength, elongation at break, and modulus, respectively, for natural rubber (NR) (see Table 5.1 for formulation). The pressure for all moldings was kept constant at 8.6 MPa, as was the molding time of 1 hour. As is evident in Figure 5.2, the strength at break of NR sintered by HPHTS increases up to a temperature of approximately 210°C. On the contrary, HTS of NR decrease in strength over all values of temperature. The most interesting point to note is that above 210°C, the HTS and sintered sheets of NR have identical strength values. An analogous conclusion is evident from the elongation at break results in Figure 5.3. Again, the sintered sheets show an increase in elongation; however, a slight difference is noted as this occurs over all temperatures. At 210°C, the

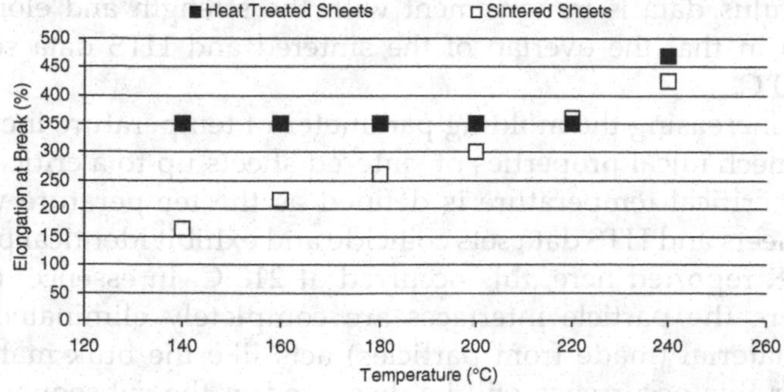

FIGURE 5.3
Elongation at break vs. temperature for heat-treated and sintered sheets of natural rubber (NR).

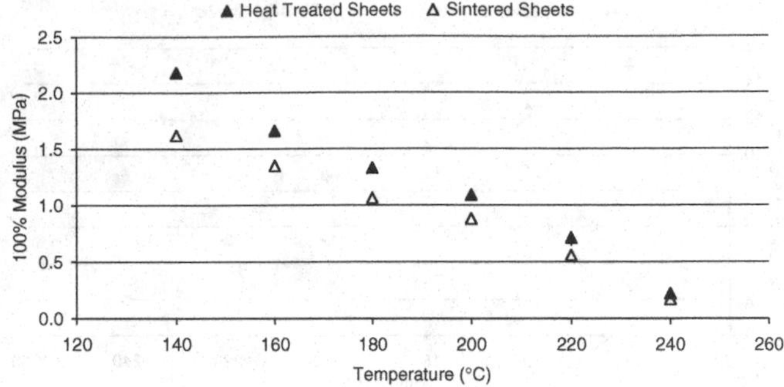

FIGURE 5.4
100% Modulus vs. temperature for heat-treated and sintered sheets of natural rubber (NR).

TABLE 5.1
Formulations for Natural (NR) and Styrene-Butadiene (SBR) Rubbers

Ingredient	NR	SBR
NR	100	
SBR		100.0
Carbon black	40	40
Paraffinic oil	10	
Naphthenic oil		8
Zinc oxide	5	5
Stearic acid	1.0	1.5
Sulfur	1.5	1.5
CBS[a]	1.5	1.5
TMTD[b]	0.2	

[a] N-cyclohexylbenzothiazole-2-sulfenamide.
[b] Tetramethylthiuram disulfide.

elongation at break falls into line with the HTS results, as the two sets of data overlap, similar to the strength at break data. Finally, in Figure 5.4 the 100% modulus data is in agreement with the strength and elongation at break data in that the overlap of the sintered and HTS data sets occurs around 210°C.

Overall, increasing the molding parameter of temperature increases the resulting mechanical properties of sintered sheets up to a critical temperature. The critical temperature is defined as the temperature where the sintered sheets and HTS data sets coincide and exhibit identical properties. For the NR reported here, this occurred at 210°C. In essence, this is the point where the particle interfaces are completely eliminated and the sintered material (made from particles) acts like the bulk material. The reasons for this occurrence will be discussed in the subsequent sections regarding backbone influence on mechanical properties and the mechanism of sintering.

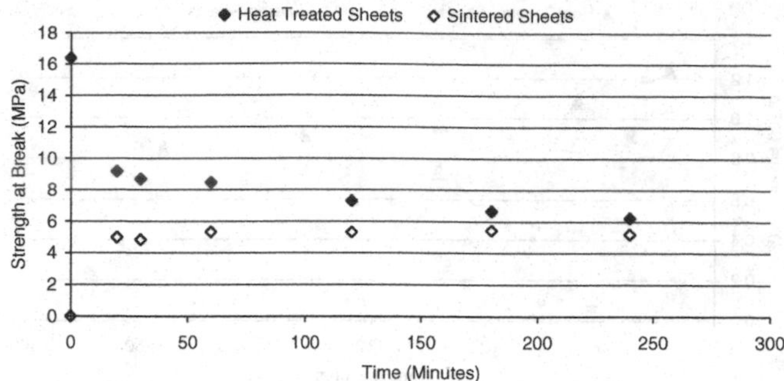

FIGURE 5.5
Strength at break vs. time for heat-treated and sintered sheets of natural rubber (NR).

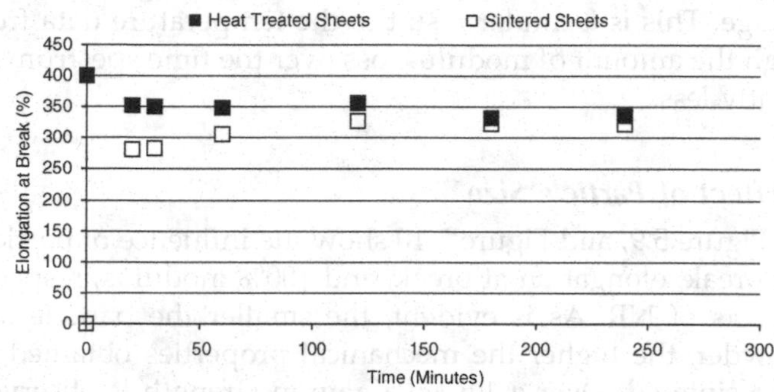

FIGURE 5.6
Elongation at break vs. time for heat-treated and sintered sheets of natural rubber (NR).

5.2.3.2 Effect of Molding Time on Mechanical Properties

Figure 5.5 and Figure 5.6 illustrate the influence of molding time on the mechanical properties of tensile strength, elongation at break, and modulus, respectively, for natural rubber. The pressure for all moldings was kept constant at 8.6 MPa, and the temperature employed was 200°C. It is evident in Figure 5.5 that time follows a nonlinear relationship with strength, unlike the temperature parameter, which was linear. The same holds true for the elongation at break in Figure 5.6. Knowing that a part sintered for zero time and will have zero strength and elongation, it appears that the parameter of time follows a logarithmic-like relationship. A similar conclusion holds true for the HTS as they decay in a logarithmic fashion. It is evident that the two sets of data are converging, and it appears that there will eventually be a "critical time," whereby the sintered sheets and HTS behave identically. It is believed that this critical time will vary as a function of temperature. Further discussion of this topic will be covered in the section on time-temperature superposition.

Figure 5.7 shows the 100% modulus as a function of time. A slight decay is evident, suggesting that the material is undergoing cross-link and back-

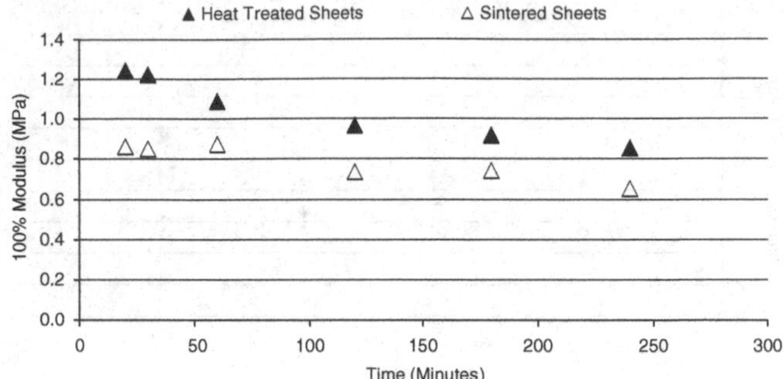

FIGURE 5.7
100% Modulus vs. time for heat-treated and sintered sheets of natural rubber (NR).

bone breakage. This is a similar result to the temperature data from Figure 5.4, although the amount of modulus loss over the time spectrum displayed is significantly less.

5.2.3.3 Effect of Particle Size

Figure 5.8, Figure 5.9, and Figure 5.10 show the influence of particle size on strength at break, elongation at break, and 100% modulus, respectively, for sintered sheets of NR. As is evident, the smaller the particle size of the starting powder, the higher the mechanical properties obtained when the particles are sintered. Over a 1.5 MPa gain in strength is observed for the smaller starting particle sized material, along with an almost 50% gain in elongation at break. The most interesting note, however, is that the moduli of materials made from the two different particle sizes are identical.

The observation that finer particles produce superior mechanical properties is consistent with all that is known about the influence of particle size on properties of composite materials. Similar observations are found in the sintering of metals and the influence of crystal size on strength. It has also

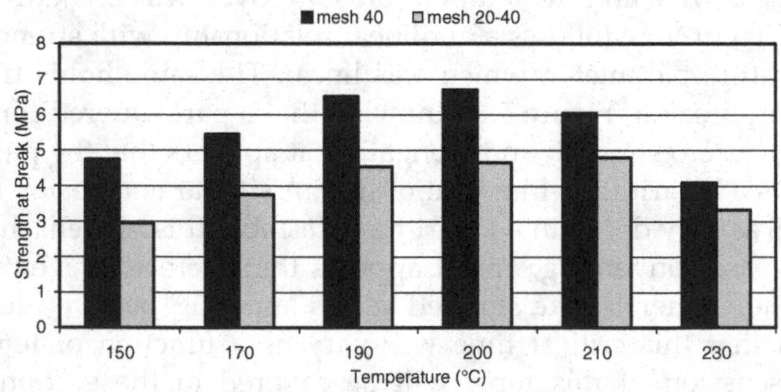

FIGURE 5.8
Effect of particle size of natural rubber (NR) powder on the strength at break of sintered sheets.

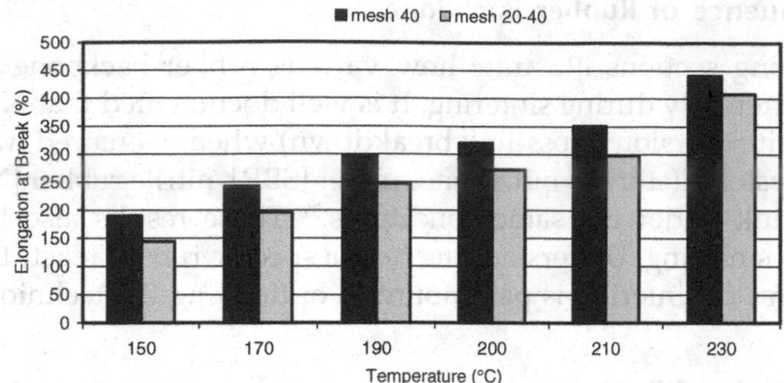

FIGURE 5.9
Effect of particle size of natural rubber (NR) powder on the elongation at break of sintered sheets.

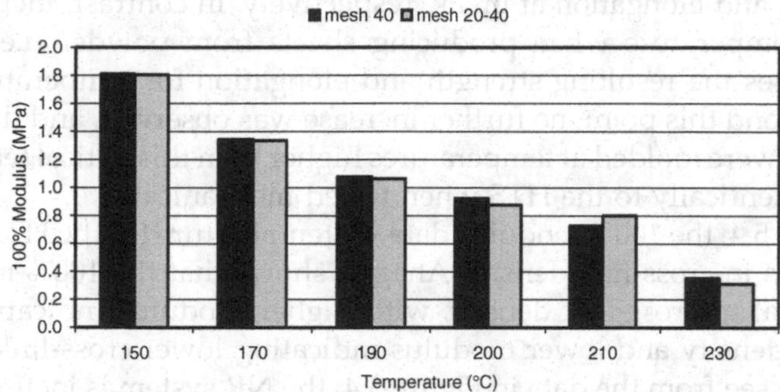

FIGURE 5.10
Effect of particle size of natural rubber (NR) powder on the 100% modulus of sintered sheets.

been shown that reinforcing rubber with smaller particles yields higher mechanical properties than a similar loading of larger particles.[30] The results in Figure 5.8, Figure 5.9, and Figure 5.10 are also in agreement with literature data regarding the sintering of polyurethane powders and with other adhesion research.[16,31] It has been postulated that smaller particles incur better adhesion when implemented into matrices, due to their higher surface area.

5.2.3.4 Molding Parameter Conclusions

It has been concluded that temperature, time, and particle size all play important roles in determining the final mechanical properties of sintered parts. Temperature has displayed a linear relationship with strength and elongation at break of sintered parts, while time appears to follow a logarithmic relationship. Using smaller particles for sintering results in a greater retention of mechanical properties, as one would expect from previous literature results. There also seems to be a time-temperature relationship; it appears that time and temperature can be written as one reduced axis.

5.2.4 Influence of Rubber Backbone

The following sections illustrate how various rubber backbone structures respond differently during sintering. It is well documented that NR systems often exhibit reversion (cross-link breakdown) when overcured, while most synthetic systems (styrene-butadiene rubber [SBR], nitrile rubber [NBR], etc.) overcross-link under the same conditions.[32] These results also have been found with sintering. Understanding how a specific rubber reacts to the high temperatures of sintering is paramount to optimizing the technique.

5.2.4.1 Natural Rubber

Figure 5.2 and Figure 5.3 showed how heat-treating vulcanized sheets (i.e., exposing vulcanized sheets to the conditions used in sintering) of NR between 140°C and 230°C causes a linear decay in the mechanical properties of strength and elongation at break, respectively. In contrast, increasing the molding temperature when producing sheets from powder (i.e., sintered sheets) raises the resulting strength and elongation for temperatures up to 210°C. Beyond this point, no further increase was observed, and the sintered sheets that were molded at temperatures higher than this critical temperature behaved identically to the HTS when tested mechanically.

In Figure 5.4, the 100% modulus data vs. temperature for the HTS explored the changes in cross-link density. Ahagon shows that the 100% modulus is proportional to cross-link density, with higher modulus indicating higher cross-link density and lower modulus indicating lower cross-link density.[33] As one can see from the data in Figure 5.4, the NR system is incurring cross-link breakdown (reversion) over the entire temperature range explored as its 100% modulus is dropping with increasing temperature. Therefore, it can be concluded that having the isoprene repeat unit along the backbone causes cross-link breakdown (reversion) when the rubber is exposed to the high temperatures associated with HPHTS.

5.2.4.2 Styrene-Butadiene Rubber

Figure 5.11 and Figure 5.12 show how heat-treating vulcanized sheets of sulfur-cured SBR (again at 8.6 MPa pressure and 1 hour; see Table 5.1 for formulation) between 160°C and 280°C causes a linear decay in the properties of strength and elongation at break, respectively, similar to HTS of NR (a slight difference is noted as NR elongation increases with strength decay). The two figures also illustrate that between 180°C and 250°C, the mechanical properties of the sintered sheets of SBR show an upward trend, until no further increase above 250°C is observed and the eventual decay of mechanical properties above ~260°C is evidenced.

Figure 5.13 shows the 100% modulus data for both the heat-treated and sintered sheets of SBR over the described temperature range (molding pressure of 8.6 MPa and molding time of 1 hour). The figure illustrates that above 220°C, the SBR system is increasing in cross-link density. This con-

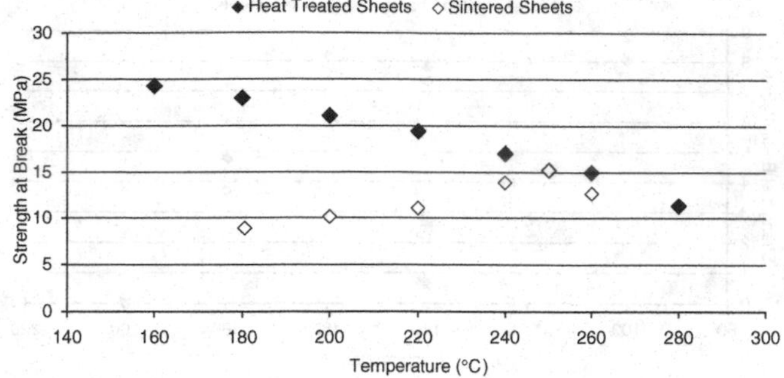

FIGURE 5.11
Strength at break vs. temperature for heat-treated and sintered sheets of styrene-butadiene rubber (SBR).

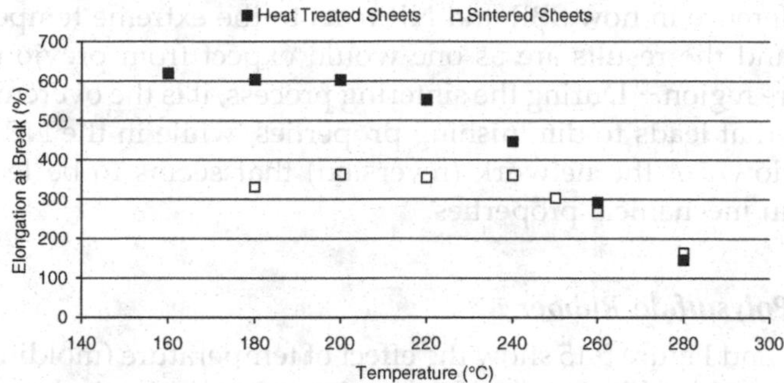

FIGURE 5.12
Elongation at break vs. temperature for heat-treated and sintered sheets of styrene-butadiene rubber (SBR).

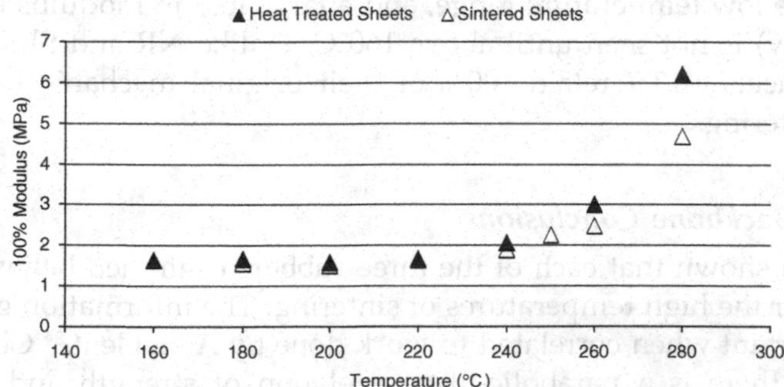

FIGURE 5.13
100% Modulus vs. temperature for heat-treated and sintered sheets of styrene-butadiene rubber (SBR).

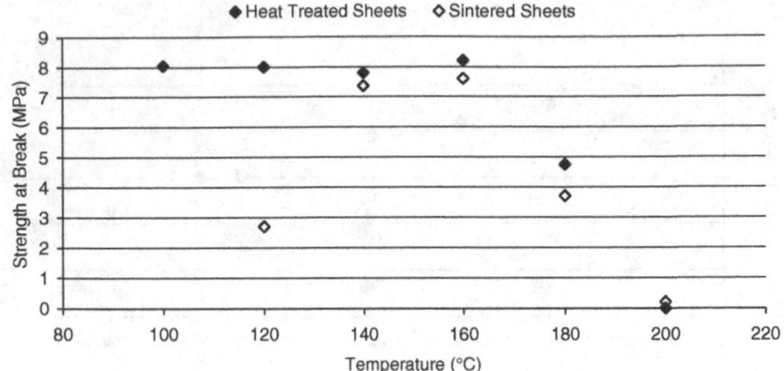

FIGURE 5.14
Strength at break vs. temperature for heat-treated and sintered sheets of polysulfide rubber (TR).

trasts with our previous data from natural rubber systems, where the 100% modulus data indicated that more cross-link breakdown (reversion), not further cross-linking, was occurring. This highlights that there is a fundamental difference in how SBR and NR react to the extreme temperatures of sintering, and the results are as one would expect from previous work in the overcure region.[32] During the sintering process, it is the overcross-linking of the SBR that leads to diminishing properties, while in the NR system, it is a breakdown of the network (reversion) that seems to be leading to a reduction in mechanical properties.

5.2.4.3 Polysulfide Rubber

Figure 5.14 and Figure 5.15 show the effect of temperature (molding pressure of 8.6 MPa and molding time of 1 hour) on the mechanical properties of strength and elongation at break, respectively, for heat-treated and sintered sheets of polysulfide rubber (TR). TR maintains its properties under heat-treatment up to about 160°C. Above this temperature, it begins to act like NR, and property decay is found. This is further illustrated in the 100% modulus data shown in Figure 5.16. Only a slight loss in modulus is registered in the low temperature range, and a large loss in modulus (i.e., cross-link density) is not seen until above 160°C. Unlike NR and SBR however, sintered sheets of TR retain 100% of their original mechanical properties during sintering.

5.2.4.4 Backbone Conclusions

It has been shown that each of the three rubbers examined behaved differently under the high temperatures of sintering. The information gathered is very important when correlated to work done by A.N. Gent.[34] Gent formulated that there is a parabolic-like correlation of strength and cross-link density. As most rubbers are formulated to yield the maximum obtainable strength when molded, any altering of the cross-link density would result in a decrease in mechanical properties. Since the modulus yields information

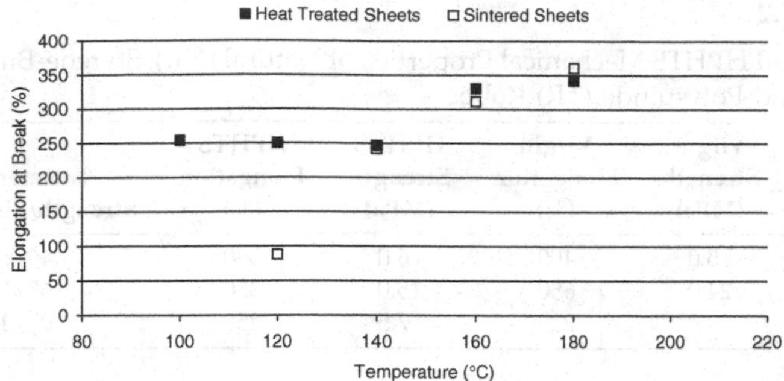

FIGURE 5.15
Elongation at break vs. temperature for heat-treated and sintered sheets of polysulfide rubber (TR).

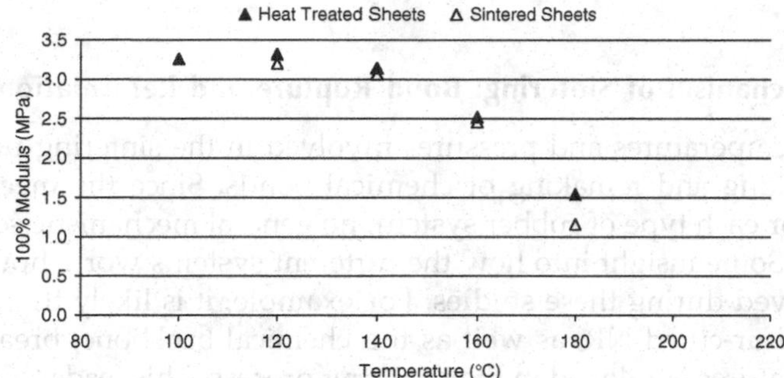

FIGURE 5.16
100% Modulus vs. temperature for heat-treated and sintered sheets of polysulfide rubber (TR).

about the cross-link density of the overall rubber network, the work presented here illustrates that each system is undergoing a different change in cross-link density. The NR and SBR systems lose mechanical properties as they incur a loss and an increase in cross-link density, respectively, and thus fall off of the peak of the parabola. TR, however, is able to retain its properties during sintering, as it is not subjected to a change in cross-link density. More details about these trends are found in a study by Williams and co-workers.[35]

Understanding how different rubbers react at high temperatures is very important when considering rubber recycling via HPHTS. Table 5.2 shows the sintered mechanical properties obtained for three types of vulcanized rubbers: natural rubber (NR), styrene-butadiene rubber (SBR), and polysulfide rubber (TR). The vulcanized properties were optimized to yield the maximum strength, and were cured at 90% of the optimum time (t_{90}). The mechanical properties of both sintered NR and SBR are less than that of the starting materials. However, the sintered TR achieves 100% recovery of the virgin properties.

In summary, studies of these systems have shown that rubbers with good mechanical properties can be obtained from waste and scrap rubber. The

TABLE 5.2

Virgin and HPHTS Mechanical Properties of Natural (NR), Styrene-Butadiene (SBR), and Polysulfide (TR) Rubbers

Rubber Type	Virgin Strength (MPa)	Virgin Elongation (%)	HPHTS Strength (MPa)	HPHTS Elongation (%)	% Retention Strength/Elongation
NR	15.0	400	6.0	330	40/85
SBR	24.0	650	15.0	350	65/55
TR	7.9	235	7.5	250	95/105

retention of properties differs, however, for each type of rubber. It appears that the retention is dependent upon many different factors, including molding conditions, particle size, and especially the chemical structure of the rubber.

5.2.5 Mechanism of Sintering: Bond Rupture and Reformation

The high temperatures and pressures involved in the sintering process can cause breaking and remaking of chemical bonds. Since the mechanism is different for each type of rubber system, no general mechanistic scheme can be drawn. Some insight into how the different systems work, however, has been achieved during these studies. For example, it is likely that the cross-links in sulfur-cured NR, as well as the chemical backbone, breakdown at the temperatures employed in the sintering process. This leads to a decrease in the maximum obtainable strength via a reversion-type mechanism. In contrast, sulfur-cured SBR systems overcross-link, probably through the residual unsaturation in the rubber. This results in lower retention of mechanical properties for these rubbers. A key finding, however, is that TR maintains a constant number of cross-links during sintering and has also been shown to recover 100% of its starting mechanical properties. This feature has made it a focal point for determining the characteristics that an ideal rubber should have if one wants to achieve full regeneration of properties during HPHTS. Further insight is available from chemical stress relaxation experiments. To this end, a discussion of the behavior of the aforementioned rubbers (NR, SBR, and TR) during chemical stress relaxation experiments is presented below.

Chemical stress relaxation, or the relaxation of mechanical stress through chemical means, was discovered in the 1940s in an investigation of cross-linked (vulcanized) rubbers including: natural rubber, neoprene, butyl rubber, styrene-butadiene copolymers, and acrylonitrile-butadiene copolymers.[26] Despite the fact that the rubbers are cross-linked and should show virtually no stress relaxation at elevated temperatures, it was discovered that at approximately 100 to 150°C there is a fairly rapid decay to zero stress at a constant strain. The stress decay is attributed to rupture of the rubber chemical network. The onset temperature of chemical interchange reactions (bond breaking and reformation of cross-link bonds) is a function of the cross-link density, and increases as the cross-link density increases. During the chemical stress relaxation experiments, it was observed that some rubbers (such as NR and butyl) became progressively softer, while other rubbers (such as SBR) became harder, even though the stress in all rubbers was decaying. Therefore, it was concluded that at elevated temperatures, two processes occurred: chemical bond scission and additional cross-linking.[26]

When a rubber is held at a constant length, the scission of the original chemical network leads to the stress relaxation. Additional cross-linking does not affect the stress of the sample held at constant length, unless there is sufficient cross-linking that a significant volume change occurs. In other words, the new network is formed in a relaxed state and therefore will not contribute to the stress. In order to separate the two reactions (chemical scission and additional cross-linking), Tobolsky developed two stress relaxation tests. He coined the terms "continuous" and "intermittent," respectively, for these two stress relaxation methods.[26]

5.2.5.1 Continuous Stress Relaxation

Continuous chemical stress relaxation provides information on the breakdown (or scission) of network cross-links. In this method, a material is set to a fixed elongation and kept at a constant temperature, and the stress is measured as a function of time. At elevated temperatures, the network undergoes chain scission and subsequent recross-linking. Any newly formed bonds, however, will be oriented in such a manner as to provide no new stress in the system (i.e., they will form in the relaxed state). As a result, the measured stress of the material will decay as a function of time as the strain is removed by bond breaking. Continuous stress relaxation provides an excellent means for understanding how quickly the old network will be eliminated during sintering. The ideal material for sintering should have a fast decay time to zero stress in a continuous relaxation experiment. This quick decay illustrates that there is a rapid bond breakage and thus high potential for eliminating the old network and efficiently forming a new one.

5.2.5.2 Intermittent Stress Relaxation

Intermittent stress relaxation provides information not only on the breakage of existing bonds, but on the formation of new bonds as well. In this method, a material is kept in the relaxed state at a fixed temperature. At various timed intervals, it is rapidly stretched to some fixed elongation and the stress at equilibrium is measured. It is then returned to its original relaxed state. This is, in essence, providing an instantaneous measurement of the modulus of the network, which directly correlates to the overall cross-link density. As the network breaks down and subsequent recross-linking occurs while the material is in the relaxed state, this method provides a measurement of net changes in the network as a function of time at a specific temperature. Thus, intermittent stress relaxation provides information about how a material will react at the extreme temperatures of sintering and the fate of the overall rubber network. An ideal rubber will maintain a constant cross-link density during sintering, and therefore will have a completely flat intermittent stress relaxation curve. A flat intermittent curve suggests that for every bond broken, one bond will form. This is important for sintering as the continuous relaxation illustrates a disappearance of the old network, while combining the two relaxations (intermittent and continuous) offers information about the formed (new) network.

5.2.5.3 Correlation to Sintering

By investigating both the continuous and intermittent stress relaxation of various rubbers, it should be possible to understand how they will behave during sintering. It is also possible to use a combination of continuous and intermittent stress relaxation experiments to mathematically determine the relative number of new bonds formed during sintering. For example, the relative continuous relaxation is equal to one minus the relative number of reactive sites that can form new cross-links. However, the relative intermittent relaxation value gives the total number of bonds in the system. Therefore, the difference between the two relative stress relaxation values is proportional to the relative amount of formed bonds/cross-links during a relaxation experiment at some fixed time and temperature. This can be expressed by:

$$N = I - C$$

where I is the intermittent stress relaxation, C is the continuous relaxation, and N is the newly formed bonds. In the ideal case, the number of newly formed bonds should be proportional to the mechanical integrity (i.e., strength) of parts produced via HPHTS. Figure 5.17 shows a plot of this data as a function of the molding temperature for NR. In this case, the number of cross-links in the system (depicted by triangles and related to the intermittent chemical stress relaxation) is decreasing with increasing temperature. The recorded behavior is identical to results depicting heat-treatment of vulcanized NR sheets over various temperatures. The squares (equal to one minus the continuous chemical stress relaxation [C]) show an increase in the number of broken bonds or reactive species in the process. The circular points depict the number of new bonds formed, N. As is shown, the number of new bonds formed follows the trends found in sintering; the strength first increases in a linear manner with temperature until a critical temperature is reached, and decreases thereafter (similar results are also shown in Figure 5.2).

The formula does not hold true for rubbers that have intermittent values that exceed one during sintering. SBR, for example, overcross-links, and therefore its intermittent stress relaxation value is greater than one, since

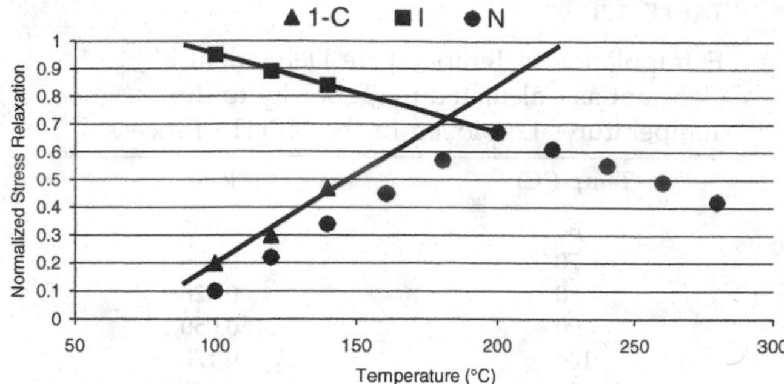

FIGURE 5.17
Correlation of new bonds formed as calculated by chemical stress relaxation to sintering results for natural rubber (NR).

more bonds form than are broken. The above expression does not account for a loss of strength due to overcross-linking, and cannot be applied in those situations. As is shown in Figure 5.17, however, it does an excellent job of predicting sintering results for stable and reverting rubbers ($I \leq 1$).

5.2.5.4 Time-Temperature Superposition

One interesting observation that comes out of Tobolsky's stress relaxation work and the above sintering results is the evidence of a time-temperature superposition, or the ability to combine time and temperature into one reduced axis. Early work by Tobolsky showed that stress relaxation of TRs performed at different temperatures could be shifted through a temperature-dependent constant so that all data fit one master curve.[26] By manipulation of Tobolsky's data, assumption of an $I = 1$ intermittent curve (valid for polysulfides) and use of the I-C equation for new bonds formed, a new graph can be formulated for polysulfide, as shown in Figure 5.18. This graph shows the predicted new bonds formed as a function of the time-temperature axis k′t, where k′ is a function of temperature and t is equal to time. Knowing that the prior section regarding sintering of TR (a similar polysulfide as that used in Tobolsky's work) was conducted over the temperature range of Tobolsky's relaxation data, a simple comparison can be made to correlate sintering to stress-relaxation and the time-temperature effect. Table 5.3 is a summary of the k′ values Tobolsky calculated and extrapolations to k′ values for 140°C and 160°C using a linear relationship predicted and verified by Tobolsky. Using the calculated k′ values and 1 hour as the time, it is predicted from Figure 5.18 that these sintering values should retain 100% of the original strength properties. Similarly, the 120°C data predict roughly 40% recovery, while below 80 to100°C, less than 15% retention is predicted. This correlates well to the measured values of 8 MPa strength for 140/160°C (out of 8 MPa starting strength, or 100% retention), 3 MPa for the 120°C data (again out of 8 MPa starting strength, or ~ 40% retention), and no retention of mechanical properties below 80°C (sintering fails to produce solid parts).

TABLE 5.3

Extrapolation of Temperature Dependent k′ Constant as Calculated by Tobolsky to the Temperatures Employed in the HPHTS Process

Temp (°C)	k′
60	0.002
70	0.005
80	0.020
90	0.050
100	0.170
110	0.400
120	0.910
140	~5
160	~25

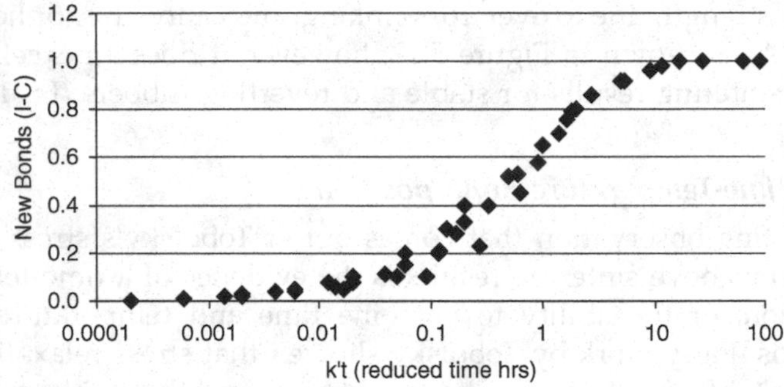

FIGURE 5.18
New bonds formed (as calculated from chemical stress relaxation results) vs. reduced time axis for polysulfide rubber (TR).

These results show that stress relaxation for polysulfides is a useful tool for predicting the retention of mechanical properties obtained via HPHTS. The results have also shown that sintering data can be plotted against a reduced time-temperature axis, since time and temperature are interchangeable.

5.2.5.5 Mechanism Conclusions

Cross-link and backbone bond breaking and subsequent reformation are the key reactions that permit sintering to occur. Chemical stress relaxation results have correlated well with sintering data, and provide an excellent tool for predicting the success of the sintering process. The ideal rubber for sintering appears to be one that has a rapid, continuous stress relaxation (i.e., a fast elimination of the old network), and a constant, intermittent stress relaxation. A constant I = 1 value is critical, as this means the newly formed network is not overcross-linking (I > 1), or reverting (I < 1), during the sintering process and thus should afford maximum property retention.

5.2.6 Use of Additives to Increase Mechanical Properties

There are many publications discussing materials that slow or stop reversion/interchange reactions, thus creating more thermally stable rubbers.[36–40] As the stress relaxation and sintering data have shown, it is probably the lowering of cross-link density in the sintered NR systems that causes the decay in mechanical properties. One possibility for increasing mechanical properties of produced parts is to incorporate reversion inhibitors during sintering, or to build back cross-link density after it has been destroyed.

5.2.6.1 Reversion in Rubbers

Reversion is a term used in the rubber industry to describe decreases in the mechanical properties (namely modulus and strength) at the end of the curing period or during service.[36,41,42] Although the complete mechanistic formulation of reversion reactions is still in debate, the general consensus is that reversion results from a degradation of load-bearing sulfur cross-links along the backbone of the rubber main chain. It is also known that certain rubbers (isoprene-containing rubbers such as NR) are more susceptible to reversion reactions than are other rubbers (SBR, butyl rubber, etc.).[32] It has been speculated, and recently demonstrated, that zinc complexes have the ability to catalyze these reversion reactions, and thus, after curing, can lower the obtained properties. Unfortunately, zinc is used in almost all sulfur-cured rubber systems due to the dramatically increased cure rate it affords; thus, it is not likely to be eliminated from vulcanization packages in the near future. In several papers, Nieuwenhuizen and co-workers demonstrate the role of zinc complexes in reversion reactions and show (through model reactions) their influence on the main rubber backbone.[36,41,42] The conclusion from the work is that zinc complexes accelerate the breakdown of sulfur cross-links, resulting in the formation of cyclic sulfur, along with diene and triene structures, along the backbone of the rubber where the sulfur cross-links had previously existed. Nieuwenhuizen and co-workers continue their work by developing a logical strategy to counteract the reversion phenomenon. They propose two approaches: first, deactivation of the zinc complex, and thus elimination of its catalytic activity that promotes reversion, and second, the use of antireversion additives such as biscitraconimides.[41] The role of the biscitraconimides is to replace broken sulfur cross-links with new, more thermally stable carbon ones. It is believed that the biscitraconimides react through a Diels-Alder mechanism to reattach the chains where cross-links had previously been, thus eliminating the loss of mechanical properties normally observed when reversion occurs.[43]

5.2.6.2 Additive Results

Although the rubbers discussed above in the sintering studies were fully cured, and the chance of reversion at room temperature or use temperatures is quite low, the potential for reversion at the high temperatures used in the

TABLE 5.4

Mechanical Properties of Sintered Sheets of Natural Rubber (NR) with Various Additives

Natural Rubber	Strength (MPa)	100% Modulus (MPa)	% Elongation
Original properties	12.9	1.8	420
Sintered Natural Rubber Powder (NR)	**Strength % Retained**	**Modulus % Retained**	**Elongation % Retained**
NR (4% adipic acid)	7	60	30
NR (1.5% CBS)	10	65	40
NR (no additives)	40	50	120
NR (2.5% benzoic acid)	60	60	140
NR (6% maleic acid)	65	60	140
NR (2.5% phthalic anhydride)	70	60	120
NR (6% maleic anhydride)	65	70	140
NR (4% phthalimide)	75	80	125
NR (4% N-methylphthalimide)	60	55	130

sintering process is significant. Furthermore, the NRs all contain the ingredients that the Nieuwenhuizen, et al. papers designate as reversion promoters. Indeed, during sintering these rubbers exhibit reversion behavior. This is not surprising considering that the operating temperature range for our process would increase the kinetics associated with the reversion reactions that the zinc complexes likely catalyze. Recently, several organic acids, "retarders" (of vulcanization) in typical vulcanization packages, have been added to sintered NR systems. It is believed that these acids react with, and neutralize, common vulcanization accelerators such as 2-mercaptobenzothiazole (MBT) and thus help slow the formation of the zinc complex [Zn $(MBT)_2$]. This complex is believed to be one of the main accelerators in cross-link formation during vulcanization.[44] Interestingly, the presence of most of these organic acids causes an increase in mechanical properties (modulus, strength, and elongation) in sintered NR systems (organic acids were powder mixed with natural rubber powder and then molded; see Table 5.4 for results).

The current understanding is that the organic acids are likely reducing the amount of the zinc complex present, and thus limiting reversion, as suggested by Nieuwenhuizen and his co-workers. In addition, acids that contained a double bond showed the largest increase in properties. It is likely that this double bond may be able to bridge a broken cross-link in a similar (yet mechanistically different) fashion, to the biscitraconimides. Contrary to conventional thinking, addition of accelerators for normal vulcanization (zinc oxide, MBT, etc.) has so far been found to decrease the mechanical properties in NR systems. A further discussion of additives is found in the work of Tripathy and fellow investigators, along with studies exploring the mechanisms of action of the various additives.[45] A brief discussion of biscitraconimide results can be found in the study by Williams and co-workers.[35]

5.3 Summary of High-Pressure, High-Temperature Sintering

The work to date on HPHTS has yielded important insight not only into the mechanism of rubber sintering, but how HPHTS can be optimized by understanding the different variables at work. Temperature and time were shown to be the most important molding parameters, each with a unique relationship to the mechanical properties obtained. Rubber backbone structure has been shown to be influential in determining property retention during sintering. Reversion and overcross-linking in NR and SBR, respectively, play important roles in diminishing sintering properties, while TR is able to recover almost all initial properties when sintered.

It has also been illustrated that a time-temperature superposition, as reported with chemical stress relaxation studies, appears to apply to sintering results as well. Chemical stress relaxation experiments correlate well with sintering data and provide a unique insight into the mechanism of sintering, especially for systems that revert or maintain constant cross-link density. It is likely that chemical stress relaxation experiments will be used to predict optimized sintering conditions in the future. Additives (specifically, organic acids) have been shown to increase property retention in a NR system, and several potential mechanisms for their effectiveness have been formulated.

In conclusion, HPHTS is a technique that can use rubber scrap in high volume as it uses 100% recycled rubber as starting material. Current limitations in processing methods hinder the development of HPHTS, but the potential exists for an economically and environmentally efficient process to be produced. A technique that reuses high volumes of scrap rubber to produce value-added products has been one of the most sought-after technologies of the past two centuries. It is believed that HPHTS offers one potential solution to this enormous environmental dilemma.

References

1. Goodyear, C., U.S. Patent 240, 1837.
2. Bradford, K. and Pierce, D.D., *Trials of an Inventor, Life and Discoveries of Charles Goodyear*, Phillips & Hunt Publishers, New York, 1866.
3. Hayward, N., U.S. Patent 1,090, 1839.
4. Goodyear, C., U.S. Patent 3,633, 1844.
5. Hancock, T., *Personal Narrative of the Origin and Progress of the Caoutchouc or India-Rubber Manufacture*, Longman, Brown, Green, Longmans, and Roberts, London, 1857.
6. http://www.rma.org/scraptires/facts_figures.html
7. Beckman, J.A. et al., Scrap tire disposal, *Rubb. Chem. Tech.*, 47, 597, 1974.
8. Myhre, M.J. and MacKillop, D.A., Rubber recycling, *Rubb. Chem. Tech.*, 75, 429, 2002.

9. Goodyear, C., British Patent 2,933, 1853.
10. Morey, C., U.S. Patent 12,212, 1855.
11. Acetta, A. and Vergnaud, J.M., Upgrading of scrap rubber powder by vulcanization without new rubber, *Rubb. Chem. Tech.*, 54, 302, 1981.
12. Accetta, A. and Vergnaud, J.M., Vibration isolation properties of vulcanizates of scrap rubber powder, *Rubb. Chem. Tech.*, 55, 328, 1982.
13. Accetta, A. and Vergnaud, J.M., Rubber recycling – upgrading of scrap rubber powder by vulcanization II, *Rubb. Chem. Tech.*, 55, 961, 1982.
14. Phadke, A.A. and De, S.K., Vulcanization of cryo-ground reclaimed rubber, *Kautschuk + Gummi Kunststoffe*, 37, no. 9, 776, 1984.
15. Phadke, A.A., Chakraborty, S.K., and De, S.K., Cryoground rubber — natural rubber blends, *Rubb. Chem. Tech.*, 57, 19, 1984.
16. Law, W.K. et al., *Polymer Recycling*, 3, 269, 1998.
17. James, O., Recycling of thermoset flexible polyurethane foams into solid polyurethane rubber, *Proceedings of the Annual Technical Conference (ANTEC) 1997*, paper no. 275, 3076, 1997.
18. Corbett, G.E. and Wadie, B.J., Preparation of elastomer sheets from flexible urethane foam formulations, *J. Cell. Plast.*, 10, 1, 26, 1974.
19. Arastoopour, H. et al., U.S. Patent 5,904,885, 1999.
20. Adhikari, B., De, D., and Maiti, S., Reclamation and recycling of waste rubber, *Prog. Polym. Sci.*, 25, 909, 2000.
21. Narkis, M. and Rosenzweig, N., *Polymer Powder Technology*, John Wiley & Sons, Chichester, 1995.
22. Morin, J.E., Williams, D.E., and Farris, R.J., A novel method to recycle scrap tires: high-pressure high-temperature sintering, *Rubb. Chem. Tech.*, 75, 955, 2002.
23. Tobolsky, A.V., Prettyman, I.B., and Dillon, J.H., *J. Appl. Phys.*, 15, 324, 1944.
24. Andrews, R.D., Tobolsky, A.V., and Hanson, E.E., *J. Appl. Phys.*, 17, 352, 1946.
25. Tobolsky, A.V., Stress relaxation studies of the viscoelastic properties of polymers, *J. Appl. Phys.*, 27, 673 1956.
26. Stern, M.D. and Tobolsky, A.V., *J. Chem. Phys.*, 14, 93, 1946.
27. Tobolsky, A.V., *Properties and Structures of Polymers*, John Wiley & Sons, New York, 1960.
28. Tobolsky, A.V. and MacKnight, W.J., *Polymeric Sulfur and Related Polymers*, Interscience Publishers, New York, 1965.
29. Aklonis, J.J. and MacKnight, W.J., *Introduction to Polymer Viscoelasticity*, Wiley-Interscience, New York, 1983.
30. Kraus, G., Reinforcement of elastomers by particulate fillers, in *Science and Technology of Rubber*, Eirich, F.R., ed., Academic Press, New York, 1978, chap. 8.
31. Anderson, L., A predictive model for the mechanical behavior of particulate composites, Ph.D. thesis, University of Massachusetts, Amherst, MA, 1989.
32. Bellander, M., Stenberg, B., and Persson, S., Crosslinking of polybutadiene without any vulcanizing agent, *Polym. Eng. Sci.*, 38, 1254, 1998.
33. Ahagon, A., Extensibility of black-filled elastomers, *Rubb. Chem. Tech.*, 59, 187, 1986.
34. Gent, A.N., Strength of elastomers, in *Science and Technology of Rubber*, Eirich, F.R., ed., Academic Press, New York, 1978, chap. 10.
35. Williams, D.E, Morin, J.E., Tripathy, A.R., and Farris, R.J., Redefining thermosets: a method of molding vulcanized rubber into new parts, *Rubber World*, 226 40, 2002.

36. Nieuwenhuizen, P.J., Haasnoot, J.G., and Reeduk, J., Homogeneous zinc (II) catalysis in accelerated vulcanization II. (Poly)olefin oxidation, dehydration, and reaction with anti-reversion coagents, *Rubb. Chem. Tech.*, 72, 15, 1999.
37. Datta, R.N., et al., Fourier transform Raman spectroscopy for characterization of natural rubber reversion and of anti-reversion agents, *Rubb. Chem. Tech.*, 72, 829, 1999.
38. Datta, R.N. and Wagenmakers, J.C., New rubber chemical for improved reversion resistance, *Kautschuk + Gummi Kunststoffe*, 49, 10, 6, 1996.
39. Schotman, A.H.M. et al., Studies on a new antireversion agent for sulfur vulcanization of diene rubbers, *Rubb. Chem. Tech.*, 69, 727, 1996.
40. Blok, E.J., Kralevich, M.L., and Varner, J.E., Preliminary studies on new antireversion agents for the sulfur vulcanization of diene rubbers, *Rubb. Chem. Tech.*, 73, 114, 2000.
41. Nieuwenhuizen, P.J., et al., Homogeneous zinc (II) catalysis in accelerated vulcanization III. Degradation modes of mono- and disulfidic cross-links, *Rubb. Chem. Tech.*, 72, 27, 1999.
42. Nieuwenhuizen, P.J., et al., Homogeneous zinc (II) catalysis in accelerated vulcanization IV. The mechanism of cross-link (de)sulfuration, *Rubb. Chem. Tech.*, 72, 43, 1999.
43. Datta, R.N., et al., Biscitraconimides as anti-reversion agents for diene rubbers: spectroscopic studies on citraconimide-squalene adducts, *Rubb. Chem. Tech.*, 7, 129, 1997.
44. Krejsa, M.R. and Koenig, J.L., A review of sulfur crosslinking fundamentals for accelerated and unaccelerated vulcanization, *Rubb. Chem. Tech.*, 66, 376, 1993.
45. Tripathy, A.R., et al., A novel approach to improving the mechanical properties in recycled vulcanized natural rubber and its mechanism, *Macromolecules*, 35, 12, 4616, 2002.

6

Powdered Rubber Waste in Rubber Compounds

Ceni Jacob and S.K. De

CONTENTS

6.1 Introduction

Rubber wastes, especially worn-out tires, occupy the lion's share among the nonbiodegradable waste materials. But the technological advance for a safe disposal and effective reuse of rubber wastes is still in its infancy. Rubber compounds with inferior mechanical properties are produced by reclaiming methods, which consist of the expensive steps of comminution and high-pressure digestion in the presence of chemicals. The incineration method for energy recovery produces environmental pollution. Pyrolysis yields materials of poor quality.

6.2 Grinding Method

The simplest approach of rubber recycling is the grinding of vulcanizates and utilization of the powdered rubber. This was first suggested by Charles Goodyear, who is also credited with the discovery of rubber vulcanization, in order to overcome the scarcity of natural rubber.[1] Later, enterprising compounders found that they could selectively use the waste rubber dust, obtained when rubber products are buffed to size, as a compounding ingredient, to reduce cost. This approach has the advantage of being environmentally friendly, with no evolution of toxic gases or chemicals. Furthermore, the material value is not debased, as it is in pyrolysis or incineration. Therefore, the focus of several studies has been on improving the cost-effectiveness of this technique. Powdered rubber has been used to improve the impact strength of plastics and the processability of rubber compounds, and also for making thermoplastic elastomeric blends.[2] There are also reports on the preparation of products, by the vulcanization of natural rubber or styrene-butadiene rubber (SBR) lattices containing finely ground scrap tires and a large amount of mineral fillers.[3]

Powdered rubber is unique as a filler because of its larger size (in the range of microns) and lower modulus, when compared to other commercial fillers used in the rubber industry. Like other fillers, the polarity (either the surface functional groups or the polar nature of the polymer) and the structure (either spongy, chain-like aggregates or free-flowing powder) of ground vulcanizates affect the physicomechanical properties of ground rubber–filled vulcanizates.[4] When added to the same base compound (that is, when the composition of the powdered rubber and the matrix are the same), powdered rubber can act as an extender to the rubber matrix, with a minimum change in compound properties. When the compositions of the powdered rubber and the matrix are different, the factors controlling the loading level for maximum performance include the compatibility of: elastomeric systems, curing systems, hardness difference, and tensile strength requirements.

6.3 Preparation of Powdered Rubber

The technique of preparation of powdered rubber greatly influences the particle size and surface topography, which in turn affects the mechanical properties of the powder-filled compositions. The effect of the grinding method on the properties of the rubber compound has been studied by several workers.[5,6] The common methods of powdering the rubber vulcanizate are described under the following subsections.

6.3.1 Ambient Grinding[7–9]

Ambient grinding consists of passing the vulcanized rubber through the nip gap of a high-powered shear mill or two-roll mill at room temperature; the number of passes determines the particle size. The higher the number of passes, the greater is the size reduction; hence, the cost of ground rubber increases as the particle size decreases. In the case of scrap tires, the metal and fiber parts are first removed and the whole tire is cut into a chip form (~5 cm). The scrap tire particle size can be further reduced by the use of granulators, cracker mills, and micromills.

6.3.2 Cryogenic Grinding[10,11]

The cryogenic grinding technique was commercialized in the late 1960s. In this process, initial particle size reduction is carried out at ambient temperature. Then, the rubber chips are cooled to sub-zero temperatures by using liquid nitrogen, and the brittle material is then crushed to tiny particles by a hammer mill. In their frozen state, the particles are sieved. The major advantage of cryogenic grinding is the high production rate. Particle size is controlled by the immersion time in liquid nitrogen and by the mesh size of the screens used in the grinding chamber of the mill.

6.3.3 Wet Grinding[9,12]

In this process, tiny rubber chips in water are passed between hard, circular grinding plates that move concurrently and are lubricated by water. In a similar process called solution grinding, the rubber chips are swollen in a solvent and then led into the gap of grinding plates. Particle size is controlled by the time spent in the grinding process.

6.3.4 Extrusion

The extrusion process, developed in Russia, produces polymer powder by employing a twin-screw extruder that imposes compressive shear on the polymeric material, at selected temperatures.[13] Later, Arastoopour and co-workers[14,15] and Khait and Torkelson[16,17] patented the improved version, solid-state shear extrusion (SSSE). In this process, vulcanized rubber chips are passed through a laboratory-size single-screw extruder, which is maintained at a constant temperature by the circulation of cold water. The channel depth of the extruder screw decreases from feed zone to the outlet of the extruder. By repeated passes through the extruder, particles of narrow size distribution and average diameter (as low as 40 to 60 μm) can be obtained.

6.3.5 Abrasion

This method is mostly confined to the preparation of tire buffings, which are obtained as a by-product of retreading. The method involves the removal of rubber particles from tire tread by an abrasion process. No systematic studies have been made on the utilization of buffings, but it has been a normal practice in industries to reuse the tire buffings to make low-grade technical products by the revulcanization of the powder. The powder can be converted into useful products by sintering at high temperature under high pressure.[18]

De and co-workers used a simple abrasion technique for the preparation of ground rubber[19,20] In this process, rectangular rubber sheets (6 to 10 mm thick) were abraded by pressing the rubber specimens against a rotating, silicon carbide abrasive wheel of a bench grinder. The powder was collected in a holder kept beneath the wheel. Particle size obtained from abrasion depends upon the grain size of the abrasive (silicon carbide) and the nature of the polymer.

6.4 Comparison of Grinding Techniques

For the production of finer mesh size powders (~ 100 μm), cryogenic grinding is more economical than ambient grinding, although production cost depends upon the liquid nitrogen consumption. Since cryogenically frozen rubber pulverizes more easily than rubber at ambient temperatures, there is less wear and tear of the machinery, and maintenance costs are significantly reduced. Gas liberation and possible ignition are eliminated in the cryogenic system, and it is easier to separate impurities like steel and other fiber materials from tires. Degradation of the rubber, due to the heat buildup during shearing, is negligible in the cryogenic grinding method, in contrast to the ambient method. Although the productivity of the wet or solution process is low, very small particle with maximum purity and in the range of 400 to 500 mesh size can be obtained.

It is claimed that the SSSE method produces rubber compounds that can be sheeted out like virgin compounds. Partial devulcanization is also believed to take place.[15,16,17] In the case of the abrasion process, particles of wider size distribution have been reported (0.4 to 150 μm), but more than 90% are finer than 10 μm. As a single-step process, the cost-effectiveness of the abrasion method is expected to be higher than the other grinding techniques.

6.5 Characterization of Powdered Rubber

Besides the composition of the powdered rubber, the other parameters affecting the performance properties are the particle size and its distribution, shape, surface topography, and the surface functionalities (that is, the presence of polar groups on the particle surface). The changes in the chemical structure of polymers in filled and unfilled vulcanizates during grinding have been studied by Izyumova and co-workers, and the changes are attributed to degradation and cross-linking.[21] Breakdown of molecular chains followed by stabilization and transformations lead to the formation of new cross-links, but the degradation may lead to reduced cross-link density. It is also found that the extent of interaction between oxygen and the ground rubber particles increases in the presence of carbon black. Several workers studied the morphology and specific area of rubber powders produced by different methods.[22,23] The scanning electron photomicrographs of the powders reveal that the cryoground rubber surface is smooth and angular, and show striated lines where fracture has occurred. In the case of ambient ground rubber, the surface is rough and convoluted. The particle size obtained by cryogenic grinding is smaller (~100 μm) and uniform, and the size distribution is narrower than that obtained by the ambient grinding technique (> 250 μm). The milling of cryoground rubber by using a colloid mill or a plate mill produces rough and convoluted surfaces.[23] Bilgili and fellow investigators[15] analyzed the particles produced by SSSE, and reported that the particle shape is irregular and the surface is rough. Due to surface oxidation, finer particles stick to the surface of the larger particles and form

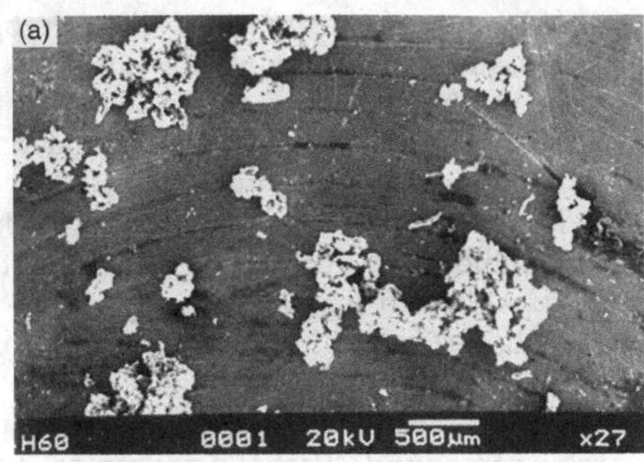

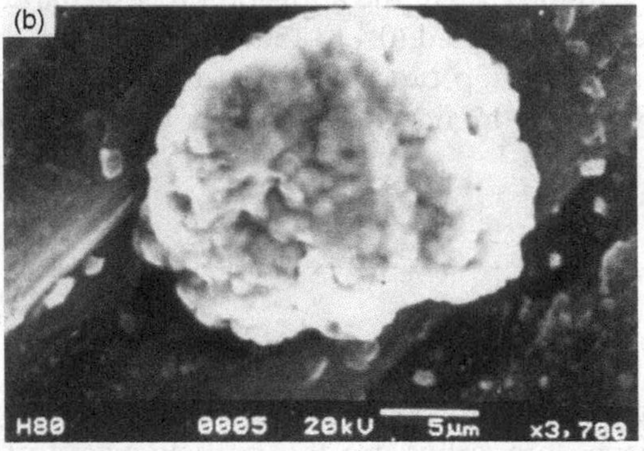

FIGURE 6.1
SEM photomicrograph of powdered (abraded) rubber[61] (a) aggregates (b) single particle.

agglomerate. Chemical analyzes reveal that some of the bonds are broken during the pulverization process and partial devulcanization takes place.

Particles obtained by the abrasion technique are found to have rough, convoluted surfaces and are irregular in shape.[19,20] The ultra-fine powdered rubber (< 10 μm) exists as aggregates (Figure 6.1a), which break down under shear. Figure 6.1b shows a typical abraded particle separated out by the ultrasonic dispersion technique. Surface oxidation may take place during abrasion, depending upon the nature of the base polymer and its aging history. While the surface area of the powder depends upon the method of production, the degree of fineness and the method of comminuting greatly affect the oxygen-containing functional groups on the surface of the rubber powder.[24]

6.6 Modifications of Ground Rubber

The most commonly used techniques of grinding are the cryogenic and ambient methods. Because it is coarse (> 250 μm), ambient ground powder will raise stress concentrations in a vulcanizate, although its rough and convoluted surface can provide interaction with the matrix. The demerit of finer cryoground rubber (> 100 μm) is that the smooth surface of the particle results in poor adhesion between the ground rubber and the matrix. Therefore, modification of the particle surface has been studied, to avoid premature failure of ground rubber–filled compositions, although it reduces the economic incentive behind recycling. Modification of the ground rubber can be done on the surface (surface functionalization) to improve the adhesion between the waste rubber and the fresh rubber phases, or it can be done in the bulk, to reduce the modulus (that is, by devulcanization).

6.6.1 Surface Modification

6.6.1.1 By Mechanical Means

The use of single- and twin-screw extruders for preparing comminuted vulcanizates combine the processes of comminution and modification due to oxidation, degradation, and secondary cross-linking. Surface roughening of cryoground particles by using a colloid mill has also been tried by some workers.[23]

6.6.1.2 By Chemical Methods

Chemical modification of the particle surface includes grafting, halogenation, oxidation with reactive gases, and coating with a cross-linkable polymer layer. Surface grafting initiated by ultraviolet light using photoinitiators is an approach to increase the activity of the ground rubber.[25] Surface grafting can also be achieved by thermally treating the ground rubber tire (GRT) with different monomers in the presence of thermo-initiators.[26] Surface chlorination of powdered rubber using trichloroisocyanuric acid has been reported to be successful.[27,28] Bauman reported the surface treatment of crumb rubber with reactive gases (e.g., mixture of halogens and oxygen), which results in the formation of polar functional groups such as carboxylate and hydroxyl.[29]

There are reports on precoating of GRT particles with trans-polyoctenamer[30] to prepare new molded products. Surface coating of ground rubber with a mixture of unsaturated curable polymer and a curing agent has also been reported to improve the performance of the ground rubber–filled composites.[31,32]

6.6.1.3 Irradiation

Corona[33] and plasma[34] treatment of ground rubber has been found to improve the adhesion between the ground rubber and fresh rubber, owing to the creation of oxidized species on the polymer surface. There are also reports on electron beam modification of ground rubber.[35]

6.6.2 Devulcanization

6.6.2.1 *Using Chemical Probes*

Several chemical probes like triphenylphosphine and sodium di-n-butyl phosphite, diphenyl disulfide, lithium aluminium hydride, and phenyl lithium are found to cleave the cross-links in a vulcanized elastomer.[36]

A process for devulcanizing and functionalizing a rubber vulcanizate by desulfurization involves suspending a rubber vulcanizate crumb in a solvent (preferably, one that swells the vulcanizate) and adding an alkali metal such as sodium to the suspension when the sulfur linkages are cleaved by sodium and the solvent gets grafted onto the crumb rubber surface.[37]

6.6.2.2 *High-Energy Excitation*

An alternative approach to devulcanization is the use of high-energy fields like ultrasonic[38,39] and microwave[40,41] to excite the material to a point of cross-link rupture.

6.6.2.3 *High-Shear Mixing*

Brown and Watson developed a mechanochemical process for devulcanization.[42] Arastoopour[14,15] and Khait[16,17] used a specially designed single-screw extruder to pulverize a variety of wastes, including rubber, by SSSE. They were able to reconstitute the rubber compound with properties similar to those of the original rubber compound, without adding any chemicals. Recently, thermochemical shear extrusion using a specially designed twin-screw extruder has also been reported to cause devulcanization of the ground tire rubber.[43]

6.6.2.4 *Microbial Desulfurization*

Sulfur-consuming bacterial genera, namely *Thiobacillus* (i.e., *T. ferroxidans, T. thiooxidans, T. thioparus*), *Sulfolobus*, and *Rhodococcus* can attack sulfur bonds in the vulcanized rubber, mainly on the surface, without affecting the rubber hydrocarbon chains, and the microbially treated (devulcanized) tire crumb can be revulcanized when added to virgin rubber.[44,45]

6.6.2.5 *High-Pressure Steam*

Several workers have used high-pressure steam autoclaving to try to devulcanize certain rubbers like natural rubber, butyl rubber, and silicone rubber.[46]

6.7 Revulcanization of Powdered Rubber and Application of Revulcanizates

Acetta and co-workers[47,48] reported that rubber powder recovered from old tires might be processed into useful products. Phadke and fellow investigators discovered that ground rubber can be revulcanized with the addition of a small amount of curatives, indicating availability of cross-linking sites in the cured particles.[49,50] It has also been found that ground rubbers vulcanize more rapidly in the presence of accelerators than do the fresh rubbers, presumably due to the formation of reactive radicals during grinding.[51] Reidel Omni Products, Inc.[52] has developed a rail crossing from scrap rubber. Kim and Lee[53] recently reported the use of a binder that coated *in situ* the GRT particles during mixing. This was followed by subsequent mixing of the powder with curatives to give a product, which could be molded to produce cost-effective rubber blocks or mats.

6.8 Mixing and Processing of Rubber Compounds Containing Powdered Rubber

Normally, powdered rubber prepared from waste rubber vulcanizates contains fillers like carbon black, silica, calcium carbonate, and clay. Therefore, powdered rubber may act as a filler in the fresh rubber compound. When more than one filler (e.g., carbon black and ground rubber) is present in the formulation, the filler that is added first to the fresh rubber may contribute more to the final properties of the vulcanizates. Swor and co-workers[54] studied the effect of time of addition of ultra-fine rubber powder (50 phr) to the Banbury cycle.At a loading of 50 phr, the effect of sequence of addition of ground rubber on the properties is negligible. Jacob and fellow investigators also made similar observations at powdered rubber loading of 50 phr in ethylene propylene diene (EPDM) compound (Table 6.1).[55]

Negmatov and co-workers[56] reported that ground rubber–filled compounds exhibit good millability.[56] When abraded powdered rubber, which exists as aggregates, is mixed with fresh rubber, it is found that the aggregates break down to particles, giving homogeneous dispersion, which is reflected in the mechanical properties of the powder-filled vulcanizates.[19] It has also been found that the addition of ground vulcanizates (particle size 400 to 500 μm) into fresh rubber during milling leads to a shift of the particle size distribution curve maximum toward the smaller particle size domain (i.e., breakdown of particles occurs).[57] A similar observation was made by Phadke and fellow investigators,[58] also in the case of natural rubber compounds containing cryoground rubber. They found that the breakdown of agglom-

TABLE 6.1

Stress-Strain Properties of EPDM Vulcanizates Containing Carbon Black (120 phr) and Powdered (Abraded) EPDM Rubber Vulcanizate as Fillers [a, b]

		Powdered Rubber Loading			
		50 phr		100 phr	
Property	F [c]	C	P	C	P
Tensile strength, MPa	6.37	6.30	6.63	6.67	5.71
Elongation at break, %	489	488	555	518	550
Modulus at 100% elongation, MPa	2.06	1.85	1.87	1.77	1.47

[a] C, carbon black added first, followed by the powdered rubber; P, powdered rubber added first, followed by carbon black.

[b] Formulations contain EPDM, 100; zinc oxide (ZnO), 4; stearic acid, 2; GPF carbon black, 120; paraffinic oil, 70; paraffin wax, 10; brown factice, 10; Tetramethyl thiuram disulfide (TMTD), 1.2; Mercaptobenzothiazole (MBT), 1.2; zinc diethyl dithiocarbamate (ZDC), 2.0; and sulfur, 1.5.

[c] Control compound with carbon black, but without powdered EPDM vulcanizate.

Source: From Jacob, C., De, P.P., and De, S.K., unpublished data.

erates is more, and hence the dispersion is better, in raw natural rubber (NR) than in NR gum compound. But in mill mixes of crumb, Bleyie showed[59] that, although the specific energy for 99.5% dispersion decreases as crumb particle size decreases from 3.2 to 0.25 mm diameter, it is still considerably greater than that needed for reinforcing carbon blacks. Phadke and coworkers[60] have compared the properties of vulcanizates prepared by mixing of ground rubber in milled sheet form (by mechanical treatment on a two-roll mill) and in the powder form. It was found that the mechanically treated powder gives better properties when blended with fresh rubber (NR compound), due to the higher breakdown of the particles and subsequent interaction with the matrix (Table 6.2).

TABLE 6.2

Effect of Mechanical Treatment of Powdered Rubber on the Properties of Powdered Rubber–Filled Compositions [a]

	Compound with 30 phr CGR[b]		Compound with 60 phr CGR	
Property	Powder Form	Milled Sheet	Powder Form	Milled Sheet
Tensile strength, MPa	12.97	18.84	14.27	20.82
Elongation at break, %	450	520	450	570
Modulus at 300% elongation, MPa	5.62	5.14	6.26	5.35
Tear strength, kN/m	48.0	54.6	52.8	49.8

[a] Formulations contain the following (in phr); NR, 100; CGR, 30/60; zinc oxide, 5; stearic acid, 2; Cyclohexyl benzothiazyl sulfenamide (CBS), 1.6; sulfur, 4.

[b] Cryoground rubber.

Source: From Phadke, A.A., Bhowmick, A.K., and De, S.K., *J. Appl. Polym. Sci.*, 32, 4063, 1986. With permission.

While studying the effect of powdered EPDM rubber (containing 120 phr carbon black and 70 phr paraffinic oil, 10 phr factice and 10 phr paraffin wax) on the properties of gum EPDM rubber compound, Jacob and co-workers[61] observed that up to 400 phr of the powdered rubber can be added easily to the fresh rubber on a laboratory-size two-roll mill (6″ × 13″ Schwabenthan, Germany). When the powdered rubber was added to the carbon black–filled EPDM compound (containing 70 phr oil) from which the powder was prepared, a drying effect was observed.[62] Earlier, Swor and fellow investigators[54] reported that the "drying" effect of ultra-fine rubber powder on sticky compounds leads to better overall processing, particularly milling. It has been found that GRT can absorb oil, the oil take-up being determined by the temperature, oil type, and particle size.[63] Ghosh and co-workers observed that silicone vulcanizate powder can be added to fresh silicone rubber compound up to a loading of 60 phr[20] and fluororubber powder can be added to a fluororubber compound up to a loading of 100 phr.[64]

As in the case of reclaimed rubber, it has been found that incorporation of small amounts of ground rubber into rubber compounds can provide processing advantages such as less air entrapment during molding, good extrudability, and less nerve during calendering. The cured rubber particles provide a path for the air to escape by bleeding air from the part. Extrudability of the compounds is improved by a decrease in die swell and extrudate distortion. It has been found that less force is required to extrude the compositions containing fine black-filled rubber particles, and the compounds containing fine particles transfer faster than other compounds during transfer molding.[54] The finer the particle size, the smoother the calendered sheets and the finer an edge that can be produced on extrusions.[65] Shrinkage is also less for compounds containing ground rubber. Several workers have found that the processability of tread compound can be improved by the incorporation of ground rubber.[66,67]

6.9 Effect of Addition of Ground Rubber on the Rheological Behavior of Fresh Rubber Compounds

The increase in Mooney viscosity by the addition of ambient ground powder is more than that by the addition of the same amount of cryoground rubber, which can be ascribed to the better interaction between the convoluted and rough surface of the spongy, ambient ground rubber and the matrix.[68]

Addition of cross-linked particles increases the viscosity at lower shear rates (e.g., Mooney viscosity; Table 6.3), but the effect is less prominent or even the reverse is observed at higher shear rates (> 1000 s^{-1}), depending upon the composition of the particles. For example, as given in Table 6.3, incorporation of fluororubber powder (containing no plasticizer) into a flu-

TABLE 6.3

Mooney Viscosity of Compounds Containing Abraded Powders[a]

Abraded EPDM Powder in Rubber Compounds	Mooney Viscosity ML(1+4)			Ref.
	G_0	G_{50}	G_{100}	
EPDM powder in gum EPDM compound[b]	35	47	61	19
EPDM powder in carbon black–filled EPDM compound[b]	24	29	37	62
EPDM powder in carbon black–filled SBR compound[c]	44	50	—	71
Abraded Specialty Powders in Gum Rubber Compounds	**Mooney Viscosity ML(1+4)**			**Ref.**
	G_0	G_{30}	G_{60}	
Fluororubber powder in fluororubber[b]	92	97	100	64
Silicone powder in silicone rubber[c]	29	46	70	20

[a] Subscript refers to the amount (phr) of ground rubber in the composition.
[b] ML(1+4)120°C.
[c] ML(1+4)100°C.

ororubber matrix[64] increases the Mooney viscosity as well as the high shear viscosity (Figure 6.2), whereas addition of EPDM-abraded powder (highly filled and oil extended) into a gum EPDM rubber compound[19,61] results in increased Mooney viscosity but lower high shear viscosity (Figure 6.3). The drop in viscosity at higher shear rates for the EPDM compound containing EPDM powder with high amount of oil and carbon black may be due to the wall slippage facilitated by the plasticizer migration. It is also found that the viscosity increases (at all shear rates) with the increasing amount of filler present in the powdered rubber (Figure 6.4). Other researchers made similar

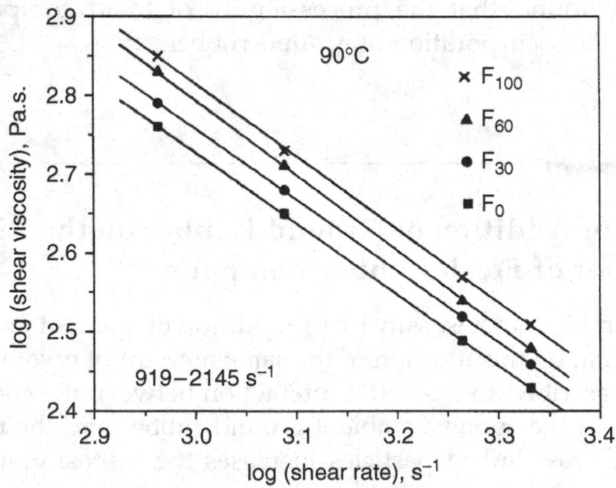

FIGURE 6.2

Apparent shear viscosity of fluororubber compound containing fluororubber powder in the shear rate range of 919–2145 s^{-1} at 90°C. [Shear viscosity is in Pa.s and shear rate is in s^{-1}.] Subscript in mix designation indicates the amount (phr) of powdered rubber in the composition.[64]

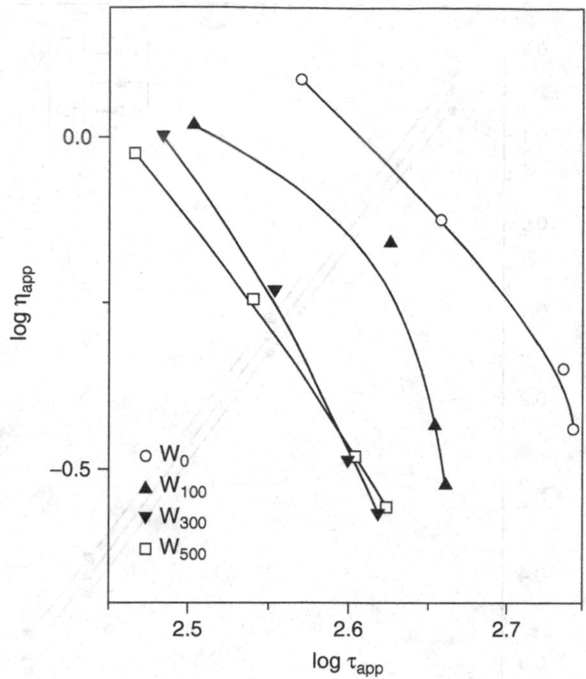

FIGURE 6.3
Plots of apparent shear viscosity of EPDM compound containing EPDM powdered rubber in the shear rate range of 306–1533 s^{-1} at 90°C. [Shear viscosity (η) is in kPa.s and shear stress (τ) is in kPa.] Subscript in mix designation indicates the amount (phr) of powdered rubber in the composition.[61]

observations on the Mooney viscosity and shear viscosity of powdered rubber–filled NR and SBR compounds.[69,70]

It is known that the extrudability of raw rubber and gum rubber compound is poor and that the addition of filler, especially reinforcing fillers like carbon black, drastically improves the extrudate characteristics. Ground rubber has been found to reduce the distortion of the extrudates.[71] Figure 6.5 shows the highly distorted extrudate of the gum SBR compound and the effect of addition of 50 phr of powdered EPDM vulcanizate on the extrudate characteristics. Addition of ground rubber to gum rubber compound broadens the shear rate range in which an undistorted extrudate can be obtained. In the case of carbon black-filled matrix compounds, addition of ground rubber slightly reduces the surface smoothness because the cross-linked particles are coarser than particles of the carbon black.[62]

Incorporation of ground rubber vulcanizate into a gum rubber compound reduces the die swell. The reduction in die swell by the addition of cross-linked particles may be due to the lower proportion of the fresh polymer chains, which can uncoil in a shear field and then recoil outside the shear field. In other words, contribution of the cross-linked polymer chains in the particles to die swell will be low due to the restriction to uncoil. A high

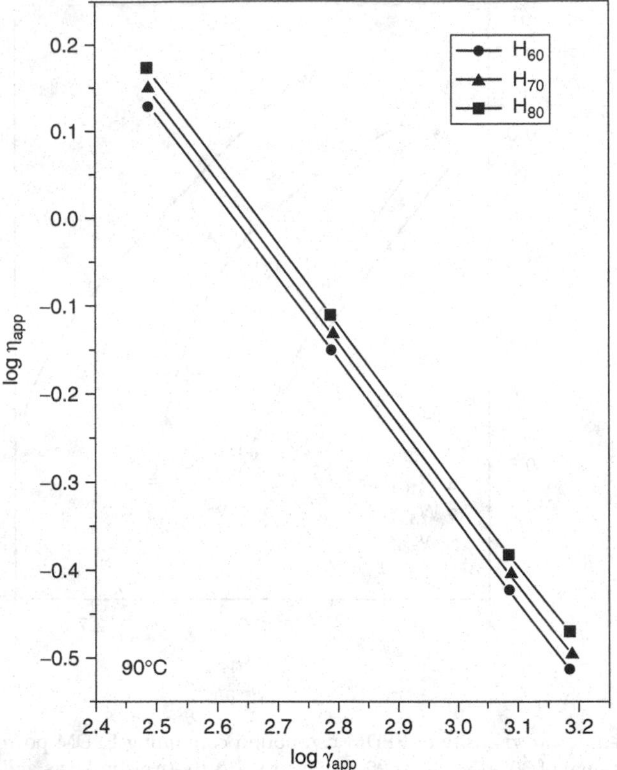

FIGURE 6.4
Effect of filler content in the powdered rubber on the apparent shear viscosity of powdered rubber–filled EPDM compound at 90°C and shear rate range of 306–1533 s^{-1}. [Shear viscosity is in kPa.s and shear rate is in s^{-1}.] Subscript in mix designation indicates the hardness (Shore A) of the precursor vulcanizate used for making the powdered rubber.[83]

degree of interaction between the polymer and the cross-linked particles, especially when the cross-linked particles are ultra-fine, highly filled, and have rough and convoluted surface, may also lead to reduced die swell. Marginal decrease in die swell has been observed with increase in filler content of particles (that is, for 8% increase in filler content, the decrease in die swell is 4% maximum in a shear rate range of 306 to 1533 s^{-1}).[55]

6.10 Effect of Addition of Ground Rubber on the Curing Characteristics of Fresh Rubber Compounds

Addition of ground rubber affects the scorch time, optimum cure time, rate of cure, and state of cure of the rubber compound, depending upon the

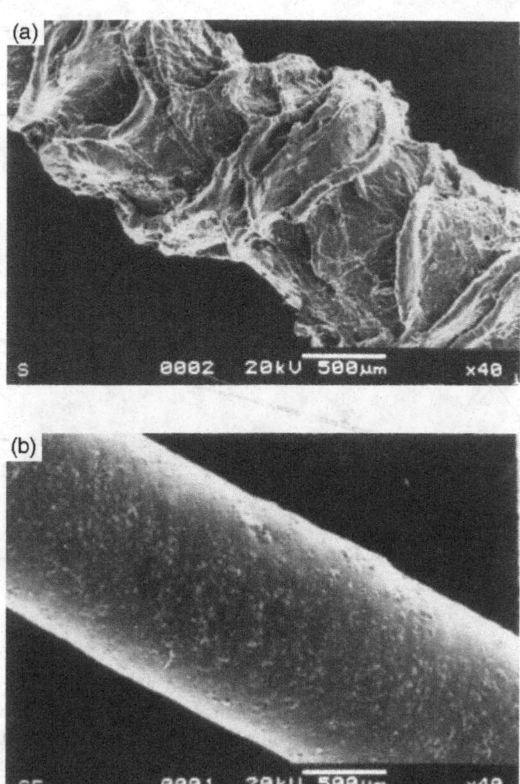

FIGURE 6.5
SEM photomicrographs of extrudates at 110°C and 613 s^{-1}. (a) Gum SBR compound and (b) powdered EPDM–filled (50 phr) SBR compound.[71]

nature of the polymer in the matrix and the characteristics of the rubber powder.[68,72,73] Accelerated-sulfur-cured particles have been found to reduce the scorch safety of the compound, presumably due to the migration of accelerator fragments from the ground vulcanizate to the matrix rubber.[60,74] According to the proposed mechanism, fresh sulfur migrates to the cross-linked particles from the matrix and cures the particles again. As a result of the second-step curing of the particles, accelerator fragments are released into the matrix from the particles. The assumption that scorch time is reduced due to accelerators is confirmed by studies using peroxide-cured particles, when no reduction in scorch time is detected.[74] It has been found that the particles reduce the scorch time even after solvent extraction of the powdered rubber, implying that the accelerators are bound to the rubber network reversibly and are released during the mixing or curing only.[74] It has also been found that the concentration of accelerators migrating to the matrix depends upon the aging conditions to which the vulcanizate is subjected before grinding. In other words, it depends upon the storage and service conditions of the postconsumer goods from which the ground rubber is

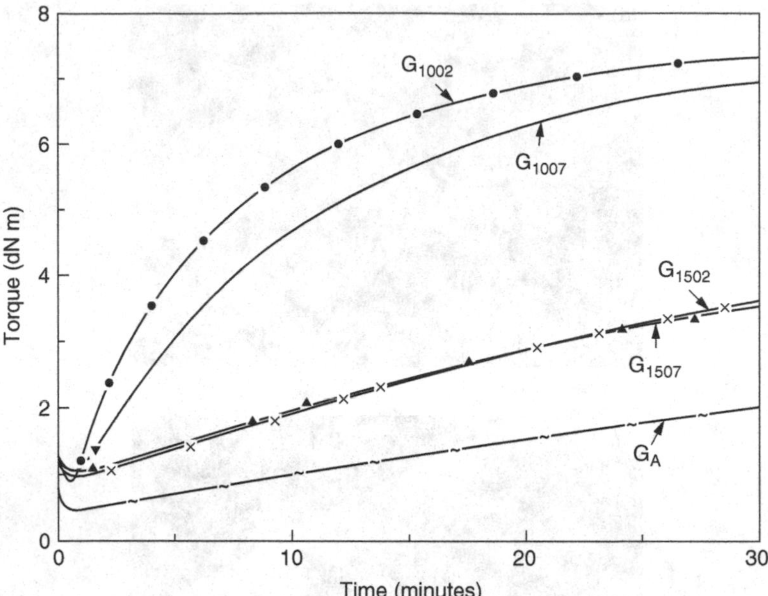

FIGURE 6.6
Rheographs of the compounds containing no additional accelerators (G_A is the control gum EPDM compound [ZnO, 5; stearic acid, 1: and sulfur, 1.5 phr] without powdered rubber; other compositions are similar to G_A, but contain 100 phr of powdered rubber subjected to different aging conditions. First three digits in the subscript of the mix number indicate the temperature of aging in degree Celsius and the fourth digit indicates the number of days of aging of the vulcanizate before grinding).[75]

prepared.[75] This is illustrated by the rheographs of gum EPDM compounds without accelerator, but with and without powdered rubber aged at different conditions. Δtorque (the difference in values of maximum (M_H) and minimum (M_L) rheometric torques [$M_H - M_L$]) values for different compounds indicate that the migration becomes less prominent with increased duration and temperature of aging of the precursor rubber sheet from which the powdered rubber is prepared (Figure 6.6).

As can be seen in Table 6.4, the change in scorch time is less in the case of filled matrix compound than in the gum matrix.[19,20,61,62,71] This may be due to the lesser flexibility of the polymer chains and the longer path length for the migration, and also to the possible adsorption of accelerators on the surface of the carbon black.[76] Furthermore, when the matrix rubber has higher affinity for curatives than the cross-linked particles, the effect is drastic, as can be seen in the case of EPDM particles in SBR matrix.[71] An increased viscosity of the compound by the addition of cross-linked particles may also have an effect on the scorch time of the compound, since higher viscosity leads to higher heat generation during shear. Another possible reason for the reduced scorch time is the ease with which the bound accelerators may form co-cross-links with the matrix rubber.

TABLE 6.4

Mooney Scorch Time (in Minutes) for Powdered (Abraded) Rubber-Filled Compositions[a]

Abraded EPDM[b] (Carbon Black-Filled) Powder in			Fluororubber Powder[c] in Gum Fluororubber (60 phr)[64]	Silicone Powder[c] in Gum Silicone (60 phr)[20]
Gum EPDM (50 phr)[19]	Filled EPDM (50 phr)[62]	Gum SBR (50 phr)[71]		
13.0 (33.0)	8.0 (9.9)	29.3 (> 90)	20.0 (16.0)	95.0 (76.0)

[a] Values in brackets stand for the corresponding compounds without powdered rubber (at 120°C for EPDM matrix and 100°C for silicone and SBR).
[b] Sulfur-cured rubber particles.
[c] Peroxide-cured rubber particles.

As seen in Table 6.4, the scorch time of vulcanizates containing peroxide-cured powdered rubber is slightly higher than the compound without powdered rubber.[20,64] This is ascribed to the dilution of vulcanizing agent (peroxide) as the amount of polymer for a fixed amount of peroxide increases in the compound by the addition of the powdered rubber.

It has also been found that the maximum rheometric torque and the Δ torque decrease with the addition of ground rubber.[61,62] Figure 6.7 shows the effect of the addition of powdered EPDM vulcanizate on the minimum rheometric torque (M_L), maximum rheometric torque (M_H), and the cure rate of an EPDM compound containing high amounts of carbon black and oil. Similar observations have been made with other rubber matrices, both gum and carbon black-filled.[19,60,71] The reasons suggested for the decreased M_H value are the plasticizer (acetone extractables) migration from the ground

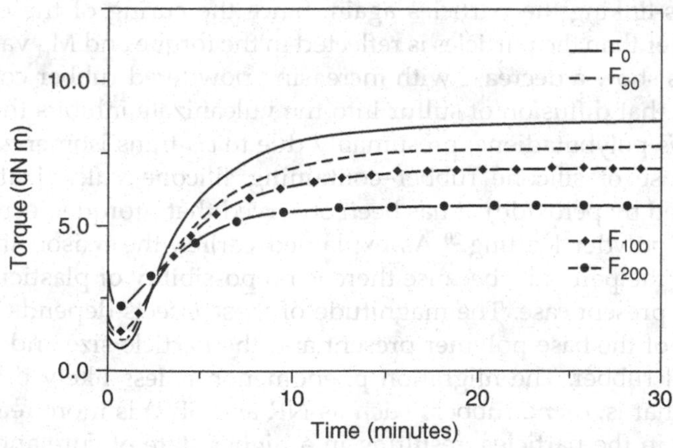

FIGURE 6.7
Effect of ground rubber concentration on the rheometer torque (F_0 is the control compound without powdered rubber; other compositions contain powdered EPDM rubber. The subscript in the mix designation indicates the amount [phr] of powdered rubber in the composition).[62]

TABLE 6.5

Cure Characteristics of NR Compounds with 40 phr Cryoground Rubber[a]

Cure Characteristics[b]	Mix Designation		
	G	G_C	G_A
M_L, N.m	1.13	1.41	1.52
M_H, N.m	9.28	8.94	9.28
M_H - M_L, N.m	8.15	7.53	7.76

[a] G is the control gum compound, G_C and G_A are similar to G but contain cryoground rubber before and after acetone extraction.

[b] Determined using Monsanto rheometer (R-100) at 150°C.

Source: From Phadke, A.A., Bhowmick, A.K., and De, S.K., *J. Appl. Polym. Sci.* 32, 4063, 1986. With permission.

rubber to the matrix rubber and migration of free sulfur from the matrix rubber to the cross-linked particles (ground rubber) because there exists a concentration gradient between the cross-linked and matrix rubber phases.[60,74] Rheometric studies with compounds containing acetone-extracted powdered rubber illustrate this. As can be seen in Table 6.5, the compound containing acetone-extracted cryoground rubber (CGR) shows higher Δtorque (M_H - M_L) than that containing the nonextracted one, which indicates the role of plasticizers in the lowering of Δtorque.[60] But the Δtorque of the compound with extracted CGR is also less than the unfilled compound, indicating the role of sulfur migration from the matrix rubber to the cross-linked particles, reducing the extent of cross-linking in the matrix rubber while cross-linking the particles again. Since the curing of the continuous matrix rather than the particles is reflected in the torque and M_H values, these parameters show a decrease with increasing powdered rubber content. It is also found that diffusion of sulfur into the vulcanizate inhibits the crystallization of *cis*-polybutadiene, presumably due to *cis*-trans isomerization.[77]

In the case of silicone rubber–containing silicone vulcanizate powder (cross-linked by peroxide), it has been observed that Δtorque decreases with increasing powder loading.[20] As explained earlier, the reason may be the dilution of the peroxide, because there is no possibility of plasticizer migration in the present case. The magnitude of these effects depends mainly on the nature of the base polymer present and the particle size and loading of the ground rubber. The migration phenomenon is less likely if the matrix polymer (that is, diene rubbers such as NR and SBR) is more reactive than the rubber in the particles, resulting in a higher state of cure and cure rate of the compound. Also, migration will be less when the particle surface area is low.

It is reported that the migration of sulfur depends on the activity of vulcanization accelerators and the catalytic effect of carbon black on vulcanization and its interaction with the accelerator.[78] Addition of accelerators, which can provide fast cure rate, can reduce the sulfur migration.[79] Use of additional sulfur can also reduce the effect, due to sulfur migration to the cross-linked particles. The actual amount of sulfur needed depends upon the curing system used. Adjustment is generally based on the rubber hydrocarbon content of the ground rubber and the amount of powdered rubber added.[54] Blooming problems have not been encountered with the extra sulfur used in powdered rubber compounds. Reinforcement of the matrix is another method. It is reported that the diffusion of sulfur is considerably lower in carbon black–filled matrix than in unfilled compounds, for the reasons described earlier in the case of accelerators.[78] Thus, migration of curatives imposes restrictions in optimizing the cure time and the curative dosage in ground rubber–filled compositions. Ishiaku and co-workers[80] studied the effect of convoluted rubber vulcanizate powder, obtained from the sanding process of polishing rubber balls, on the properties of NR compound. It was observed that scorch time and cure time decrease and the tendency toward reversion increases with increasing powder concentration.

6.11 Effect of Addition of Ground Rubber on the Mechanical Properties of Rubber Vulcanizates

There are several reports on mechanical properties of ground rubber–filled rubber vulcanizates. The factors affecting the performance of a vulcanizate containing powdered rubber are the particle size, hardness, and aging condition of the precursor rubber article; filler content in the particle; and the particle-matrix rubber interaction. Rothemeyer[22] discussed the effects of grinding techniques on the particle size, structure, and distribution of powders obtained from waste rubber, as well as the effects of different powders on the physical properties of the resulting vulcanizates.

Even at 20 phr loading of ground rubber (ambient-ground or cryoground) drastic deterioration in mechanical properties of the rubber vulcanizate is observed. Nah and fellow investigators studied the effect of polynorbornene vulcanizate powder (up to 40 phr) on the properties of NR and polybutadiene rubber (BR). The addition of rubber powder to a low-strength BR compound did not induce any significant changes in the tensile strength. However, in the case of high-strength NR compound, the tensile strength decreased significantly upon addition of the powder.[81] Jacob and co-workers, while studying the effect of carbon black–filled NR and EPDM-abraded powders (< 50 μm) in the respective gum matrices, observed that the tensile strength of NR remains constant, whereas a reinforcing effect is observed in the case of low-strength EPDM rubber at a loading of 50 phr of powdered rubber.[19,82] The abraded EPDM powder shows reinforcing effect in the SBR

TABLE 6.6
Physical Properties of Vulcanizates Containing 50 phr Abraded Powders[a]

Physical Properties	NR Powder (50 phr) in Gum NR Compound[82]	EPDM Powder (50 phr) in Gum EPDM Compound[19]	Filled EPDM Compound[62]	Gum SBR Compound[71]
Tensile strength, MPa	20.1	3.60	6.30	4.66
	(20.8)	(1.50)	(6.37)	(1.59)
Elongation at break, %	868	305	488	428
	(988)	(157)	(489)	(167)
Modulus at 100% elongation, MPa	0.81	1.31	1.85	1.16
	(0.85)	(1.16)	(2.05)	(1.17)
Tear strength, kN/m	31.7	16.5	27.0	19.3
	(29.8)	(8.9)	(32.4)	(11.6)

[a] Values in parentheses stand for the samples without powdered rubber.

matrix also. However, the effect is less prominent in the NR matrix (Table 6.6). While the modulus (at 100% elongation) of NR vulcanizate remains almost constant on addition of 50 phr powdered NR vulcanizate (i.e., from 0.85 MPa to 0.81 MPa), the modulus of NR vulcanizate containing powdered EPDM vulcanizate increases slightly (i.e., from 0.88 MPa to 0.98 MPa) at the same loading.[71,82] This may be due to the higher filler content in EPDM powdered rubber (46%) when compared to the NR vulcanizate powder (31%). Studies on the EPDM matrix rubber containing EPDM powdered (abraded) rubber with different filler contents also demonstrate that the mechanical properties of the vulcanizates show improvement as the filler content in the powdered rubber increases (Table 6.7).[83] It has also been found that the gum powdered rubber cannot reinforce the matrix, indicating the vital role of the filler present in the particles in the matrix reinforcement (Table 6.8).[83] Jacob and fellow investigators also studied the effect of abraded EPDM powder loading on the vulcanizate properties of filled EPDM, NR, and SBR rubbers (Table 6.6 and Table 6.9).[62,71] It is known that NR degrades easily under shear and heat during abrasion, but the stress-strain properties of powdered NR vulcanizate–filled gum NR compound indicate that the degradation has negligible effect on the properties.[82] Furthermore, the ultra-fine size (~20 μm) and the convoluted and rough surfaces of the particles (higher surface area) provide high interaction with the matrix rubber.

Burford and Pittolo[84] studied the effect of hardness of the rubber powder (ambient-ground, using shear mill equipped with 1-mm size screen) on the failure properties of unfilled and carbon black–filled butadiene rubber compounds. Hardness of the powdered rubber has been varied by varying the filler content or the dosage of the curing agent. It is found that increasing hardness of the rubber powder causes progressive reduction in tensile strength and elongation at break, but increase in modulus. Kim studied the fracture behavior of crumb rubber–filled elastomers under a tensile load.[85] He reported that, while in the case of unfilled vulcanizate the fracture starts

TABLE 6.7

Effect of Filler Content in the Powdered Rubber on the Properties of Powdered Rubber–Filled Vulcanizates

Property	Mix Designation			
	H[a]	H_{60}	H_{70}	H_{80}
Tensile strength, MPa	1.50	5.71	6.63	7.04
Elongation at break, %	157	410	405	345
Modulus at 100% elongation, MPa	1.17	1.35	1.53	1.89
Modulus at 300% elongation, MPa	—	3.86	4.54	5.92
Tear strength, kN/m	8.9	18.8	23.0	29.8
Hardness, Shore A	45	47	50	53
Tension set (at 100% elongation), %	b	2	2	2
Hysteresis loss, ($\times 10^{-6}$), J/m^2	0.003	0.007	0.010	0.015
Abrasion loss, cc/hr	0.35	0.95	0.86	0.78

[a] Formulation H contains EPDM, 100; ZnO, 5; stearic acid, 1; TMTD, 1; MBT, 0.5; and sulfur, 1.5. In addition to the above, H_{60}, H_{70}, and H_{80} contain 100 phr of powdered (abraded) EPDM vulcanizate prepared from window seals of hardness 60, 70, and 80 Shore A, respectively, and filler content of 46, 48, and 54%, respectively.

[b] Could not be determined due to low elongation at break of the sample.

Source: From Jacob, C., Ph.D. thesis, 2002. With permission.

TABLE 6.8

Effect of Filler Content in the Ground Rubber on the Mechanical Properties of Ground Rubber–Filled Rubber Vulcanizates[83,a]

Property	Mix Designation		
	G_0[b]	G_{100}[c]	C_{100}[c]
Tensile strength, MPa	1.50 (1.27)	5.50 (5.43)	1.63 (1.61)
Elongation at break, %	157 (107)	415 (229)	234 (186)
Modulus at 100% elongation, MPa	1.16 (1.27)	1.33 (2.10)	1.04 (1.15)
Tension set at 100% elongation, %	—	2	2

[a] Values in parentheses stand for the aged samples (150°C for 72 hr).

[b] G_0 contains EPDM, 100; ZnO, 5; stearic acid, 1; TMTD, 1; MBT, 0.5; and sulfur, 1.5 (in phr). In addition to the above, G_{100} and C_{100} contain 100 phr of powdered EPDM rubber.

[c] Mix C_{100} is equivalent to G_{100}, except that the W-EPDM used here is gum powder.

Source: From Jacob, C., Ph.D. thesis, 2002. With permission.

TABLE 6.9

Mechanical Properties of Powdered (Abraded) Rubber–Filled Rubber Vulcanizates Containing Carbon Black[a,b]

Property	Mix Designation						
	EPDM Matrix[62]			SBR Matrix[71]		NR Matrix[71]	
	F_0	F_{100}	F_{200}	S_0	S_{50}	N_0	N_{50}
Tensile strength, MPa	6.37	6.67	6.33	7.90	11.51	19.12	21.09
Elongation at break, %	489	518	472	280	410	697	735
Modulus at 100% elongation, MPa	2.06	1.77	1.69	2.55	2.37	1.65	1.70
Tear strength, kNm^{-1}	32.4	25.6	22.8	42.1	38.3	62.4	48.8
Hardness, Shore A	57	54	51	56	57	45	47
Abrasion loss, cc/hr	2.5	2.8	3.1	0.93	1.06	1.78	1.99
Resilience, %	43	47	46	79	72	84	79
Hysteresis loss, ($\times 10^{-6}$), J/m^2	0.036	0.027	0.026	0.011	0.014	0.006	0.011
Tension set at 100% elongation, %	4	4	4	0	2	2	3

[a] The subscript in the mix designation indicates the amount (phr) of powdered EPDM rubber in the compound.

[b] Formulations: F series contains EPDM, 100; ZnO, 5; stearic acid, 1; N660 carbon black, 120; paraffinic oil, 70; paraffin wax, 10; brown factice, 10; TMTD, 1.2; MBT, 1.2; ZDC, 2.0; and sulfur, 1.5; S series contains SBR, 100; ZnO, 5; stearic acid, 2; N550 carbon black, 30; aromatic oil, 3; TMTD, 0.5; Mercapto benzothiazyl sulfenamide (MBTS), 1.5; and sulfur, 2.5; and N series contains NR, 100; ZnO, 4; N550 carbon black, 30; aromatic oil, 3; stearic acid, 2; CBS, 1.75; and sulfur, 1.75.

from the surface, it initiates from a cavity around the particle in the rubber powder–filled vulcanizates. While the fracture surface of gum EPDM vulcanizate shows a fish-tail feature, characteristic of brittle fracture, powdered rubber–filled vulcanizate shows the characteristics of carbon black–reinforced matrix (Figure 6.8).[61]

Klingensmith and co-workers[9] reported that cryogenically ground SBR and EPDM vulcanizates, when incorporated in the respective rubber compounds, results in deterioration of mechanical properties. The extent of deterioration increases with increase in particle size[86–88] and loading of ground rubber.[89] Moreover, with increasing aging severity of the precursor rubber article (used in making the powder), the deterioration in properties becomes prominent (Table 6.10).[75] This can be due either to the degradation of the rubber in the particles, or to the increased cross-link density or hardness of the particles, which facilitates an easy pull out from the matrix rubber under a tensile load. Swor and fellow investigators also studied the effect of loading of fine rubber crumb, prepared by mechanical means, on the properties of SBR compound.[54] At high loadings (i.e., 150 phr) the drop in properties is drastic. According to Fujimoto and co-workers,[90] sulfur in the rubber matrix phase diffuses into the powder phase, lowering the matrix cross-link density, which in turn affects the mechanical properties. Makarov and fellow investigators suggested that when both the ground rubber and the matrix rubber have identical types of cross-linking (i.e., either monosulfide or polysulfide

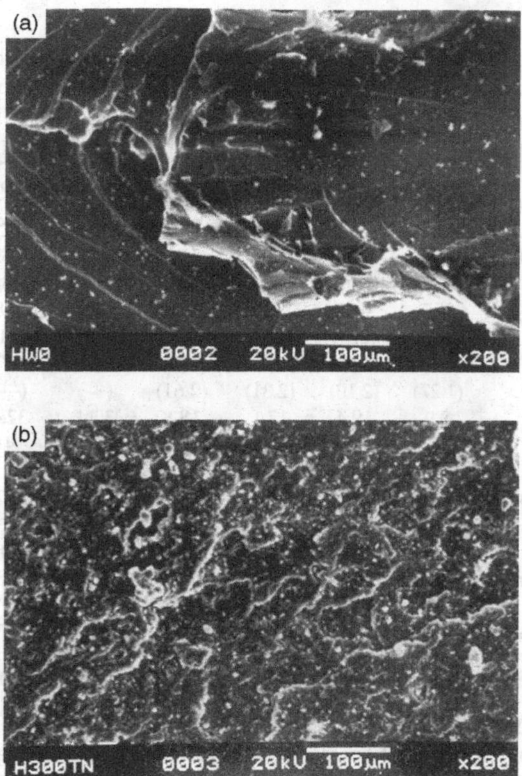

FIGURE 6.8
SEM photomicrographs of the tensile fractured surface of vulcanizates. (a) Gum vulcanizate, (b) with powdered rubber.[61]

cross-links), the resulting vulcanizates have increased tensile strength, fatigue life, and wear resistance.[91]

The cross-link density and modulus of the crumb (powder)-filled material is a function of the previous history of vulcanization and composition of the powdered rubber, as well as that of the matrix. When the crumb is added to NR compound, stress-strain behavior changes and at intermediate strains (150 to 300%), the powder stiffens the compound.[23] Powdered EPDM vulcanizate has been incorporated up to a loading of 500 phr in gum EPDM compound, and the mechanical properties of the vulcanizates indicate that the optimum loading is 300 phr.[61] Although 100 phr of powdered rubber contributes to a slight increase in modulus (at 100% elongation) of the vulcanizate, further addition up to 500 phr results in negligible changes only. At higher elongation (i.e., at 300% elongation) the modulus increases continuously, implying that the contribution of carbon black content in the particles to the modulus is evident only at higher strains. The tension set at 100% elongation for the vulcanizates is low and is independent of the powder loading (Table 6.11). The stress-strain curves for these vulcanizates are given

TABLE 6.10

Effect of Aging Conditions of the Precursor Vulcanizate Used for the Preparation of Powdered Rubber on the Properties of Powdered Rubber–Filled Rubber Vulcanizates[a]

	Mix Designation							
	Gum EPDM Matrix Compounds				Carbon Black–Filled Matrix Compounds			
Property	G	G_{1002}	G_{1007}	G_{1507}	F	F_{1002}	F_{1007}	F_{1507}
Tensile strength,[b] MPa	1.50 (1.27)	5.50 (5.52)	4.83 (4.73)	4.10 (3.50)	6.37 (7.57)	6.67 (6.30)	6.62 (7.51)	5.10 (7.02)
Elongation at break,[b] %	157 (107)	415 (229)	391 (177)	294 (129)	489 (88)	518 (94)	535 (129)	387 (109)
Modulus at 100% elongation,[b] MPa	1.16 (1.27)	1.30 (2.10)	1.29 (2.31)	1.46 (2.61)	2.06 (—)	1.77 (—)	1.81 (5.93)	1.78 (6.60)
Tear strength, kN/m	8.9	19.4	17.1	18.8	32.4	32.4	30.5	27.2
Tension set at 100% elongation, %	[c]	2	2	2	4	4	4	2
Hysteresis loss, $\times 10^{-6}$, J/m^2	0.003	0.008	0.008	0.011	0.036	0.027	0.030	0.030
Hardness, Shore A	45	45	46	47	57	54	56	54
Abrasion loss, cc/hr	0.35	1.1	1.0	1.2	2.5	2.8	2.7	2.8
Rebound resilience, %	75	67	66	63	43	47	46	49
Heat buildup, °C	1	6	5	8	15	17	16	15

[a] Formulation G contains EPDM, 100; ZnO, 4;stearic acid, 1; TMTD, 1, MBT, 0.5; and sulfur 1.5. Formulation F contains EPDM, 100; ZnO, 4; stearic acid, 1; GPF carbon black, 120; paraffinic oil, 70; TMTD, 1.2; MBT, 1.2; ZDC, 2.0; and sulfur 1.5. G_{1002}, G_{1007}, and G_{1507} and F_{1002}, F_{1007}, and F_{1507} contain 100 phr of powdered rubber in addition to the ingredients in G and F respectively. First three digits in the subscripts of mix designation indicate the temperature of aging in degree Celsius and the fourth digit indicates the number of days of aging.

[b] Values in parentheses stand for the samples after aging (150°C/ 3 days).

[c] Could not be determined due to low elongation at break.

Source: From Jacob, C., et al.[75] (communicated to *Kautschuk Gummi Kustst.*) With permission.

in Figure 6.9. When the powdered EPDM vulcanizate is added as a filler to a highly filled compound, marginal decrease in the modulus is observed, as can be seen in Table 6.9, which may be due to lesser interaction between the matrix and the powdered rubber than that between the matrix and fresh carbon black.[62]

Although the addition of ground rubber in a rubber matrix decreases the tensile strength, the trouser tear strength increases.[74] It is proposed that the ground rubber acts as a stress-raising flaw in a tensile test, but promotes crack tip blunting in the tear test. This can be due to the fact that a stress concentration can lead to a lower strength in tensile test, whereas a stress raiser, even a cavity, can blunt the propagating tear tip in a tear test. The higher force fluctuations in ground rubber–filled vulcanizate indicate a higher stick-slip tendency. Gibala et al. attribute the higher tear strength to the greater maximum tear force fluctuations. In the case of powdered rubber–filled gum EPDM vulcanizate, it is found that the tear strength (angle specimens) and the tensile strength continuously increase up to a loading of

TABLE 6.11

Effect of Powdered Rubber (Abraded) Loading on the Mechanical Properties of Powdered Rubber–Filled Vulcanizates[a,b]

Property	Mix Designation			
	W_0	W_{100}	W_{300}	W_{500}
Tensile strength, MPa	1.50 (1.26)	5.66 (5.65)	11.50 (10.33)	9.61 (11.31)
Elongation at break, %	157 (107)	411 (264)	649 (335)	544 (346)
Modulus at 100% elongation, MPa	1.17 (1.20)	1.35 (1.95)	1.31 (2.56)	1.40 (2.86)
Modulus at 300% elongation, MPa	—	3.86 (—)	4.08 (9.13)	4.59 (9.51)
Heat build-up (ΔT) at 50°C, °C	1	10	12	13
Hysteresis loss (×10^{-6}), J/m^2	0.003	0.007	0.010	0.012
Rebound resilience, %	75	70	67	64
Tension set (at 100% elongation), %	c	2	2	2
Tear strength, kN/m	8.9	18.8	27.3	26.7

[a] Values in parentheses stand for the aged specimens (150°C for 72 hr).
[b] Formulations contain EPDM, 100; ZnO, 5; stearic acid, 1; TMTD, 1; MBT, 0.5; and sulfur, 1.5 (in phr), in addition to the powdered EPDM rubber. The subscript in the mix designation indicates the amount of powdered rubber (in phr) present in the composition.
[c] Could not be determined; the sample broke.

Source: Jacob, C. et al., *Rubber Chem. Technol.*, 76, 36, 2003. With permission.

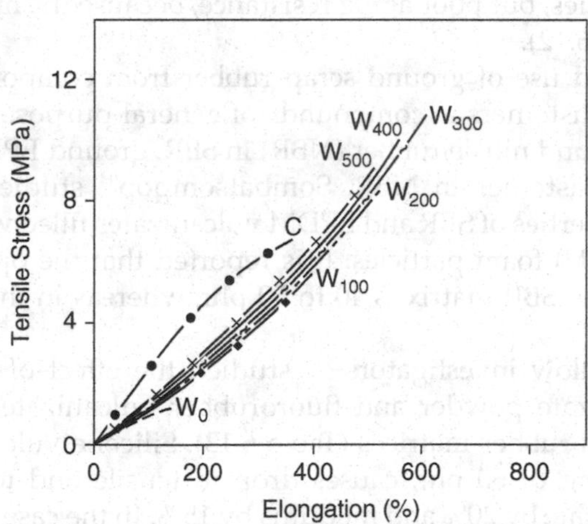

FIGURE 6.9
Stress-strain plots of powdered EPDM–filled EPDM vulcanizates. The subscripts in the mix designation indicate the amount (phr) of powdered rubber in the composition.[61] C = vulcanizate for making powdered rubber.

TABLE 6.12

Properties of NR Vulcanizates Containing 30 Phr of GRT[a]

Physical Properties	G_0[b]	G_1[c]	G_2[d]
Tensile strength, MPa	14.0	4.2	8.0
	(8.8)	(2.5)	(2.5)
Elongation at break, %	1175	620	860
	(770)	(360)	(400)
Tear strength, kN/m	28.2	18.5	21.1
	(20.3)	(10.6)	(9.7)

[a] Values in parentheses stand for the aged samples (100°C for 36 hours).
[b] Compound without GRT.
[c] Particle size 52–72 mesh (300–215 μm)
[d] Particle size 100–150 mesh (150–100 μm)

Source: From Naskar, A.K., Pramanik, P.K., Mukhopadhyay, R., De., S.K., and Bhowmick, A.K., *Rubber Chem. Technol.*, 73, 902, 2000. With permission.

300 phr (Table 6.11). The increase in tearing energy with the addition of ground rubber may be attributed partly to a less cross-linked matrix, due to sulfur migration from the matrix rubber to the cross-linked particles.[61]

Naskar and co-workers[92] studied the effect of GRT of different particle sizes on the properties of NR compound. It is found that smaller particles contain less amount of polymer (i.e., NR, BR, SBR) but higher amount of fillers (i.e., carbon black, silica) and metals (i.e., copper, manganese, iron). Hence, the NR compounds containing smaller GRT particles show higher physical properties, but poor aging resistance, because the metals act as pro-oxidants (Table 6.12).

Fesus[93] studied use of ground scrap rubber from compounds of higher-cost, specialty elastomers in compounds of general-purpose rubbers. Examples include ground nitrile rubber (NBR) in SBR, ground EPDM in SBR, and ground fluoroelastomer in NBR. Sombatsompop[94] studied the dynamic mechanical properties of SBR and EPDM vulcanizates filled with cryoground polyurethane (PU) foam particles. It is reported that the optimum amount of particles in the SBR matrix is 40 to 50 phr, whereas in the EPDM matrix it is 30 phr.

Ghosh and fellow investigators[20,64] studied the effect of the addition of silicone vulcanizate powder and fluororubber vulcanizate powder in the respective virgin rubber matrices (Table 6.13). Silicone vulcanizate powder, even at a loading of 60 phr, causes drop in tensile and tear strengths of silicone vulcanizate by 20% and modulus by 15%. In the case of fluororubber, the tensile and tear strengths drop by 15% with the addition of 100 phr of fluororubber vulcanizate powder. It is evident that the fall in properties is marginal when abraded particles are used, owing to the very fine particle size and rough surface obtained by the grinding technique.

TABLE 6.13

Mechanical Properties of Powdered (Abraded) Rubber–Filled Compositions [a]

Properties	Silicone Rubber Compounds[20,b]			Fluororubber Compounds[64,c]		
	S_0	S_{30}	S_{60}	F_0	F_{30}	F_{100}
Tensile strength, MPa	7.9	6.6	6.3	6.3	7.6	7.6
Elongation at break, %	212	190	179	834	803	726
Modulus at 100% elongation, MPa	4.1	3.9	3.5	0.8	0.9	1.0
Tear strength, kN/m	24.1	21.0	19.1	23.2	21.8	19.4
Hardness, Shore A	74	72	71	40	41	44
Tension set at 100% elongation, %	4	5	6	8	10	8
Hysteresis loss, ($\times10^{-6}$), J/m^2	0.060	0.055	0.054	0.010	0.012	0.011

[a] The subscripts in the mix designation indicate the amount of powdered rubber (phr) in the composition.

[b] Silicone rubber compounds contain silicone rubber, 100; Dicumyl peroxide, 2; and powdered silicone rubber vulcanizate (variable).

[c] Fluororubber compounds contain fluororubber, 100; Dicumyl peroxide, 2; Triallylcyanurate, 5; $Ca(OH)_2$, 5; and powdered fluororubber vulcanizate (variable).

Addition of ground rubber (filled and unfilled) reduces the abrasion resistance of the vulcanizates, which may be due to the reduced cross-link density of the rubber matrix, facilitating an easy pull out of the particles under the continuous shear force (Table 6.9 and Table 6.10). The effect is pronounced as the particle size increases.[95] When the powdered rubber is added to a gum rubber compound, the hardness of the resulting vulcanizate increases due to the carbon black filler present in the ground vulcanizate. But when powdered rubber is added to a highly filled compound, the hardness decreases. Resilience of the vulcanizates decreases when the powdered rubber is added to the gum compound, whereas it increases in the case of filled compounds. A similar trend is found in the case of hysteresis loss (Table 6.9 and Table 6.10). The contrasting behavior of the gum and highly filled matrices may be due to the lesser interaction of the cross-linked particles with the matrix rubber in the presence of a large amount of finer reinforcing carbon black already present in the filled matrix. Fatigue resistance of a carbon black–filled rubber vulcanizate is improved by the addition of cured particulate rubber, when the rubber crumb is added at 10 to 40 phr after the carbon black addition is completed.[96,97]

In highway testings, it is observed that ultra-fine rubber powder–filled tire tread compounds exhibit marked improvements in their resistance to groove cracking, compared to those without powdered rubber.[56] Increase in dynamic flexing strength by the addition of ground vulcanizate has also been reported.[98,99] Effect of the degree of mechanical degradation of sulfur vulcanized butadiene rubber wastes on the tensile strength and fatigue strength of tread rubber containing 10% waste has been studied by Solov'ev and co-workers.[99] Adhesive interaction of vulcanizate particles with the rubber matrix is increased by the polymer degradation.

The physical properties of EPDM foam and SBR foam containing ground rubber (i.e., SBR-natural rubber blend, as in waste tire) and carbon black as a filler were studied by Seo et al.[100,101] The blowing ratio, tensile strength, and elongation decrease rapidly at > 30 phr of ground rubber in EPDM. EPDM foams showed excellent thermal stability. In the case of SBR foam, for the same amount of filler, ground rubber is better than $CaCO_3$ in terms of tensile strength and elongation at break of the compounds. The fracture surface becomes rougher as the amount and particle size of the ground rubber increase.

6.12 Effect of Addition of Powdered Rubber on the Properties of Rubber Blends

Ghosh and co-workers[102] studied the replacement of individual rubbers with the corresponding vulcanizate powders in blends of silicone rubber and fluororubber (Table 6.14). Due to the difference in viscosity of the rubbers, the morphology of the blends without powdered rubber is that of a disperse fluororubber phase in a continuous silicone matrix. When the fluororubber is replaced by its powdered rubber (abbreviated as FVP), negligible difference in properties occur, whereas the drop in properties is higher when the continuous silicone rubber is replaced by its vulcanizate powder (abbreviated as SVP). As in the case of single matrix, the addition of cross-linked particles results in lowering of maximum rheometric torque. Processability

TABLE 6.14

Mooney Viscosity and Mechanical Properties of the Blends Containing Silicone Rubber Powder (SVP) and Fluororubber Powder (FVP)

Properties	Ratio of Silicone Rubber/SVP/Fluororubber/FVP		
	100/0/100/0[a]	50/50/100/0[b]	100/0/50/50[c]
ML(1+4) 120°C	42	89	44
Tensile strength, MPa	6.6	5.4	7.6
Elongation at break, %	302	301	313
Modulus at 100% elongation, MPa	3.5	2.5	3.3
Tear strength, kN/m	28.3	27.2	26.0
Hardness, Shore A	70	62	69
Tension set at 100% elongation, %	8	7	8

[a] Dicumyl peroxide, 4 phr; Triallylcyanurate 5; and calcium hydroxide [$Ca(OH)_2$], 5.
[b] Dicumyl peroxide, 3 phr; Triallylcyanurate 5; and calcium hydroxide [$Ca(OH)_2$] 5.
[c] Dicumyl peroxide, 3 phr; Triallylcyanurate 2.5; and calcium hydroxide [$Ca(OH)_2$] 2.5.

Source: From Ghosh, A., Antony, P., Bhattacharya, A.K., Bhowmick, A.K., and De, S.K., *J. Appl. Polym. Sci.* 82, 2326, 2001. With permission.

studies using Monsanto Processability Tester at higher shear rates indicate that the effect of fluororubber replacement by FVP on the shear viscosity is marginal, whereas the shear viscosity increases with increasing level of silicone rubber replacement by SVP.

References

1. Isayev, A.I., Rubber recycling, in *Rubber Technologist's Handbook,* S. K. De and J. R. White. eds., Rapra Technology Ltd., Shawbury, Shropshire, U.K., 2001, chapter 15.
2. De, S.K., Re-use of ground rubber waste – a review, *Prog. Rub. Plast. Technol.,* 17(2), 113, 2001.
3. Anonymous, Goodyear Tire and Rubber Co., Akron, USA, Process of preparing low cost utility pads from scrap rubber, *Res. Discl.,* 136, 21, 1975.
4. Grebekina, Z.I., Zakharov, N.D., and Makarov, V.M., Study of the effect of the nature of rubber phases on properties of rubber-ground vulcanizate dispersed systems, *Izv. Vyssh. Uchebn. Zaved, Khim. Khim. Tekhnol.,* 23(9), 1161, 1980.
5. Drozdovskii, V.F., Grinding of rubber-containing wastes and the properties of ground vulcanizates, *Kauchuk i Rezina,* 1, 36, 1993.
6. Slusarski, L. and Sendlewski, R., Effect of grinding methods on scrap rubber powder properties, *Gummi Asbest Kunstst,* 34(12), 783, 1981.
7. Drozdovski, V.F., Production of comminuted vulcanisates, *Prog. Rub. Plast. Technol.,* 14, 116, 1998.
8. Adhikari, B., De, D., and Maiti, S., Reclamation and recycling of waste rubber, *Prog. Polym. Sci.,* 25, 909, 2000.
9. Klingensmith, W. and Baranwal, K., Recycling of rubber — an overview, *Rubber World,* 218(3), 41, 1998.
10. Stark, J., in paper "The Tire Cycle Solution: Minnesota's Answer to the Scrap Tire Problem," Tire Technology Conference, Clemson University, Oct. 28–29, 1987.
11. Leyden, J., Cryogenic processing and recycling, *Rubber World,* 203(6), 28, 1991.
12. Rouse, M.W., Development and application of superfine tire powders for compounding, *Rubber World,* 206(3), 25, 1992.
13. Kulakova, G.P., Zelenetskii, S.N., Prut, Z.V, and Enikolopyan, N.S., Properties of finely ground rubber vulcanizate, *Kozh.-Obuvn. Promst,,* 6, 40, 1989.
14. Bilgili, E., Arastoopour, H., and Bernstein, B., Pulverization of rubber granulates using the solid state shear extrusion process. Part II. Powder characterization, *Powder Technology,* 115(3), 277, 2001.
15. Bilgili, E., Arastoopour, H., and Bernstein, B., Analysis of rubber particles produced by the Solid State Shear Extrusion Pulverization process, *Rubber Chem. Technol.,* 73, 340, 2000.
16. Khait, K. and Torkelson, J.M., Solid-State Shear Pulverization of Plastics: A Green Recycling Process, *Polymer Plastics Technology Engineering,* 38(3), 445, 1999.
17. Khait, K., "Novel Used Tire Recovery Process for Value-added Products," presented at the 145th ACS Rubber Division Meeting, Chicago, Illinois, April 19–22, 1994.

18. Williams, D.E., Morin, J.E., and Farris, R.J., A novel method to recycling scrap tires: High pressure, high temperature sintering, *Rubber Chem. Technol.*, 75, 955, 2002.
19. Jacob, C., Bhowmick, A.K., De, P.P., and De, S.K., Recycling of EPDM waste. I. Effect of ground EPDM vulcanizate on properties of EPDM rubber, *J. Appl. Polym. Sci.*, 82, 3293, 2000.
20. Ghosh, A., Rajeev, R.S., Bhattacharya, A.K., Bhowmick, A.K., and De, S.K., Recycling of silicone rubber waste: effect of ground silicone rubber vulcanizate powder on the properties of silicone rubber compound, *Polym. Eng. Sci.*, 43, 279, 2003.
21. Izyumova, V.I., Zakharov, N.D., Kostrykina, G.I., and Shakh-Paron'yants, A.M., Study of mechanochemical changes in the structure of cis-polybutadiene-based vulcanizates during grinding, *Izv. Vyssh. Uchebn. Zaved., Khim. Khim. Tekhnol.*, 24(5), 634, 1981.
22. Rothemeyer, F., Reprocessing of rubber wastes, *Kautsch. Gummi. Kunstst.*, 46, 356, 1993.
23. Burford, R.P. and Pittolo, M., Characterization and performance of powdered rubber, *Rubber Chem. Technol.*, 55, 1233, 1982.
24. Solov'ev, M.E., Zakharov, N.D., and Goncharenko, T.G., Effect of the method of grinding of vulcanizates on the nature of particle surfaces, *Izv. Vyssh. Uchebn. Zaved., Khim. Khim. Tekhnol.*, 26(3), 356, 1983.
25. Fuhrmann, I. and Karger-Kocsis, J., Promising approach to functionalisation of ground tyre rubber – photochemically induced grafting, *Plast. Rubber Comp.*, 28(10), 500, 1999.
26. Pramanik, P.K. and Baker, W.E., Toughening of ground rubber tire filled thermoplastic compounds using different compatibilizer systems, *Plast. Rubber Comp. Process. Appl.*, 24, 229, 1995.
27. Kim, J.K. and Burford, R.P., Study on powder utilization of waste tires as a filler in rubber compounding, *Rubber Chem. Technol.*, 71, 1028, 1998.
28. Naskar, A.K., De, S.K., and Bhowmick, A.K., Surface chlorination of ground rubber tire and its characterization, *Rubber Chem. Technol.*, 74, 645, 2001.
29. Bauman, B.D., High-value engineering materials from scrap rubber, *Rubber World*, 212(2), 30, 1995.
30. Mahlke, D., Recycling of waste rubber by surface modification, *Kautschuk Gummi Kunstst.*, 46, 889, 1993.
31. Fesus, E.M. and Eggleton, R.W., Recycling rubber products sensibly, *Rubber World*, 203(6), 23, 1991.
32. Dierkes, I.W., Solutions to the rubber waste problem incorporating recycled rubber, *Rubber World*, 214(2), 25, 1996.
33. Pramanik, P.K. and Baker, W.E., LLDPE composites filled with corona discharge treated ground rubber tire crumb, *J. Elast. Plast.*, 27, 253, 1995.
34. Xu, Z., Losure, N.S. and Gardner, S.D., Epoxy resin filled with tire rubber particles modified by plasma surface treatment, *J. Advanced Materials*, 30(2), 11, 1998.
35. Rajalingam, P., Sharpe, J., and Baker, W.E., Ground rubber tire/thermoplastic composites: Effect of different ground rubber tires, *Rubber Chem. Technol.*, 66, 664, 1993.
36. Warner, W.C., Methods of devulcanization, *Rubber Chem. Technol.*, 67, 559, 1994.
37. Myers, R.D. and MacLeod, J.B., Rubber devulcanization process, US Patent 5798394, 1998.

38. Isayev, A.I., Chen, J., and Tukachensky, A., Novel ultrasonic technology for devulcanization of waste rubbers, *Rubber Chem. Technol.*, 68, 267, 1995.
39. Isayev, A.I., Kim, S.H., and Levin, V.Yu., Superior mechanical properties of reclaimed SBR with bimodal network, *Rubber Chem. Technol.*, 70, 194, 1997.
40. Parasiewiez, W., "Microwave recycling of rubber wastes," paper presented at 7th International Seminar of Elastomers, Bangkok, 16–17 December 1998.
41. Fix, S. R., Microwave devulcanization of rubber, *Elastomerics*, 112 (6), 38, 1980.
42. Brown, C.J. and Watson, W.F., Recycling of vulcanized factory waste, *Rubber World*, 218(2), 34, 1998.
43. Maridass, B. and Gupta, B.R., Recycling of waste tire rubber powder: devulcanization in a counter-rotating twin screw extruder, *Kautschuk Gummi Kunstst.*, 56(5), 232, 2003.
44. Kim, J.K. and Park, J.W., The biological and chemical desulfurization of crumb rubber for the rubber compounding, *J. Appl. Polym. Sci.*, 72, 1543, 1999.
45. Bredberg, K., Christiansson, M., Bellander, M., Stenberg, B., and Holst, O., Properties of rubber materials containing recycled microbially devulcanized cryo-ground tire rubber, *Prog. Rub. Plast. Technol.*, 17(3), 149, 2001.
46. Schaefer, R.A. and Berneking, R.J., "Light Colored Reclaim," presented at the Fall ACS Rubber Division Meeting, Atlanta, Georgia, Oct. 7–10, 1986.
47. Acetta, A. and Vergnaud, J.M., Vibration isolation properties of vulcanizates of scrap rubber powder, *Rubber Chem. Technol.*, 55, 328, 1982.
48. Acetta, A. and Vergnaud, J.M., Upgrading of scrap rubber powder by vulcanization without new rubber, *Rubber Chem. Technol.*, 54, 301, 1981.
49. Phadke, A.A., Bhattacharya, A.K., Chakraborty S.K., and De, S.K., Studies of vulcanization of reclaimed rubber, *Rubber Chem. Technol.*, 56, 726, 1983.
50. Phadke, A.A. and De, S.K., Vulcanization of cryo-ground reclaimed rubber, *Kautsch. Gummi Kunstst.*, 37, 776, 1984.
51. Khvedchenya, O.A., Kostrykina G.I., and Zakharov N.D., Effect of mechanical grinding of SKD and BSK rubbers on the structure and properties of their vulcanizates, *Kauch. Rezina*, 6, 23, 1981.
52. Riedel Omni Products Inc., reported by Shaw, D., Rail crossings use scrap rubber, *European Rubber J.*, 176(3), 14, 1994.
53. Kim J.K. and Lee S.H., New technology of crumb rubber compounding for recycling of waste tires, *J. Appl. Polym. Sci.*, 78, 1573, 2000.
54. Swor, R.A., Jenson, L.W., and Budzol, M., Ultra-fine recycled rubber, *Rubber Chem. Technol.*, 53, 1215, 1980.
55. Jacob, C., De, P.P., and De, S.K. (unpublished data).
56. Negmatov, S.S., Umarova, M.A., Kh. Matkarimov, S., Kh. Nazirbekov, R., and Kobelev, I.A., Feasibility of using ground wastes of mechanical rubber goods in compositions of vulcanizates, *Uzb. Khim. Zh.*, 2, 59, 1986.
57. Markov, V.V., Kvardashov, V.P., Maloshuk, Yu S., Polyakov, O.G., Vil'nits, S.A., and Karmilina, L.I., Additional dispersion of ground vulcanized rubbers in rubber mixtures, *Tr. Mosk. Inst. Tonkoi Khim. Tekhnol.*, 5(1), 119, 1975.
58. Phadke, A.A., PhD thesis, "Studies on reclaimed rubber with special reference to cryo-ground rubber," chapter 5(C), Indian Institute of Technology, Kharagpur, India, 1985.
59. Bleyie, P.L., Influence of morphology and particle size of powdered rubber on mill processing, *Rubber Chem. Technol.*, 48, 254, 1975.
60. Phadke, A.A., Bhowmick, A.K., and De, S.K., Effect of cryoground rubber on properties of NR, *J. Appl. Polym. Sci.*, 32, 4063, 1986.

61. Jacob, C., Bhowmick, A.K., De, P.P., and De, S.K., Utilization of powdered EPDM scrap in EPDM compound, *Rubber Chem. Technol.*, 76, 36, 2003.
62. Jacob, C., Bhowmick, A.K., De, P.P., and De, S.K., Studies on ground EPDM vulcanisate as filler in window seal formulation, *Plast. Rubber Comp.*, 31(5), 212, 2002.
63. Koutsky, J., Clark, G., and Klotz, D., The Use of Recycle Tyre Rubber Particles for Oil Spill Recovery, *Conservation and Recycling*, 1, 231, 1977.
64. Ghosh, A., Bhattacharya, A.K., Bhowmick, A.K., and De, S.K., Effect of ground fluororubber vulcanizate powder on the properties of fluororubber compound, *Polym. Eng. Sci.*, 43, 267, 2003.
65. Klingensmith, B., Recycling, production and use of reprocessed rubbers, *Rubber World*, 203(6), 16, 1991.
66. Zhang, P., Yang, H., Zhu, W., Zhao, M., Zhao, J., and Zhao, S., Morphological structure of ground rubber and its application in tread compound, *Luntai Gongye*, 18(3), 152, 1998.
67. Pavlovskii, L.L., Kuznetsova, O.P., Kumpanenko, E.N., and Prut, E.V., Properties of tread rubber compounds filled with ground vulcanizate, *Proizvodstvo i Ispol'zovanie Elastomerov*, 8, 18, 1992.
68. Gibala, D., Laohapisitpanich, K., Thomas, D., and Hamed, G.R., Cure and mechanical behavior of rubber compounds containing ground vulcanizates. Part II – Mooney viscosity, *Rubber Chem. Technol.*, 69, 115, 1996.
69. Changwoon, N. and Shinyoung, K., Rheological and mechanical properties of styrene-butadiene rubber compounds containing rubber powder vulcanizates, *J. Polym. Eng.*, 17(4), 323, 1997.
70. Changwoon, N. and Hyeonjae, K., Effect of loading and particle size of rubber powder vulcanizate on physical properties of natural rubber compound, *Pollimo (Korean)*, 21(4), 648, 1997.
71. Jacob, C., De, P.P., and De, S.K., Studies on powdered EPDM rubber waste as a filler in Natural rubber and SBR compounds, *Kautschuk Gummi Kunstst* (communicated).
72. Kim, S.W., Lim, H.S., Kim, D.J., and Seo, K.H., Cure behaviors and physical properties of EPDM vulcanizates containing ground rubber, *Pollimo*, 21(3), 401, 1997.
73. Zhao, S., Zhou, Y., and Bai, G., Study on rheological-mechanical properties and morphology of waste rubber powder/SBR compounds, *Hecheng Xiangiiao Gongye*, 16(3), 152, 1993.
74. Gibala, D., Thomas, D., and Hamed, G.R., Cure and mechanical behavior of rubber compounds containing ground vulcanizates. Part III – tensile and tear strength, *Rubber Chem. Technol.*, 72, 357, 1999.
75. Jacob, C., De, P.P., De, S.K., Gong, B., and Bandhyopadhyay, S., Powdered rubber filled EPDM compound: Effect of aging of precursor rubber sheet (used in making the powder) on the properties of EPDM compound (communicated to *Kautschuk Gummi Kustst*).
76. Sharapova, L.N., Chakanova, A.A., Zakharaov, N.D., and Borisova, E.Yu, Role of sulfur diffusion in vulcanizate structure formation in heterogeneous rubbers containing ground vulcanizates, *Izv. Vyssh. Uchebn. Zaved., Khim. Khim. Tekhnol.*, 25(9), 1135, 1982.
77. Kostrykina, G.I., Solov'ev, M.E, and Zakharaov, N.D., Effect of a ground vulcanizate on the crystallization of cis-butadiene rubbers, *Izv. Vyssh. Uchebn. Zaved., Khim. Khim. Tekhnol.*, 24(7), 900, 1981.

78. Sharapova, L.N., Chekanova, A.A., Kostrykina, G.I., and Zakharov, N.D., Change in structural parameters of rubber-ground vulcanizate heterogeneous systems during vulcanization, *Kauch. Rezina,* 12, 6, 1983.
79. Sharapova, L.N., Chekanova, A.A., Zakharov, N.D., and Lyapina, L.A., Effect of carbon black type on vulcanizate structure and properties of rubber-ground vulcanizate systems in presence of different accelerators, *Kauch. Rezina,* 11, 12, 1983.
80. Ishiaku, U.S., Chong, C.S., and Ismail, H., Cure characteristics and vulcanizate properties of a natural rubber compound extended with convoluted rubber powder, *Polymer Testing,* 19(5), 507, 2000.
81. Nah, C., Cho, S., Lee, K.J., Jeon, D.J., Cho, C.T., and Kaang, S., Mechanical properties of rubber compounds containing polynorbonene vulcanizate powder, *Pollimo* (Korean), 20(4), 547, 1996.
82. Jacob, C., De, P.P., and De, S.K., Utilization of powdered natural rubber vulcanizate as a filler, *Rubber World,* 227(2), 43, 2002.
83. Jacob, C., Ph.D. thesis, "EPDM Recycling: Utilization of Powdered Rubber Vulcanizate in Polymer Matrices," chapter 6, Indian Institute of Technology, Kharagpur, 2002.
84. Burford, R.P. and Pittolo, M., The mechanical properties of rubber compounds containing soft fillers. Part 1. Tensile properties and fracture morphology, *J. Mat. Sci.,* 19, 3059, 1984.
85. Kim, J.K., Fracture behavior of crumb rubber-filled elastomers, *J. Appl. Polym. Sci.,* 74, 3137, 1999.
86. Cheater, G., Recycling progress reviewed, *Eur. Rubber J.,* 161(4), 11, 1979.
87. Dempster, D., Crumbed scrap proves successful in Canadian road surfaces, *Eur. Rubber J.,* 159(9), 87, 1977.
88. Burgoyne, M.D. and Evans, R.J., The use of reground scrap as a filler – extender, presented at 112th meeting of the Rubber Division, ACS, Cleveland, Ohio, Oct. 4–7, 1977, abstract in *Rubber Chem. Technol.,* 51, 385, 1978.
89. Han, M.H., "Ground Rubber Criteria for Use in Tire Stock Compounds," presented at the 7th *International Seminar on Elastomers,* Bangkok, Thailand, Dec. 16–17, 1998.
90. Fujimoto, K., Nishi, T., and Okamoto, T., Studies on structure and physical properties of vulcanizates containing comminuted vulcanizates. VI. Influence of the type of crosslinks on the physical properties of the composites, *Nippon Gomu Kyokaishi,* 54(5), 310, 1981.
91. Makarov, V.M., Zakharov, N.D., Gracheva, G.N., and Vil'nits, S.A., Effect of the character of the vulcanization structure of ground vulcanised rubbers on the properties of vulcanizates containing them, *Kauch. Rezina,* 10, 21, 1975.
92. Naskar, A.K., Pramanik, P.K., Mukhopadhyay, R., De, S.K., and Bhowmick, A.K., Characterization of ground rubber tire and its effect on natural rubber compound, *Rubber Chem. Technol.,* 73, 902, 2000.
93. Fesus, E., "Scrap Rubber: a Compounding Tool," paper presented at Rubber Divn., ACS, spring meeting, Chicago, Illinois, April 19–22, 1994.
94. Sombatsompop, N., Dynamic mechanical properties of SBR and EPDM vulcanisates filled with cryogenically pulverised flexible polyurethane foam particles, *J. Appl. Polym. Sci.,* 74, 1129, 1999.
95. Solov'ev, M.E., Zakharov, N.D., Ovchinnikova, V.N., and Goncharenko, T.G., Effect of the nature of the surface of ground vulcanizate on the properties of rubbers containing it, *Kauch. Rezina,* 6, 11, 1982.

96. Grebenkina, Z.I., Makarov, V.M., Zakharov, N.D., and Vil'nits, S.A., Rubber mixture, Russian Patent SU 590316 19780130, 1978.
97. Kirillov, A.A., Zakharov, N.D., and Neienkirkhen, Yu. N., Safronov, V.I., Effect of the size of particles of ground vulcanizate on the properties of the rubber containing it, *Kauch. Rezina*, 6, 16, 1979.
98. Pavlovskii, L.L., Kuznetsov, O.P., Arzumanova, Ya S., and Prut, E.V., Ground vulcanizates in formulations for electrically conducting rubbers, *Proizvodstvo i Ispol'zovanie Elastomerov*, 10, 8, 1992.
99. Solov'ev, M.E. and Zakharov, N.D., Effect of degree of destruction of milled vulcanizate on the properties of rubbers containing it, *Kauch. Rezina*, 8, 15, 1986.
100. Lim, J.C., Lim, H.S., and Seo, K.H., Cure behavior and physical properties of SBR foam containing ground tire waste, *Komu Hakhoechi*, 32(2), 105, 1997.
101. Seo, K.H., Lim, J.C., and Lim, H.S., Effect of ground rubber on mechanical properties of EPDM foam, *Eylasutoma* (Korean), 35(2), 132, 2000.
102. Ghosh, A., Antony, P., Bhattacharya, A.K., Bhowmick, A.K., and De, S.K., Replacement of virgin rubbers by waste ground vulcanizates in blends of silicone rubber and fluororubber based on tetrafluoroethylene/propylene/vinylidene fluoride terpolymer, *J. Appl. Polym. Sci.*, 82, 2326, 2001.

7

Rubber Recycling by Blending with Plastics

D. Mangaraj

CONTENTS

7.1 Introduction

Except for thermoplastic elastomers (TPEs) and thermoplastic vulcanizates (TPVs), all rubber products are thoroughly cross-linked to provide the needed property profile including strength, modulus, and resilience. As a result, vulcanized rubber products do not dissolve in a solvent or melt at high temperature and as such cannot be reprocessed to a desired shape. Recycling of vulcanized rubber waste therefore poses a serious problem, resulting in costly landfills, devastating tire-fires, and habitats for mosquito breeding.[1] The environmental hazards of tire dumps in industrialized countries in Europe and the U.S. are a serious problem that begs for immediate attention. Myhre and MacKillop[2] have recently reviewed the problem and have discussed the recycling technology being currently practiced.

It is estimated that approximately three to five billion tires are piled across America, and the amount is increasing every year by 250 to 275 million pounds. In the year 2000, approximately 276 million scrap tires were generated, of which 273 million were used in various markets. Of this, 61.6% were used for tire-derived fuel (TDF), 14% for civil engineering applications, 8.9% for ground rubber, 3.9% for being cut and punched, and 3.4% for agricultural/miscellaneous applications.[2] The cost of energy recovery from tires is high, and the volume of retreading is in decline. Rubberized asphalt, although superior to regular asphalt in performance on a long-term basis, costs twice as much and has not reached popularity on a large scale.[1] It is therefore important to develop technologically sound and cost-effective methods for recycling rubber from tire scrap.

The best way to recycle rubber products would be to devulcanize and reuse them in the rubber industry. Although processes for devulcanization, including chemical, thermal, thermomechanical, and ultrasonic, have been worked out,[2] they are costly and not suitable for commercial application, particularly in manufacturing highly engineered products like tire. Since 70% of all rubber produced in the world goes for tire manufacturing, the avenue for recycling rubber to manufacture non-tire rubber products is very limited. As such, past efforts to recycle rubber by blending it with virgin rubber have met very limited success. The other alternative is to blend the crumb or ground rubber with a material having the ability to flow under heat and pressure, so that it can be shaped into useful articles at a reasonable cost. This can be accomplished by mixing finely ground rubber with plastics, along with necessary additives. Hence, rubber recycling, by blending with thermoplastic and thermoset polymers, has been a subject of interest for quite some time and has been used for manufacturing a variety of consumer products. In this chapter, rubber-plastic blends will be reviewed as a cost-effective process for rubber recycling, using both vulcanized and devulcanized rubber. The advantages of using rubber-plastic blends will be discussed, along with the need for compatibilization and the methods used for achieving that. Past work on product and process development by blending ground rubber with plastics will be reviewed, and suggestions for the future will be provided.

The following is a list of abbreviations often used in this chapter:

- EPDM — ethylene-propylene-diene-monomer rubber
- GRT — ground rubber from waste tires
- HDPE — high-density polyethylene

- IIR — butyl rubber
- LDPE — low-density polyethylene
- NBR — nitrile rubber
- NR — natural rubber
- PE — polyethylene
- PP — polypropylene
- PVC — polyvinyl chloride
- SBR — styrene-butadiene rubber

7.2 Advantages of Rubber Recycling by Blending with Plastics

The key to success in recycling is reducing cost. This is achieved through continuous processes, such as extrusion or injection molding of thermoplastics and reaction-injection molding (RIM) and pultrusion of thermosets. Rubber goods, however, are manufactured largely by batch processes. Blending vulcanized rubber powder, or grits, with plastics provides an economic route for recycling rubber. The essential steps for getting ground rubber from tire waste (GRT), such as cutting, metal and fiber removal, and size reduction, are common preliminary steps prior to both blending and devulcanization. Since devulcanization requires additional energy and processing, the cost of GRT is substantially lower than that of devulcanized rubber.

Addition of rubber to plastics also improves some of the key properties of the plastics, particularly impact resistance and "good feel" behavior. Engineering plastics such as nylon, polyester (PE), polycarbonate, styrene-acrylonitrile copolymer, and alloys of polyphenylene-oxide can be blended with recycled rubber powder to improve their impact resistance. Low-cost thermoplastic vulcanizates can be produced by blending GRT with commodity thermoplastics such as PE, polypropylene (PP), and polyvinyl chloride (PVC).

During the last two decades, plastics processing technology has seen tremendous progress. Blending and alloying of dissimilar plastics, reactive processing, mixing and compounding in twin-screw extruders, computer-assisted processing and quality control, gas- and water-assisted injection molding, and in-mold painting and decorating are examples of the large array of such new developments.[3] Similar progress has been made in processing of thermoset plastics, including RIM, pultrusion, etc. The recent developments in processing wood flour–based composites and natural fiber–reinforced composites are noteworthy.[4] Net result is that plastics processing cost has gone down substantially and a variety of fillers, additives, and reinforcements, including polymeric ones, are available. These can be incorporated into plastics to improve the properties, without appreciable addition to the cost. This knowledge bank can be profitably used to develop useful, marketable products from ground-rubber/plastic blends at a reasonable cost.

A variety of engineering and commodity plastics are commercially available, covering a wide range of cost. Thermoplastics such as PE, PP, and PVC are not only cheap, but also are available in a wide range of melt index and micro-structure, which can be used for blending with recycled rubber. Polyolefins produced by metallocene and single-site catalyst technology are somewhat elastomeric in nature[5] and possibly are more suitable for blending with rubber than their conventional counterparts. In addition, a substantial quantity of commodity and engineering plastics is available as recyclate from municipal solid waste and manufacturing waste. This can be used to blend with scrap rubber, to further reduce product cost and alleviate the current environmental problems.

Our knowledge of blending and alloying of plastics and elastomers has grown immensely in the last two decades.[6] The principles underlying polymer-polymer compatibility have been understood very well, and practical methods for compatibilization, interface modification, and rheological optimization have been worked out.[7]

In recent years, TPEs and TPVs have replaced conventional rubbers in a variety of applications including appliance, automotive, medical, engineering, etc. TPEs are made by copolymerization and by blending thermoplastics with a rubbery component. TPVs, on the other hand, are made by dynamically vulcanizing the rubber component in a rubber/thermoplastic blend during mixing. Both materials are processed like thermoplastics and are recyclable. Whereas blending of thermoplastics with devulcanized rubber may provide novel TPEs, blending of very fine ground rubber particles from tire waste with a suitable thermoplastic, along with adequate compatibilization, will lead to development of novel, low-cost TPVs. TPVs are currently finding wide application in automotive, electric, appliance, tools, and a variety of personal products due to their easy processability, elastic properties, and soft-feel appeal.[8] The increasing application of TPVs in the automotive industry is due to faster production cycles compared to vulcanized rubber, lower part weight, elimination of scrap through the use of regrind, reduced material consumption, and better environmental compliance. In the past, TPVs were largely used for under-the-hood parts due to their resistance to oil and grease and thermal resistance over a temperature range of – 40^{0}C to 150^{0}C. Such parts include rack and pinion boots, air-management ducts, tie rod end seals, brake cable covers, steering column covers, distributor wire boots, seat belt sleeves, and windshield washer reservoir seals. TPVs have also been successfully used in glass and metal encapsulation by overmolding. The soft, rubbery feel of TPVs has opened up new interior applications such as air bag covers, floor mats, cup holders, and closeout seals.

Automotive shredder residue (ASR) is rich in both thermoplastics and rubber. It may be possible to blend ASR without separating it into individual components,[9] and thereby reduce the cost of ASR recycling.

In short, there is a great scope for recycling both devulcanized rubber and ground rubber by blending with plastics. Whereas devulcanized rubber will act as a second polymeric component, vulcanized ground-rubber particles will act as a low-cost organic filler or extender. Polymeric in nature, ground-rubber particles provide additional advantage over inorganic fillers such as carbon black, talc, or silica for bonding better with matrix polymer — after suitable chemical or mechanical treatment or in the presence of a compatibilizer.

7.3 Compatibilization

The key to success in blending rubber with plastics is to compatibilize the two components to get satisfactory performance. Since most polymers, including elastomers, are high-molecular-weight compounds and are thermodynamically immiscible with each other, their blends undergo phase separation, with poor adhesion between the matrix and dispersed phase. The properties of such blends are often inferior to those of the individual components. At the same time, it is often desirable to combine the processing and performance characteristics of the two polymers, to develop useful products in a cost-effective manner.

Miscibility between the two polymers is best accomplished either by reducing the heat of mixing (ΔHm) or by making it negative. Heat of mixing is reduced when the two polymers are similar and their cohesive energy densities are close to each other. For regular solutions, $\Delta Hm = K(\delta_1 - \delta_2)^2$, where δ represents the solubility parameter (square root of cohesive energy density) and K is a constant. In other words, a blend can be technologically compatible, if the two polymers have similar intermolecular forces or if they interact or react at the phase interface. Hence, compatibilization may be described as a process that reduces the enthalpy of mixing or making it negative. This is best accomplished either by adding a third component, called a compatibilizer, or by enhancing the interaction of the two component polymers, chemically or mechanically.[10]

The role of the compatibilization is to:

- Reduce interfacial energy and improve adhesion between phases.
- Achieve finer dispersion during mixing. In blending recycled rubber with plastics, smaller particle size of the ground rubber and deagglomeration during mixing play a major role.
- Stabilize the fine dispersion against agglomeration during processing and throughout the service life.

The ultimate objective is to develop a stable morphology that will facilitate smooth stress transfer from one phase to the other and allow the product to resist failure under multiple stresses.

In general, compatibilization of two dissimilar polymeric materials is carried out either by mechanical or by chemical methods, or by both simultaneously.

7.4 Mechanical Compatibilization

The first step in blending two polymers is mixing the components into a homogenous mass. Both single-screw and twin-screw extruders are used for blending thermoplastics in their melt phase along with additives such as fillers, plasticizers, stabilizers, pigments, etc. Elastomers are blended in internal mixers such as Banbury® or on open roll mills. In both cases, the materials are exposed to shearing stress. The balance between drop breakup and the coalescence process determines the size of the dispersed phase. The need for reducing potential energy initiates the agglomeration process, which is less severe if the interfacial tension between the two components is minimal. A small amount of compatibilizer acts like a solid emulsifier when added, and reduces interfacial tension. This, in turn, stabilizes the primary droplets, thereby reducing the size of the dispersed phase. It has also been established that better dispersion is achieved when both components have similar viscosity. The rheological condition for forming a co-continuous phase is:

$$\eta_1\phi_2/\eta_2\phi_1 = 1$$

where η_1, η_2 are viscosities and ϕ_1 ϕ_2 are weight fractions of the two components in the blend.[11] A co-continuous phase provides the special morphology, where the two phases behave in tandem and exhibit the best properties of the two components. On the other hand, if the viscosity of the minor phase is high, it does not get broken down into small, dispersed particles. Mangaraj and co–workers[12] have shown that the component with high viscosity can be better dispersed if it is premixed with a plasticizer to provide lower viscosity during blending. They obtained a fine dispersion of Nylon 66 in its blend with high-molecular-weight polycarbonate by plasticizing the latter with polycaprolactone.

The cooling rate of the blend also influences the particle size. Whereas rapid cooling provides small particles, slower cooling allows ripening (agglomeration) and generates large particles.[13]

The above principles apply only to mixing of un-cross-linked polymers and may not apply to mixing ground rubber with thermoplastics. Use of ground rubber of fine particle size and mixing under high shear conditions may be required to get uniform blend of GRT with thermoplastics.

7.4.1 Chemical Compatibilization

Chemical compatibilization is aimed at reducing interfacial tension and is carried out in both reactive and nonreactive mode.[14] In nonreactive mode, an external plasticizer or a polymeric material such as a copolymer (preferably a block copolymer, which has chain units similar to both components of the blend) is added to the blend to achieve compatibilization. Di-block copolymers are preferred over single polymers or copolymers, although in many instances tri-block copolymers (such as styrene-ethylene-butylene-styrene) have also been used,[15] as have random and graft copolymers. Block and graft copolymers reduce the interfacial tension by spreading at the interface and mixing with both phases through their component parts, which are similar to one phase or the other. In reactive mode, block and graft polymers, formed *in-situ* during mixing of the two components, assist in compatibilizing the blend.

7.4.2 Nonreactive Compatibilization

Block and graft copolymers, available commercially or prepared prior to blending, are often used for compatibilization. The theoretical and practical aspects of nonreactive compatibilization have been discussed by Asaletha and Thomas,[16] in their study on compatibilization of natural rubber (NR)/ polystyrene (PS) blends. Graft copolymer of styrene onto NR was used as a compatibilizer. The latter was prepared by mixing an emulsion of styrene monomer with NR latex and exposing the mixture to γ radiation from cobalt 60 source. The workers monitored the effect of molecular weight of the homopolymer and graft copolymer, concentration of graft copolymer, and the sequence of addition on compatibilization efficiency. The increase in the size of the dispersed phase resulted in the decrease of tensile strength. Modulus increased with the increase in compatibilizer concentration, reaching maximum at a critical volume fraction, ϕ_c, which is sufficient to saturate the interphase surfaces, as noted by Noolandi and Herg.[17] The experimental and calculated values of ϕ_c were found to be comparable and were within a range of 0.005 to 0.02, showing that a small amount of compatibilizer is needed to provide adequate compatibility. The investigators also found that the critical surface area (Σ_c) of the copolymer compatibilizer required to saturate the interface is higher for a two-step process, where the copolymer is preblended with the dispersed phases, than when it is mixed with both components in one step; it is also greater, the higher the molecular weight of the compatibilizer. Further reduction of interfacial tension improved compatibilization. In the case of a block copolymer, the chemical structures of the individual blocks also play an important role. The closer the chemical structure of the blocks to the chemical structures of the blend components, the better the compatibilization.

Reiss and co-workers have shown that, for a polyisoprene/polystyrene blend, block copolymers provide better compatibilization than graft copolymers.[18] Solubilization of compatibilizer by individual phases takes place when

TABLE 7.1

Commercially Available Compatibilizers for Plastic and Rubber Blends

Compatibilizer	Manufacturer	Trade Name
1. Ethylene acrylic acid copolymer	DuPont	Surlyn 1652
2. Ethylene acrylic acid copolymer	Dow Chemical	Primachor
3. Maleic anhydride grafted PP and PE	Crompton	Polybond
4. Propylene maleic anhydride copolymer	Eastman Kodak	Epolene
5. EPDM, maleic anhydride copolymer	DuPont	Fusabond and MF 274D
6. SEBS, maleic anhydride copolymer	Shell Chemical	Kraton 1901X
7. EPDM, maleic anhydride copolymer	Crompton	Royaltuf
8. EVA, grafted maleic anhydride	DuPont	Fusabond MG 423D
9. HDPE, grafted maleic anhydride	DuPont	Fusabond MB 100D
10. LLDPE, grafted maleic anhydride	DuPont	Fusabond MB 110/MB 226D
11. Ethylene octene copolymer, maleic anhydride	DuPont	Fusabond MN 493D
12. Maleic anhydride grafted polybutadiene	Ricon	

Source: Mangaraj, D., *Comprehensive Polymer Science,* (Supp. 2), Allen, G., Aggarwal, S.L., and Russo, S., eds., Elsevier (Pergamon), London, 1996, p. 605. With permission.

the molecular weights of the blend components are comparable or smaller than the molecular weight of corresponding block copolymer in the compatibilizer. Teysie and co-workers,[19] who examined the compatibilizing action of many copolymers, concluded that the structure and the molecular weight of the copolymer control the efficiency of compatibilization; they also found that tapered block copolymers are more effective as compatibilizer than linear block copolymers. Boudreau[20] has shown that both impact resistance and yield strain improved appreciably when a small amount of ethylene vinyl acetate copolymer (EVA) was added to recycled PET/high-density polyethylene (HDPE) blends (recyclates obtained from municipal solid waste). He suggested that EVA acts like an adhesive at the interface of PE and PET domains.

A number of commercially available compatibilizers are often used to prepare polymer blends. They are listed in Table 7.1 with trade names and manufacturers.[21]

7.4.3 Reactive Compatibilization

Compatibilization that is carried out during the blending process by adding a reactive material as a blend component, is classified as reactive compatibilization. The classic example of the use of reactive compatibilization is in DuPont's super-tough nylon, where nylon is blended with maleic anhydride–grafted ethylene-propylene-diene monomer rubber (EPDM) to improve toughness.[22] The maleic anhydride moieties of the latter react with amine end groups of nylon at phase interface, forming a copolymer of nylon-EPDM that compatibilizes the two phases.

Blending of nylon with functionalized elastomers has been successfully carried out by several authors, including Ide and Hassegawa, Han, Cimino and co-workers, and McKnight and co-workers.[23] Recently, Scott and Macosko[24] have carried out an in-depth investigation of maleic acid (MA)-grafted EPDM/Nylon 6 blend. The increase in torque and temperature during blending indicated the existence of chemical reaction, leading to higher molecular weight, greater viscosity, and heat buildup. Exotherm due to mechanical work (mixing) contributes less than 5% of the total temperature buildup. The rest is due to exotherm of the coupling reaction. The morphology of the blend was examined by scanning electron microscopy (SEM) for both nonreactive and reactive blends using samples microtomed at – 100C with a diamond knife, followed by removal of rubber by exposing the fractured samples to hot xylene. SEM micrographs for nonreactive and reactive blends with 20% and 50% EPDM showed that the particle size reduction was appreciably greater in the reactive blend than in the nonreactive blend.

Rajalingam and Baker[25] have used precoating of GRT with functional polymers to improve the compatibility and impact resistance of GRT/PE blends. They used four different polymers: ethylene acrylic acid copolymer, SEBS terpolymer, EVA, and maleic anhydride–grafted PE as coupling agents. The GRTs were precoated with coupling agents in a Haake-Buchler Rheomix batch mixer under high shearing conditions for 3 minutes at 180C and 150 rpm. Coated and uncoated GRTs were blended with linear low-density polyethylene (LLDPE) and HDPE at 180C and 100 rpm. Test specimens were prepared by injection molding and were tested for impact resistance by a Rheometrics RDT-5000 instrument. Control blends of LLDPE and HDPE with functional polymers were made under similar conditions. The concentration of the functional polymer varied from 0 to 15 wt% for LLDPE. Almost all blends showed maximum impact energy at a concentration of 6.7 wt% of the reactive polymer. The blend containing maleic anhydride–grafted SEBS showed maximum increase. However, addition of the same coupling agents with HDPE showed either zero or a small improvement in impact energy, as shown in Table 7.2.

TABLE 7.2

Effect of Functional Polymers on the Impact Energy of Treated PE Blends

	LLDPE		HDPE	
Blend Composition	**Impact Energy (J)**	**% Increase**	**Impact Energy (J)**	**% Increase**
LLDPE	14.9	—	20.6	—
PE/ethylene coglycidyl MA	16.6	4.1	20.6	—
Ethylene covinyl acetate	16.8	5.5	20.6	0
SEBS	18.4	15.0	21.9	6.4
Maleated SEBS	18.7	17.0	21.2	3.2
Maleated PE	16.9	6.2	20.8	1

LLDPE, linear low-density polyethylene; MA, maleic anhydride; PE, polyethylene; SEBS, styrene-ethylene-butylene-styrene block copolymer.

Source: Rajalingam, P. and Baker, W.E., *Rubber Chem. Technol.*, 65, 1992, p. 908. With permission.

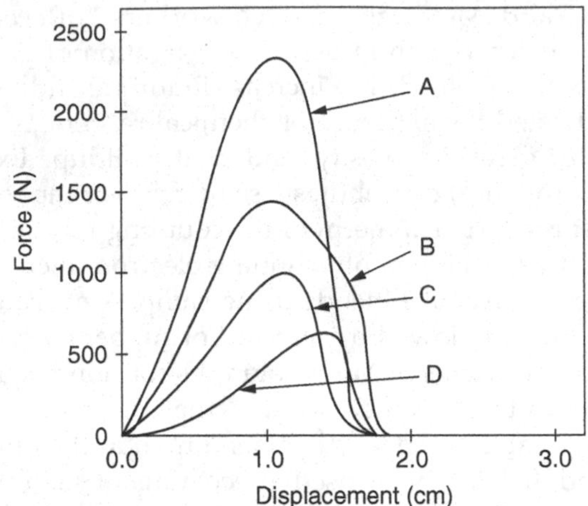

FIGURE 7.1
Force vs. displacement curves for the impact failure. (A) HDPE, (B) LLDPE, (C) LLDPE/GRT, (D) HDPE/GRT composites. (From Rajalingam, P. and W.E. Baker, W.E., *Rubber Chem. Technol.*, 65, 1992, p. 908. With permission.)

It is obvious that functional polymers do not improve, or improve only by a small percent, the impact energy of HDPE, whereas the improvement for LLDPE is substantial, as much as 17% with maleated SEBS. The force displacement curves are given in Figure 7.1, for LLDPE and HDPE, along with their blends with GRT. The initial slope, representing the stiffness of the blend, decreases slightly on addition of GRT to LLDPE, but decreases considerably for HDPE.

Table 7.3 gives the impact energy and percent of improvement for a series of blends containing 56% PE, 40% GRT, and 4% functional polymers. It is obvious that adding GRT to PE decreases the impact energy by about 50% in the case of LLDPE and about 75% in the case of HDPE. Addition of functional polymers (6 wt%) to the blend increases the impact energy for both thermoplastics by 40 to 60%, to a greater extent for HDPE compared to that for LLDPE. It was concluded that the improvement in impact energy for HDPE is largely due to increase in ductility, where as that for LLDPE, a fairly a ductile plastic, is due to improved compatibility.

Earlier, Duheime and Baker[26] had used amine-terminated nitrile rubber and ethylene-acrylic acid copolymer (EAA) to improve the impact properties of GRT/PE blend, with limited success. Precoating GRT with EAA overcame this difficulty, probably due to enhanced interaction.[27] Parmanik and Baker[28] blended GRT with waste plastics and found that the composites behaved like impact-resistant plastics. Rodriguez[29] found that addition of vinyl silane improved the mechanical properties of GRT-polymer composites.

Reactive compatibilization can also be carried out by adding one or more monomers and initiator to the blend. These can react with one or both phases, providing a graft copolymer *in-situ* that acts as a compatibilizer. Beaty and

TABLE 7.3
Impact Energy of PE/GRT Blend

Composite	LLDPE Impact Energy (J)	LLDPE % Increase	HDPE Impact Energy (J)	HDPE % Increase
PE	14.9	—	20.6	—
PE/GRT (60:40)	7.9	—	5.7	—
PE/GRT/EVA	12.5	58	9.9	74
PE/GRT/Maleated EVA	12.5	58	9.2	62
PE/GRT/SEBS	10.6	34	9.0	58
PE/GRT/Maleated SEBS	11.9	51	9.2	62
PE/GRT/Maleated PE	11.3	43	7.8	37

EVA, ethylene vinyl acetate copolymer; GRT, ground rubber tire.

Source: Rajalingam, P. and Baker, W.E., *Rubber Chem. Technol.*, 65, 1992, p. 908. With permission.

co-workers[30] added methyl methacrylate and peroxide to waste plastics (containing PE, PP, PS, and PET). The graft copolymer formed *in-situ* homogenized the blend very effectively. Alternatively, one of the blend components can be grafted with a suitable monomer prior to blending, to enhance compatibility. However, when Adam, Sebenik, and co-workers[31] added ethyl acrylate (EA)-grafted ground polybutadiene rubber to virgin polybutadiene, they could not achieve adequate vulcanization under normal conditions. Therefore, additional cross-linking was provided to restore the physical properties. It was inferred that grafting consumed some of the residual double bonds in polybutadiene, leading to a decrease in cross-linking sites and inefficient vulcanization. The resulting blend had characteristics of acrylic rubber (AR), such as good resistance to oil. This was attributed to the presence of poly-ethyl-acrylate side-chains on ground styrene-butadiene rubber (SBR) surface. On the other hand, when the EA-grafted poly-butadiene granulate was added to AR, the properties of the AR, including the resistance to oil, did not change appreciably, due to improved compatibilization (Table 7.4).

Brutlett and Lindt,[32] in a very interesting article, "Reactive Processing of Rubbers," discussed various methods of functionalizing rubbers, including

TABLE 7.4
Properties of AR Containing Nongrafted and EA-Grafted SBR Granulate

Properties	0 pGRT	20p NGr	20 pGr.(NW)	20p Gr.(W)	20p Gr(γ radn)
Hardness (Shore, A)	65	68	65	67	60
Tensile strength, Nmm^{-2}	12	8.4	9.4	8.8	8.9
Elongation at break, %	150	150	230	170	170
Swelling in oil, %	18.5	23.6	20.3	19.3	19.2

Gr, grafted; NGr, nongrafted; NW, nonwashed; p, phr; radn, radiation; W, washed.

Source: Adam, G., Sebnik, A., Usredkar, U., et al., *Rubber Chem. Technol.*, 64, 1991, p. 133. With permission.

maleation (thermal and grafting), Diels-Alder reaction, halogenation, electrophilic and nucleophilic reactions, and grafting. They discussed the necessary conditions for successfully carrying out these reactions with ground rubber particles and using the products to blend with thermoplastics. However, the level of grafting should be carefully selected, taking into consideration the characteristics of the blend components and the property enhancement desired.

Reactive compatibilization has been extensively used in the development of thermoplastic elastomers, particularly TPVs.[33] Coran and Cheng[34] have successfully used phenolic curative in the dynamic vulcanization of nitrile rubber (NBR)/PP and EPDM/PP blends. In this case, the phenolic resin reacts with PP *in-situ* and the product participates not only in the curing of EPDM, but also in compatibilizing PP with the rubber. This, in turn, reduces the particle size of the dispersed phase and provides a continuous morphology. The strength of the TPV depends largely on particle size of the dynamically vulcanized rubber particles and the degree of cross-linking of the rubber (Figure 7.2).

Impact modification of epoxy compounds using carboxy-terminated nitrile rubber (CTBN) and amine-terminated nitrile rubber (ATBN) is another example of reactive compatibilization. In this case, the carboxylic and amino groups react with epoxy groups, thereby compatibilizing the polar epoxy resin with nitrile rubber and simultaneously, participating in the cross-linking process of the epoxy resin.[35]

Polymer molecules often undergo fracture under mechanical stress, producing free radicals. When a polymer mixture is subjected to such stress

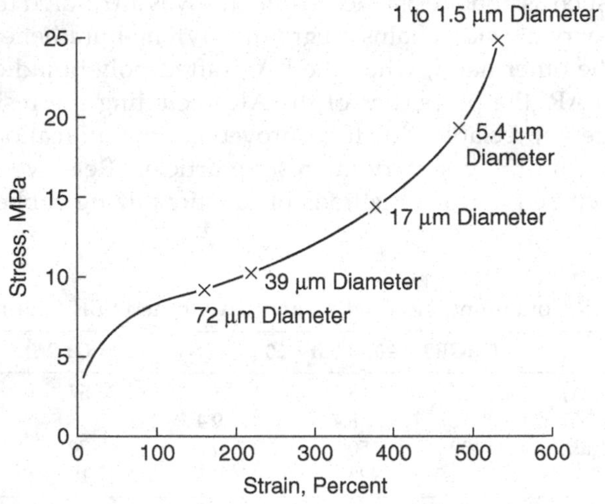

FIGURE 7.2
Effect of rubber particle size on stress-strain properties of TPV. (From Coran, A.Y. and Patel, R., *Rubber Chem. Technol.*, 56, 1983, p. 1045. With permission.)

reactions, the macro-radicals formed *in-situ* react with each other and form block copolymers, which compatibilize dissimilar polymeric materials.[36] Bilgill, Arastoopour, and Bernstein[37] have used this process to recycle solid waste containing high- and low-density PE and PET.

Stress reaction also takes place under ultrasonic irradiation of a polymer mixture. When polymeric materials, especially elastomers, are subjected to high-frequency ultrasound, cavitations occur, leading to high local stress and high temperature. This leads to rupture of polymer molecules, resulting in the formation of macro-radicals, which subsequently react with each other and compatibilize the blend. Isayev and Hong[38] have used this technique to devulcanize rubber and to prepare rubber-rubber and rubber-thermoplastic blends.

In a recent article, Isayev[39] discussed continuous mixing and compounding of polymer-filler and polymer-polymer mixtures using ultrasonic waves. He has studied the effects of ultrasonic treatment on the rheology and mechanical properties of rubber-plastic and rubber-rubber blends. Polyolefins such as HDPE and PP were blended with NR, SBR, and ethylene-propylene-diene rubber (EPDM). The 50/50 blends were first fed into a twin-screw extruder at a feed rate of 1.0 gm per minute. The extruder temperature varied from 140 to 160C for HDPE blends, and from 165 to 190C for PP blends. The products were dried and subsequently fed into a single-screw extruder with an ultrasonic attachment. A 20 kHz ultrasonic horn operated by a 300 kw power supply was attached to the extruder barrel. The gap between the horn and the screw was set at 2 mm and the feed rate was maintained at 0.63g/s. The treatment for HDPE/rubber blends was carried out at 150C and the treatment for PP/rubber blends at 190C.

The blends were characterized with respect to their surface characteristics, flow behavior, and physical performance. The viscosities of the ultrasonically treated blends were lower than those of untreated blends; viscosity decreased with amplitude of the ultrasonic horn, indicating that polymer chain break-ups are due to ultrasonic cavitations. Table 7.5 gives the tensile strength, elongation at break, and Young's modulus of untreated and treated blends. Significant increase in all three properties in the case of treated blends indicates the existence of enhanced intermolecular interaction and possibly, formation of block copolymers by the interaction of different types of macro-radicals formed *in-situ*.

This was confirmed by solvent extraction of the treated and untreated blends, followed by the study of their surface characteristics by atomic force microscopy. The copolymers formed during ultrasonic treatment of rubber/plastic blends provided improved adhesion between the phases, leading to better dispersion and smaller particle size of the dispersed phase.

The improved compatibilization is reflected in higher impact energy of ultrasonically treated blends, compared to that of the untreated blends. Similar property enhancement has been observed when thermoplastic blends, rubber blends, and silica-filled rubber compounds were subjected to ultrasonic treatment.

TABLE 7.5

Mechanical Properties of Untreated and Ultrasonically Treated Blends

Blend	Tensile Strength, MPa	Elongation at break, %	Young's Modulus, MPa	Toughness MPa	Impact Energy, Joule
PP/NR					
Untreated	8.39	38.9	191.0	2.99	2.75
Treated	11.27	111.7	250.0	12.19	5.89
PP/EPDM					
Untreated	9.87	29.1	229.0	2.68	8.40
Treated	10.51	87.6	261.0	9.00	9.46
HDPE/NR					
Untreated	4.92	189.0	75.1	8.17	6.09
Treated	5.71	300.6	126.5	16.00	8.12
HDPE/EPDM					
Untreated	4.74	88.2	103.0	4.24	8.50
Treated	4.84	116.0	109.0	5.76	9.63
HDPE/SBR					
Untreated	5.33	15.4	59.0	0.59	7.53
Treated	5.46	28.7	101.0	1.32	8.76

Source: Isayev, A.I., Paper #108, presented at the ACS Rubber Division Meeting, October, Pennsylvania, 2002. With permission.

The characteristics of the GRT particles, including the particle size and specific surface area, are very important for the performance of the rubber/plastic blends. They are also important for maintaining product consistency and marketing. The American Society for Testing and Materials has developed tests and standard characteristics for GRT, which not only help recyclers, but also with trading of GRT in the stock market.[40]

7.4.4 Surface Characteristics of Ground Rubber and Their Influence on GRT/Plastic Blends

The surface characteristics of GRT particles depend largely on the process used for size reduction, which is carried out in three different ways: dry grinding, wet grinding, and cryogenic grinding. The gross morphology and the surface chemistry of the ground rubber depend largely on the intensity of stress, temperature, and the surrounding atmosphere. Rajalingam and coworkers[41] have studied the effect of different grinding processes on the impact resistance of ground rubber/LLDPE blends. The SEMs of 0.35 mm ground rubber particles show that the surface of the particles produced by

TABLE 7.6

Surface Area of Differently Ground GRT

Mode of Grinding	Surface Area $m^2/gm \times 10^{-1}$
Ambient (1)	4.40
Ambient (4)	1.88
Wet ambient (7)	1.95
Cryogenic (3)	1.35
Cryogenic (5)	1.42

ambient dry grinding and ambient wet grinding is covered with porous, rough nodules. On the other hand, the particles produced by cryogenic grinding are smooth and angular, indicating brittle fracture of the rubber below glass transition temperature. In addition, the surface area (Table 7.6) of the differently ground rubber are different, the ambient ground one having the maximum surface area.

The impact resistance of a filled composite also depends on particle size of the rubber particles. In an earlier investigation, the Rajalingam and co-workers found that the reduction of particle size from 0.35 mm to 0.1 mm increased the melt viscosity and impact strength by about 20%, as well as tensile elongation by 40 to 50%.

In a subsequent study on GRT/LLDPE blend, the authors found a 17% increase in impact energy, but a decrease in melt flow index from 2.7 to 1.7 g/10 minutes, when the particle size of GRT was reduced from 0.6 mm to 0.074 mm. Addition of EAA to the blend increased both impact energy and melt flow index (MFI), due to improved compatibilization. The impact energies of all composites containing different amounts of ambient-ground GRT are similar and slightly higher than cryoground GRT–based composites at a particular particle size (Table 7.7). The ambient-ground GRT blends also exhibit better processability. This may be due to their greater surface area, porous rough nature, and greater fiber content.

The ground GRT samples were subjected to plasma treatment, corona treatment, and electron beam irradiation. ESCA studies showed that electron

TABLE 7.7

Effect of Grinding Type on Impact Energy

		Impact Energy	
GRT Type	**Grinding Mode**	**Sieved**	**Unsieved**
GRT 1	Ambient	10.4	9.7
GRT 2	Ambient/Cryo	10.6	10.7
GRT 3	Cryo	10.0	9.2
GRT 4	Ambient	7.9	7.4
GRT 5	Cryo	10.6	10.2
GRT 6	Cryo	10.6	9.9
GRT 7	Wet ambient	12.5	12.0

TABLE 7.8

Effect of GRT Surface Treatment on the Impact Energy of the Blends

	Impact Energy (J)			
Treatment	None	Plasma	Corona	Electron Beam (10 kGy)
GRT 1	9.7	10.2	9.9	12.9
GRT 2	10.7	10.5	10.9	13.1
GRT 7	12.0	—	—	13.6

GRT, ground rubber from waste tires.

beam–irradiated composites showed a higher percentage of oxygen on the surface, similar to the surface area of wet-ground GRT. There is also some indication of formation of ether (– O-CH_2) groups and/or ester groups (–CO-O) groups on GRT surface. The MFI of the composites based on irradiated blends is higher than the nonirradiated ones and is further enhanced by the addition of a coupling agent. It is suggested that low-molecular-weight species formed during irradiation also act as lubricants and contribute to the increase in MFI. The impact energy of differently irradiated compounds is given in Table 7.8. It is obvious that electron beam irradiation provides greater impact energy for the composites than either plasma or corona treatment, indicating the importance of surface chemical reaction to the coupling action between GRT and the thermoplastic.

The authors also studied the effect of different tire sources and grinding techniques on impact energy. Thermoplastic blends with GRT from passenger tire and truck tire waste showed minimum impact energy. Addition of 4% coupling agent improved impact energy substantially. For similar particle size distribution, the ground truck tires are superior to ground passenger tires, and the ambient ground truck tires in wet conditions are better than truck tires ground by other methods, in so far as impact energy of the blends is concerned.

Earlier, De and co-workers[42] had developed a size reduction process based on abrading waste rubber parts. Surface characteristics of particles obtained by this process were fibrilar and porous. As a result, the ground rubber particles could be successfully incorporated in new rubber formulations without appreciable loss of performance.

Ground rubber particles are often surface treated with reactive chemicals such as ozone, chlorine, or sulfur dioxide, either to make their surface polar or to generate reactive groups, which improve their prospect of compatibilization in reactive or nonreactive mode. Rubbers containing double bonds can be epoxidized, maleated, and oxidized, forming reactive groups, which help in their compatibilization with other plastics and rubber.[43] Rubber Research Elastomerics has developed a process for surface treating ground rubber with a special functional polymer that cross-links with virgin rubber during processing and as such, can load 50 to 85% ground rubber without significant loss in performance.[44] Verdestein of Belgium has developed a continuous process of waste tire pulverization, purification, and regeneration (possibly with a proprietary surface treatment) and markets the product under the trade name Surcum®. It can be used to make rubber products with properties similar to those obtained from virgin rubber.[45] Surcum® powder with particle size between 42.5 and 159 micrometers provides significant increase in tensile strength and elongation at break, compared to rubber powder without surface treatment. It is reported that the activation process involves the formation of a surface layer of a cross-linkable polymer which, in the presence of rubber curatives, helps bond the powder to the matrix.

Composite Particles, Inc. markets Vistamer R™, a surface-modified crumb rubber that can be used for producing molded goods with superior properties at a reduced cost.[46] Goldsmith and Eagleton[47] have developed a proprietary surface treatment, Tire-Cycle®, for GRTs to improve their compatibility with virgin rubber. Radiation, both γ and electron beam, has been used to change GRT surface characteristics and to compatibilize GRT blends with plastics.[48]

Composites made from blends of GRT powder with LDPE/EVA and LDPE/PP/EVA have been studied.[49] It is claimed that the presence of EVA at the thermoplastic/rubber interface increases the adhesion between the matrix and the powder and thereby increases the deformability of the rubber/plastics blend.

Ground rubbers powdered by cryogenic grinding and ambient grinding methods have been blended with thermoplastics and have been used for treating sports grounds.

World Tech Management, Inc. claims to have found a way of chemically linking passenger tire scrap rubber with plastics, without the need for bonding agents or additives.[50] The crumb rubber from passenger tires is chemically linked with either recycled PP or HDPE. The development and testing of the product, trade named Xylex, is discussed. The company is moving toward commercialization.

Kim and Burford[51] did a comprehensive study on the effect of surface treatment and the use of compatibilizers on the morphology and physical properties in blending ground tire crumb with natural rubber and nitrile rubber, containing silica filler. Carbon black-filled compounds were not used in order to match modulus of the crumb and the nonvulcanized rubber and to facilitate morphology study. The crumb rubber was surface chlorinated before blending. An external compatibilizer, poly-octenemer from Struktol, was also used in this study. It was found that whereas both surface chlorination and compatibilizer addition helped in improving the performance of NBR/NR blend, there was no significant effect on the performance of NR/crumb rubber blend. It was concluded that since the surface of NR is very similar to that of ground rubber, there was no net improvement in performance by adding chlorinated crumb to NR. However, the surface of NBR is more polar than that of ground rubber. Chlorination of the crumb, therefore, helps the compatibility by improving polarity and providing higher intermolecular interaction between the crumb and NBR. Addition of Struktol compatibilizer helps in matching the viscosity of the two components and, thereby, further improves compatibilization.

Oldfield and co-workers[52] found that greater compatibility is obtained by surface chlorinating crumb rubber, using trichloro-isocyanourate. Lawson and fellow investigators[53] found that surface-chlorinated crumb has higher surface energy and better adhesion to hetero-polar systems. Bauman[54] investigated the effect of surface modification of crumb rubber by active gases. Pittolo and co-workers[55] used surface grafting of styrene monomer and surface chlorination to modify the surface of ground rubber, to improve its compatibility with plastics.

7.5 Product and Process Development

7.5.1 Rubber-Thermoplastic Composites

7.5.1.1 Thermoplastic Elastomers

Coran's[56] idea that a blend of thermoplastic and dynamically vulcanized rubber provides the best of two worlds, namely excellent rubbery behavior with processability of thermoplastics, has inspired many authors to develop similar products by replacing (fully or partially) the rubber phase with vulcanized rubber powder, including GRT. The need for chemically bonding the plastic to the rubbery phase has been overcome by various methods, including the use of special cross-linking agents, surface functionalization, and reactive surface treatment.

De and co-workers[57] have carried out substantial fundamental work in this area at the Indian Institute of Technology, Kharagpur. Their initial attempt was to develop a thermoplastic elastomer by blending GRT and acrylated HDPE. GRT obtained from Recovery Technology, Inc. in Ontario, Canada, was maleated in the presence of maleic anhydride and dicumyl peroxide in Brabender Plasticorder at 160C and 60 rpm. Acetone-extracted GRT was treated in the same manner. The surface-treated GRTs were characterized by x-ray photoelectron spectroscopy, diffuse reflectance infrared Fourier transform spectroscopy (DRIFT), and differential scanning calorimetry (DSC). The decrease in the intensity of the peak at 1402 cm^{-1} in the DRIFT spectra, corresponding to the – C-H deformation of > C = CH, indicated the grafting of maleic anhydride to the rubber hydrocarbon in GRT. Spectroscopic evidence also confirmed that the rubber particles are chemically bonded not only to maleic anhydride, but also to the zinc ions (from zinc oxide and stearic acid used in rubber compounding) in GRT through ionic interaction. The hydrophilic character of GRT also increased on maleation, leading to an increase in surface energy from 26.75 to 28.42 mN/m. The high temperature transition observed by DSC was attributed to the dissociation of ionic clusters arising from the zinc salt mentioned earlier.

TABLE 7.9

Physical Properties of the 60/40 Rubber/Plastic Blends Containing GRT and m-GRT

GRT Type	Tensile Strength, MPa	Elongation at Break %	Young's Modulus, MPa	Work of Rupture (kJ/m²)	Tear Strength, kN/m	Tension Set (%)
GRT	8.2 (7.1)	152 (203)	36.6 (22.3)	4.5 (5.3)	56.6 (60.8)	24 (26)
m-GRT	10.3 (9.0)	158 (190)	41.2 (37.2)	5.9 (6.2)	62.5 (62.0)	24 (26)

Properties in parentheses are those for acetone-extracted GRT.

Source: Naskar, A.K., De, S.K., Bhowmick, A.K., *J. Appl. Polym. Sci.*, 84, 2002, p. 370. With permission.

Both GRT and maleated GRT were blended with EPDM rubber in 30:68 ratios, in a Brabender Plasticorder for 5 minutes at 30°C. The product was then milled in a two-roll mill at 2 mm nip setting, sheeted out, and then blended with 40 parts of acrylated HDPE and 0.6 parts of dicumyl peroxide at 160C for 2.5 minutes. The sheets were subsequently remixed in Brabender Plasticorder at 160°C and were molded into test specimen by compression molding. The physical properties of 30/68/40 EPDM/GRT/HDPE blend using GRT and maleated GRT are given in Table 7.9.The values in parentheses are for blends in which GRT was not acetone extracted and were used, as such, for maleation and for blending.

It was found that tensile strength, elongation at break, modulus, work of rupture, and tear strength were higher when maleated GRT was used in the blend, in place of untreated GRT. The elongation at break was comparable. This shows that maleation helps in improvement of physical properties, probably due to greater interaction between the rubber and the plastic phase. The properties of blends containing GRT not subjected to acetone extraction, and the corresponding maleation product, are somewhat lower than the blends with extracted GRT and their maleation product, except for the work of rupture and elongation at break.

It was suggested that acetone extraction, in general, removes the plasticizer and processing oil present in GRT, which results in lowering the elongation at break, tear strength, and work of rupture. Reprocessing of the blends did not change the physicals to an appreciable extent, showing that the blends are truly thermoplastic elastomer.

Naskar, Bhowmick, and De[58] studied the possibility of replacing EPDM with GRT in a thermoplastic elastomer based on dynamically vulcanized EPDM and acrylated HDPE (AHDPE). T hey studied three series of formulations. In the first series, they varied the amount of dicumyl peroxide (DCP) in 60:40 blend of rubber with plastics, keeping a 1:1 ratio between EPDM and GRT. The GRT was a cryogenically ground rubber obtained from Recovery Technology, Ontario, Canada. In the second series, they varied the AHDPE concentration from 0 to 100, keeping a 1:1 ratio between EPDM and GRT and 1 part of DCP. In the third series, they varied the EPDM-to-GRT

TABLE 7.10

Physical Properties of Dynamically Vulcanized 60:40 Rubber/Plastic Bends

GRT/EPDM Ratio	Tensile Strength, MPa	Elongation at Break,%	Tear Strength, kN/m	Toughness $J/m^2 \times 10^{-3}$	Tension set %	Hardness Shore A
60:0	3.5	200	44.5	2.7	42	69
50:10	8.5	388	72.4	12.0	20	80
40:20	6.5	260	63.8	7.3	22	80
30:30	7.2	247	64.3	6.3	24	81
20:40	6.5	155	47.1	3.8	28	83
10:50	5.7	85	38.0	1.8	—	8
0:60	5.2	43	25.8	0.7	—	88

Source: Naskar, A.K., Bhowmick, A.K., De, S.K., *Polym. Eng. Sci.*, 41, 2001, p. 1087. With permission.

ratio at a constant rubber-to-plastic ratio of 60:40 and constant DCP content. EPDM/GRT mixtures were prepared by mixing them in a Brabender Plasticorder at 60 rpm and a two-roll mill. The mixture was then added to HDPE and further mixed in the Plasticorder, running at 60 rpm at 160°C. DCP was added after 2.5 minutes. The blend was taken out to the two-roll mill after 1.5 minutes, sheeted, and remixed in the Plasticorder for an additional 2 minutes at 160°C. The sheets were compression molded and cured in a hydraulic press for 2 minutes at 170°C. The cured compounds were characterized for their mechanical properties, thermal behavior (by differential scanning colorimetry [DSC] and dynamic mechanical thermal analysis [DMTA]), rheology, and morphology.

Analysis of the physical properties showed that the best combination of tensile strength, elongation at break, and toughness was obtained when one part of DCP was used and the rubber-to-plastic ratio was maintained at 60:40. Table 7.10 gives the physical properties of the dynamically vulcanized blends of AHDPE/GRT/EPDM as a function of GRT/EPDM ratio for 60:40 rubber/plastic blends. Only the hydrocarbon content of GRT was taken into consideration in calculating the ratio of GRT to EPDM. Based on the results of physical properties, it was concluded that virgin EPDM in AHDPE/EPDM blend–based TPV can be replaced by GRT up to 50% without any appreciable loss of properties.

Scanning electron micrographs showed that GRT particles are dispersed in the EPDM matrix. HDPE forms a continuous phase only after its concentration reaches about 35%. At loadings greater than 40%, there was a decrease both in elongation at break and toughness. Post yield deformation was exhibited at still higher concentration due to toughening of the brittle plastic by rubber particles. Dynamic mechanical analysis showed two transition temperatures, one at 95°C, characteristic of the melting transition of HDPE, and the other at – 36°C, the Tg of pure EPDM and GRT. Whereas the glass-rubber transition temperature did not change with blend composition, Tm of the blend decreased with decrease in HDPE concentration. The dynamic mechanical behavior of the blends corresponded to the prediction based on theoret-

TABLE 7.11

Mechanical Properties of 70/30 EPDM/PP Blends Containing Vulcanized EPDM

Property	Mix Numbers						
Blend No.	F_0	F_{10}	F_{20}	F_{30}	F_{35}	F_{40}	F_{45}
Tensile strength (MPa)	4.8	4.3	4.2	4.2	4.9	5.5	6.1
Elongation at break (%)	220	183	215	216	205	239	249
100% Modulus (MPa)	3.5	3.8	3.3	3.0	3.6	3.8	3.9
Tear strength, kNm^{-1}	30.2	25.8	26.4	24.5	29.3	29.0	35.0
Tension set (100% elongn.)	14	26	18	14	14	14	14
Toughness (Jm^{-2})	3323	2794	3079	3136	3093	4039	4516
Hysteresis loss, $Jm^{-2} \times 10^6$	0.078	0.101	0.078	0.084	0.084	0.092	0.088

Source: Jacob, C., De, P.P, Bhowmick, A.K., De, S.K., *J. Appl. Polym. Sci.*, 41, 2001, p. 1087. With permission.

ical models. Transmission electron microscopy (TEM) photomicrographs showed that at 40% loading, HDPE formed a continuous phase, and EPDM with encapsulated GRT particles tended to form a co-continuous phase.

Jacob, De, Bhowmick, and De[59] studied the possibility of replacing virgin EPDM with waste EPDM vulcanizate powder (r-w-EPDM) in a thermoplastic elastomer composition, based on PP and EPDM. The ratio of r-w-EPDM to virgin EPDM varied from 0:100 to 45:55, keeping the total rubber content constant. Dynamic vulcanization of the PP/rubber blend was carried out at 180°C in the presence of one part of dicumyl peroxide per hundred parts of rubber in a Brabender Plasticorder. The dynamically vulcanized blends were then compression molded at 200°C and 5 MPa pressure and tested for their morphology, fractography, and dynamic mechanical behavior. The mechanical properties of the blends are given in Table 7.11. The subscripts give the parts of r-w-EPDM in 100 parts of total EPDM.

It is evident that the properties dropped initially on replacing virgin EPDM with vulcanized waste, but recovered after about 35% replacement, and then exceeded the control at 45%. It was suggested that the properties of the blended thermoplastic elastomer depend on the morphology, degree of cure, and the filler content. The initial lowering of strength was ascribed to a smaller degree of cure (as evidenced by a Monsanto Rheometric study), which was later offset by a higher loading of fillers present in r-w-EPDM. The degree of crystallinity of PP decreased from 46% to 11% in regular TPVs based on PP and EPDM. In the present case, it dropped to 11% and then increased to 12% at 40% loading of r-w-EPDM. Dynamic Mechanical Analysis showed that $(\tan \delta)_{max}$ decreased with the increase in r-w-EPDM content. However, the glass transition temperature of PP was not affected.

The authors also studied the cure characteristics of the compounds, with and without waste rubber. The degree of cure was slightly higher in compounds containing waste EPDM compared to those containing virgin EPDM.

It was suggested that carbon black and plasticizers in virgin EPDM–based compounds absorb a certain amount of DCP during initial mixing at 70°C, thereby decreasing its concentration at the time of dynamic vulcanization at 180°C. It was also determined that the accelerator migration from waste EPDM plays only a minor role in the dynamic vulcanization process. It was concluded that waste vulcanized EPDM can be substituted for virgin EPDM up to 45% in the preparation of dynamically vulcanized PP/EPDM–based TPV, without any loss of performance.

Anandhan, Bandyaopadhyay, De, Bhowmick, and De[60] have recently completed a comprehensive study on the development of a new thermoplastic elastomer based on nitrile rubber and styrene-acrylonitrile copolymer (SAN). They have shown that as much as 50% of nitrile rubber (based on rubber hydrocarbon content) can be replaced by ground nitrile rubber waste (r-w-NBR) without affecting the physical properties. They prepared nitrile rubber sheets based on a seal formulation and cured it for 6.75 minutes at 150°C. The sheets were aged for 48 hours and 160 hours (to simulate rubber waste), after which they were abraded against a rotating silicon carbide wheel of a mechanical grinder at room temperature, to produce r-w-NBR. Their swelling coefficients were measured in methyl ethyl ketone to characterize the extent of aging. It was assumed that the cross-link density will change on aging, and that will be reflected in the change in swelling coefficient. The waste rubber powder (r-w-NBR) was characterized in respect to particle size, morphology, and density. The blends of NBR and SAN were prepared in a Brabender Plasticorder with cam type rotors. SAN was dried at 80°C in vacuum for 3 hours prior to mixing. The TPV was prepared from a 70/30 blend of nitrile rubber and SAN copolymer containing zinc oxide, stearic acid, mercaptobenzothiazole (MBT), tetramethyl thiuram disulfide (TMTD), and sulfur. Virgin nitrile rubber in the blend was gradually replaced by r-w-NBR. Blending was done at 180°C in Brabender Plasticorder followed by sheeting in a cold two-roll mill. The sheeted compounds were cured in a hydraulic press at 5 MPa pressure and 210°C temperature. The cured samples were characterized with respect to their mechanical properties, morphology (by TEM), surface characteristics (by atomic force microscopy; AFM), and reprocessability by repeated extrusion. Mechanical properties of the blends containing different amounts of 96-hour-aged r-w-NBR are given in Table 7.12. Subscripts in the table note the percentage of virgin rubber replaced by waste rubber in the blend. The tensile strength, toughness, tear strength, and 100% modulus increased up to 45% r-w-NBR, after which they declined drastically. The elongation at break decreased gradually up to 45% and then dropped abruptly with further replacement of NBR with r-w-NBR. Hysteresis increased gradually up to 30% replacement and then remained constant. These results show that virgin NBR in a NBR/SAN blend can be replaced up to 45% without loss of properties. When similar blends were prepared containing 168-hour-aged NBR powder, the mechanical properties went down to small extent. This is probably due to higher amount of degradation, as evident from higher swelling coefficient in MEK.

TABLE 7.12

Mechanical Properties of NBR/SAN Blends with Different Amounts of r-w-NBR

Mechanical Property	Blends							
	W_0	W_{20}	W_{30}	W_{40}	W_{45}	W_{50}	W_{60}	W_{70}
Tensile strength, MPa	11.9	12.5	14.3	13.6	15.5	11.3	10.3	10.0
Elongation at break, %	267	262	235	251	224	203	141	114
100% modulus, MPa	6.3	7.5	9.0	8.0	9.9	8.2	9.2	8.8
Tear strength, kNm^{-1}	60.9	62.6	50.2	63.6	73.8	60.0	58.6	54.4
Tension set @ 100% elongn. %	24	26	20	24	24	24	26	26
Hysteresis loss, $Jm^{-2} \times 10^{-6}$	0.111	0.134	0.167	0.151	0.185	0.158	0.173	—
Toughness (Jm^{-2})	8664	9812	9969	9896	10292	6806	4685	3609
Swelling coefficient (dioxane) (%)	355	341	333	323	314	302	291	279

Source: Anandhan, S., De, P.P., Bhowmick, A.K., De, S.K., Bandyaopadhyay, S., *J. Appl. Polym. Sci.*, (in press). With permission.

Transmission electron micrographs of the blends showed that SAN formed the continuous phase and the virgin NBR–encapsulating r-w-NBR particles formed the dispersed phase. The surface topography, as seen from AFM pictures, revealed that the surface of the blends were very rough, with appreciable amount of bumps and holes. The surface roughness, arising out of this type of topography, led to increase in tensile strength and 100% modulus. The hysteresis values followed the same trend as roughness, possibly because of increased molecular friction. The potential for the migration of residual curatives from the vulcanized powder to the virgin rubber was investigated by Monsanto Rheometric studies. It was found that such migration was very small, and fresh curatives need be added to NBR for satisfactory vulcanization.

Weber, Mertzel, and co-workers [61] have described a process for preparing a thermoplastic composition that contains a blend of ground rubber (particle size, 80 mesh or smaller) and thermoplastics, along with a compatibilizer. Earlier, the same authors had described the results of a study on a ground rubber/polyolefin composition that provided good mechanical strength, comparable to those of thermoplastic elastomers. Vulcanized EPDM rubber, cryogenically ground to fine particles by a slurry process (developed by Rouse)[62] was mixed with polyolefin-type plastics or polyolefin EPDM based

TABLE 7.13

Typical Example of TPE Blend Composition

Ground EPDM, 200 mesh	100 parts (by weight)
Polypropylene (Rexene W110)	60
Flexon 885, Paraffinic oil	30
Silica, FK 500LS (Degussa)	3
Antioxidant	0.2
Lubricant	0.4

EPDM, ethylene-propylene-diene monomer rubber (vulcanized).

Source: Rouse, M.A., U.S. Patent #5,411,215, May, 1995.

TABLE 7.14

Physical Properties of TPE Blend and Santoprene W233

Property	TPE Blend	Santoprene
Hardness, Shore A	72	72
Tensile strength, psi	803	727
Elongation at break %	226	651
Tear strength, kN/m	31	32
Compression set, % (22 hr at 70°C)	56	53

Source: Rouse, M.A., U.S. Patent #5,411,215, May, 1995.

TPEs and TPVs, along with a compatibilizer-cum-plasticizer (Flexon paraffinic oil)[63] in an extruder to make pellets. The latter could be processed by a variety of thermoplastic processing techniques, including compression/injection molding and extrusion. Recycled PE and PP scrap was also used in this process. A typical example of a blend composition is given in Table 7.13. The properties are given in Table 7.14, along with those of a commercial thermoplastic vulcanizate, Santoprene, for comparison. It may be seen that all the properties of the blend are comparable to those of Santoprene, except elongation at break.

Burgoyne, Fisher, and Jury[64] have patented a process in which ground SBR (from GRT) is blended with virgin rubber and a styrene-based thermoplastic, in the presence of a compatibilizer and vulcanizing agents, to give a homogenous mass that can be molded and vulcanized to give useful rubber products. Styrene-based thermoplastics include polystyrene, acrylonitrile-styrene-butadiene terpolymer (ABS), acrylonitrile styrene acrylate (ASA), and SAN. A variety of uncured rubbers, such as SBR, NR, butyl rubber, polybutadiene, nitrile rubber, and silicone rubber, could be used in this process. The homogenizer-cum-compatibilizer used in the patent includes hydrocarbon resins such as poly-octenemer, fatty acid esters, and Struktol 40 (produced by Struktol Canada Inc., Scarborough, Ontario, Canada). Electron micrographs show good homogenization. Compression set and tensile strength of the vulcanized blends are slightly superior to Santoprene and Multibase thermoplastic elastomers. Tear strength is comparable. The elongation at break is somewhat inferior. However, these compositions pass the

3-hour, 100°C fogging test required for automotive specification, whereas both Santoprene and Multibase fail. This process allows incorporation of high levels (up to 70%) of ground rubber (GRT) compared to typical use of about 10% ground rubber in GRT/virgin rubber blends. The amount of GRT is adjusted to provide necessary levels of compression set, surface smoothness, etc. The rubber-thermoplastic compositions are particularly suitable for making seals, seal extensions, and other automotive components.

Johnson[65] has worked out a process for blending finely ground vulcanized rubber scraps, including scrap tires with thermoplastic resins and/or SEBS copolymers to produce a variety of rubbery compositions, of different hardness from 60 Shore A to 40 shore D.

Jury and Chien,[66] of NRI, patented a process for making a TPE from scrap-tire rubber powder, polyolefin, uncured rubber (styrenic thermoplastic elastomer), and a vinyl homo or copolymer. The 20 to 100 mesh GRT powder, free of metal and fiber contaminants, was used. The high-molecular-weight olefin polymers could be PE, PP, poly-isobutene, or their copolymers. Whereas the styrenic block copolymer helps in compatibilizing the ground rubber containing substantial amount of SBR, (from tire-tread compound), the vinyl polymer helps in bonding the vulcanized rubber particles with nonvulcanized rubber (or block copolymer) and polyolefin. The composition is premixed in an internal mixer at about 275 to 400°F. The blended mass is pelletized or powdered for processing by extruder or injection molding. It may contain small amounts of plasticizers, processing aids, mold-release agent, and antioxidants. The resulting product is a thermoplastic, general-purpose elastomer, trade named Symar T, with good fluid resistance suitable for replacing thermoset rubbers in automotive and other applications. Typical examples of blend composition and properties are given in Table 7.15 and Table 7.16.

Haber[67] has recently reviewed the rubber recycling processes practiced at NRI. Both postconsumer passenger car tire waste and tire plant waste are used. The rubber is mainly a mixture of NR and SBR. Their process includes three stages, namely separation of steel cords and fibers, granulation, and thermomechanical reactivation. Products of each stage are marketed for

TABLE 7.15

Typical Composition of the Blend

Micro tire crumb	51%
Polyethylene	16%
Polypropylene	8%
EVA	11%
SBS TPE	12%
Antioxidant	1%

EVA, ethylene-vinyl acetate copolymer; SBS, styrene-butadiene-styrene block copolymer.

Source: Jury, J.R. and Chien, A.W.V., U.S. Patient #6313183, July 7, 2001.

TABLE 7.16

Typical Properties of the Processed Blend

Hardness	77 Shore A
Sp. gravity	1.04
Tensile strength	5.2 MPa
Elongation at break	260%
100% Modulus	3.5 MPa
Compression set	60% at 23C
Tear strength	32 Kg/m at 23°C

Source: Jury, J.R. and Chien, A.W.V., U.S. Patent #6313183, July 7, 2001.

different applications. The proprietary thermomechanical modification process uses both heat and intensive mixing to devulcanize GRT and to reactivate the surface of the rubber particles.

NRI produces two types of materials from recycled rubber, namely Symar D and Symar T. Symar D is a rubber product, which is further compounded with virgin rubber to produce molded goods. The uncured sheets produced in the process are used for producing sheeted, compression-molded, or injection-molded products, which are subjected to batch or continuous vulcanization. They also market a masticated rubber compound made from postindustrial and postconsumer scrap rubber containing short fibers, tire cords, and processing oil. The compound is processed at high temperature and pressure, and the molded parts are usually black. The automotive application of the masticated rubber compound includes splash shields, close-cut and sight shields, headlamp seals, spare tire protector, and fuel tank pads, etc. These products successfully compete with products made from alternative materials such as ground rubber dispersed in EVA matrix or bonded by polyurethane. They are, however, superior to thermoplastic-based compounds in respect to their thermal resistance.

The second line of product, Symar T, is a thermoplastic elastomer. As described earlier, this is based on a blend of reactivated rubber and a thermoplastic. It is produced in two grades: Symar T 200 and Symar T 700. Symar T 200 is used for under-the-hood applications, where color and surface appearance are not important. As a result, more than 50% of postconsumer passenger car–tire recyclate can be used in producing this grade. The first commercial use of this product is a 2-mm-thick, 3-in by 24-in lower radiator–to-fascia seal for Daimler-Chrysler's 1999 Jeep Grand Cherokee sport utility vehicle. Symar T processes like a conventional TPE on standard injection molding machines. It is also used for other automotive applications such as sound dampening and under-the-hood parts, air deflectors, radiator seals, and sill covers. Table 7.17 gives the typical properties of Symar T 200, along with specifications. Symar T 700, containing 30% recyclate, is typically black and textured and is designed for exterior applications such as air dams, site shields, and sill covers, where the color and appearance, although critical, are not specified.

TABLE 7.17

Typical Properties of Symar T 200

Property	Value	Specification
Tensile strength (MPa)	6	5.5
Tear strength (kN/m)	50	40
Hardness (Shore A)	90	90
Compression set (%)	77	80 max
Specific gravity	1.0 + 0.02	1.0
Recycled rubber content %	50	50 max
Low temp. resistance	No cracking	Pass (at –40°C)

Source: Haber, A., Proceedings of Conference on Plastics Impact on Environment, GPEC 2202, Detroit, Michigan, Feb 13-14, 2002, Paper #17, p. 149 (CD-ROM, 012). With permission.

Pillai and Chandra[68] have described a unique process in which the vulcanized rubber is dynamically devulcanized in the blending process and then revulcanized in final processing. The three-step process includes mixing on an open roll mill, followed by mixing in an internal mixer and then in an extruder. The ground scrap rubber is blended with epoxidized NR and stearic acid or zinc stearate on an open roll mill. Fifty to eighty parts of this mix are then blended with 50 to 15 parts of a thermoplastic resin in an internal mixer in the presence of 1 to 5 parts of zinc oxide, 1 to 8 parts of EVA, and 0.2 to 2 parts of a peptizer. Mixing is carried out at around 350°F for 4 to 10 minutes, during which the scrap gets devulcanized. The product is then mixed with an accelerator package containing 0.2 to 1.0 parts of dibenzothiazole disulfide and a similar amount of diphenyl guanidine, along with 3 to 5 parts of zinc stearate, and blended with a thermoplastic polyolefin such as PE or PP. The mixing process is carried out for 1 to 2 minutes in an extruder and then dumped into a rotary pelletizer. The viscosity of the load goes down and then up, showing that at first, the scrap rubber gets devulcanized, and then mixes with the thermoplastic, giving a homogenous mass of higher viscosity. The rubber particles dispersed in a thermoplastics matrix get revulcanized under the influence of an accelerator package. Typical scraps used in this process include roofing membranes, NR gloves, truck tire retread, passenger tire crumb, auto-window seals, etc. The composition and properties of the thermoplastic elastomer-like blends are given in Table 7.18 and Table 7.19, respectively, and compared with those of conventional TPVs. The TPVs made by this process, from recycled rubber scrap, appear to have better strength and stiffness and relatively lower compression set than conventional TPVs made from virgin rubber.

Liu, Mead, and Stacer[69] have carried out a systematic investigation to produce novel thermoplastic elastomers and toughened plastics by blending recycled rubbers with virgin PP of different molecular weights. The recycled rubber compounds employed were EPDM, SBR, and SBR/NR blends obtained from sources such as shoe soles, roofing membranes, and scrap tires. Rubber powders of five different particle size and PPs of five different

TABLE 7.18

Example of Typical Blend Compositions

Base composition: EPDM roofing membrane scrap, 60 parts by weight.	
Epoxidized natural rubber	1.80
Stearic acid	0.70
Final composition: base composition	62.50
Polyethylene (virgin)	40
Zinc oxide	3
Diphenyl guanidine	0.25
Dibenzo-carbazole disulfide	0.50
Ethylene vinyl acetate copolymer	3.0

Source: Pillai, C.R.P. and Changra, H., U.S. Patent #6,313,183 BI, November 6, 2001.

TABLE 7.19

Properties of the Thermoplastic Elastomer Product

Blend Composition	Tensile Strength, psi	Elongation at Break %	Modulus at 50% Strain, psi	Compression Set (% at 23°C)	Hardness Shore A
EPDM scrap/PE	1660	170	1630	78	93
Under-hood body					
Plug scrap/PP	2600	410	2080	89	93
Window seal scrap/PP	1650	110	1580	58	92
Truck tire scrap/PP	1600	120	1280	40	92
Passenger tire scrap/PP	1490	90	1440	59	86
Santoprene grade101 73	820	120	710	40	77
Santoprene grade101 75	830	220	630	29	80

EDPM, ethylene-propylene-diene monomer rubber.

Source: Pillai, C.R.P. and Chandra, H., U.S. Patent 6,313,183 BI, November 6, 2001.

molecular weights were used in this work. Tertiary butyl hydroperoxide along with octyl-phenol-formaldehyde resin were used for reactive compatibilization. Mixing was carried out using a Plasticorder at 200°C and 30 rpm for 30 to 45 minutes until the torque became stable.

The additives were introduced 2 minutes after the addition of the rubber. Blends designed for making thermoplastic elastomers were prepared using the dynamic vulcanization procedure used by Coran and Patel.[33] Compression-molded specimens were used for testing. Crystallinity, dynamic mechanical properties, and apparent viscosity were measured for each composition. Dynamic Mechanical Analysis (DMA) study indicated phase separation for all blends. The DMA peak at – 55 to – 56°C corresponded to the glass transition temperature of the rubbers (EPDM, SBR, etc.) and the peak close to 0°C corresponded to the glass transition temperature of PP. The blends exhibited pseudo-plastic behavior in all cases. Apparent viscosities were measured using a capillary rheometer at shear rates varying between

10 to 5000 per second. The results were fitted into power law equation $\eta = k(\Upsilon/\Upsilon_0)^{n-1}$, where η is apparent viscosity, γ is shear rate, n is the power law exponent, and k is the consistency index. In general, n decreased and k increased with rubber content, consistent with the general theory of rheology of suspensions. The rubber particles acted as suspensions in a continuous phase of PP melt. However, the increase of viscosity was not as pronounced as would have been expected in the case of rigid fillers, due to the deformability of rubber particles. The increase of the exponent n with smaller rubber particle size suggested greater interfacial contact area between the two phases. The smaller value of n at high concentrations of rubber particles indicated shear rate sensitivity of the blend.

Mechanical properties went down with increase in powder content, but rebounded substantially on the addition of a reactive compatibilizer, due to interphase mixing and some cross-linking. Octyl-phenol formaldehyde (OPF) and tertiary butyl hydroperoxide were added to the blend during mixing of PP with rubber powder, to carry out the reactive blending. It was postulated that OPF grafted to both PP and EPDM, thereby enhancing the compatibilization between the thermoplastic and rubber phases. Residual peroxide helped to further cure EPDM. Both of these factors contributed to improved strength. Reactive blending increased stress-bearing capability of all the blends (except for those at 10% rubber). Tensile strength increases as high as 80% were obtained at the highest rubber concentration. Elongational capabilities of 105 to 110% were obtained for reactive blends with 50% and 80% rubber content. At low levels of rubber, the strain capability more than doubled by reactive processing. The increase in both stress and strain led to the increase of strain energy density and provided rubber-toughened plastics. Preliminary blending of maleated PP with EPDM powder did not provide significant improvement in properties, and the reaction took more time to complete. The increase in tensile strength and elongation at break at different rubber content with and without reactive compatibilizer is illustrated in Figure 7.3. A suggested mechanism for interfacial reaction between the phenolic additive and SBR is illustrated in Figure 7.4.

When property enhancement was studied as a function of particle size, it was noted that there was no significant difference in tensile strength of blends containing different size particles. However, the elongation at break increased at all rubber concentrations by more than 40% for blends containing smaller particles. The blend with smaller particles also showed lower consistency index.

Both tensile strength and elongation at break increased with increase in the molecular weight of the PP used. The increase in elongation at break is more dramatic, 50 to 500% at lower rubber content, and a modest 200 to 400% at the highest rubber content. This increase was ascribed to lower crystallinity in the high-molecular-weight PP phase, due to inclusion of rubbery materials and reactive blending. Measurement of crystallinity of blends using differential scanning calorimetry supported this speculation. The inclusion of the rubbery phase probably hinders the individual PP

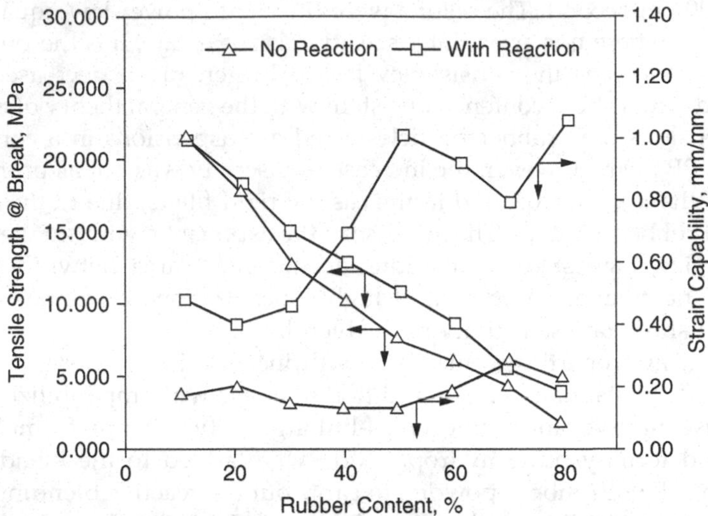

FIGURE 7.3
Tensile strength and strain capability of EPDM/PP blends with and without reactive compatibilization (From Liu, H., Mead, J.L., and Stacer, R.G., *Rubber Chem. Technol.*, 75(1), 2002, p. 49. With permission.)

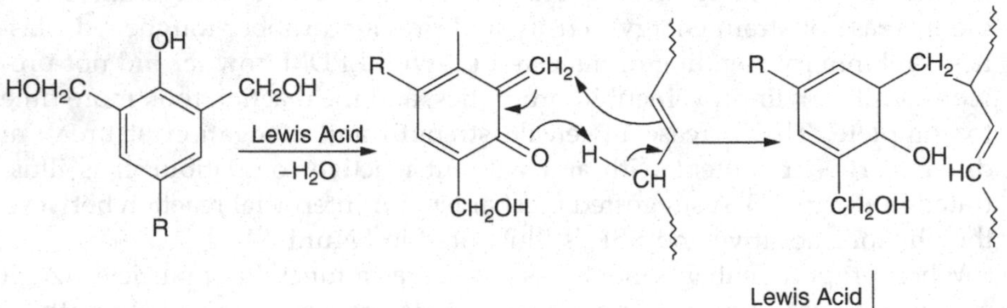

FIGURE 7.4
Possible mechanism for the compatibilization of SBR/PP blend by phenolic resin. (From Liu, H., Mead, J.L., and R. G. Stacer, *Rubber Chem. Technol.*, 75(1), 2002, p. 49. With permission.)

molecules from participating in the growth of crystallites. Similar results were obtained when PP was blended with NR/SBR blends.

In short, the authors (Liu, Mead, and Stacer[69]) concluded that mechanical properties of the recycled rubber/plastic blends, without adequate compatibilization, are not high enough for practical applications. Better properties are obtained by using rubber particles of smaller size and high molecular weight thermoplastics, and with adequate compatibilization.

Both ambient and cryogenically ground scrap tire rubber were blended with virgin and recycled PP to provide blends that behave like dynamically vulcanized blends. It was claimed that the product could be used in applications such as artificial pitches for sports facilities.[70]

7.5.1.2 Roofing and Miscellaneous Products

It is reported that waste polymers such as vulcanized rubber scrap (mainly from scrap tires) and recycled LDPE, PP, or PVC can be used for blending, individually or in combination with virgin polymers and end-of-life recycled polymers.[71] The preparation of these compositions and their use in the manufacture of porous hoses, automotive parts, flooring, traffic control and security elements, motorway sound barriers, and packing for wastewater treatment have been noted.

Astro-Valcour, Inc. of Glen Falls, New York, recycles PE scrap materials from packaging converters and fabricators and combines them with postconsumer scrap and virgin PE to manufacture protective cushioning products such as bubbles, planks, and foam. Partially devulcanized rubber waste can be used in this process.[72] It is reported that plastics-rubber blends surpass paper-based cushioning products in shock absorption, fatigue endurance, and thickness loss (compression set) under load.

Doan[73] disclosed a process for manufacturing street signs and other similar products made from reground waste rubber, SBR, and polyolefins such as PE and PP. First, the ground rubber was surface plasticized in a twin-screw extruder at high temperature and pressure and subsequently blended with polyolefin or acetal resin. A large amount of ground rubber, as high as 75%, can be used in this process.

Mertzel, Weber, and co-workers[74] have patented a composition containing ground vulcanized rubber waste and a thermoplastic, preferably a polyolefin, in the presence of a compatibilizer (based on paraffinic oil such as Flexon, of Exxon Chemical Company). Fillers such as silica, antioxidants, and lubricants are also added. Mixing is carried out either in a batch or continuous process. The blend is pelletized for subsequent molding and for making products such as seals, seal extensions, and other automotive components.

Ferro Corp. is reported to produce filled and reinforced PP compounds that contain at least 25% postconsumer recycled material and are aimed at the automotive market.[75] Highway guardrail posts are manufactured from GRT, recycled PE oil jugs, and herbicide containers, in a technology partnership between Amity Plastics and Alberta Research Council.

World Tech Management, Inc. has found a method for chemically bonding passenger tire scrap rubber to plastics without the need for additives. They have chemically linked crumb rubber with PP or HDPE.[76]

Howard and Coran[77] have reported a study aimed at developing useful products by blending GRT with HDPE. The effects of particle size and GRT concentration on the mechanical and rheological properties of the blends were discussed. The composition of a rubber/plastic blend, based on semi-crystalline HDPE (obtained from recycled milk jugs) and a rubber binder based on EPDM, was demonstrated.

Bennett's[78] patented reactive compounding technology enables chemical binding between different polymers. The company has recently achieved appreciable success in developing thermoplastic-elastomer blends and has entered the thermoplastic-elastomer market with two series of products. While propyrene SP (SBS /PP blends) are based on virgin raw materials, Bennetire is a TPE based on SBR/NR grind, originating from truck tires.

Nagayasu, Orisaki, and fellow investigators[79] have patented a process in which fine particles obtained by crushing rubber waste are mixed with rough, larger particles obtained from thermoplastic waste and fed into an injection molding machine to provide an ebony (wood)–like product that is hard and tough with high impact resistance and good weather and water resistance.

Rubber scrap prepared in particulate form from worn-out tires was segregated and sieved to provide particles of suitable size. The latter was used as filler for scrap PE and PVC to manufacture traffic-control equipment such as road guides, road emergency supports, speed bumps, and guiding rails.[80] Plastic scraps were obtained from cable production plants and were used without removing metal contaminants. Reclaimed PE from other sources was also used. Paraffin oil was used as a compatibilizer, EVA as impact modifier, and phospho-gypsum as filler. The compositions were optimized for providing the necessary performance characteristics and for processing by injection molding, extrusion, and compression molding.

There have been several attempts to produce roofing tiles and roofing membranes by blending thermoplastics and rubber waste. U.S. Patent 4,028,288 describes blending of ground tire scrap, including fibrous tire cords, with thermoplastics such as PE, PP, and polystyrene under heat and pressure to make roofing products.[81] Butilino[82] describes the admixture waste rubber, plastics, and fibers with thermoplastics such as PE in an extruder to provide roofing products. Roofing membranes have been produced by dispersing crumb rubber in polyurethane matrix.[83]

Roofing products have also been developed by blending rubber scrap with polysulfide rubber. The composition is used to cast roofing membranes that are subsequently cured *in-situ*.[84] Pulverized waste tires were blended with a limited amount of recycled thermoplastics and additional ingredients such as starch, acetone, caustic soda, glass wool, and ammonium phosphate to provide better moldable roofing singles.[85]

Attempts have been made to use chemical bonding to improve the cohesive strength of the crumb rubber–based products. Sulfur and zinc-stearate

TABLE 7.20

Properties of GRT/SBS Composite

Properties	Flow Direction	
	Parallel	Normal
Tensile strength (psi)	775	875
Elongation at break (%)	275	525
100% Modulus (psi)	630	275
300% Modulus (psi)		650
Tear resistance (p/i)	195–200	200–235
Elongation (%)	13	25

Source: Doan, R.C., U.S. Patent #4,9999970,043, February 27, 1989.

have been used to achieve better properties of rolled products, suitable for roofing.[86] Doan[87] has patented a process in which approximately 70% ground rubber is mixed with styrene-butadiene–styrene block copolymer such as Kraton D-3202 in a twin-screw extruder and subsequently maintained at high pressure and temperature for some time to provide a moldable product. It is presumed that the rubber crumb gets surface plasticized by the thermoplastic elastomer and gets rejuvenated in the subsequent heating under pressure, to flow better in the injection-molding process. This process claims to use the highest recyclate content without losing much in performance. Table 7.20 provides the properties of the injection-molded product in parallel and normal direction to flow.

Johnson[88] has described a process for making extruded roofing products containing large amounts of crumb rubber, polyolefin, and a functionalized olefin copolymer containing long-chain unsaturated mono- or di-carboxylic units, vinyl esters, and acid anhydrides. Rosenbaum[89] has designed an extruder with a special transition nozzle to produce articles from blends of the whole ground rubber crumb, containing metals and tire cord, and thermoplastics, especially polyolefins and high-impact polystyrene. Recently, Blalock and Nelson[90] patented a process for manufacturing roofing shingles from ground rubber waste and thermoplastics, which are lightweight and have outstanding strength and durability, together with excellent temperature stability, weatherability, and resilience. The products are particularly suitable for use in southern parts of the U.S., which experience severe hailstorms. The authors used branched polyolefins containing hexene and octene co-monomer, such as Engage 8150 (Du Pont-Dow Elastomers) and Exact (Exxon) EPDM, along with fillers and processing oil for blending with crumb rubber in an internal mixer and extruding the blend to produce continuous sheets. The latter are then made into roofing shingles. Table 7.21 presents typical composition of their blend.

The final products have excellent visual appearance and pass the usual pull test and tear test required for roofing membrane application. They provide low-cost, efficiently manufactured, polymer-based products such as roofing shingles. The shingles are weatherproof, inexpensive to manu-

TABLE 7.21

Material Composition of the Blend

Engage 8150 polyolefin	50 parts by weight
Silicate clay	100
Regenerated crumb	100
Nordel EPDM 2470	50
Sundex oil	100
Carbon black	100

Source: Blalock, C. and Nelson, L., U.S. Patent #6,194,519 BI, February 27, 2001.

facture, and durable compared to conventional shingles. The rubber material can be most efficiently provided by the use of crumb rubber particles from spent automobile tires and other recycled rubber debris that are readily available at very low cost. The products have an average life far greater than that of current commercial roofing shingles, which are made using fiberglass or asphalt.

Styrene-butadiene rubbers are often used in the production of tire treads, especially in passenger tires. Burgoyne, Fisher, and Jury[91] of National Rubber Technology, Canada, have patented a process of recycling SBR crumb rubber by blending it with virgin SBR and polystyrene. Recycled polystyrene obtained from styrofoam products is especially suitable for this purpose because of its high surface-to-volume ratio and low cost. A homogenizing agent such as Strucktol 40 MS is added to the blend. A cure package containing suitable vulcanizing agents, as well as processing aids and fillers are added to provide a strong, well-cross-linked product. Blending is carried out in an internal mixer below 275°F. When scanning electromicrographs of thin sections from the vulcanized product are compared with the micrographs of nonvulcanized compound, the vulcanized product appears to be very homogenous with very fine particle size for the dispersed phase, compared to that in the nonvulcanized product. Use of polystyrene helps the compound to be stiff and retain stiffness up to 150°F. This provides a method for making stiff rubber products such as mud flaps, which need to maintain stiffness at moderately high temperature.

A new U.S. start-up tire recycling firm, Re-Engineered Composite Systems, LLC, has signed an exclusive worldwide licensing agreement with the University of Massachusetts to develop a patent-pending blending technology that grafts thermoplastic resins onto the surface of vulcanized rubber particles.[92]

Khait[93] used the solid-state extrusion process (SSEP) for the pulverization of waste rubber and for mixing the pulverized rubber with recycled LLDPE, LDPE, and HDPE, in the presence of a compatibilizer, EVA. Subsequently, the author customized a Berstorff twin-screw extruder to use it for SSEP pulverization of plastic waste for recycling with rubber scrap.[94] Scrap gets softened as it passes through the barrel. Then it is conveyed into a cooling zone, where the material is pulverized under a combination of shear stress and high pressure. The kneading disks present in this zone carry out max-

TABLE 7.22
Physical Properties of APTR/Postconsumer Thermoplastic Blends

Ingredients	Tensile Strength MPa	Elongation at Break (%)	Tear Strength Kg/cm	Hardness Shore D
ATPR/LLDPE-P/EVA	5.92	26	4248	40
ATPR/LLDPE/EVA-P	6.21	34	4338	40
APTR/HDPE-P/EVA	8.82	33	7051	50
ATPR/HDPE/EVA-P	9.03	28	6873	50
ATPR/LDPE-R/EVA	5.86	100	5248	34
ATPR/LDPE/EVA-P	6.75	168	5480	35

Source: Khait, K., ACS Rubber Division Meeting, Louisville, Kentucky, October 8 – 16, 1996. With permission.

imum pulverization, reducing the scrap into particles of 10 to 40 mesh. It was stated that powders of desired size could be obtained by varying the equipment design and the processing parameters. Fluffs of PP, HDPE, and LLDPE, when subjected to this process, produced elongated particles of greater surface area. The molded plaques from the powder exhibited higher tensile strength, indicating improved compatibilization.

Khait used a modified Berstorff co-rotating twin-screw extruder to carry out SSEP of ground rubber waste and thermoplastics.[95] In one set of experiments, she pulverized all the base polymers, including ground rubber, and added EVA before molding. In the other, she pulverized all the components simultaneously, including EVA. Most particles had a diameter between 100 to 250 microns and a cauliflower-type open morphology, suitable for blending. All samples were mixed in a Brabender Prep Mixer using cam blades at 180°C and 80 rpm for 5 minutes and compression molded at 220°C. The physical properties of the different blends are given in Table 7.22. APTR represents the pulverized rubber powder and P at the end means pulverization of all components carried out simultaneously.

The results clearly indicate that the blends have appreciable strength, particularly tear strength, and all the properties are better when all the ingredients are pulverized in one step. This further indicates that the coupling agent probably acts better when subjected to SSEP. The blends were subsequently converted into heavy mats, with large, sharp knobs by compression molding, indicating that the mold flow of the blend was appreciably good.

Rihai and co–workers[96] have used solid-state shear extrusion (SSSE) for recycling waste rubber with thermoplastics such as polyethylene. In this process, the materials are simultaneously subjected to high shear and high pressure, much below their melting point. The elastic deformation under the action of high pressure and high shear releases energy that is spent in creating new surfaces. When this process is carried out in a twin-screw extruder, the materials are pulverized. When a few parts of PE are added to the scrap rubber, the surface area of the rubber particles increases and then sharply decreases at about 5% PE. The roundness of rubber particles undergoes a similar change during blending of rubber scrap with plastics.

TABLE 7.23
Mechanical Properties of PVC/GNBR at Different Composition

Composition (by wt.) PVC/GNBR	Tensile Strength Kg/cm^2	Elongation at Break, %	Impact Energy J/cm	Flex Crack Resistance No. Cycles to Crack
100/00	550	25	0.40	1
80/20	238	180	1.75	56
60/40	162	270	9.0	100,000
40/60	107	360	Pass*	Greater**
20/80	41	525	Pass*	Greater**

Note: GNBR, ground nitrile rubber; PVC, polyvinyl chloride.

* Without break.

** Greater than 100,000.

Source: Tipanna, M. and Kale, D.D., *Rubber Chem. Technol.*, 70, 1997, p. 815. With permission.

Printing rollers used in lithographic or offset printing are made out of NBR. The rubber is usually removed by grinding, and the metal mandrel is reused in the manufacture of new rollers. Tipanna and Kale[97] blended finely ground nitrile rubber (GNBR) from old printing rollers with PVC (weight average mol. wt. 140,000, poly-dispersity Mw/Mn = 2.5) and studied the physical properties of the blends as a function of rubber loading. The average size of the rubber particles were from 100 to 300 μm. The acrylonitrile content of NBR was 21.8%. Dibutyl phthalate (five parts), barium cadmium thermal stabilizer (two parts), and zinc oxide (one part) were also added to the compound. Blending was carried out in a Haake Rheomix at 15 rpm. The rubber content was changed from 10 to 90%. The physical properties of the different blend compositions and flex crack resistance are given in Table 7.23.

It appears that the ground waste NBR is compatible with PVC and acts like a solid plasticizer. This confirms earlier observations that PVC is compatible with SAN and with NBR over a certain range of nitrile content.[98,99] This type of compatibilization has been attributed to the similarity of the solubility parameters of the two component polymers and the existence of specific interaction between them.

Zytek and Zollinger[100] blended nonvulcanized and vulcanized NBR with PVC and saw no significant difference between the physical properties of the blends. It was concluded that, although vulcanization restricts molecular mobility, it does not affect intermolecular interaction.

Automotive companies are promoting the reuse and recycling of plastics and rubber. Ford Motor Company is working with TPE supplier Syntene Company (Richmond, Indiana) to recycle tires into various automotive parts such as brake pedal pads.[101] Tires are ground into fine powder, magnetically separated to remove steel and other metals, screened, and sorted. The resulting fine particles are then mixed with plastic, turned into pellets, and molded into vehicle parts. The pads contain at least 50% recycled tire content. Chrysler is working with National Rubber Company (Toronto, Canada) to produce steering shaft seals and fender liners from recycled rubber. The company is developing polymer-grade recycled rubber for use in auto parts.

Rouse[102] has discussed the use of rubber obtained from postconsumer rubber products as a compounding ingredient in the manufacture of EPDM hose. The USCAR Vehicle Recycling Development Center separated the scrap hoses, converted them into fine particles, and separated the particles into different grades. The processability and properties of the blends containing the different grades of recycled rubber powder are being investigated.

Rumber Materials is promoting its new product called Rumber Composite, a blend of ground rubber tires with PE or PP.[103] The material is offered as an impact modifier to extend durability and promote both tear and crack resistance. It has been used in making road signs, buckets and feeders for livestock, plastic pallets, and dustbins.

Plastic crossties or sleepers now are challenging the wooden ties long favored by railroads.[104] The Chicago Transit Authority (CTA) has installed some 20,000 plastic ties and is ordering 10,000 more. Plastic ties should account for 25-30% of all of CTA's ties in 10 years. Plastic ties do not rot or splinter over time. Some producers even offer 50-year warranties on the life of plastic ties. All the commercialized plastic ties contain at least 50% recycled HDPE and other materials, including ground rubber waste. With a standard-size plastic tie typically weighing around 200 pounds, increased penetration of these ties in railroad application would create substantial demand for both plastic and rubber waste.

A sealant that has received approval from Miami-Dade County's Office of Code Compliance has been introduced by Urecoats Industries, Inc. It is made from Urecoats 100 fluid; compounded mainly from urethanes, crumb rubber from recycled tires, and asphalt. The sealant is resistant to oils and chemicals.[105]

In a recent study, typical hospital waste consisting of polyolefins was mixed with crumb rubber and characterized for mechanical properties. Various coupling agents were used as a means of combining the materials. The blend with the highest mechanical properties was specified for use as different types of value-added products.[106]

7.5.1.3 Rheological Studies

It is important to have a thorough understanding of the rheological behavior of rubber/plastic composites as a function of temperature and pressure, to develop appropriate processing methods for thermoplastic/GRT blends. Although there are studies in this area for polymer/hard-filler composites, very little is known about systems in which fillers are soft and are likely to deform at high temperature.

Recently, Naskar, De, and Bhowmick[107] have studied the thermorheological behavior of thermoplastic elastomer blends based on EPDM and EAA, (GRT), and maleic anhydride–grafted GRT (m-GRT) to determine the influence of maleation on the GRT, particularly the rheological characteristics of the blend. Higher storage and loss modulus and shear viscosity were noted

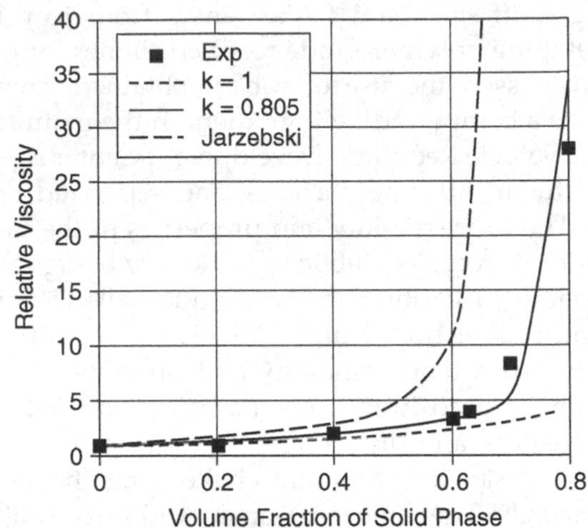

FIGURE 7.5
Relative viscosity of GRT/LDPE composite at 190°C.(From Sbraski, I. and Bhattacharya, S.N., Proceedings of International Conference, Calcutta, India, 1997, p. 64. With permission.)

for the m-GRT–based TPE compared to the GRT-based TPE composition. An interaction between m-GRT and HDPE was evident from the loss tangent spectra of the blends, and also from the die swell of the melt extrudates. The m-GRT–based TPE produced smoother extrudates at low shear rates than the blend containing non-maleated GRT. However, at higher shear rates, both compositions showed melt fracture.

Sbraski and Bhattacharya[108] have studied the rheology of ground rubber/LDPE composites at very high concentrations of ground rubber. The blends were prepared by mixing the powder with LDPE in a Haake Rheocord, and the rheological measurements were carried out using a high-pressure capillary rheometer. Figure 7.5 shows plots of relative viscosity as a function of volume fraction of the solid phase. The relative viscosity, i.e., the ratio of the viscosity of the composite to that of LDPE, showed little variation with shear stress up to a volume fraction of 0.4. However at higher volume fractions (0.5 and 0.6), the shear viscosity was very high. Further, it exhibited a yield stress at low shear rates. The maximum packing fraction was found to be approximately 0.65 and did not change much either with shear stress or with particle size.

Different theoretical models have been developed to predict relative viscosity as a function of shear rate of filled composites, assuming different values for packing fraction,φ_m, and power law constant, k. Sbraski and coworkers[109] have recently developed a model for shear rheology of composite melts, where the particles are deformable, poly-dispersed, and irregularly shaped. The experimental results of ground rubber/LDPE blend conform to this model for a k = 0.805 and for a higher value for φm, close to that of k. The higher value of k has been ascribed to the deformability of GRT particles at high temperature. It is evident from Figure 7.6 that Sbraski's model can

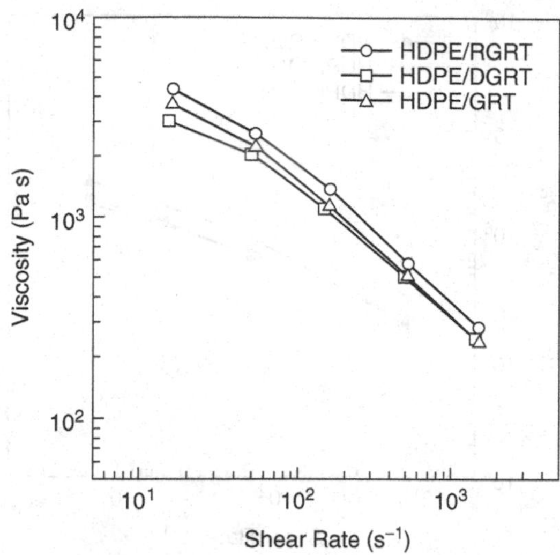

FIGURE 7.6
Viscosity of GRT/HDPE blends as a function of shear rate. (From Hong, C.K. and Isayev, A.I., Paper # 107, ACS Rubber Division Meeting, Orlando, Florida, 1999 and *J. Elastomer and Plastics*, 33(1), 2001, p. 47. With permission.)

successfully predict the relative viscosity of GRT/thermoplastic composites as a function of shear rate.

Vetkin, Goncharov, and Shilov[110] studied the rheological properties (consistency factor and flow index) of composites containing LDPE and crumb rubber from tire curing bags, using a capillary rheometer. The effects of temperature and blend composition on rheological characteristics were determined.

7.5.1.4 Processes

Isayev and Hong,[111] in recent years, have used the ultrasonic devulcanization process for developing compatibilized rubber/plastic blends from recycled rubber waste and polyolefins. In an interesting series of experiments, Hong and Isayev used three different combinations of mixing and vulcanization procedures, i.e., (1) by mixing GRT with HDPE pellets in a twin-screw extruder followed by ultrasonic devulcanization in an extruder, provided with an ultrasonic horn and a die, and subsequent revulcanization in an internal mixer to provide revulcanized Blend (RGRT); (2) by mixing HDPE pellets and ultrasonically devulcanized rubber tire (DGRT), followed by revulcanization in an internal mixer; and (3) by mixing HDPE pellets with GRT, followed by revulcanization in an internal mixer. They studied the flow characteristics using a capillary rheometer, morphology by xylene etching followed by SEM, and physical properties including impact behavior by ASTM methods. The viscosity of HDPE increased on adding GRT, DGRT, and RGRT and decreased with increase in shear rate (Figure 7.6). However, after revulcanization, all three blends show almost identical behavior.

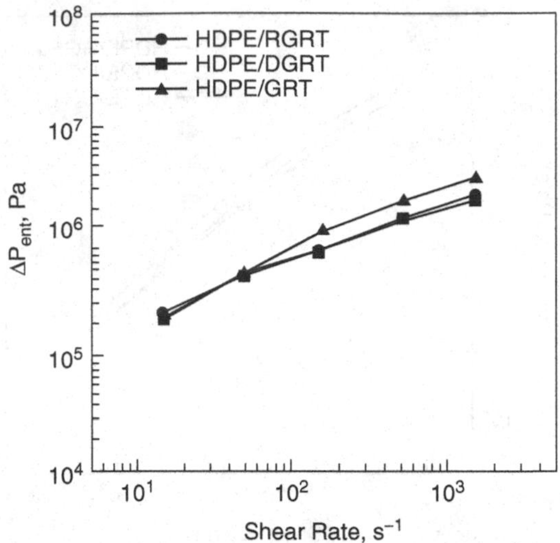

FIGURE 7.7
Pressure drop at die entrance of GRT/HDPE blends. (From Hong, C.K. and Isayev, A.I., Paper # 107, ACS Rubber Division Meeting, Orlando, Florida, 1999 and *J. Elastomer and Plastics*, 33(1), 2001, p. 47. With permission.)

The pressure drop at the entrance of the rheometer, which represents melt elasticity, is the lowest for HDPE/GRT blend but increases with addition of RGRT (Figure 7.7). This shows that the melt elasticity increases with increasing interaction between the plastic and rubber as more and more ground rubber gets devulcanized and dynamically revulcanized in the process. Scanning electron micrographs of the blends show that the particle size of the rubber domain is smaller for HDPE/RGRT than for HDPE/GRT and HDPE/DGRT blends. Further, whereas the surface of HDPE/GRT blends is smooth, that of HDPE/RGRT blends is rough and porous, which facilitates better mixing between the plastic and the rubber phases.

The above results show that simultaneous devulcanization and blending provide better opportunity for plastic rubber interaction than plain blending or blending after devulcanization. This is reflected in the physical properties of the blends. The impact resistance, tensile strength, and elongation at break of the HDPE/RGRT blend (mixing by twin-screw extruder) are far better than those of the other two blends. Addition of coupling agents such as maleated PP or chlorinated PE did not improve the mechanical properties appreciably. This shows that blending with a twin-screw extruder, followed by exposure to ultrasonic waves, increases the plastic-rubber interaction, leading to the formation of some interfacial bonds that provide compatibilization and co-continuous morphology.

The tensile strength, elongation at break, and impact energy (area under the force displacement curves for falling dart experiments) for the three blends are shown in Figure 7.8, Figure 7.9, and Figure 7.10, respectively. It

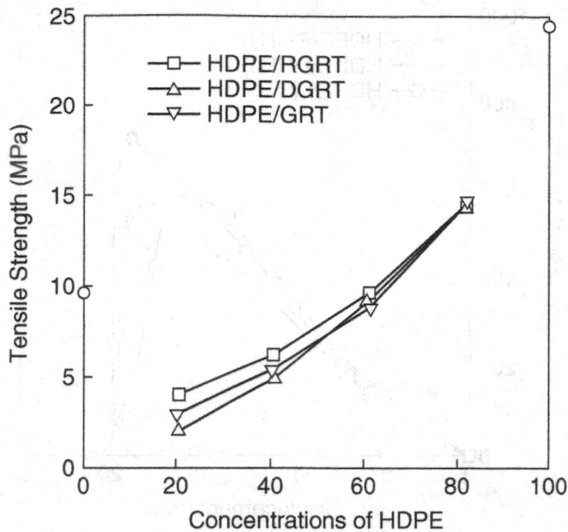

FIGURE 7.8
Tensile strength of HDPE/GRT blends. (From Hong, C.K. and Isayev, A.I., Paper # 107, ACS Rubber Division Meeting, Orlando, Florida, 1999 and *J. Elastomer and Plastics*, 33(1), 2001, p. 47. With permission.)

is obvious that the HDPE/RGRT blend has the highest impact resistance among the three blends. The physical properties, including impact resistance of RGRT/HDPE blends at different concentrations of HDPE, has also been studied. Based on these results, it may be inferred that blending with a twin-screw extruder followed by exposure to ultrasonic waves increases the plas-

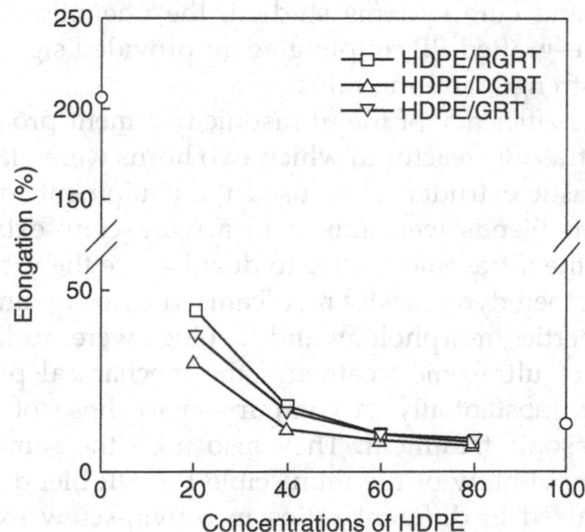

FIGURE 7.9
Elongation at break of GRT/HDPE blends. (From Hong, C.K. and Isayev, A.I., Paper # 107, ACS Rubber Division Meeting, Orlando, Florida, 1999 and *J. Elastomer and Plastics*, 33(1), 2001, p. 47. With permission.)

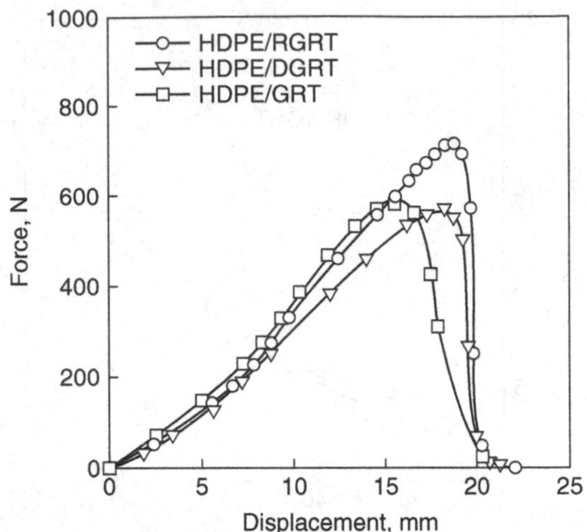

FIGURE 7.10
Force displacement curves for GRT/HDPE blends. (From Hong, C.K. and Isayev, A.I., Paper # 107, ACS Rubber Division Meeting, Orlando, Florida, 1999 and *J. Elastomer and Plastics*, 33(1), 2001, p. 47. With permission.)

tic-rubber interaction and possibly the formation of some interfacial bonds and block copolymers at the interface, which leads to improved physicals.

Luo and Isayev[112] have blended DGRT (ultrasonically devulcanized crumb rubber) with PP. The blend was subjected to both conventional and dynamic vulcanization processes in the presence of sulfur and phenolic cross-linking agents. The blends had better impact strength than PP. Among the various compatibilizers and cure systems studied, the phenolic cure system and maleic anhydride–grafted PP coupling agent provided significant improvement in tensile strength and modulus.

To improve the efficiency of the ultrasonic treatment process, Isayev and Oh[113] used an ultrasonic reactor in which two horns were placed in a slit die attached to a plastic extruder. They used the equipment to blend PP with GRT. Initially, the blends were mixed in a twin-screw extruder and then passed through the ultrasonic reactor to devulcanize the rubber. The resulting product was then dynamically revulcanized by using an internal mixer. Mechanical properties, morphology, and rheology were studied. Under optimal conditions of ultrasonic treatment, the mechanical properties of the blend improved substantially in comparison to those of the blends not exposed to ultrasonic treatment. They also used the same equipment to improve the compatibility of the immiscible PP/NR blends. The two components were mixed in different ratios in a twin-screw extruder and the blends were passed through the ultrasonic reactor. Mechanical properties, rheology, and morphology of the blends were studied using SEM and AFM. Under optimal conditions of ultrasonic treatment, the mechanical properties of the blends improved substantially. The AFM studies indicated development of interfacial layers, interfacial roughening, and improved interfacial adhesion. The untreated blends exhibited weak adhesion and delamination at the interface.

Australia's CSIRO claims to have developed a technique in which crumb rubber from scrap tires can be blended with virgin rubber and plastic resins to achieve savings in the amount of materials used.[114] The technique involves the chemical modification of the surface of crumb rubber using a "bridging molecule" to couple it with virgin material. Researchers at CSIRO have devised a treatment that allows significant proportions of rubber crumb to be incorporated into many plastics products. The process allows production of as-new quality tires using up to 50% recycled rubber content. The patented process involves the chemical modification of the molecular surface of crumb rubber to bridge or couple it with the virgin material. Brief details are given of potential applications. It is reported that in Australia, approximately seven to eight million tires are discarded each year, which either go to landfill or are burned.[115] The new patented technology for surface treatment by CSIRO's Building, Construction Engineering Department, will help in using the scrap tires to produce useful polymeric composites for many civil and industrial applications.

A new way of recycling vulcanized rubber, the Cycer Method, makes it possible to manufacture products that contain up to 80% recycled rubber.[116] Rolf Degerman invented the method while he was working with the production of puncture-proof tires for wheelchairs. The new invention consists of forming a material composition using recycled vulcanized rubber as the main component. It also includes a thermoplastic in powder form and a quantity of expandable microspheres. When this material composition is heated in a sealed unit, internal pressure is generated and the softened plastic is integrated with the rubber. Subsequently, the die form is cooled down and the finished product is lifted out of the form. The three main components of the system include a mixing station, a form-filling device, and an oven heated to 150°C. To commercialize the method, a new company has been formed in Sweden, Cicero Scandinavia, and a factory is currently being built to fully demonstrate the merits of the Cycer Method in an industrialized setting.

7.5.2 Rubber-Thermoset Composites

The building blocks of most thermoset materials, such as epoxy or urethane, are liquid. Hence, dispersion of finely ground rubber in the liquid matrix is easier and more uniform. The ground rubber is not likely to interfere with the cross-linking reactions, which are used to solidify the thermosets. The thermoset matrix shrinks, not only due to cooling to room temperature, but also due to cross-linking reaction. Hence, compatibilization is very important to provide a co-continuous morphology. This is accomplished by the suitable choice of cross-linking agents, which are similar in polarity to the thermoset matrix and have long, nonpolar chains for miscibility with rubber.

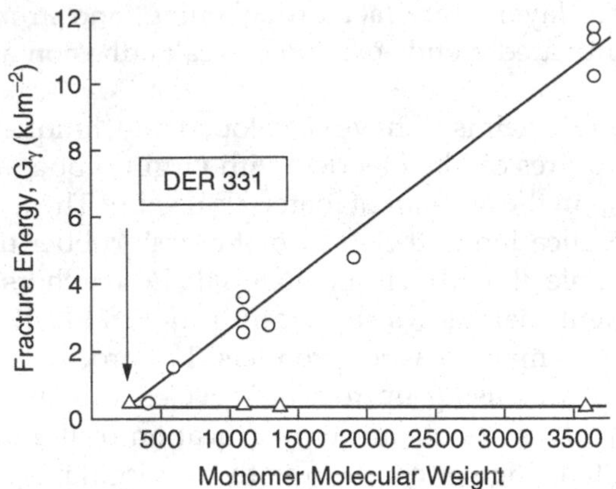

FIGURE 7.11
Impact energy vs. molecular weight between cross-links. ○, modified; △, unmodified. (From Ballester, J.L., Department of Theoretical & Applied Mechanics, Cornell, University, Ithaca, NY, private communication.)

As mentioned earlier, ground rubber waste has been used for impact modification.[117] A combination of finely ground rubber (up to 25%), carboxy-terminated nitrile rubber with Dow epoxy 331, and Jeffamine hardener was used to prepare the blend which was cured for 12 hours at room temperature and 1 hour at 100°C. Toughness increased with increasing molecular weight of the epoxy monomer and cross-linking agent, since together, they determine the chain length between cross-links. Figure 7.11 illustrates the influence of epoxy molecular weight on the fracture energy of the cross-linked unmodified and rubber-modified epoxies. Several mechanisms, including cracks impinging and crack blunting at ground rubber particles, shear yielding, crazing, debonding, and particle tearing, have been put forward for the toughness improvement of ground rubber-thermoset plastic blends. Properties can be adjusted by altering surface treatments and/or rubber content to meet the requirements of different applications. Other potential applications of this technology include its use in polymer concrete structures.

Ballester[118] has successfully used ground rubber from scrap tires to modify epoxy resin to improve its toughness, with minimal impact on strength and stiffness. This makes the material more attractive for high-volume uses such as railroad cross ties.

The use of car-tire regrind as the toughening phase in epoxy resins was also investigated by using different surface-modification techniques, to improve the compatibility at the interface between the rubber particles and the epoxy matrix.[119] Silane coupling agents, plasma surface modification, and acrylic acid with a small amount of benzoyl peroxide initiator were used to modify the surface properties of rubber particles. Ground rubber from used tires, with or without surface treatment, was mixed with epoxy resin and the mixture was poured into molds, to obtain test specimens. After curing, mechanical tests were performed and the fractured surfaces of the specimens were examined by SEM. Changes in mechanical properties were followed as a function of volume percentage of rubber particles and types of surface treatment. In general, both tensile properties and impact resistance decreased with increase in rubber content. Only small increases in fracture toughness were observed for all surface-treated specimens.

Civelli[120] has patented a process in which a polymerizable liquid such as epoxy resin is blended with rubber crumbs having coarse, siliceous grains. The liquid slurry is cast into a sheet-like configuration. It is heated for sufficient time to provide a sheet-like product that can be used for pavements, driveways, flooring tiles, or roofing shingles. The product is said to have excellent weather, wear, and chemical resistance. The surface characteristics of GRT are important to the performance of these sheets.

A noncellular polyurethane composite material is produced by mixing properly dried ground rubber with a suitable polyol and an isocyanate prepolymer, to form a precursor blend that can be used to make solid tires, shoes and shoe soles, surface coatings, and other applications. The products are strong and wear resistant. Henrich[121] has patented a process in which rubber powders can be blended with thermoplastic materials such as PE, PP, thermoplastic polyurethane, and polyvinyl compounds and extruded into sheets of desired thickness for applications such as roofing tiles.

Synthetic sports pitches are widely replacing traditional gravel, sand, and cement surfaces. In many parts of the world, leisure centers and clubs have installed synthetic sports ground for their own flood-lit sports facilities for mini-football, netball, tennis, hockey, and many other sports activities.[122,123] Being resilient, these synthetic sports pitches reduce the number of accidents and are usable in all types of weather conditions. In the U.K., irrespective of the type of surface used, the installation of playground surfaces must meet the requirements of BS 5696. In the U.S., the safety surfacing system must comply with the Americans with Disabilities Act (ADA) by providing the handicapped with a smooth, cushioned access route to buildings and equipment. The International Amateur Athletics Association has also laid down stringent specifications for durability, strength, and safety of the sports pitches; all these need be considered in designing synthetic sports pitches.

Currently, synthetic pitches are made out of ground rubber from discarded vehicle tires and single-component, medium-viscosity polyurethane prepolymer based on methylene diisocyanate (MDI).[123] Tire waste is preferred in place of EPDM, SBR, or NR, to keep costs low. Spent tires in the form of regrind, shreds and buffing, and mixes of various types and grades are being used. However, depending on the use of the surface, virgin materials may be blended with recycled rubber. EPDM, SBR/NR blends, and NBR are preferred for professional sports surfaces. Recently, the Field TurfEEC company has developed new pitches with longer turfs by using hybrid fibers made from copolymers and PE/PP blends and filling the turf with recycled rubber crumbs. Whereas the hybrid fibers provide softness to the turf, the recycled rubber crumbs provide a comfortable surface to slide across.

The production wastes from fertilizer phospho-gypsum plants are blended with virgin and recycled polymers[124] to provide novel polymeric compositions that are used to make porous pipes, vehicle parts, flooring, road-safety and traffic-control equipment, sound-absorbing motorway screens, packing in biological wastewater treatment and biodegradable flower pots, and for buckets, fencing, and similar profiles.

7.5.3 Miscellaneous

Green-Man Technology's purchase of Cryopolymers, Inc. includes two minority equity interests, Cryopolymer Leasing, Inc., and Cryopolymer Management, Inc.[125] The acquisition enlarges Green-Man's operations with the addition of a fully operational, modern processing plant that produces finely ground rubber crumb for sale to manufacturers of rubber and plastics products or to be used by Green-Man Technology, whose main product is a proprietary blend of recycled tire rubber and plastics called GEM-Stock.

Shahidi, Arastoopour, and Teymour[126] have discussed the use of ground rubber particles for surface coatings in a recent presentation. They changed the hydrophobic surface character of rubber particles to make it hydrophilic by adding polyacrylic acid, thereby forming an amphiphilic "particulate phase semi-interpenetrating polymer network" (PPSIPN). The composite is used as an additive to waterborne polymeric coatings or paints, to improve their impact resistance and adhesion to steel substrates. Reduced hardness of these coatings also makes them suitable for sports surfacing applications.

Microbial degradation of tire rubber has been tried both for disposal of waste tires and for surface modification of crumb rubber. Yeast appeared to metabolize the rubber without sticking to it, and the treated particles appeared to be more hydrophilic due to the presence of carboxylic groups on the surface.[127] This facilitates blending the treated particles with polar materials like clay, and possibly polar plastics such as epoxy and polyurethane. Species of bacteria that are capable of breaking sulfur-sulfur bonds and carrying out chain scission to devulcanize rubber have been identified.[128] Recently, De and co-workers[129] have developed a novel process for reclaiming rubber waste by using a vegetable product and di-allyl sulfide. Battelle Pacific Northwest National Laboratory at Richland, Washington, has developed a process for bacterial degradation and devulcanization of waste rubber.[130]

Talibani and Hanelaw[131] studied the performance of new cement additives, including rubber powder obtained from tire scrap. Proper combination of Ironite sponge, Anchorage clay, and rubber powder eliminated the micro-annulus and micro-fracture of the cement. They provided an experimental approach to select the right type and right particle size of rubber powder that can be used as an additive to cement in the production of cement pipes.

7.6 Conclusions and Recommendations

The need for recycling used tires is very high. With increasing use of automobiles and truck transportation, particularly in countries with large populations like China and India, the environmental problem associated with tire dumps and tire fires will be severe. In addition, disposal of waste tires by land-filling will not only consume large sections of valuable land and the associated cost of transportation, but also will waste substantial amounts of organic structural materials that are costly and have limited availability. Therefore, recycling of waste rubber from used tires is very important, and the pace of development of necessary technologies needs to be accelerated.

Recycling of waste tire rubber can best accomplished by blending with thermoplastic and thermoset materials in a cost-effective and efficient manner. With increasing use of plastics in applications from packaging to building and construction, the potential market for products made from GRT/plastics blend is great. Further, blending of GRT with recycled high-density PE from milk jugs and low-density PE from grocery bags to develop marketable products will simultaneously help the recycling of municipal solid waste.

The last few years have seen a tremendous progress in tire-rubber recycling, particularly by blending with plastics. Methods for size reduction, compatibilization, and cost-effective processing of GRT/plastic blends have been worked out. A variety of methods including surface modification, surface functionalization, and surface treatment by functional polymers of GRT, have been developed to achieve high levels of interfacial adhesion and interphase compatibility between rubber and plastics. The prospect of using bacterial degradation to depolymerize rubber and induce surface functionality is quite feasible. Therefore, it is hoped that waste-tire recycling will soon be a reality, not only to eliminate environmental hazards, but also to conserve and enrich our material resources.

Suitable methods for size reduction and standardization of particle size, along with increasing knowledge of surface chemistry of GRT particles and cost-effective processes for devulcanization, will add to the development of many new products from blends of tire waste with thermoplastics and thermosets, including thermoplastic elastomers. It is possible to get improved properties by blending partially devulcanized GRT with plastics. Ultrasonic-assisted, continuous devulcanization and blending with plastics, once the hardware problems are solved, will provide GRT/rubber blends and thermoplastic vulcanizates with enhanced properties.[111,112] The work of Pillai and co-workers, on sequential devulcanization and revulcanization by tweaking rubber chemistry, provides an excellent method for recycling GRT.[68]

The prospect of making low-cost thermoplastic elastomers, both TPEs and TPVs, by blending thermoplastics with GRT is very alluring because of their high demand in the automotive, electronic, and appliance industries. Recyclability of TPEs and TPVs also provides incentive because of the potential for repeated use and environmental compliance. Additional benefits will be obtained when such TPEs or TPVs can be compounded with biodegradable polymers and biofibers from renewable resources.

It has been demonstrated that the impact resistance of many engineering plastics, both thermoplastics and thermosets, can be greatly enhanced by blending with GRT after surface treatment or in the presence of a suitable compatibilizer. Since impact resistance is an important requirement for many products, ground-up tire waste can replace costly impact modifiers currently being used. They can be combined with nano-fillers to provide improved impact resistance and barrier properties.

Further, the fact that GRT powder is a soft, low-cost organic filler, with the potential for surface modification by using a variety of chemical methods, provides an opportunity to blend it with various thermoplastics and thermoset materials, to meet different processing and product requirements. It will be very productive to focus more of our intellectual and financial resources toward: improving and standardizing procedures for GRT production, grinding used tires into fine particles with specific surface characteristics, and developing cost-effective methods to change their surface chemistry in order to blend them with thermoplastic and thermoset plastics.

Finally, there is a great potential for recycling GRT in civil-engineering applications, including rubberized asphalt. Although asphalt is not considered a polymer, it is an organic material, like thermoplastics. Our knowledge gained from rubber-plastic blends can be successfully used to lower the cost and improve the properties of rubberized asphalt for large-scale application in building and in repairing highly needed infrastructure throughout the world. This will provide a large potential for rubber and plastics recycling, while solving some of our difficult environmental problems.

References

1. Mangaraj; D., *Proceedings,* International Conference on Rubbers, Calcutta, India, 1997, p. 61.
2. Myhre, M. and MacKillop, D.A., *Rubber Chem. Technol.,* 75 (2), 2002, p. 429.
3. Anon., *Plastics Technology,* June 2003, p. 46. Anon, *Plastics Additives and Compounding,* June 2002, p. 24; W.A. Ploski, Paper #142, ACS Rubber Division Meeting, Orlando, Florida, September 1999.
4. Prieto, A. and Jana, S.C., *Proceedings,* Soc. Plast. Eng., ANTEC 2002, p. 2164; Misra, M., Mohanty, A.K., Drazl, L.T., ibid, p. 383.
5. Vaidya, U.I., *Proceedings,* International Conference on Rubbers, December 12–14, Calcutta, India, p. 147.
6. Paul, D.R. and Barlow, J.W., *J. Macromol. Sci. Rev. Macromol. Chem.,* C18, 1980, p. 109; Mangaraj, D., Markham, R.L., Heggs, R., et al., Battelle Memorial Institute, "Multiclient Report on Polymer Blends and Alloys," 1986.

7. Markham, R.L., *Elastomer Technology Handbook,* Cheremisinoff, P., ed, CRC Press, Boca Raton, Florida, Ch. 2; Mangaraj, D., Parson, A., Markham, R., "Battelle Multi-client Report on Compatibilization of Polymer Blends," 1994.
8. Raeder, C.P. and Weglin, R.C., *Proceedings,* Soc. Plast. Eng., ANTEC 1994, p. 3026; Gordon, J.J. and Lemieux, M.A., *Rubber World,* August 1999, p. 40.
9. Mangaraj, D., "Battelle Multi-Client Report on Recycling Plastics Waste from Automotive Industry," 1988; Deanin, R.D. and Curry, M.J., "Secondary Reclamation of Plastics Waste Report, Phase 1," 1987. Crawford, W.J., Polymer Preprint, 24(2), 1983, p. 432.
10. Mangaraj, D. *Rubber Chem. Technol.,* 75, July-August, 2002, p. 365; Berstoff, R.U., *Rubber World,* May 2002, p. 36.
11. Miles, I.S. and Zurek, A., *Polym. Eng. Sci.,* 28, 1988, p. 796.
12. Mangaraj, D., Heggs, R., et al., *Plast. Eng.,* 44(6), 1988, p. 29.
13. Sunderarajan, U. and Macosko, C.W., *Macromolecules,* 28, 1995, p. 2697.
14. Ayla, A., Hess, W.M., and Scott, C.E., *Rubber Chem. Technol.,* 64, 19, 1992.
15. Mangaraj, D. and Markham, R., *Proceedings,* Royal Society of Chemistry Symposium on Chemical Aspects of Plastics Recycling, Manchester, England, 1996.
16. Asaletha, R., Thomas, S., and Kumaran, M.G., *Rubber Chem. Technol.,* 68, 671, 1995.
17. Noolandi, J. and Hong, K.M., *Macromolecules,* 17, 531, 1984.
18. Reiss, G., Kohler, J. et al., *Macromol. Chem.,* 101, 58, 1967.
19. Teysie, P., Fayat, R., and Jerome, R., *Polym. Eng. Sci.,* 27, 1987, p. 328.
20. Boudreau, K.A., *Proceedings,* Soc. Plast. Eng. ANTEC 1991, p. 2562.
21. Mangaraj, D., *Comprehensive Polymer Science,* 2nd Suppl., Allen, G., Aggarwal, S.L.,and Russo, S., eds., Elsevier (Pergamon), London, 1996, p. 605.
22. Epstein, B.R., Lantham, A.A., et al., *Elastomerics,* 119 (10), 1987, p. 10.
23. Ide, F. and Hassegawa, A., *J. Appl. Poly. Sci.,* Part B. *Polym. Physics,* 32, 1994, p. 205; Han, C.H., *Advances in Chemical Series,* No. 206, p. 171; Cimino, C.H., Capella, F., et al., *Polymer,* 27, 1985, p. 1874; McKnight, W.J., Lenz, R.W., et al., *Polymer,* 27, 1985, p. 1874.
24. Scott, C.E. and Makosco, C.W., *Proceedings,* Reactive Processing International Polymer Processing Expo. 27, 328, 1987.
25. Rajalingam, P. and Baker, W.E., *Rubber Chem. Technol.,* 65, 1992, p. 908.
26. Duheime, R.M. and Baker, W.E., *Plastics, Rubber Composites, Processing and Applications,* April 15, 1991, p. 87.
27. Oliphant, K. and Baker, W.E., *Polym. Eng. Sci.* 26, 1986, p. 1455.
28. Parmanik, P.K. and Baker, W.E., *Plastics, Rubber Composites Processing and Applications,* 24, 1995, p. 229.
29. Rodriguez, E.L., *Polym. Eng. Sci.,* 28, 1988, p. 1455.
30. Beaty, C., Paper presented at ACS Meeting, New York City, 1992.
31. Adam, G., Sebnik, A.,and Osredkar, U., et al., *Rubber Chem. Technol.,* 64, 1991, p.133.
32. Brutlett, D.J. and Lindt, J.T., *Rubber Chem. Technol.,* p. 411.
33. Coran, A.Y. and Patel, R., *Rubber Chem. Technol.,* 56, 1983, p. 1045.
34. Coran, A.Y. and Cheng, O., *Rubber Chem. Technol.,* 70, 1998, p. 281.
35. Riew, C.K. and Gilham, J.K., eds., *Rubber Modified Thermoset Resins, Advances in Chemistry,* 208, 1984; Bagheri, R., Williams, M.A., Pearson, R.A., *Polymer Engineering and Science,* 37 (2) February 1997.

36. Khait, K., *Proceedings,* Soc. Plast. Eng. ANTEC, 1994, p. 3006.
37. Bilgili, E., Arastoopour, H., and Bernstein, B., *Rubber Chem. Technol.,* 73, 2001, p. 340.
38. Isayev, A.I. and Hong, C.K., *Proceedings,* Soc. Plast. Eng. ANTEC 2002, p. 1334.
39. Isayev, A.I., Paper #108, ACS Rubber Division Meeting, Philadelphia, October 2002.
40. Baranwal, K.C., *Proceedings,* International Conference on Rubbers, Calcutta, India, 1997, p. 56.
41. Rajalingam, P., Sharpe, J., and Baker, W.F., *Rubber Chem. Technol.,* 66, 1993, p. 664.
42. Phadke, A.A., Chakraborty, S.K., and De, S.K., *Rubber Chem. Technol.,* 57, 1984, p. 19; Jacob, C., De, P.P., Bhowmick, A.K., and De, S.K., *Plast. Rubber Composites,* 31 (4), 2002; *J. Appl. Polym. Sci.,* 82, 2001, p. 3293.
43. Baumann, B., Paper presented at ACS Rubber Division Meeting, Akron, Ohio, 1996.
44. Ballman, B. and Spalding, B.J., *Chemical Week,* December 1987, p. 48.
45. Verdestein Rubber Recycling, "Product Bulletin on Surcum®", Dierkes, W., *Rubber World,* 214 (20), 1996, p. 25.
46. Anon., *Rubber Recycling Innovation,* October 1995, p. 7.
47. Feus, E., Paper #22, Rubber Division, ACS, April 1994.
48. Singh, A., Int. Conf. on Adv. Additives, Modifiers, Poly. Blends, Miami, Florida, 1992; Eaves, M., *J. Adhesives,* 5 (1) 1973; Xanthos, D.W. Yu and Gogos, C.G., *Polym. Mater. Sci. Eng.* 67, 1992, p. 313.
49. Serenko, O.A., Avinkin, O.A., et al., *Polym. Sci.,* Series A, 43 (2), 2001, p. 129.
50. Anon, *Scrap Tire News,* 14 (6), June 2000, p. 11; Anon, *Plastics News, U.S.A.* 11(32), September 1999, p. 11; *Rubber and Plastics News,* 29 (4), September 1999, p. 70/3.
51. Kim, J.K. and Burford, R.P., *Rubber Chem. Technol.,* 71, 1028, 1998.
52. Oldfield, D. and Symer, T.E.M., *J. Adhesives,* 1983, p. 77.
53. Lawson, P., Kim, K.J., and Fritz, T.L., *Rubber Chem. Technol.,* 56, 1995, p. 25.
54. Bauman, B.D., Paper #35, Rubber Division, ACS Meeting at Chicago, Illinois, April 19–22, abstract in *Rubber Chem. Technol.,* 67, 1994, p. 775.
55. Pittolo, M., Kim, J.K., and Park, J.W., Sixth Intl. Symposium on Elastomers, Koycongiu, Korea, 1996, p. 101; Pittolo, M. and Burford, R.P., *Rubber Chem. Technol.,* 58, 1997, p. 1965.
56. Coran, A.Y., *Thermoplastic Elastomers,* Chapter 7, Legge, N.R., Holden, G., Schroeder, H.E., eds., Hanser, New York, 1987.
57. Naskar, A.K., De, S.K.,and Bhowmick, A.K., *J. Appl. Polym. Sci.,* 84, 2002, p. 370.
58. Naskar, A.K., Bhowmick, A.K., and De, S.K., *Polym. Eng. Sci.,* 41, 2001, p. 1087.
59. Jacob, C., De, P.P., Bhowmick, A.K., and De, S.K., *J. Appl. Polym. Sci.,* 82, 2001, p. 3304.
60. Anandhan, S., De, P.P., Bhowmick, A.K., De, S.K., and Bandyaopadhyay, S., *J. Appl. Polym. Sci.* (in press).
61. Weber, R.G., Mertzel, E.A., and Citracel, L., et al., U.S. Patent #6,384,145, May 7, 2002.
62. Mertzel, E.A., Weber, R.G., and Citracel, L., U.S. Patent #6,015,861, January 18, 2000.
63. Rouse, M.A., U.S. Patent # 5,411,215, May, 1995.
64. Burgoyne, M.D., Fisher, J.F., and Jury, J.R., U.S. Patent #55,102,109, April 23, 1996.
65. Johnson, L.D., U.S. Patent #5,157,082, November 13, 1992.

66. Jury, J.R., and Chien, A.W.V., U.S. Patent 6,313,183, July 7, 2001.
67. Haber, A., *Proceedings,* Conference on Plastics Impact on Environment, GPEC 2002, Detroit, Michigan, February 13–14, 2002, Paper #17, p. 149 (CD- ROM, 012).
68. Pillai, C.R.P. and Chandra, H., U.S. Patent #6,313,183 BI, November 6, 2001.
69. Liu, H., Mead, J.L., and Stacer, R.G., *Rubber Chem. Technol.,* 75(1), 2002, p. 49.
70. Anon., *Scrap Tire News,* 14 (6), June 2000, p. 11.
71. Kowalska, E., Wielgosz, Z., and Pelka, J., *Proceedings,* Adcon World, Berlin, October 8–9, 2001, Paper #39, p. 9.
72. Information from Astro Vulcor, Inc., P.O. Box 148, Glen Falls, NY 12801.
73. Doan, R.C., U.S. Patent #5,733,943, February 7, 1996.
74. Mertzel, E.A., Weber, R.G., and Citracel, L., U.S. Patent #6,015,861, January 18, 2000.
75. LeGault, M., *Canadian Plastics,* 57 (10), October 1999, p. 28.
76. Begin, S., *Rubber and Plastics News,* 29 (4), September 20, 1999, p. 70.
77. Howard, H. and Coran, A.Y., *Rubber and Plastics News,* 30(24), June 25, 2001, p. 12.
78. Kramer, J., van der Groeb, B., and Billiet, J., *Rubber Tech. Intl.,* 1998, p. 70.
79. Nagayasu, N., Orisaki, O., et al. U.S. Patent #4,795,603, October 8, 1987.
80. Kowalska, E. and Wielgosz, Z., *Polymer Recycling,* 3(1): 61, 1997/98.
81. Turner, J.E., U.S. Patent #4,028, 288, June 7, 1977.
82. Butilino, European Patent #0401885.
83. McClellan, T.R., U.S. Patent #5,385,953, October 18, 1993.
84. Kiser, W., U.S. Patent #5,580,638, December 1996; Kiser, W., U.S. Patent #5,525, 399, June 11, 1996.
85. Lee, W., U.S. Patent #5,635,551, June 3, 1997.
86. Frankland, W.H., U.S. Patent #4,244,841, March, 24, 1980.
87. Doan, R.C., U.S. Patent #4,970,043, February 27, 1989.
88. Johnson, L.D., U.S. Patent #5,157,082, November 13, 1990.
89. Rosenbaum, J.E., U.S. Patent #5,523, 328, May 6, 1994.
90. Blalock, C. and Nelson, L., U.S. Patent #6,194,519 BI, February, 27, 2001.
91. Burgoyne, M.D., Fisher, J.F., and Jury, J.R., U.S. Patent #5,385,953, April 23, 1996.
92. Anon; *Scrap Tire News,* 16 (11), November 2002, p. 7.
93. Khait, K., *Proceedings,* Soc. Plast. Eng. ANTEC 1995, p. 2066.
94. Khait, K., Rubber Division Meeting, ACS, Paper #53, Cleveland, Ohio, 1995.
95. Khait, K., ACS Rubber Division Meeting, Louisville, Kentucky, October, 8–16, 1996.
96. Rihai, A., Li, J., Arastoopour, J., Ivanov, G., et al.; *Proceedings,* Soc. Plast. Eng. ANTEC 1993, p. 891.
97. Tipanna, M. and Kale, D.D., *Rubber Chem. Technol.,* 70, 1997, p. 815.
98. Kim, J.H., Barlow, J.W., and Paul, D.R., *J. Polym. Sci. Polym. Phys.* 21, 1983, p. 367.
99. Coleman, M.M., Serman, C.J., Bhagwager, D.E., and Painter, P.C., *Polymer,* 31, 1990, p. 1187.
100. Zytek, P. and Zollinger, P., *J. Polym. Sci.,* Part A-1, 6, 1988, p. 467.
101. Lang, N., *Waste Age,* January, 1995, p. 77.
102. Rouse, M.A., Paper #19, presented at the ACS Rubber Division Meeting, *Abstracts,* (69), p. 141.
103. Powers, R.L., *Rubber and Plastics News,* 2, 20 (15), April 1999, p. 6.
104. Rosenzweig, M., *Modern Plast. Intl.,* 32 (11), November 2002, p. 50.
105. Anon, *Plastics in Building and Construction,* 25(7), April, 2001, p. 4.

106. Kennerknecht, B.,*Proceedings,* Soc. Plast. Eng. ANTEC Conference 1998, (3), p. 3384.
107. Naskar, A.K., De, S.K., and Bhowmick, A.K., *Polymer and Polymer Composites,* 10 (6), 2002, p. 427.
108. Sbraski, I. and Bhattacharya, S.N., *Proceedings,* International Conference, Calcutta, India, 1997, p. 64.
109. Bhattacharya, S.N. and Sbraski, I., *Rubber News,* 37 (12), 1998, p. 23.
110. Vetkin, Y.A., Goncharov, G.M., and Shilov, M.O., *Intl. Polym. Sci. and Technol.,* 25 (12), 1998, p. 79.
111. Hong, C.K. and Isayev, A.I., Paper #107, ACS Rubber Division Meeting, Orlando, Florida, 1999; *J. Elastomer and Plastics,* 33 (1), 2001, p. 47.
112. Luo, T. and Isayev, A.I., *J. Elastomer and Plastics,* 30 (2), April 2, 1998, p. 133.
113. Oh, J.S. and Isayev, A.I., ACS Rubber Division Meeting, Fall, Cleveland, Ohio, October, 2001.
114. Harper, J., *Urethanes Technol.,* 16 (5), October/November 1999, p. 22; *Plastics News (USA),*11 (32), September 1999, p. 11.
115. Wu, D.Y., *Packaging Digest,* 36 (6), June 1999, p. 42.
116. Anon, *Scrap Tire News,* 15 (3), March 2001, p. 5.
117. Ballester, J.L., Department of Theoretical & Applied Mechanics, Cornell University, Ithaca, New York, private communication.
118. Ballester, J.L., *Proceedings,* GPEC 2002, Plastics Impact on Environment, Detroit, Michigan, February 13–14, 2002, Paper #18, p. 159, CD Rom-012.
119. Sipahi-Saglam, E., Kanyak, C., Akovali, G., et al., *Polym. Eng. Sci.* 41 (3), March 2001, p. 514.
120. Civelli, H.A., U.S. Patent #5,258,222, November 2, 1993.
121. Henrich, R., U.S. Patent #6,255,391 B1, July 3, 2001.
122. Sen, A., *Popular Plastics and Packaging,* 43 (3), March 1998, p. 67.
123. Kowlaska, E. and Weilgosz, Z., *Polymer Recycling,* 3 (1), 1997–1998, p. 61.
124. Kowalska, E., Weilgosz, G., and Pelka, J., *Polymer and Polymer Composites,* 10 (1), 2002, p. 83.
125. Anon, *Scrap Tire News,* 12 (1), January 1998, p. 4.
126. Shahidi, N., Arastoopour, H., and Teymour, F., Paper #88, ACS Rubber Division Meeting, Cleveland, Ohio, October 14–17, 2003.
127. Beckman, J.A., Crane, G., Key, E.L., and Laman, J.R., *Rubber Chem. Technol.,* 47, 1974, p. 574.
128. Holst, G., Stenberg, B., and Christianson, M., *Biodegradation,* 9, 1998, p. 301.
129. De, P., Ghosh, A.K., et al., *Polymer Recycling,* 4 (3), 10998–10999, p. 151.
130. Vijayendran, B., Senior Research Leader, Battelle Memorial Institute. Private communication.
131. Talibani, S. and Hanelaw, G., *Energy Sources,* 21 (1-2), 1999, p. 232.

8

Strategies for Reuse of Rubber Tires

D. Raghavan

CONTENTS

The abatement of environmental problems is one of the biggest challenges of technology today. Strategies for reuse of waste polymeric material can lead to a direct reduction of the disposable waste in landfills. Among the several polymeric materials, waste tire in landfill is a major concern because yearly, in the U.S., 279 million tires are discarded with another 300 million stockpiled in large, open areas. Since waste rubber is poorly degradable in a landfill environment, material and energy recovery are better alternatives to disposal. Solutions have centered around retreading and deriving fuel from rubber tires, incinerating rubber tires for pyrolyzed gas and carbon black, using rubber tires in marine environments as barrier reefs, mixing the shredded rubber tires with asphalt, and mixing rubber particles in concrete mixes. In the first part of this chapter, we review the various solutions in some detail. Recent advances in both material mixture composition and mixing technique for preparing rubberized asphalt that have led to better material performance will also be discussed. We present observations on the use of tire particles as replacement aggregates for low-strength concrete material. The importance of surface treatment of powdered tire rubber to increase its adhesion to cement paste is examined. Finally, novel applications for scrap tire as additives for commingled plastics are also described.

8.1 Introduction

The use of natural rubber in making useful materials dates back to the early 1800s. Several close chemical replicas of natural rubber emerged during the mid-1940s and they were referred to as "synthetic natural rubber." One of these is, in fact, synthetic cis 1,4 polyisoprene. The synthetic rubbers differ in their properties from natural rubber. During the mid-1940s, synthetic elastomers started replacing natural rubber *(Hevea braziliensis)* in many applications because of the limited availability of natural rubber and because of economics. Since then, there has been a steady introduction of next-generation synthetic elastomers for increasing the durability of tire products. Typical formulation for tire tread is given in Table 8.1.[1] To realize their full potential, all the rubber compounds have to be cross-linked. The traditional cross-linking method is based on heating the rubber compound with a sulfur-based chemical and other vulcanizing agents. This often results in a polymer network with various cross-linking structures including monosulfidic, disulfidic, or polysulfidic cross-link units. The sulfur may be bonded to the polymer chain via intranetwork/internetwork linkages, or may exist in the free form. The chemistry of vulcanization can be exceedingly complicated and the associated chemistry often affects overall mechanical properties.

Cross–linked rubber is widely used in automobile and truck tires, mats, battery covers, hoses, gaskets, shoe soles, dock rubber, shims, etc. Among the broad use of synthetic rubber, automobile and truck tires represent the bulk of the rubber use. According to the International Institute of Synthetic Rubber Products, the North American demand for synthetic rubber is 2.2 million metric tons.[2] Sheets of rubber and layers of fiber-reinforced rubber are placed in a manner to obtain different properties in each of the three principal directions of tire structure. The structure and material used in the design of a tire can be vastly different from the exterior tread to the interior lining of the tire.

For example, sidewalls of the tire are designed to resist curb scrub, crack propagation, flexing, and attack by ozone in air, while rubber treads are designed to primarily resist abrasive wear. In fact, most modern tires are carefully designed to meet this diverse requirement and thereby extend the life expectancy of tires. Despite the technological revolution in the formula-

TABLE 8.1
Representative Tire Formulation in Parts by Weight

Components	Wt% of Compound
Polybutadiene	~40
Natural rubber or butadiene/styrene copolymers	~50
Processing oil	~9
Carbon black	~52.5
Zinc oxide	~4
Stearic acid	~1
Sulfur	~2
Accelerator	~1
Antioxidant	~2
Protection wax	~2

Source: Adapted from Gent, A.N. and Livingston, D.I., Automobile tires. In: *Concise Encyclopedia of Composite Materials*, Kelly, A., Cahn, R.W., Bever, M.B. (eds.), Pergamon Press, 1989.

tion chemistry of rubber tires, they wear upon repeated use, rendering the tires nonfunctional for their intended use. These tires are accumulated in large quantities and represent a growing disposal problem.

8.2 Disposal of Tires

Discarded vulcanized rubber tires now account for 3% of the weight of all municipal refuse and are one of the fastest growing forms of refuse. The tonnage is expected to increase annually at ~2%, as it has for the past 100 years. It is approximated that 67% of disposed rubber is primarily automobile and truck tires. It is widely reported that, in the U.S., about 279 million tires are discarded in the landfill every year, with another 300 million stockpiled in large open areas.[3] Disposing of tires in open areas is not only unacceptable, but tires in open areas are major breeding grounds for mosquitoes because they hold water. Furthermore, tires in open areas are possible fire hazard material and can potentially cause significant damage to the land, water, and air neighboring the site. Therefore, disposal of tires in open areas is not only a cause of environmental concern, but also represents a health and fire hazard.[4,5] Often, most discarded tires are stored in individual sites before being transported to landfill. Given the enormous volume of refuse tires, the collection and transportation of rubber tires from individual sites to landfills is becoming a practical and an economic issue.[6] Unlike many biodegradable materials, rubber tires in a landfill environment are resistant to degradation. Because of poor degradability of rubber and the cost of maintaining the landfill site, the practice of disposal of whole tires in many landfills is no longer a viable choice. As an alternative, a few landfill sites accept shredded tires. Tire shredding adds additional economic burden to the disposal of tires in landfill, given that shredding of tires requires usage of specialized equipment and costs about $65 to $85 per ton.[7]

8.3 Solutions to the Rubber Disposal Problem

The short-term solution to tire disposal is to design rubber tires that are stronger, more resilient, and more durable under adverse climatic conditions. With the constant demand for replacing older automobile tires, the rubber industry has an incentive to formulate durable rubber tires that can withstand severe environmental conditions. Use of improved vulcanization systems and compounding of elastomers has allowed rubber vulcanizates to be formulated so they don't deteriorate or lose their integrity when wet or frozen. The cross-linking of polymer chains covalently makes vulcanized rubber resistant to chemical, microbial, and mechanical degradation. Often, a large amount of carbon black is added to improve the abrasion resistance of rubber. Carbon black-reinforced tire prolongs the tire life expectancy.[1] Currently, there is an ongoing revolution in the tire industry to formulate rubber compounds with a life expectancy of at least 80,000 miles. For example, tire manufacturers are addressing the rolling resistance without sacrificing traction of the tire by resorting to solution styrene-butadiene rubber or isoprene terpolymers instead of conventional emulsion styrene-butadiene rubber.[2] To protect the rubber vulcanizates from attack by molds and to minimize their susceptibility to fungal attack, vulcanization accelerators (derivatives of sulfonamide) are added in the vulcanization system. Also, aging of vulcanized rubber is overcome by the addition of antioxidants (e.g., a derivative of amine) to the rubber formulation. The chemistry of the vulcanization of rubber and tire formulation contributes toward making the tires last longer.

Despite the recent success in extending the life expectancy of tires, the rubber industry and solid-waste management industry are faced with a challenge to address the long-term problem of discarded tires. The challenge is to recycle, reuse, reduce, or recover rubber in an environmentally benign, energy-efficient, and cost-effective manner. The long-term solutions so far advanced have been (a) retreading and splitting, (b) reclaiming, (c) combustion for deriving fuel and energy, (d) pyrolysis for producing carbon black, (e) use in barrier reefs and erosion control structures, (f) use in construction materials such as mixing crumb rubber with asphalt or concrete.

8.3.1 Retreading and Splitting

In retreading, tire carcass is passed through a recapping system to introduce new tread to the tire. Tire splitting involves removal of the bead wire by cutting or stamping, and the tread is cut and peeled off the tire carcass. The remaining carcass, which is a tough, durable, and fabric-reinforced sheet, is divided into three sections. The three sections are planed to uniform thickness and then placed in a press, where they can be die cast into desired shapes for use as gaskets, seals, doormats, and automobile tailpipe insulators.[8] However, retreading and splitting are done for a relatively small percent of the tires. Only 20% of discarded tires are retreaded, while only 5% of the discarded tires are reclaimed as alternative material.[9]

8.3.2 Rubber Reclaiming

Another alternative to disposal of tires in landfill is the recovery of material and energy from scrap tires. Rubber reclaiming involves rupturing the covalent carbon-sulfur bond existing in the vulcanized rubber. The reclaiming process involves either chemical or thermal approaches. The use of steam pressure to devulcanize rubber material is a well-known technique. In most instances, rubber reclaiming can be highly complicated, cumbersome, and costly. The process also produces an inferior-quality rubber that is difficult to use with pristine rubber in the manufacture of tires. In addition, with the availability of less expensive plastics and oil-extendable rubbers, there is less need for the use of reclaimed rubber, except for more costly polymer, e.g., silicones[10] On the other hand, rubber reclaiming can become an extremely attractive method when the fuel and petroleum-derivative costs for polymer increase.

Microbial devulcanization is another promising way to increase the recoverability of rubber material. An obstacle to achieving biodevulcanization is the presence of toxic rubber additives.[11] Recent studies by Bredberg and coworkers[12] have shown that fungus growth on rubber detoxifies the rubber material. Growth of *Thiobacillus ferroxidans* has been noticed on detoxified rubber, suggesting the possibility of sulfur oxidation from rubber. We conducted a similar study to Bredberg's to evaluate the microbial desulfurization of rubber vulcanizate that is free of biocidal compound.[13,14] The two parameters that were monitored during biodesulfurization were oxygen uptake and sulfate ion formation. Oxygen is consumed by microbial cells during the oxidation of sulfur in rubber; there is also formation of sulfate ions in the solution. Figure 8.1a and Figure 8.1b show both the oxygen consumption by microbial cells and sulfate ion formation in the solution, during desulfurization of rubber by *Thiobacillus ferroxidans and Thiobacillus thioxidans.*[13,14] The oxygen consumption by microbial cells and the sulfate ion formation were found to be proportional to the total sulfur content in vulcanized rubber. Results of this work corroborate the literature finding that rubber vulcanizates that are free of antioxidants are susceptible to microbial growth.[15]

8.3.3 Pyrolysis

Pyrolysis is a process in which the chemical bonds in rubber are completely broken by heat. Scrap tire pyrolysis has been the subject of many investiga-

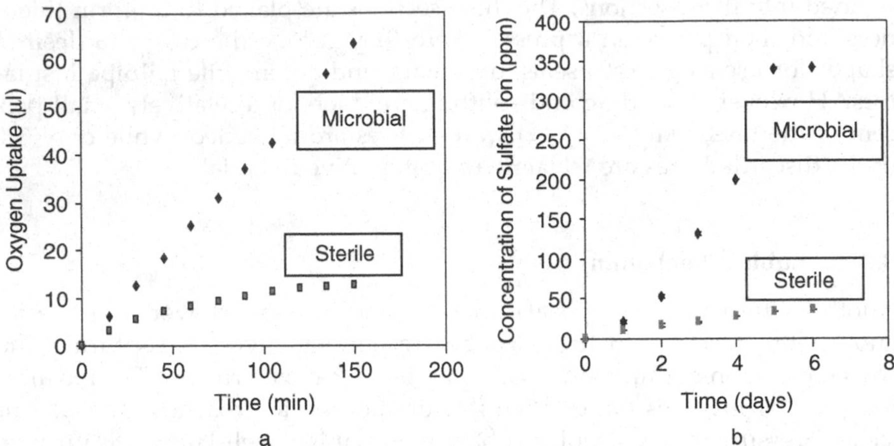

FIGURE 8.1
Measurement of (a) oxygen uptake by microbial cells and (b) sulfate ion formation during biodesulfurization of rubber.

tions because of the by-products generated during the process.[16–18] During the pyrolysis process (500 to 900°C), the rubber is broken down to yield oil, gas, fiberglass, steel, and carbon black. The collected gas is essentially a mixture of ethylene, propylene, butane, butylenes, and other lower-molecular-weight hydrocarbon gases. The carbon black is separated from inorganic contaminants such as fiberglass and steel. The gains made in oil, gas, and fuel recovery by burning the tires are highly sensitive to the economic fluctuation surrounding the oil and gas market. Attempts at producing carbon black and liquid hydrocarbon from tires have met with limited success.

While carbon black from rubber tires eliminates shredding and grinding costs, the carbon black is much more expensive than that obtained by traditional methods; it is also inferior and of less reliable quality than carbon black from petroleum oils.[16] The highly unpredictable composition of carbon black makes the wider use of carbon black in mechanical goods less likely. An alternative use of carbon black made from tires is as activated carbon.[19]

8.3.4 Tire-Derived Fuel

The most prevalent use of scrap tires is the use of whole or partially shredded tires in the manufacture of Portland cement. Tires have approximately three times the energy of municipal solid waste. Recently, tires have been used as partial substitutes for coal (1 lb of tire is roughly equivalent to a BTU value generated from 5/4 lb of coal). The use of tires could translate to a 25% reduction in the amount of coal consumption by cement industries, and could be beneficial if there is a coal shortage or if the cost of coal-derived fuel becomes expensive.[20] Use of rubber tires would also save the cost associated with coal cleanup (to lower nitrogen and sulfur content). Another benefit in the use of whole tires as fuel is the consumption of steel belt, carcass, and sidewall in the kiln. Alternatively, recycled rubber can serve as a good energy source for the generation of electricity. The energy produced from scrap tires can have a major market in the pulp and paper industry and also the industrial/municipal sector. Based on current estimates, it is projected that by 2005, 56 million tires will be consumed by cement kilns, 51 million tires by paper and pulp industries, and 47 million tires by coal-firing power plants.[20] However, there is poor acceptance of the use of tires in cement kilns, the paper and pulp industry, and power plants because it would require the installation of a new feed system for accepting scrap rubber as fuel source. The high cost of such an installation makes the use of scrap rubber less attractive than burning natural gas, oil, or coal as fuel substitutes.[21,22] Other barriers to the wider use of tires for fuel include possible unreliability of tire supply, increased cost associated with meeting stricter emission standards, and overcoming public perception of burning tire.[23,24] Residues from burning tires have been reported to form a char that increases combustible heat loss and tends to clog furnace grates.

8.3.5 Reefs and Erosion Structures

Scrap whole tires have been used for artificial fishing reefs, oyster beds, and as a floating breakwater. Goodyear has about 2,000 fishing reefs of old tires and one of the longest reefs in Florida (it stretches 2.4 km) is made of 3×10^6 tires.[11,25] Studies by the U.S. Bureau of Reclamation have demonstrated that fishes and aquatic macroinvertebrates are attracted to these structures, increasing the species density neighboring to the costal region. However, the use of tire reefs in aquatic environments that have smaller volumes of water, e.g., a canal, have raised water quality concerns. Analysis of the tire leachate for both organic and inorganic compounds showed that zinc (751 μg/L compared to 8.7 μg/L in lake water) was the primary toxicant, followed by copper, with trace amounts of organic compounds.[26] Although the amount of zinc leached from tire reefs in small bodies of water is considerably high, the amount does not raise acute or chronic toxic issues in of itself, provided large bodies of water are in contact with a tire reef.[27]

8.3.6 Rubberized Asphalt

The addition of ground tire rubber to asphalt has been of great interest to both the tire and paving industries. If only a small amount of waste tires can be added to the pavement mix, the asphalt industry can absorb a considerable fraction of the steady waste rubber stream. This is because the annual asphalt concrete production roughly amounts to 500 million tons. Two different processes (wet and dry) are used in the formulation of rubberized asphalt.[28] Both processes begin with the use of crumb rubber, which is made by grinding the rubber into fine particles (1/4 inch to 40 mesh). In the dry process, the tire rubber particle are premixed with aggregate prior to blending the mix with asphalt aggregate. Within the dry process, two systems are popular: the PlusRide and the TAK system. In the wet process, asphalt cement is preblended with tire rubber particles before introducing it into the hot mix. The rubber is mixed with the asphalt at high temperatures, 300-400°F for 1/2 to 1 hour. During the blending process, the rubber particles swell and soften by absorbing the aromatic oils from the asphalt, providing the asphalt mixture additional elasticity and durability. During the mixing of crumb rubber tire–modified asphalt, there is a slow phase separation of the components, since the crumb rubber is usually denser than the hot asphaltic phase.[29] The carbon black in the crumb rubber particle causes the crumb rubber to slowly sediment at high temperature. An approach to improve the uniform dispersion of the additive in the asphalt mixture is the efficient mixing of the rubber and asphalt. The percentage of rubber in the binder strongly affects characteristics and performance of the asphalt material.[30] Typically, rubber constitutes about 3% of the mix by weight, but it can vary depending on the specific application of asphalt material.

Usage of scrap tire rubber in asphalt has been demonstrated to have potential benefits, such as the improvement in the short and long term performance of rubberized asphalt. Some of the swollen rubber acts to thicken the material by increasing the viscosity of the mix, while the aggregate helps stop cracks from propagating. The rubber improves asphalt ductility, skid resistance under icy conditions, crack resistance, and noise reduction.

The benefits from using scrap rubber in asphalt, however, are strongly influenced by a number of factors such as the amount of rubber used in the binder, the mix temperature, and the duration of the blending process.[31] For example, high-temperature rheological measurements of the loss modulus and complex modulus of rubberized asphalt have shown only a marginal improvement in properties when large quantities of crumb rubber are added to the asphalt binder. The disadvantages of using rubber-modified asphalt concrete over conventional concrete are thus far unclear due to the lack of results available. Processing and performance issues have slowed the wider acceptance of rubber tires in large-scale asphalt pavement. Tests have shown that, for the construction of roadways, the price of using rubber-modified asphalt is almost 50% higher than that of using conventional asphalt.[32] Part of the cost can be offset by the extended period of performance provided by rubberized asphalt over the conventional asphalt. Another disadvantage of rubber-modified asphalt is the possible harmful effects to the environment. During processing of asphalt, toxic emissions may be released into the air or absorbed into the surrounding ground, polluting water supplies. Another potential disadvantage is the inability to recycle the rubberized asphalt. Asphalt free of rubber can be recycled as aggregate in common asphalt,[33] but the lack of such an approach to recycle rubberized asphalt is considered a disadvantage. Because the issues of cost, emissions, and recyclability are yet unresolved, the use of rubberized asphalt in pavement has met with considerable resistance.

Even if these issues are dealt with satisfactorily, some reports have indicated that the asphalt industry can absorb only 30 to 40% of the scrap tires generated.[34] Furthermore, automobiles driven on asphalt are far less efficient than those driven on concrete.[35] Hence, there is a need for development of innovative materials using scrap tires. The growth in rubberized concrete technology will open new avenues for high-volume use in the transportation and construction industries and will have a significant impact on the solid-waste management industry.

8.3.7 Rubberized Concrete

During the past decade, considerable efforts have been made to develop rubberized concrete.[9,36–61] There are many advantages to the addition of crumb rubber to conventional cement-concrete structures, including increased crack resistance and shock wave absorption, reduced heat conductivity, reduced noise levels, lighter weight, and increased resistance to acid rain. Other properties have also been reported to have been improved, e.g., freeze-thaw resistance and impact resistance.[9] Possible uses of the rubberized concrete would be in sub-bases for highway pavements, highway medians, sound barriers, and other transportation structures.[45]A new class of material, rubberized concrete is processed by mixing rubber with concrete. Concrete is typically a mixture of Portland cement, fine aggregate, coarse aggregate, air, and water. Aggregates typically amount to 80% of the concrete volume. Although aggregates are chemically inert, they are known to exert a major influence on the overall performance of concrete. Primarily, the work on rubberized concrete has centered on the use of shredded rubber as an aggregate, with the exception of a more recent study where shredded rubber was used as a replacement of cement in concrete.[46]

Until now, most of the research on uses of shredded rubber in concrete has been conducted using coarse or fine granular rubber. The maximum size and grading of rubber granules used by various investigators varied considerably. Particle sizes used have ranged from 0.06 to 2 mm in diameter. It is worth mentioning that the rubber source and grinding process can alter the fiber content of shredded rubber, its shape, and its texture.[58-60] Results of testing have shown that the compressive and flexural strengths of concrete decreased with the addition of granular rubber. The size of rubber granules has appeared to have a major influence on the compressive strength of concrete. Topcu's[37] and Eldin and Senouci's investigations[36] have shown that the addition of coarse-graded rubber particles lowered the strength of concrete more than the fine-graded rubber particles did. Tests by Ali and coworkers[61] reported opposite results for rubberized concrete with coarse-rubber particles and fine-rubber particles. Possible reasons for the variation in reported strength could be the rubber source, rubber geometry, and initial preparation of shredded rubber for use in concrete.

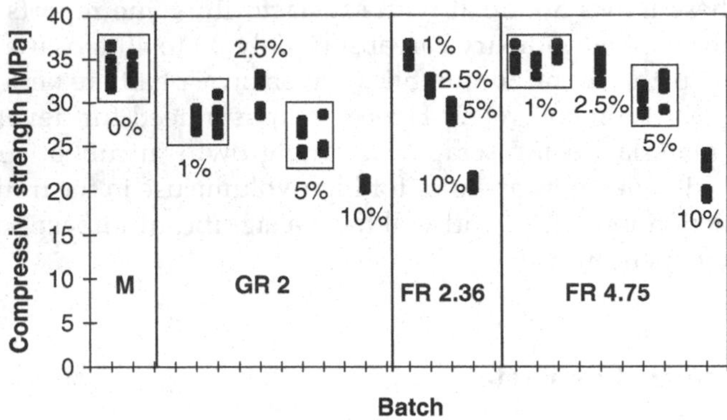

FIGURE 8.2
Compressive strength of mortar cubes.

In our study, we soaked and thoroughly washed the rubber with water in an attempt to free the rubber of contaminants before use in cementitious material. Figure 8.2 shows the individual compressive strengths of mortar cube specimens for the different mixtures. Rubber fibers that passed through the 9.72 mm sieve (#4) and were retained by the 4.75 mm sieve (#6) were designated as FR 9.72; the rubber that passed through the 4.75 mm sieve (#6) and were retained by the 2.36 mm sieve (#8) were designated as FR 4.75; and the rubber fibers that passed through the 2.36 mm sieve (#8) and were retained by the 1.18 mm sieve (#16) were designated as FR 2.36. GR 2 represents granular rubber of approximately 2 mm diameter. Figure 8.3 shows the pictorial representation of the mechanically ground tire rubber fibers. Notice that the particle size of scrap tire prepared by mechanical grinding could range from several inches to a fraction of an inch. There are two methods of grinding rubber tires. The cryogenic method is based on freezing the rubber below its Tg and fracturing the specimen in a hammer mill. This method limits the particle size distribution of ground rubber tires. We adopted the low-cost mechanical grinding procedure of shredding rubber tire for the work described below.

In Figure 8.2, we show the strength of mortar mixtures for several compositions. The boxes surrounding the points correspond to more than one batch of sample tested. Details of the work can be found in Reference 42. Within the batch, coefficient of variation ranged from 2 to 9%. We noticed that increasing the content of rubber decreased the compressive strength of mortar. At 1, 2.5, and 5% rubber by mass of cement, fibrous rubber showed a smaller reduction in the compressive strength of mortar than did granular rubber. However, the trend did not hold true with the addition of 10% rubber by mass of cement.

To examine whether there were statistically significant differences in the average compressive strength due to the geometry of rubber or to a combination of the amount of rubber and the geometry of rubber, an analysis of

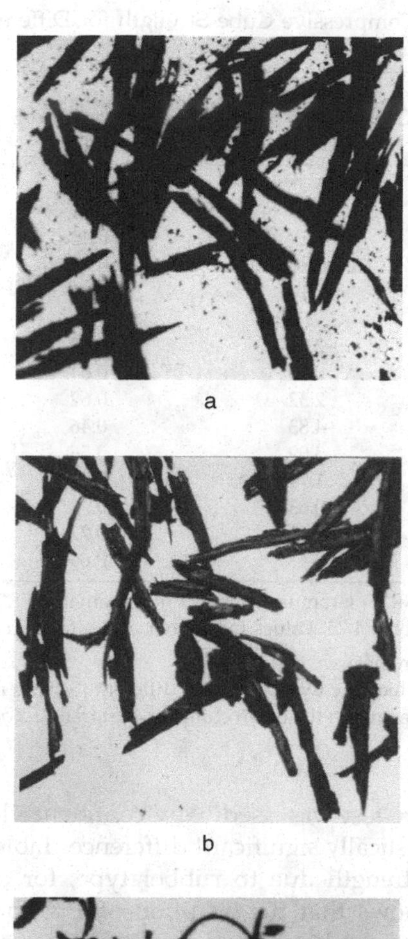

a

b

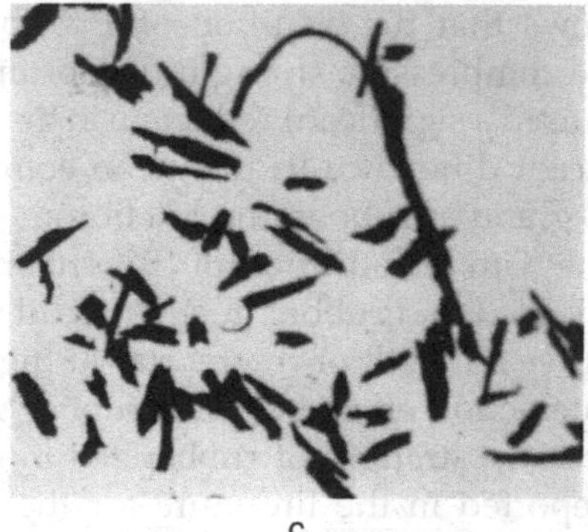

c

FIGURE 8.3
Pictures of the rubber fibers (a) FR 9.72, (b) FR 4.75, and (c) FR 2.36 used in the preparation of rubber-filled mortar.

TABLE 8.2

Differences in Mean Compressive Cube Strength for Different Rubber Types and Contents

Rubber Type and Content	Difference in Compressive Strength (MPa)	Standard Error (MPa)	Confidence Level** (%)
N – GR2 (1)	5.6***	0.44	≈100
N - FR2 (1)	–1.50	0.61	*
N - FR4 (1)	–1.19	0.43	*
GR2 (1) - FR4 (1)	–6.80	0.42	≈100
GR2 (1) - FR2 (1)	–7.10	0.61	≈100
FR4 (1) - FR2 (1)	–.0.31	0.59	*
GR2 (2.5) - FR4 (2.5)	–2.71	0.62	99.4
GR2 (2.5) - FR2 (2.5)	–0.39	0.64	*
FR4 (2.5) - FR2 (2.5)	2.32	0.62	96.3
GR2 (5) - FR4 (5)	–4.83	0.46	≈100
GR2 (5) - FR2 (5)	–3.09	0.56	≈100
FR4 (5) - FR2 (5)	1.74	0.55	*
GR2 (10) - FR4 (10)	–0.63	0.73	*
GR2 (10) - FR2 (10)	–0.15	0.77	*
FR4 (10) - FR2 (10)	0.48	0.69	*

Note: N = No rubber, GR2 = Granular rubber 2 mm diameter, FR2 = Fibrous rubber 2.36, FR4 = Fibrous rubber 4.75. Values for rubber percentage are shown in parentheses

* No difference between means.

** Confidence level at which the two means are different (Scheffé method).

*** N - GR2 (1) = the difference in the mean compressive strengths of mortar with no rubber and 1% GR 2.

variance and post-hoc test was used. Any confidence level below 95% was equivalent to no statistically significant difference. Table 8.2 summarizes the difference in cube strength due to rubber type, for different amounts of rubber. The table shows that the addition of 1% fibrous rubber did not significantly affect the compressive strength of mortar, but the addition of 1% granular rubber caused a significant reduction in the compressive strength.

There was a significant difference in the mean compressive strengths of mortar containing 5% granular rubber and 5% fibrous rubber. However, the trend did not hold true with the addition of 10% rubber by mass of cement. Overall, the results revealed that rubber geometry did not have a significant effect on the compressive strength of the mortar at high wt% of rubber, but there was a strong interaction effect between rubber geometry and rubber content on the compressive strength of rubberized mortar.

Since, the results reported in the literature for the determination of the compressive strength of rubberized concrete have included cylindrical and cube specimens, we compared the strength of the cylindrical and cube specimens of rubberized mortar. Figure 8.4 shows the individual compressive strengths of the cylindrical mortar specimens. No replicate batches were included in these tests. Within the batch, coefficients of variation ranged from 1.4 to 4.5%. As was the case with cube strength, it is shown in Figure

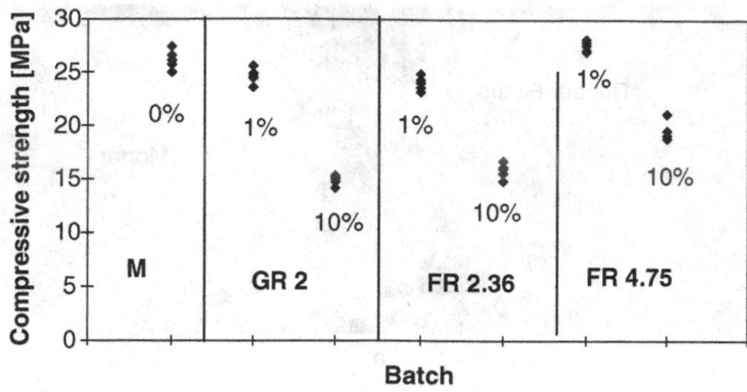

FIGURE 8.4
Compressive strength of mortar cylinders.

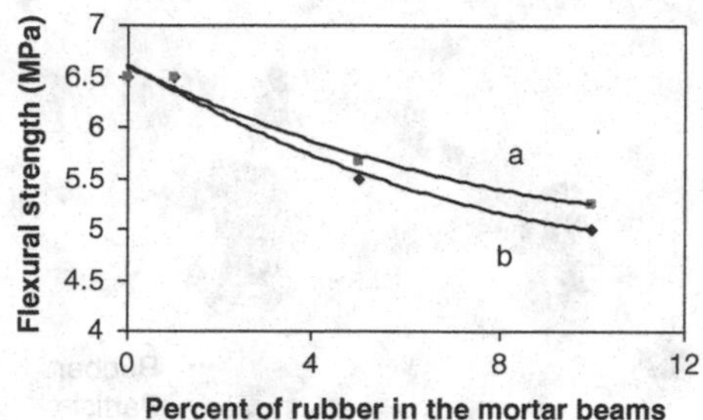

FIGURE 8.5
Flexural strength of (a) FR 4.75 and (b) GR 2 rubber-filled mortar beams.

8.4 that increasing the content of rubber decreased the cylinder compressive strength of mortar. Just as generally reported in the literature, we found that the compressive strength of mortar obtained from the cube test is slightly higher than that obtained from cylinder tests.

Figure 8.5 shows the flexural strength of the mortar beams as a function of rubber geometry and mass percent of rubber. The results of the flexural strength tests appear to indicate that rubberized mortar containing FR 4.75 may perform slightly better than that containing GR 2.[42] Furthermore, our results showed that the flexural strength of the mortar beams decreased with addition of rubber. The reduction of strength of mortar containing rubber is due to the replacement of load-carrying material by the low-modulus rubber aggregate material, or the weak rubber-cement interface, or their combined effect, which may be the reasons for premature failure of the composite.[7,36,56]

We also noticed that the beam specimen with rubber fibers was able to withstand some load, even when it was cracked. This was probably due to the bridging of cracks by the fibers. The specimens did not physically separate into two pieces under flexural loading. Figure 8.6a shows a fractured mortar specimen containing fibrous rubber. It can be seen that the mortar matrix failed, while the rubber fibers bridged the crack and prevented catastrophic failure of the specimen during the test. Unlike the bridging of

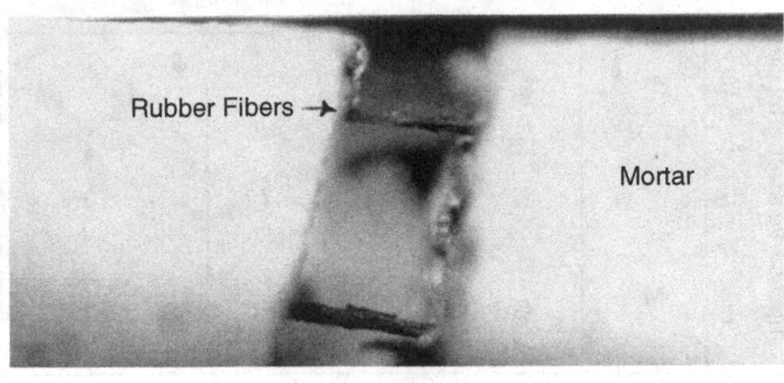

a

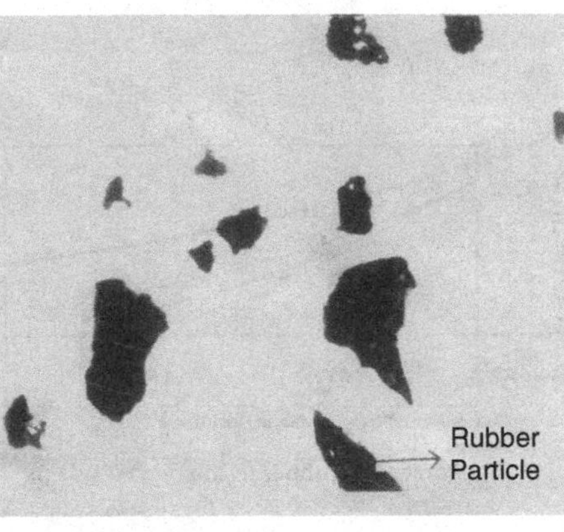

b

FIGURE 8.6
Fracture specimen of (a) fibrous rubber-filled cement specimen and (b) granular rubber-filled cement specimen.

cracks by the rubber fibers in fibrous, rubber-filled mortar when the specimen was loaded to peak load, the specimen with granular rubber broke into two halves when the peak load was reached (Figure 8.6b). Microscopic examination of the fracture surface of the granular rubber-filled mortar specimen showed rubber particles distributed in the matrix (Figure 8.6b). A close examination of the optical micrograph showed that that fracture occurred at the rubber/cement interface. The pull out characteristics of the granular rubber particles from the mortar matrix are consistent with widely reported findings that rubber/cement interface is weak.[35]

8.4 Rubber Treatment

Several attempts have been made to strengthen the rubber/cement interface by washing the granular rubber particles prior to their use in rubberized concrete.[36–40,48,50,61] Eldin and Senouci[36] showed that by washing the rubber particles with water, the cleaned rubber particles adhered to the cement matrix. Rostami and co-workers[50] also showed that washing the rubber particles with water and carbon tetrachloride, and water and latex admixture, can improve the adhesion of rubber to cementitious matrix. Chemical treatment of rubber with dilute nitric acid, sulfuric acid, and NaOH has also been investigated for improving the adhesion of rubber to cement.

The treatment of rubber with nitric acid was expected to chemically oxidize rubber and introduce polar groups, to improve the adhesion of the rubber particles with cement matrix. Contrary to expectation, nitric acid treatment of rubber was found to decrease the strength of rubber-filled cementitious matrix.[47] Lepore and Tantala[49] reported opposite results for when the rubber particle was treated with sulfuric acid. Segre,[51,52] using a combination of chemical techniques and surface-probing methods, demonstrated that the hydrophilicity of the rubber surface can be greatly improved by soaking the rubber in acidic or basic medium. The zinc stearate commonly found in rubber mixture diffuses to the surface and renders the rubber hydrophobic. By treating the rubber with acid, zinc stearate can be readily hydrolyzed (about 2 wt%) to stearic acid; by treating the rubber with base, the zinc ions can be replaced by sodium ions of NaOH (base medium), to form readily soluble sodium stearate. The water contact angle of the acid- or base-treated rubber was found to be lower (i.e., more hydrophilic) than untreated rubber. Addition of acid- or base-treated rubber to cement paste was found to improve its flexural strength and fracture energy, compared to that of untreated rubber-filled cement paste.

An alternate approach to improve rubber/cement adhesion is to make molecular contact between hydrophilic clay and hydrophobic rubber, with the use of coupling agents. Organofunctional alkyl trialkoxy silane coupling agents are commonly used to promote interfacial bonding of organic phase with inorganic material. The coupling agent should possess pendant groups that will open the double bonds of rubber as well as bond with cement.

Figure 8.7 is an illustration of the structure of gamma mercapto trimethoxy silane (GMPTS) coupling agent–bonded cement and recycled rubber interface. At pH 4.0, the coupling agent hydrolyzes to form methanol and trihydroxy silane derivative. The hydroxyl group of the silane agent can participate in the bond formation with the cement surface. The hydration of calcium silicate (primary component of Portland cement) yields a saturated solution of Ca^{+2}, SiO_4^-, and OH^-. Upon equilibration, the solution forms calcium silicate hydrate gel and calcium hydroxide.[62] The metal hydroxide bonds with the hydroxyl group of the silane agent to form a M-O-Si bond. At elevated temperatures, mercapto group (–SH) of GMPTS can be involved in the bond formation with the unsaturated groups in vulcanized rubber.

To establish that indeed, mercapto group of GMPTS is involved in the bond formation at elevated temperature, X-ray Photoelectron Spectroscopy (XPS) measurements were performed on the model system (polybutadiene and GMPTS coupling agent). A shift in S(2p) peak of the coupling agent was

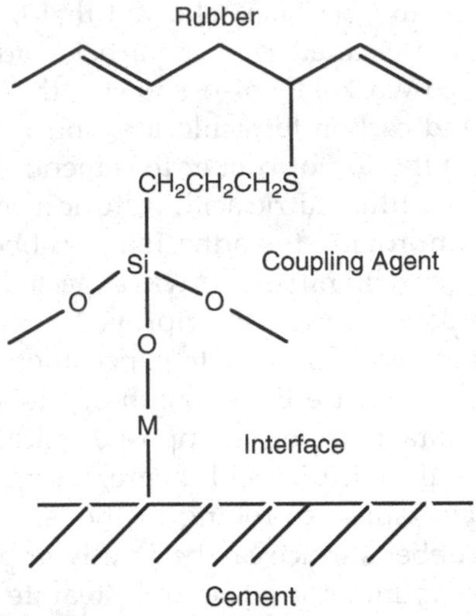

FIGURE 8.7
Schematic representation of rubber-coupling agent–cement interface.

noticed when the GMPTS was allowed to react with polybutadiene at elevated temperature. The shift may be a result of chemical bonding of thiol group with the unsaturation in polybutadiene.

A similar conclusion was drawn by Chung and Hong, who used vinyl triethoxy silane coupling agent to chemically bond rubber to aggregates.[53] They showed that the interfacial adhesion of rubber and aggregate could be improved when a recipe of rubber, vinyl trimethoxy silane coupling agent, sulfur, and zinc oxide was mixed with aggregate at 160°C.

An attempt was made to obtain qualitative measure of the bonding of rubber to cement surface by observing the peeled rubber surface in bonded and nonbonded composite. Figure 8.8 shows the effect of a coupling agent on the fractured surface of rubber/cement composite. On peeling the rubber from the cement, the rubber surface had several gray spots of cement particles adhering to it. The localized patches of gray spots were negligible on the rubber surface without the coupling agent. The presence of cement grains in the rubber sample is an indication of cohesive failure in the cement.

The extent of bonding between the rubber and the coupling agent–treated cement paste can be expressed as peel strength. Figure 8.9 is a plot of the peel strength results for various conditions. Details of peel strength measurement and calculation can be found elsewhere.[54,55,62] We noticed poor bonding for composite specimens with a coupling agent that were compression molded at room temperature. As expected, it appears that the coupling agent did not react with rubber at room temperature. When a rubber/cement composite specimen with a coupling agent was compressed and cured at elevated temperature (~150°C for 1 hour), there was a measurable degree of

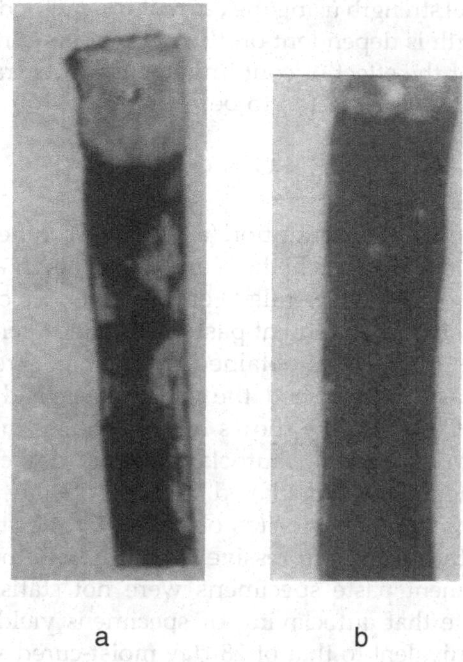

FIGURE 8.8
Picture of peeled rubber surface in (a) bonded and (b) unbonded rubber/cement composite.

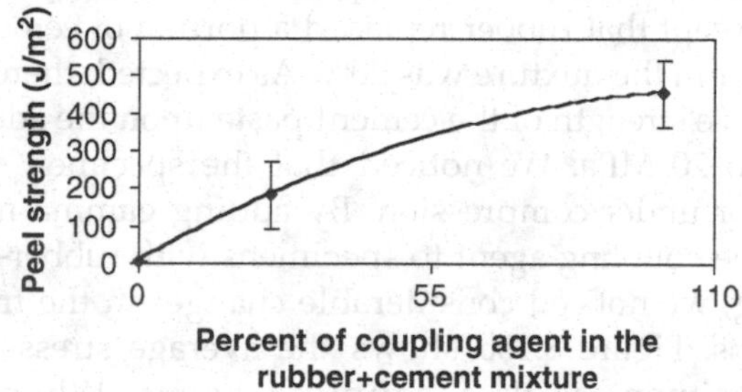

FIGURE 8.9
Plot of peel strength data of rubber-filled cement paste as a percent of coupling agent concentration.

bonding. The peel load-displacement profile shows two distinct regions: (1) linearly increasing region followed by (2) a plateau region. The load at the plateau region was used in the determination of peel strength. For composite specimens with a layer of undiluted coupling agent (100%), the peel strength was ~455 Jm^{-2} and for composite specimens with a layer of diluted coupling agent (25%), the peel strength was ~182 Jm^{-2}. These values are in general agreement with the reported adhesion energy for an elastomer layer bonded to a glass substrate by a vinyl triethoxy silane coupling agent.[63] In the absence of a coupling agent, the composite showed such poor bonding that it could not be tested for peel strength using the current test method. The data suggest that the peel strength is dependent on the mass of bonding coupling agent. The relationship for the effect of coupling agent concentration on the interfacial peel strength, G, is expected to be:

$$G = G_0 \times C/C_0 \tag{8.1}$$

where G_0 is 455 Jm^{-2} at C_0 concentration (as received) of the coupling agent.[64]

In an attempt to determine whether improvements in adhesion translate to desirable failure of the rubber-filled cement paste, we compared the fracture behavior of rubber-filled cement paste and silane agent–treated, rubber-filled cement paste. Initially, we obtained the average stress-strain curve of hydrated cement paste. As expected, the cement paste had high strength and low failure strain. Cement paste shows minor changes in fracture behavior with preparation conditions (i.e., autoclaving or 28-day curing of the paste specimen). In general, the autoclaved and the 28-day cured specimens showed a brittle fracture behavior (as observed by the stress-strain curve). The differences in the mean compressive strengths of autoclaved and 28-day moisture-cured cement paste specimens were not statistically significant. Our results indicate that autoclaving of specimens yielded cement pastes with a strength equivalent to that of 28-day moist-cured strength. This is in general agreement with the data reported in the literature for 28-day aerated and autoclaved cementitious specimens.[62,63]

A rubber-filled cement paste was prepared similarly to the hydrated cement paste, except that rubber replaced a portion of cement. The volume fraction of rubber in the mixture was 20%. As expected, there was a decrease in the compressive strength of the cement paste upon the addition of rubber, from 35 MPa to 20 MPa. We noticed that the specimen exhibited brittle fracture behavior under compression. By adding gamma mercapto propyl trimethoxy silane coupling agent to specimens with rubber-cement mixture and autoclaving, we noticed considerable changes in the fracture behavior of the specimens. Figure 8.10b shows the average stress-strain curve for composite specimens with a coupling agent. When the coupling agent–treated rubber cementitious composite specimen was aged for 28 days, we found no substantial improvement in the fracture stress and strain of the composite, and the behavior was similar to the 28-day cured rubber cement mixture with no coupling agent. When the specimen was autoclaved, the composite was able to withstand a fraction of the ultimate load for long periods of time (observed as a tail in the stress-strain curve). The observation of a large tail at 150°C and 170°C autoclaved rubber cement with bonding agent is the evidence of greater ductility.

The fracture surfaces of rubber-filled cement composite and rubber-bonded cement composite were examined by scanning electron microscopy.[55] The micrograph showed torn rubber particles in the cement matrix. Unlike untreated rubber particle, where we noticed a particulate pull out, we observed tearing of the rubber particulate in treated rubber cement compos-

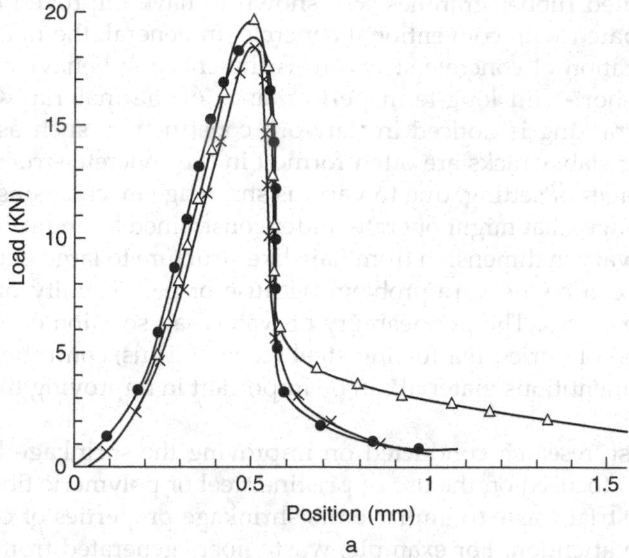

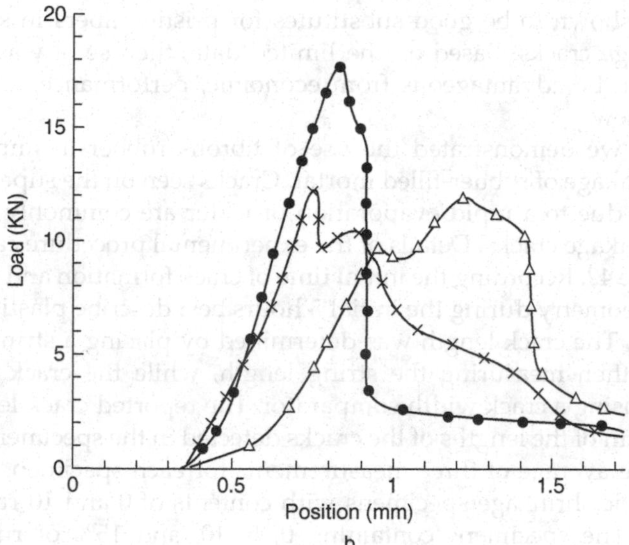

FIGURE 8.10
Average stress-strain graph of (a) rubber-filled cement paste and (b) coupling agent–treated rubber-filled cement paste. ●, 25°C; ×, 150°C; △, 170°C.

ite, suggesting the role of the coupling agent in strengthening the interface. The post-crack strength is improved by switching from granular rubber to fibrous rubber.

Until now, most of the research on uses of rubber particles in concrete has been conducted to evaluate the strength of rubberized concrete and adhesion of the rubber/cement interface. The exception is a study[35] where concrete with elongated rubber granules was shown to have improved crack resistance, compared with conventional concrete. In general, the main cause for the deterioration of concrete structure is the cracking behavior that determines the short- and long-term performance of the material. Commonly, shrinkage cracking is noticed in flatwork construction, such as pavement and parking slab. Cracks are often formed in the concrete structure during the early stages of setting, due to various shrinkage mechanisms (drying or thermal change) that might operate under constrained boundary conditions. Cracks can vary in dimension from hair-like structure to large cracks. Cracks in concrete can be a severe problem because of permeability of salt water through the cracks. The permeability of water/salt solution could promote the corrosion of buried reinforcing steel rebars.[65] Thus, control of shrinkage cracks in cementitious material can be important in improving the longevity of material.

In the past, research conducted on improving the shrinkage behavior of concrete has focused on the use of pristine steel or polymeric fibers.[66–68] The use of industrial waste to improve the shrinkage properties of concrete has drawn little attention. For example, waste fibers generated from the carpet industry amount to 2 million tons per year, and fibers from rubber tire cords have been shown to be good substitutes for pristine fibers in suppressing the shrinkage cracks. Based on the limited data, the use of waste fibers in concrete can be advantageous from economic, performance, and disposal points of view.

Recently, we demonstrated the use of fibrous rubber in improving the plastic shrinkage of rubber-filled mortar. Cracks seen on the superficial layer of concrete due to a rapid evaporation of water are commonly referred as plastic shrinkage cracks. Details of the experimental procedure can be found in Reference 42. Recording the initial time of crack formation and monitoring the crack geometry during the initial 3 hours best describe plastic shrinkage of concrete. The crack length was determined by placing a string along the crack and then measuring the string length, while the crack width was measured using a crack width comparator. The reported crack length represents the sum of the lengths of the cracks detected in the specimen. The crack width is the average of three measurements for each specimen. Figure 8.11 shows plastic shrinkage specimens with contents of 0 and 10 rubber fibers in mortar. The specimens containing 0, 5, 10, and 15% of rubber fibers cracked, while those containing 1% of polypropylene fibers did not crack in the first 3 hours.[42] In the cracked specimens, the crack was always observed over the central stress raiser. The addition of fibrous rubber resulted in a smaller and a noncontinuous crack. Apparently, crack propagation was arrested several times due to interference by the fibrous rubber. The fibrous particles, despite the weak bonding of rubber, provided sufficient restraint to prevent the crack from progressing.

To quantify the plastic shrinkage cracking of rubber containing mortars, the width of the cracks and the time of appearance of first crack were noted; the results are shown in Figure 8.12. After 3 hours, the mortar specimen

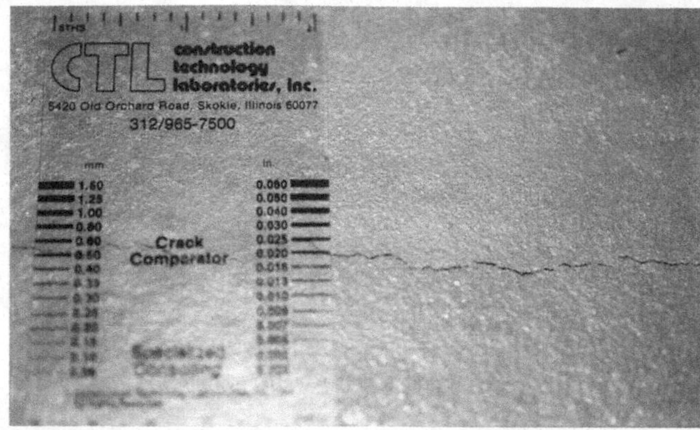

a

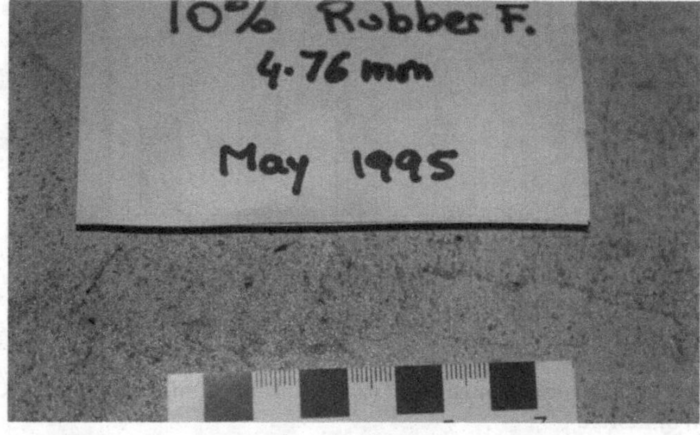

b

FIGURE 8.11
Picture of the specimen after testing for plastic shrinkage (a) plain mortar (b) 10% fibrous rubber-mortar.

(control) developed a crack having an average width of about 0.9 mm, while the average crack width for the specimen with 5% fibrous rubber was about 0.4 mm. It was found that the onset of cracking was delayed by the addition of fibrous rubber; the mortar without fibers cracked within 30 minutes, while the specimens with 5% and FR 4.75 fibrous rubber cracked after 1 hour. The crack length, the crack width, and the time of failure of the rubber-containing mortar were dependent on the geometry of the rubber in the mortar.

Figure 8.13 summarizes the flow characteristics of mortar mixtures. The details of the test can be found elsewhere.[42,69] The VeBe time is a measure of the workability of the mixture. A shorter VeBe time represents a more workable mixture. The reported results are averages of two or three tests. The control mixture and the mixture with 10% fibrous rubber had similar VeBe times, while the mixture with 1% fibrous rubber had a lower VeBe time. The VeBe time for 1% polypropylene fibers was much larger (241

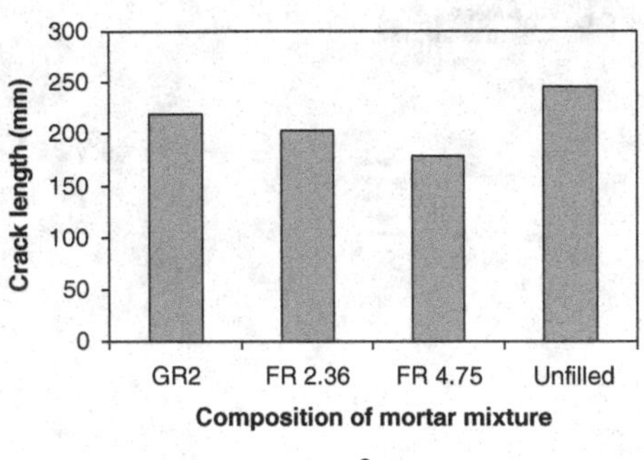

FIGURE 8.12
Results of (a) crack width measurements and (b) time of appearance of initial crack as a function of composition of mortar.

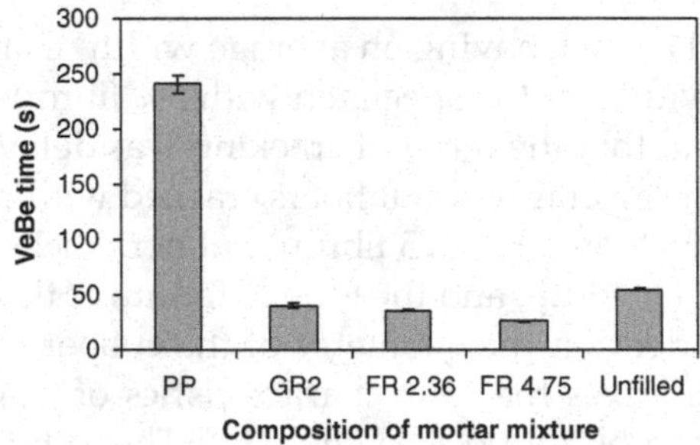

FIGURE 8.13
Flow characteristics (VeBe time) of mortar mixture.

seconds). It is believed that the effects of the polypropylene fibers in preventing the free flow of the mixture are attributable to their greater number and high aspect ratio.

8.5 Miscellaneous Applications

Molded plastic products can also be manufactured by using blends of fine-ground rubber and plastics. The molded product could be used to produce artificial timbers for landscaping and to extrude more durable slates for use in roof shingles. The tire content of the blend would protect the plastic from UV light deterioration.

Civil-engineering applications, such as light landfill cover and potential landfill drainage layer, are also attractive applications for shredded rubber material. Many local, state, and federal officials have been reluctant to use the shredded rubber tires because of lack of data that address the long-term effects of possible leachate from various recycled tire compounds on air, water, and soil.

8.6 Summary

The development of rubberized concrete requires consideration of the chemistry of rubber and cement, as well as geometric effects of shredded rubber on the mechanical and durability characteristics of the overall concrete. The post-crack strength is improved by switching from granular rubber to fibrous rubber. It is expected that with further development and optimization of technical issues and improved design procedures, many, if not most, applications of rubber in low-strength concrete can be very interesting. It is believed that, if tires are used in the above-mentioned markets, then the annual scrap tire demand will exceed the disposed refuse.

Acknowledgments

The author would like to take this opportunity to acknowledge H. Huynh from Howard University, C. Ferraris from NIST, and R. P. Wool from University of Delaware. Financial support from NIST, IBM, and NSF is recognized. I would also like to thank C. Moreland from Michelin Americas Research Corporation for proofreading the manuscript.

References

1. Gent, A.N. and Livingston, D.I., Automobile tires. In: *Concise Encyclopedia of Composite Materials,* Kelly, A., Cahn, R.W., and Bever, M.B. (eds.), Pergamon Press, 1989, p. 21.
2. Tullo, A.H., Synthetic rubber, *Chemical & Engineering News,* 81 (15), 23, 2003.
3. Anonymous, New directions. *Adv. Matls. Proc.* 135 (1), 21, 1989.
4. Eldin, N.N. and Senouci, A.B., Rubber-tire particles as concrete aggregate, *ASCE J. Mat. Civ. Eng.* 5, 478, 1993.
5. Thompson, W.H., Studies on *Aedus triseriatus* — the LaCrosse virus carrier, *Wis. Acad. Rev.,* 31, 64, 1984.
6. Serumgard, J.R. and Blumenthal, M.H., A practical approach to managing scrap tires, *MSW Management,* 48 (September 1993).
7. Eldin, N.N. and Senouci, A.B., Observations on rubberized concrete, *Cem. Concr. Agg.* 15 (1), 74, 1993.
8. Singh, S.S., Innovative application of scrap tires, *Wisc. Prof.Engineers,* 14, 1993.
9. Paul, J., Rubber reclaiming, *Encycl. Polym. Sci. Eng.,* 14, 787, 1985.
10. Paul, J., Rubber, *Encycl. Chem. Technol.,* 19, 1002, 1982.
11. Rodriguez, F., The prospects of biodegradable plastic, *Chem. Technol.,* 50, 409, 1971.
12. Bredburg, B.E., Andersson, E. L., and Holst, O., Microbial detoxification of waste rubber material by wood rotting fungi, *Bioresource Technol.,* 83 (3), 221, 2002.
13. Raghavan, D., Guay, R., and Torma, A.E., A study of biodegradation of polyethylene and biodesulfurization of rubber, *Appl. Biochem. Biotechnol.,* 24, 387, 1990.
14. Torma, A.E. and Raghavan, D., Biodesulfurization of rubber, *Proceedings,* Bioprocess Engineering Symposium, Hochmuth, R.M., ed., American Society of Materials Engineering, New York, 1990, 81.
15. Thaysen, A.C., Bunker, H.J., and Adams, M.E., Rubber acid damage in fire hoses, *Nature,* 155, 322, 1945.
16. Sverdrup, E.F. and Wendrwow, B.R., Rubber reclaiming, *Encyl. Polym. Sci. Technol.,* 14, 787, 1985.
17. Barbooti, M.M., Hassan, E.B., and Issa, N.A., Thermogravimetric and pyrolytic investigation on scrap tires, *J. Petro. Res.,* 8 (2), 229, 1989.
18. Bouvier, J.M., Carbel, F., and Gelus, M., Gas-solid pyrolysis of tire wastes — kinetics and material balance of batch pyrolysis of used tires, *Resour. Conserv.,* 15 (3), 205, 1987.
19. Wojtowicz, M.A., in *Workshop on National Cooperation Strategies for Reuse of Waste Rubber Tires in Infrastructure,* Raghavan, D., Sabnis, G.M., and Ahmad, S., eds., NIST, Gaithersburg, Maryland, (1994).
20. Sullivan, J.J., in *Workshop on National Cooperation Strategies for Reuse of Waste Rubber Tires in Infrastructure,* Raghavan, D., Sabnis, G.M., and Ahmad, S., eds., NIST, Gaithersburg, Maryland, (1994).
21. O'Keefe, W., Stone, clay and glass: fuel price shift puts fully researched tire burning on hold at cement plant, *Power,* 128 (10), 115, 1984.
22. Lee, B., Research in progress in developing tire-concrete, *Proceedings,* ACI Spring Convention, ACI, Salt Lake City, Utah, 1995.

23. Lemieux, P., Lutes, C.C., and Santoianni, D.A., Emissions of organic air toxics from open burning: a comprehensive review, *Progress in Energy and Combustion Science*, 30(1), 1, 2004.
24. Dose, G.J., Elfving, D.C., and Link, D.J., Zinc in foliage downwind from a tire-burning power plant, *Chemosphere*, 31(3), 2901, 1995.
25. *Rubber World*, 67 (October, 1978).
26. Nelson, S.M., Mueller, G., and Hemphill, D.C., Identification of tire leachate toxicants and a risk assessment of water quality effects using tire reefs in canals, *Bull. Environ. Contam. Toxicol.*, 52, 574, 1994.
27. "Waste Tires in Sub-Grade Road Beds," Minnesota Pollution Control Agency (1990), Waste Tire Management Unit, St. Paul, Minnesota.
28. Takallou, H.B. and Takallou, M.B., Recycling tires in rubber asphalt paving yields cost disposal benefits, *Elastomeric*, 123 (7), 19, 1991.
29. Morrison, G.R., Lee, N.A., and Hesp, S.A.M., Recycling of plastic and rubber tire waste in asphalt pavements, *MRS International Meeting on Materials and Processes for Environmental Protection*, (344), Materials Research Society, Pittsburgh, Pennsylvania, 189, 1994.
30. Sainton, A., Advantages of asphalt rubber binder for porous asphalt concrete, *Transp. Res. Rec.*, 1265, 69, 1995.
31. Blumenthal, M.H., Experience smooths a bumpy road, *Solid Waste Technol.*, January-February 1995.
32. McQuillen Jr., J.L., Takallou, H.B., and Hicks, R.G., Construction of rubber modified asphalt pavement, *J. Transp. Eng.*, 114 (3), 259, 1988.
33. Anonymous, What's the future for rubberized asphalt?, *Biocycle*, 32, 63, 1991.
34. Anonymous, Recycled tires down the road, *Biocycle*, 34, 9, 1993.
35. Frankowski, R., Rubber-Crumb-Reinforced Cement Concrete, U.S. Patent, #5,391,226, 1995.
36. Eldin, N.N. and Senouci, A.B., Measurement and prediction of the strength of rubberized concrete, *Cement and Concrete Composites*, 16 (4), 287, 1994.
37. Topcu, I.B., The properties of rubberized concrete, *Cement Concrete Res.*, 25 (2), 304, 1995.
38. Topcu, I.B., Assessment of brittle index of rubberized concrete, *Cement Concrete Res.*, 27 (2), 177, 1997.
39. Topcu, I.B. and Avcular, N., Collision behavior of rubberized concrete, *Cement Concrete Res.*, 27 (12), 1893, 1997.
40. Topcu, I.B. and Nuri, A., Analysis of rubberized concrete as a composite material, *Cement Concrete Res.*, 27 (8), 1135, 1997.
41. Khatib, Z.K. and Bayomy, F.M., Rubberized portland cement concrete, *J. Mater. Civ. Eng.*, 11 (3), 206, 1999.
42. Huynh, H., Raghavan, D., and Ferraris, C.F., "Rubber Particles From Recycled Tires in Cementitious Composite Materials," National Institute of Standards and Technology, NISTIR 5850, NIST, Gaithersburg, Maryland, 1996, 1–20.
43. Raghavan, D., Huynh, H., and Ferraris, C.F., Workability, mechanical properties, and chemical stability of a recycled tyre rubber-filled cementitious composite, *J. Mat. Sci.*, 33, 1745, 1998.
44. Shutov, F. and Volfson, S.,New principle of plastic waste recycling: solid state shear extrusion, *Polym. Matl. Sci. Eng.*, 67, 404, 1992.
45. Goldstein, H., Not your father's concrete, *Civ. Eng.* 65 (5), 60, 1995.

46. Chung, K-H. and Hong, Y-K, Scrap tire/aggregate composite: composition and primary characterizations for pavement material, *Polym. Composites,* 23 (5), 852, 2002.
47. Lee, B.I., Burnett, L., Miller, T., Postage, B., and Cuneo, J., Tyre rubber/cement matrix composite, *J. Mater. Sci. Lett.* 12 (13), 967, 1993.
48. Fattuhi, N.I. and Clark, L.A., Cement-based materials containing shredded scrap truck tyre rubber, *Const. Bldg. Matls.* 10 (4), 229, 1996.
49. Lepore, J.A. and Tantala, M.W., Behavior of rubber induced concrete (RIC), In *Concrete Institute of Australia – Concrete 97,* 1997, pp. 623–627.
50. Rostami, H., Lepore, J., Silverstraim, T., and Zandi, I., Use of recycled tire rubber in concrete, In: *Proceedings, International Conference Concrete 2000,* UK, 1993, pp.391–399.
51. Segre, N., Monteiro, P.J.M., and Sposito, G., Surface characterization of recycled tire rubber to be used in cement paste matrix, *J. Colloid Interface Sci.,*248, 521, 2002.
52. Segre, N. and Joekes, I., Use of tire rubber particles as addition to cement paste, *Cement Concrete Res.,* 30, 1421, 2000.
53. Chung, K-H. and Hong, Y-K., Introductory behavior of rubber concrete, *J. Appl. Polym. Sci.,* 72, 35, 1999.
54. Raghavan, D., Tratt, K., and Wool, R.P., Recycled rubber in cement composites, *MRS International Meeting on Materials and Processes for Environmental Protection,* (344) Materials Research Society, Pittsburgh, Pennsylvania, 177 (1994).
55. Raghavan, D., Tratt, K., and Wool, R.P., Interfacial bonding of recycled rubber-cement composite, *Process Adv. Mater.,* 4, 203, 1994.
56. Raghavan, D., Study of rubber-filled cementitious composites, *J. Appl. Polym. Sci.,* 77, 934, 2000.
57. Krysztafkiewicz, A., Jesionowski, T., and Rager, B., Reinforcing of synthetic rubber with waste cement dust modified by coupling agents, *J. Adhesion Sci. and Technol.,* 11 (4), 507, 1997.
58. Sherwood, P.T., The use of waste and recycled materials in roads, *Proc. Instn. Civ. Engrs. Transp.* 111, 116, 1995.
59. Prasher, C.L., *Crushing and Grinding Process Handbook,* John Wiley & Sons Limited, New York, 1987.
60. Rouse, M.W., Development and application of superfine tire powders for rubber compounding, *Rubber World,* 204 (4), 1992.
61. Ali, N.A., Amos, A.D., and Roberts, M., Use of ground rubber tires in Portland cement concrete, In: *Proceedings of International Conference Concrete 2000,* University of Dundee, UK, 1993, pp. 379–390.
62. Taylor, H.W., *Cement Chemistry,* Academic Press, New York, (1990).
63. Mindness, S. and Young, J.F., *Concrete,* Prentice-Hall, New Jersey (1981).
64. Ahagon, A. and Gent, A.N., Effect of interfacial bonding on the strength of adhesion, *J. Polym. Sci.,* 13, 1285, 1975.
65. Moreno, M., Morris, W., Alvarez, M.G., and Duffó, G.S., Corrosion of reinforcing steel in simulated concrete pore solutions: effect of carbonation and chloride content, *Corrosion Science,* 46(11), 2681, 2004.
66. Soroushian, P., Khan, A.,and Hsu, J-W., Mechanical properties of concrete materials reinforced with polypropylene or polyethylene fibers, *ACI Mater. J.,* 89 (6), 535, 1992.

67. Wu, H.C., Lim, Y.M.,Li, V.C., and Foremsky, D.J., Utilization of recycled fibers in concrete, In: *4th International Engineering Conference Materials for the New Millennium*, Chong, K.P., ed., ASCE, Washington, DC, 1996, 799.
68. Wang, Y., Cho, B.S., and Zureick, A.H., Fiber reinforced concrete using recycled carpet industrial waste and its potential use in highway construction, In: *Proceedings of the Symposium on Recovery and Effective Reuse of Discarded Materials and By-Products for Construction of Highway Facilities*, Denver, Colorado, October 19–22, 1993.
69. Huynh, H., "Mechanical and Chemical Properties of Rubber-Cement Composites," MS thesis, Howard University, 1996.

9

Ultrasonic Devulcanization of Used Tires and Waste Rubbers

A.I. Isayev and Sayata Ghose

CONTENTS

9.1 Abstract

A recent method for recycling thermosets is with the application of ultrasound, by which waves of certain levels, in the presence of heat and temperature, can rapidly break down the three-dimensional structure of cross-linked rubber, making it reprocessable. This chapter provides an up-to-date account of ultrasonic devulcanization of rubbers, including the history of the technology, available reactors, extensive experience accumulated with various types of rubber, attempts to develop mechanisms of devulcanization, and models to theoretically describe the process and a possibility of its scale–up. The chapter gives a comparative analysis of process characteristics, rheological and mechanical properties, structural transformations, curing behavior and nuclear magnetic resonance (NMR) relaxation, and diffusion processes taking place in unfilled and filled rubbers during their ultrasonic treatment. Properties of blends of devulcanized and virgin rubbers are also presented. Sulfur- and peroxide-cured rubbers are considered, along with details of degradation mechanisms of the rubber network by ultrasonic waves. Future directions and further development of ultrasonic devulcanization technology are also discussed.

9.2 Introduction

One of the most crucial problems faced by man in this era of rapid development and economic growth is the problem of waste management, including management of used tires and other rubbers. According to the recent (2001) statistics of the Rubber Manufacturers Association, the total number of tires scrapped annually in the U.S. is 281 million, and the approximate number of tires in the stockpiles is 300 million.[1] Of these 300 million, about 115 million scrap tires are being used to make tire-derived fuels (TDFs), 40 million are used for civil-engineering application, 33 million are processed into ground rubber, and only 8 million are made into new products. Apart from tires, other rubber wastes have also become a growing concern today. Over 350 million pounds of rubber are scrapped annually from the production of non-tire goods in the form of runners, trims, and pads.[2]

Until quite recently, the utilization of polymeric wastes was considered a task for select engineers only; the majority of scientists were indifferent to it. However, the situation changed dramatically over the last 30 years. Synthetic polymers exceeded the volume of production and use of traditional materials like metals, glass, ceramics, wood, and paper. In contrast to their natural counterparts, these polymers are neither reusable nor biodegradable, and are rather stable to natural factors.

The major problem encountered with the recycling of thermosets is that, unlike thermoplastics, they cannot be reprocessed simply by the application of heat. This is due to the presence of their cross-linked structure and three-dimensional network. To recycle thermosets, the cross-links in the network structure have to be broken before the material can be reprocessed; this is the inherent difficulty in the recycling of thermosets. Thus, many of the advantages of cross-linked polymers, such as their variety, stability, and ability to form spatial molecular networks, turn out to be the factors hindering the recycling of the used articles.

The process of recycling would, then, involve the conversion of the three-dimensional, insoluble, and infusible thermoset to a soft, tacky, processable, and vulcanizable product that simulates many of the properties of the virgin compound. Recovery and recycling of rubber from used and scrap rubber products would save precious petroleum resources and also solve the scrap/waste rubber disposal problem.

Reclaimed rubber is the product that results when waste vulcanized scrap rubber is treated to produce a material that can be easily processed, compounded, and vulcanized with or without the addition of either natural or synthetic rubbers. Recycled rubber can be generalized to include any rubber waste that has been converted to an economically useful form, such as reclaimed rubber, ground rubber, or reprocessed synthetic rubber.[3]

Incineration and landfilling, the two most common methods of recycling, generate environmental pollution; therefore, the time has come to search for alternate methods of recycling rubber wastes. Numerous techniques have been proposed for recycling waste rubbers, including methods like catalysis,[4] mechanical,[5,6] microwave,[7] biotechnological,[8] and solid-state shear pulverization.[9,10] An extensive compilation of numerous chemical reclaiming methods has been done,[11] and extensive reviews on the various methods of rubber recycling have been discussed.[2,12,13]

One of the recent methods of recycling of thermosets has been by the application of ultrasound. Ultrasonic waves of certain levels, in the presence of heat and temperature, are able to rapidly break down the three-dimensional structure of the cross-linked rubber. This devulcanized rubber is soft and reprocessable. The advantages of using ultrasound are that the process is continuous, it occurs within seconds or less, and it does not require the use of any chemicals.

The main objective of this chapter is to give a detailed account of the process of devulcanization of filled and unfilled rubbers by the application of high-power ultrasound. Devulcanization has been defined as the process of cleaving, totally or partially, the poly-, di-, and monosulfide cross-links that were formed during the initial vulcanization.[11] However, this is not viable and hence, in the present study, devulcanization is defined as the process of cleaving of cross-links, along with some main chain breakage, to produce useful materials that can be revulcanized and shaped into products.

9.3 Development of Technology

9.3.1 Introductory Remarks

Numerous publications over the years have been devoted to the study of the effect of ultrasound on polymer solutions[14–19] and on polymer melts during extrusion.[20–25] Significant efforts have also been made to understand the mechanism of the effect of ultrasound on fluids[26,27] and degradation of polymers in solution.[28]

The application of ultrasonic waves to the process of devulcanizing rubber is a novel and attractive field of study. It was originally thought that rubber is vulcanized, rather than devulcanized, by ultrasound.[29] Rubber devulcanization by using only ultrasonic energy was first discussed by Okuda and Hatano.[30] It was a batch process in which a small piece of vulcanized rubber was devulcanized at 50 kHz ultrasonic waves after treatment for 20 minutes. The process claimed to break down carbon-sulfur bonds and sulfur-sulfur bonds, but not carbon-carbon bonds. The properties of the revulcanized rubber were found to be very similar to those of the original vulcanizates.

During the last decade, a novel, continuous process has been developed; many studies have focused on the ultrasonic devulcanization of various rubbers as a suitable way to recycle used tires and waste rubbers.[31-72] This technology is based on the use of high-power ultrasound, along with an extrusion equipment. The ultrasonic waves of certain levels, in the presence of pressure and heat, can quickly break up the three-dimensional network in cross-linked rubber. The process of ultrasonic devulcanization is very fast, simple, efficient, and solvent and chemical free. Devulcanization occurs within seconds and may lead to the preferential breakage of sulfidic cross-links in sulfur-cured, vulcanized rubbers. The process is also suitable for de-cross-linking of peroxide-cured rubbers and plastics.

9.3.2 Equipment

The devulcanization process is typically carried out by means of ultrasonic devulcanization reactors. So far, three types of devulcanization reactors based on single-screw extruders have been developed. A schematic diagram of the various devulcanization reactors suitable for carrying out this process is shown in Figure 9.1. The first is a coaxial reactor,[33,34] the second is a barrel reactor,[53] the third is a grooved barrel reactor.[65] The coaxial devulcanization reactor that was initially developed consists of a 38.1 mm single rubber extruder with the L/D = 11 and a coaxial ultrasonic die attachment, as shown in Figure 9.1a.[33] The barrel has three temperature control zones, equipped with electrical heaters and fans. The ultrasonic unit is composed of a 3.0 kW ultrasonic power supply, an acoustic converter, a 1:1 booster, and a 76.2 mm diameter cone-tipped horn. The horn vibrates longitudinally with a frequency of 20 kHz and varying amplitudes ranging from 5 to 10 microns. The ultrasonic waves propagate perpendicular to the flow direction in the die gap. The ultrasonic unit is mounted on four rigid tie bars, fixed to the extruder flange. It moves easily along the bars and is readily fixed by special pins. The ultrasonic energy consumed during experiments is measured by a watt meter attached to the ultrasonic unit.

Both the die and the horn have sealed inner cavities for cooling water stream. The die and horn are cooled down with tap water, to reduce the heat buildup caused by dissipation of the ultrasonic energy in the rubber, so that the degradation of rubber due to the high temperature can be minimized. The screw speed is variable, from 20 to 80 rpm. The cooling water flow rate for both die and horn is 0.09 m^3/hour. A flush-mounted pressure gage and thermocouple are attached to the barrel, to monitor the pressure and the temperature of the rubber at the entrance to the die.

The convex tip of the horn matches the concave surface of the die, so that the clearance between the horn and the die is uniform. The clearance is controlled and varied from 0.25 to 5.0 mm. A uniform or variable die gap can be used. The rubber flows through the clearance and is devulcanized under the action of ultrasonic waves, propagating perpendicular to the flow direction.

Later, the barrel (Figure 1b) and the grooved barrel (Figure 1c) ultrasonic reactors were developed. In the barrel reactor, two ultrasonic, water-cooled horns of rectangular cross sections were inserted into the barrel through two ports.[53] The clearance existing between the horns and the barrel was sealed by Teflon gaskets. Also, two restrictors made of bronze were placed in the barrel. These restrictors blocked the flow of rubber and forced the rubber to flow through the gap created between the rotating screw and the tip of the horn. The clearance can be varied from 0.25 to 5.0 mm. In the devulcanization section, the larger diameter of a rotating shaft provided converging flow of the rubber to the devulcanization zone. The latter may enhance the devulcanization process.

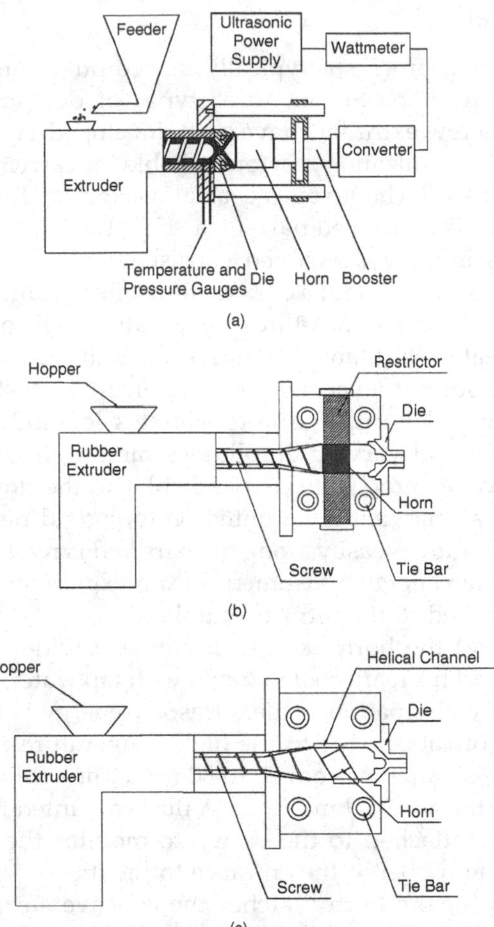

FIGURE 9.1
Schematic representation of the coaxial (a), barrel (b), and grooved barrel (c) reactors.

Figure 9.1c shows the grooved barrel reactor.[65] In the grooved barrel ultrasonic reactor, two helical channels were made on the barrel surface (grooved barrel). Rubber flows into the helical channel and passes through the gap created between the rotating shaft and the tip of the horns, where devulcanization takes place. Two ultrasonic horns of rectangular cross section (38.1 × 38.1 mm^2) are inserted into the barrel through two ports. The axes of the horns are perpendicular to the barrel and thus allow the imposition of longitudinal waves perpendicular to the flow direction. The horns are cooled by tap water. The gap can vary from 0.25 to 5.0 mm. The clearance between the horns and the ports of the barrel are sealed by 40 wt% graphite-filled polyimide gaskets. Two pressure transducers are placed in the barrel in front of and behind the devulcanization section. The latter allows the measure of pressure difference in the devulcanization section. There are two thermocouples, one attached to a temperature controller maintaining the barrel temperature and the other connected to a temperature-measuring device. The screw diameter is 50.8 mm in the devulcanization zone; this larger diameter is intended to provide a converging flow of rubber to the devulcanization zone. A circular die is attached to the extruder. The diameter of the circular die can be varied to vary the pressure in the devulcanization zone. Also, the die can be changed depending on the shape of the final extrudate.

The extruder of the reactors is fed by means of feeders using chunks of rubber or rubber particles of a given mesh size. The screw of the extruder takes this material, compresses it, and transfers it to the devulcanization zone. In this zone, resulting from compression and extension actions of ultrasound waves, a breakage of chemical bonds takes place due to a fatigue-like effect. The first action of the compression-extension waves occurs at the entrance edge of the horn, initiating the devulcanization process. In this area, significant mechanical energy is generated by the ultrasonic waves, leading to extension-contraction stresses on molecular chains; these stresses, in turn, cause the breaking of chemical bonds. Apparently, these high stresses were first initiated by the acoustic cavitation in the area surrounding the entrance edge of the horn. The cavitation is generated in the nonhomogeneous media consisting of packed particles and voids present in any rubber. Also, the nonuniformity of the material density throughout the rubber facilitates the dissipation of ultrasonic energy. The devulcanization propagates further downstream as the material moves from the edge of the horn toward its end. It should be mentioned that the effect of devulcanization greatly depends on the thickness of rubber layer in the devulcanization zone. Typically, at gap sizes higher than about 5 mm, devulcanization does not occur at all.

9.3.3 Application of Technology

Amplitude of the ultrasonic waves, pressure in the devulcanization zone, and residence time are the main operating parameters affecting the degree and rate of devulcanization. If the amplitude is too small or the pressure is too low, no devulcanization takes place. The process is very fast; the average residence time in the order of seconds. The effect of the temperature is complicated by its influence on the rubber viscosity and, therefore, on the pressure in the treatment zone. Typically, the temperature was found to have little effect on the efficiency of devulcanization due to the fact that ultrasonic waves generate tremendous dissipation of energy in rubber, leading to a heat-up of the sample. Thus, as stated above, the process requires cooling of the horn and the die to avoid excessive degradation of the main chains.

Under the license from the University of Akron for the ultrasonic devulcanization technology, NFM Inc. of Massillon, Ohio, has built a prototype of the machine for ultrasonic devulcanization of tire and rubber products.[66] It was reported that retreaded truck tires containing 15 wt% and 30 wt% of ultrasonically devulcanized carbon black–filled styrene-butadiene rubber (SBR) had passed the preliminary dynamic endurance test.[67]

Extensive studies on the ultrasonic devulcanization of rubbers[31–72] and some preliminary studies on ultrasonic de-cross-linking of cross-linked plastics[32] were carried out. It has been shown that this continuous process allows the recycling of various types of rubbers and thermosets. As the most desirable consequence, ultrasonically devulcanized rubber becomes soft, which enables this material to be reprocessed, shaped, and revulcanized in very much the same way as the virgin rubber. This new technology has been used successfully in the laboratory to devulcanize ground tire rubber (GRT),[33,34,38,40,53,66,67] unfilled and filled natural rubber (NR),[47,51] guayule rubber,[68] unfilled and filled SBR,[33,35–45] unfilled and filled silicone rubber,[46,48,52,59–61, 63] unfilled and filled ethylene-propylene diene monomer (EPDM) and EPDM roofing membrane,[58,62,65] unfilled polyurethane rubber (PU),[64] butadiene rubber (BR),[69] fluoroelastomer, ethylene vinyl acetate foam, and cross-linked polyethylene.[31,32] After revulcanization, rubber samples exhibit good mechanical properties, which in some cases for unfilled vulcanizates, are comparable to or exceed those of the virgin vulcanizates.[43,48,58]

Ultrasonic devulcanization studies were concerned with finding the effect of processing parameters (e.g., the pressure, power consumption, die gap, temperature, flow rate, and ultrasonic amplitude) on devulcanization, structural changes occurring in various rubbers, rheological properties and curing kinetics of devulcanized rubbers, and mechanical properties of revulcanized rubbers, and also the effect of the design of the devulcanization reactor. These are discussed in the following sections.

9.4 Process Characteristics

9.4.1 Processing Parameters

Processing parameters varied or measured in the devulcanization experiments can be divided into two groups: (1) independently varying, or controlled parameters, and (2) dependent parameters. The first group includes the barrel temperature, the clearance, the ultrasonic amplitude, and the flow rate. The average residence time of the rubber in the treatment zone is a derivative of two independent factors: the clearance (gap) in the devulcanization zone and the rubber flow rate. Therefore, the residence time falls in the first group (controlled parameters) as well. The second group (dependent parameters) includes temperature and pressure of the rubber entering the devulcanization zone and the ultrasonic power consumption. The dependent parameters are affected not only by controlled parameters, but also by each other. Interrelations of the processing parameters are discussed below. During devulcanization, the ultrasonic power consumption and the die pressure

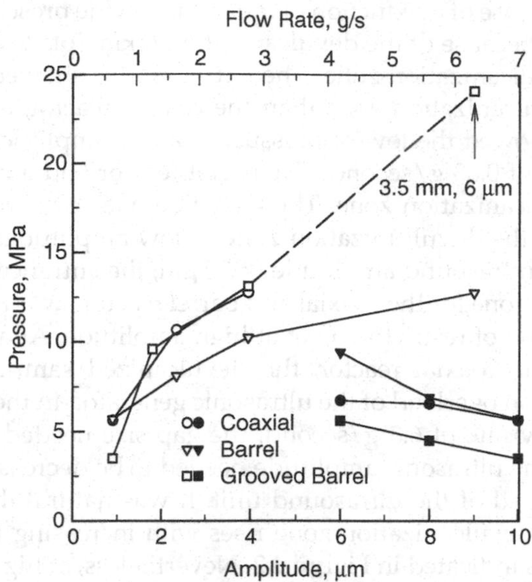

FIGURE 9.2
Entrance pressure of devulcanization zone vs. ultrasonic amplitude at flow rate of 0.63 g/second (solid symbols) and vs. flow rate at an amplitude of 10 µm (open symbols) for 30 mesh GRT at a clearance of 2 mm and a barrel temperature of 178°C. (Adapted from Yun, J., Oh, J.S., and Isayev, A.I., *Rubber Chem. Technol.*, 74, 317–330, 2001; and Yun, J., Ph.D. dissertation, University of Akron, 2003.)

are parameters governing the process. Comparative analysis is made using coaxial, barrel, and grooved barrel reactors during devulcanization of 30 mesh GRT.[53,70] In the barrel reactor, the devulcanization zone was located in the barrel of the extruder, where additional shearing occurs because of screw rotation. In the coaxial reactor, the devulcanization zone was located at the exit of the extruder. In the grooved barrel reactor, the devulcanization zone is located along the helical channel in the barrel.

9.4.2 Die Pressure and Power Consumption

9.4.2.1 Tire Rubber

As described above, the die pressure is a dependent parameter since it is affected by the controlling parameters. Also the pressure depends on the type of the reactor. Figure 9.2 shows the entrance pressure at the devulcanization zone vs. amplitude of ultrasound at a flow rate of 0.63 g/second, and the entrance pressure of devulcanization zone vs. flow rate (die characteristics) at an amplitude of 10 µm and a clearance of 2 mm for various reactors during devulcanization of GRT.[53,70] In all the reactors, the entrance pressure of the devulcanization zone was substantially reduced as the amplitude of ultrasound was increased. Ultrasound facilitated the flow of rubber through the gap not only because of a reduction of the friction in the presence of ultrasonic waves, but also because of the devulcanization taking place as GRT particles entered the devulcanization zone. The barrel reactor showed a higher pressure in the devulcanization zone than the coaxial reactor, and the grooved barrel reactor showed the lowest pressure at a low amplitude of ultrasound and a flow rate of 0.63 g/second. The barrel reactor had a converging zone before the devulcanization zone. The GRT flow was essentially blocked by the restrictor of the devulcanization zone at low amplitudes of ultrasound. However, at an ultrasound amplitude of 10 μm, the entrance pressure of the devulcanization zone for the coaxial and barrel reactors was almost the same, due to a reduction of restrictor effect at high amplitude. At a flow rate of 6.3 g/second for the coaxial reactor, the devulcanized sample could not be obtained due to an overload of the ultrasonic generator. In the grooved barrel reactor, at a flow rate of 6.3 g/second, the gap size needed to be increased to 3.5 mm and the ultrasonic amplitude needed to be decreased to 6 μm, also due to an overload of the ultrasound unit. It was natural that the entrance pressure of the devulcanization zone rises with increasing flow rate for all three reactors, as indicated in Figure 9.2. Nevertheless, at high flow rates and at an ultrasound amplitude of 10 μm, the barrel reactor had lower entrance pressure at the devulcanization zone than that of the other reactors. The difference in die characteristics among three reactors having a clearance of 2 mm in the devulcanization zone is possibly related to the difference in power consumption and the difference in shearing conditions experienced by the rubber. In the barrel reactor and grooved barrel reactor, the GRT in the devulcanization zone is subjected to a pressure and drag flow, while in the coaxial reactor it is subjected to a pressure flow alone.

Power consumption is also dependent on the processing parameters. It is affected by the design of the reactor, in addition to independent processing parameters. Figure 9.3 presents the power consumption density vs. amplitude of ultrasound at a flow rate of 0.63 g/second and power consumption density vs. flow rate at an ultrasonic amplitude of 10 μm for various reactors during devulcanization of GRT.[70] The power consumption density was calculated based on the measured power consumption by dividing it by the cross sectional area of the horn. The measured ultrasonic power consumption was expended due to dissipation losses and breakage of bonds, which led to devulcanization. It was not possible to estimate experimentally what part of power is consumed by devulcanization alone. In addition, the power expended on heat dissipation in the material and the power transmitted by the traveling wave through the rubber could not be separated. The only measurable losses were the initial power consumption of the acoustic system, when the horn works without loading. In obtaining Figure 9.3, these losses were subtracted from the total power consumption. In all three reactors, the power consumption density was increased as the amplitude of ultrasound and flow rate were increased. However, the power consumption density of the barrel reactor was significantly higher than that of the coaxial reactor and the grooved barrel reactor. Nevertheless, it was found that the

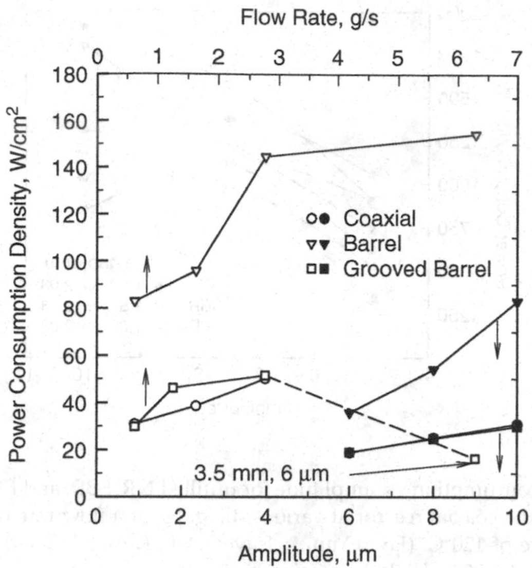

FIGURE 9.3
Power consumption density vs. ultrasonic amplitude at flow rate of 0.63 g/second (solid symbols) and vs. flow rate at an ultrasonic amplitude of 10 μm (open symbols) for 30 mesh GRT at a clearance of 2 mm and a barrel temperature of 178°C. (From Yun, J., Ph.D. dissertation, University of Akron, 2003.)

sample exiting the barrel reactor indicated less devulcanization, as shown in the next section. Evidently, the barred reactor had significant energy losses due to installation of seals around the horns in the extruder. In fact, Teflon®, which was used as a sealing material, was gradually softened during the devulcanization experiment, possibly indicating the energy losses. On the other hand, in the grooved barrel reactor, the clearances between the horns and the ports of the barrel were sealed by two gaskets made of 40 wt% graphite-filled polyimide. This material provides enhanced resistance to wear and friction, as well as improved dimensional and oxidative stability. This is attributed to lesser energy losses in the grooved barrel reactor when compared to the barrel reactor. In addition, an abnormally low power consumption density was observed at the flow rate of 6.3 g/second in the grooved barrel reactor. However, more work is needed to fully understand this phenomenon. The mean residence time of the devulcanization zone that is inversely proportional to the flow rate of GRT is calculated. Of note, the mean residence time that could be achieved in the devulcanization zone for the barrel reactor and grooved barrel reactor was as low as 1.2 seconds and 1.3 seconds, respectively.

9.4.2.2 *Various Types of Rubbers*

Power consumption also depends on the type of rubber being devulcanized. Thus, a comparative analysis of power consumption is also made for devul-

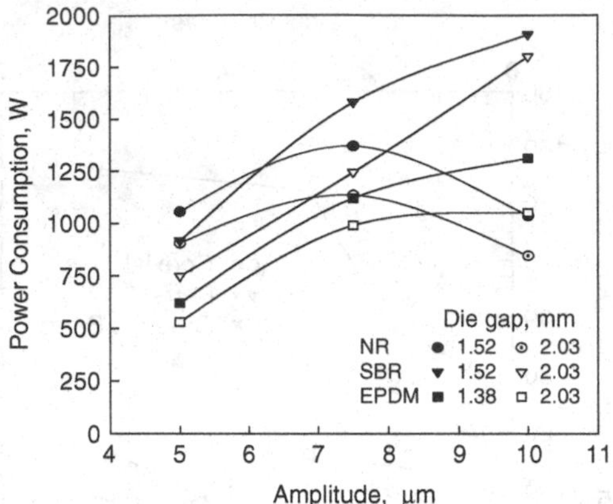

FIGURE 9.4
Ultrasound power consumption vs. amplitude for unfilled NR, SBR, and EPDM rubbers during devulcanization in the coaxial reactor at various die gaps, at a flow rate of 0.63 g/second and a barrel temperature of 120°C. (From Yun, J., Isayev, A.I., Kim, S.H., and Tapale, M., *J. Appl. Polym. Sci.*, 88, 434–441, 2003. With permission.)

canization of unfilled NR, SBR, and EPDM rubbers in the coaxial reactor.[62] Figure 9.4 shows the comparison of ultrasonic power consumption vs. ultrasonic amplitude for these rubbers during the devulcanization at various die gaps.[62] Ultrasonic energy loss and breakage of bonds leading to devulcanization was the main expenditure for the measured ultrasonic power in the coaxial reactor. In case of SBR and EPDM rubber, as expected, ultrasonic power consumption was increased with amplitude of ultrasound due to an increase of the strain amplitude experienced by the rubbers. In contrast, in the case of NR, ultrasonic power consumption showed a maximum value at 7.5 μm. This indicates that more energy was transmitted into the rubber material with an increase of amplitude during devulcanization of SBR and EPDM rubbers. But, for NR, it is believed that the imposition of ultrasonic waves leads to simultaneous breakage and reformation of bonds in the material. In this case, while devulcanization was the dominant phenomenon between 5 and 7.5 μm, some revulcanization occurred with increasing ultrasound intensity when the amplitude was increased from 7.5 to 10 μm. The latter accounts for a reduction in the power consumption at an amplitude of 10 μm. At the same processing conditions, the higher ultrasonic power consumption leads to a higher degree of devulcanization. Based on a comparison of the power consumption values of SBR and EPDM, it is inferred that SBR is easier to devulcanize than EPDM. The differences in devulcanization among NR, SBR, and EPDM are possibly due to the different chemical structure of the polymer chains and the thermomechanical stability. In addition, a decreasing gap size leads to a higher ultrasonic power consumption, which is evidently related to the higher degree of devulcanization. With a decreasing gap size, the rubber experiences higher strain amplitude at a constant ultrasonic amplitude.

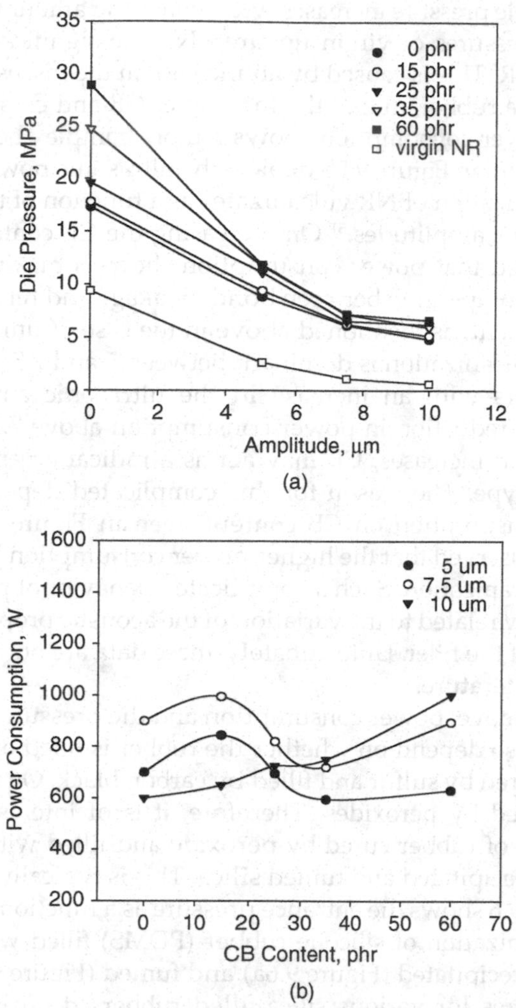

FIGURE 9.5
Die pressure vs. amplitude at various CB contents (a) and power consumption vs. CB content at various amplitudes (b) for NR vulcanizates devulcanized at a flow rate of 0.63 g/second, a clearance of 2.54 mm, and a barrel temperature of 120°C using the coaxial reactor. (From Hong, C.K. and Isayev, A.I., *J. Appl. Polym. Sci.*, 79, 2340–2348, 2001. With permission.)

Die pressure and ultrasonic power consumption are dependent not only on the type of rubber; they also vary upon the addition of fillers to a rubber (Figure 9.5).[51] Figure 9.5a gives the dependence of the die entrance pressure on ultrasonic amplitude during treatment of virgin uncured and unfilled NR and devulcanization of unfilled and carbon black (CB)-filled NR vulcanizates, using the coaxial reactor.[51] Similarly to GRT, the die pressure for NR decreases as the amplitude of the ultrasound is increased. This is due to the combined effect of the softening of rubber, due to devulcanization in the die gap, and reduction in friction between the particles and the die walls, due to ultrasonic vibrations. The die pressure increases with increasing amounts of CB loading. Also, the die pressures of virgin uncured NR are significantly lower than those of cured NR. This is caused by an increase in the viscosity or resistance to the flow of the rubber, due to the loading of CB and cross-linking.

Ultrasonic power consumption shows a more complex behavior with CB loading. In particular, Figure 9.5b depicts the ultrasonic power consumption during devulcanization of NR vulcanizates as a function of the carbon black loading at various amplitudes.[51] On increasing the CB content from 0 to 25 phr, it is observed that power consumption shows a maximum at 7.5 μm. This is due to a competition between bond breakage and reformation during ultrasonic treatment, as mentioned above in the case of unfilled NR. In this case, while devulcanization is dominant between 5 and 7.5 μm, some revulcanization occurs with an increase in the ultrasonic amplitude, which accounts for the reduction in power consumption above 7.5 μm. However, as the CB content increases, CB may act as a radical acceptor of a special polyfunctional type. The reason for this complicated dependence of ultrasonic power consumption on CB content, seen in Figure 9.5b, is unclear. However, it is observed that the higher power consumption leads to a higher degree of devulcanization. Such a complicated behavior of power consumption is apparently related to the variation of the acoustic properties of rubbers upon addition of the filler. Unfortunately, these data are not readily reported in the existing literature.

As indicated above, power consumption and die pressure vary with types of rubber and also depend on whether the rubber is filled or unfilled. These rubbers were cured by sulfur and filled by carbon black. On many occasions, rubbers are cured by peroxides. Therefore, it is of interest to look at the devulcanization of rubber cured by peroxide and filled with other types of fillers such as precipitated and fumed silica. This is typically done in silicone rubbers. Figure 9.6 shows die entrance pressure as a function of an amplitude during devulcanization of silicone rubber (PDMS) filled with various concentrations of precipitated (Figure 9.6a) and fumed (Figure 9.6b) silica.[71] Die entrance pressures for various silica-filled rubbers decrease as amplitude increases, indicating that devulcanization is facilitated at a higher intensity of applied ultrasound. In addition, die pressures increase with concentration of the silica filler. Initial entrance pressures for fumed silica-filled PDMS are higher than those of precipitated silica-filled samples since fumed silica acts as stronger reinforcing filler than precipitated silica. Upon imposition of ultrasound on fumed silica-filled PDMS, the pressure decreases to values similar to that of precipitated silica-filled PDMS. Reduction of the die pressure is higher in the case of fumed silica-filled PDMS. In addition, Figure 9.7 shows the power consumption as a function of amplitude during devulcanization.[71] Power consumed during devulcanization increases with amplitude and filler concentration. For an unfilled PDMS system, power consumption at some conditions is higher than that of precipitated silica-filled materials. Apparently, ultrasound breaks both chemical cross-links between PDMS chains and physical cross-links between rubber chains and

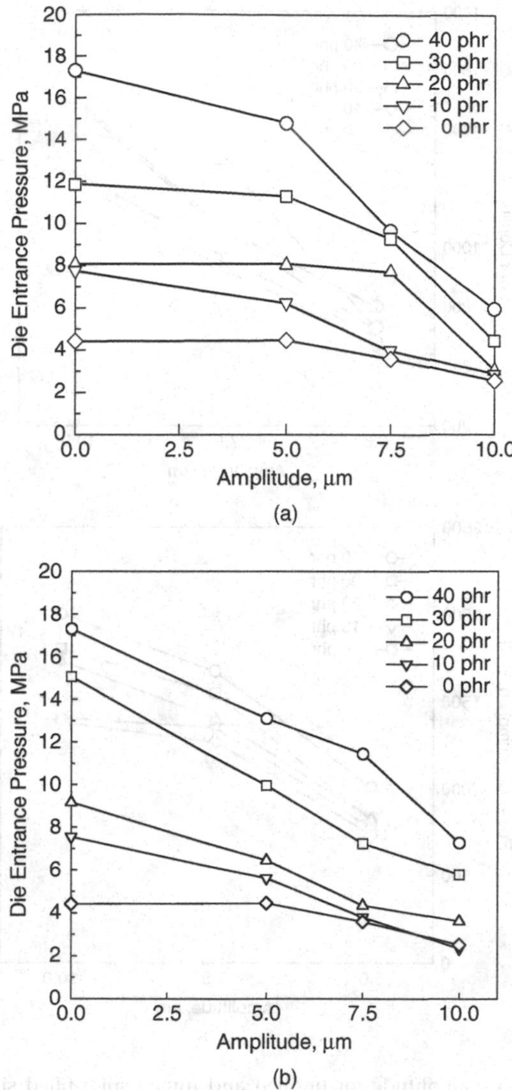

FIGURE 9.6
Power consumption vs. amplitude for both unfilled and precipitated (a) and fumed (b) silica-filled silicone rubber in the coaxial reactor at a flow rate of 0.32 g/second and a gap of 0.63 mm. (From Shim, S.E., Ph.D. dissertation, University of Akron, 2001.)

fillers. Considering that more energy is needed to break chemical cross-links than physical ones, most of the energy produced by ultrasound may be consumed to break the cross-links of unfilled silicone rubber. This phenomenon was also observed in the devulcanization of filled SBR.[72] However, the power consumption for fumed silica-filled PDMS shows significantly higher values than those of precipitated silica-filled PDMS, indicating more energy is required to break the fumed silica-filled PDMS network.

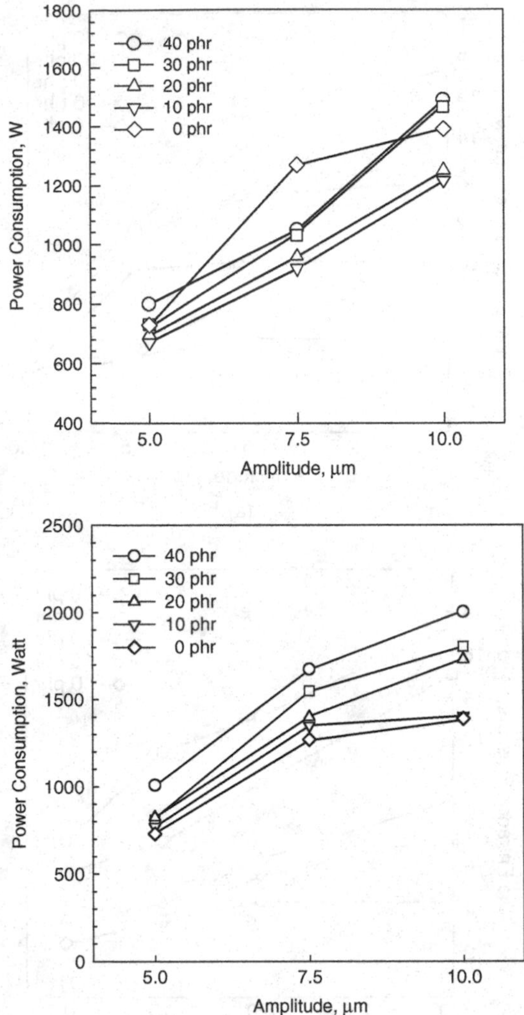

FIGURE 9.7
Power consumption vs. amplitude for unfilled and fumed silica-filled silicone rubber in the coaxial reactor at a flow rate of 0.32 g/second, a gap of 0.63 mm, and a barrel temperature of 180°C. (From Shim, S.E., Ph.D. dissertation, University of Akron, 2001.)

9.5 Cure Behavior

9.5.1 Tire Rubber

Ultrasonically devulcanized rubber can be revulcanized by adding curing ingredients to the samples and compounding them. Typically, revulcanization was done using the same recipe as that of the original rubbers. Figure 9.8

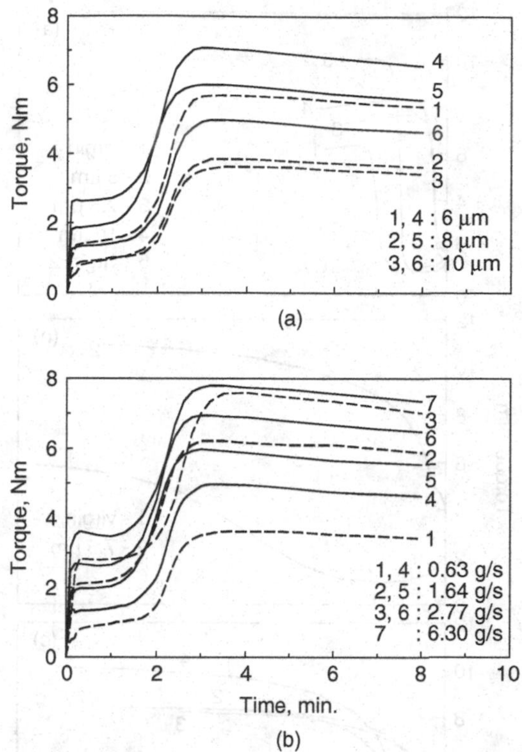

FIGURE 9.8
Comparison of cure curves of DGRT 178°C obtained at various amplitudes with a flow rate of 0.63g/second (a) and at various flow rates with an amplitude of 10 μm (b) prepared at gap of 2 mm using the coaxial (dash lines) and barrel (solid lines) reactors at a barrel temperature of 178°C. (From Yun, J., Oh, J.S., and Isayev, A.I., *Rubber Chem. Technol.*, 74, 317–330, 2001. With permission.)

shows the comparison of the cure curve of devulcanized GRT (DGRT) obtained at various amplitudes at a flow rate of 0.63 g/second (a) and various flow rates at an amplitude of 10 μm (b) at a cure temperature of 178°C.[53] Curing occurs fast; the torque attained a minimum at about 1 minute and a maximum at about 3 minutes and then decreased slightly. In general, fast revulcanization is an attribute of reclaimed rubber.[73] As the ultrasound amplitude increased, the minimum and maximum torque decreased. A comparison with the virgin tire compound could not be made since the composition and recipe of the GRT was unknown. Generally, the higher the degree of devulcanization, the lower were the minimum and the maximum torques. Thus, cure curves were also an indirect indication of the degree of devulcanization.

9.5.2 Various Types of Rubbers

It is of interest to see how different types of devulcanized rubber undergo revulcanization. In this regard, Figure 9.9 represents the cure curves of

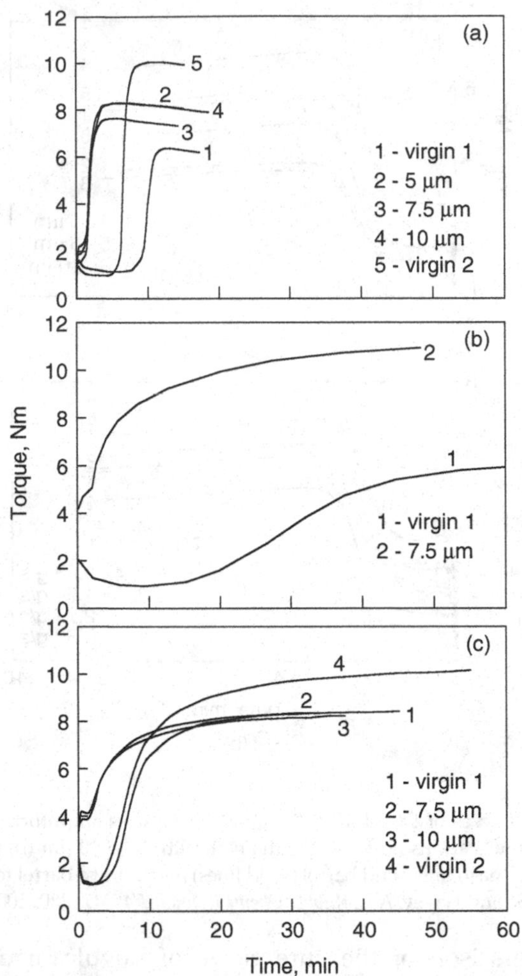

FIGURE 9.9
Cure curves for virgin and devulcanized NR (a), SBR (b), and EPDM (c) rubbers at gaps of 2.54, 1.52, and 1.01 mm, respectively, obtained at different amplitudes at a flow rate of 0.63 g/second and a barrel temperature of 120°C in the coaxial reactor. (From Yun, J., Isayev, A.I., Kim, S.H., and Tapale, M., *J. Appl. Polym. Sci.*, 88, 434–441, 2003. With permission.)

unfilled virgin and devulcanized NR (a), SBR (b), and EPDM (c) rubbers obtained at various amplitudes and gap sizes at a flow rate of 0.63 g/second and a cure temperature of 160°C.[62] These curves were obtained for the devulcanized samples that had the best millability and reprocessability. It should be noted that the gap size during devulcanization was different for different rubbers. Gap sizes for NR, SBR, and EPDM were 2.54 mm, 1.52 mm, and 1.01 mm, respectively. This was also the indirect indication of the ability of particular rubbers to be devulcanized, because a higher gap size reflects the ease of devulcanization. It can be observed that the devulcanized

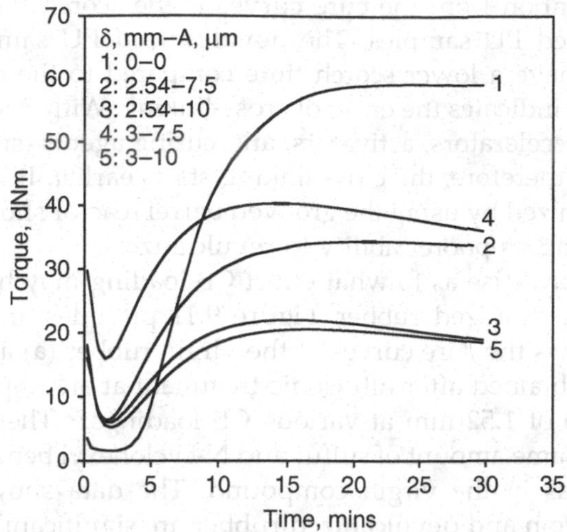

FIGURE 9.10
Cure curves for unfilled polyurethane rubber devulcanized at a temperature of 120°C, a feed rate of 1.26 g/second, and various gap sizes and amplitudes using the coaxial reactor. (From Ghose, S. and Isayev, A.I., *J. Appl. Polym. Sci.*, 88, 980–989, 2003. With permission.)

rubbers cure faster than virgin rubbers. The shortness or absence of the induction period indicates that cross-linking reactions start immediately upon heating. This could be explained by the presence of residual curative in the devulcanized sample, which is a typical characteristic of reclaimed rubber.[73] Among these rubbers, EPDM has the lowest amount of double bonds. During revulcanization, the final torque of devulcanized NR and SBR is higher than that of their virgin counterparts, while for virgin and devulcanized EPDM, it is about the same. This is apparently related to the degree of unsaturation of rubbers. In fact, due to this reason, an addition of twice the amount of curatives in virgin NR (virgin 2), having a high degree of unsaturation, can cause significant increase in the final torque. But in EPDM, due to its low degree of unsaturation, little increase is observed in virgin 2 in comparison with virgin 1.

Another rubber that has a low degree of unsaturation like EPDM rubber is PU rubber. Figure 9.10 shows the cure curves of unfilled devulcanized PU rubber obtained from the coaxial reactor.[64] It is seen that with increase in the amplitude, there is a decrease in the maximum torque value that does not reach the torque value of the original PU rubber. Also, the devulcanized samples cure at a faster rate than the virgin sample, with a substantially lower final torque. PU rubber has a degree of unsaturation even lower than that of EPDM. This explains the lower ability of devulcanized PU rubber to revulcanize. The minimum torque values for a particular feed rate and gap size show a decrease with increase in amplitude. A lower minimum torque indicates a lower viscosity of the devulcanized sample. At higher amplitudes, there is greater devulcanization of the sample and hence, the viscosity drops.

Another observation from the cure curves is the scorch time of the virgin and devulcanized PU samples. The devulcanized PU samples, similar to other rubbers, have a lower scorch time compared to the original sample. The scorch time indicates the onset of cross-linking. With devulcanized samples, there are accelerators, activators, and curing agents (sulfur) remaining in the samples; therefore, the cross-linking starts earlier. It is noted that PU rubber devulcanized by using the grooved barrel reactor showed even lower torque, indicating its poorer ability to revulcanize.

A question may arise as to what effect CB loading may have on the cure behavior of devulcanized rubber. Figure 9.11 provides an answer to this question. It shows the cure curves of the virgin rubber (a) and the devulcanized SBR (b) obtained after ultrasonic treatment at an amplitude of 10 μm and a clearance of 1.52 mm at various CB loadings.[72] These rubbers were cured with the same amount of sulfur and N-cyclohexyl benzothiazole sulfenamide (CBS) as in the virgin compound. The data show that the cure behaviors of virgin and devulcanized rubber are significantly different. The dependencies are in good agreement with data obtained on unfilled SBR. In contrast to the virgin samples, cure curves of devulcanized samples show a slight tendency to minimize or no minimum in torque during the course of heating the samples in a rheometer. This minimum is typically observed during curing of virgin rubber, and is related to a decrease of viscosity upon heating. This tendency is also maintained for filled SBR at CB loading of 15 phr. However, in the case of filled SBR compounds with a CB content of 30 phr and higher, there are no minima, and the final torque in all SBR samples is higher than that of the virgin rubber.

9.6 Gel Fraction and Cross-Link Density

9.6.1 Tire Rubber

Devulcanized rubbers typically consist of some amount of sol and gel of reduced cross-link density in comparison with the virgin vulcanizates. The cross-link density can be characterized by the average molecular mass of the network chain, i.e., the average chain length between two network junctions. Figure 9.12 presents the gel fraction (a) and cross-link density (b) vs. the ultrasound amplitude for DGRT and revulcanized GRT (RGRT) using three reactors at a flow rate of 0.63 g/second.[53] Both the gel fraction and cross-link density decreased substantially during the ultrasound treatment. All of the DGRT samples were cured by using the same cure recipe. The devulcanized GRT, having higher gel fraction and cross-link density, had a correspondingly higher gel fraction and cross-link density on revulcanization. This observation is in agreement with the minimum and maximum torque data depicted in Figure 9.8. One could observe that DGRT and RGRT of the barrel reactor had higher gel fraction and cross-link density than those of

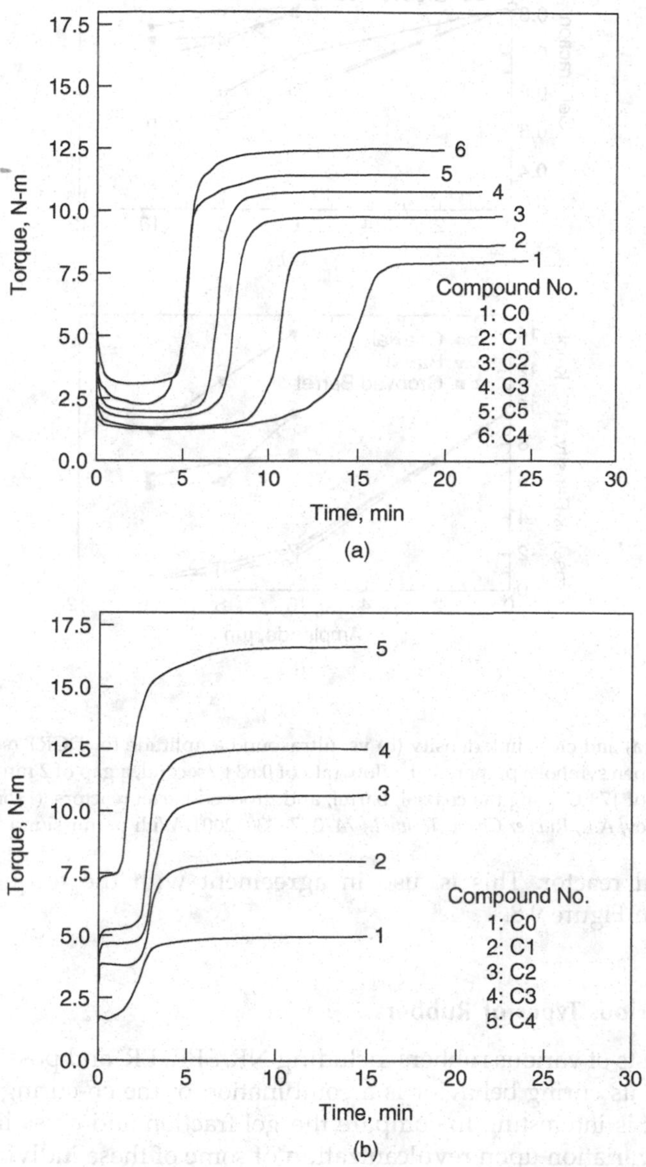

FIGURE 9.11
Cure curves of filled virgin (a) and devulcanized (b) SBR of various carbon black content at 160°C obtained at a flow rate of 0.63 g/second, an amplitude of 10 μm, a gap of 1.52 mm, and a barrel temperature of 120°C using the coaxial reactor; C0 – 0 phr, C1 – 15 phr, C2 – 25 phr, C3 – 35 phr, C4 – 60 phr. (From Kim, S.H., Ph.D. dissertation, University of Akron, 1998.)

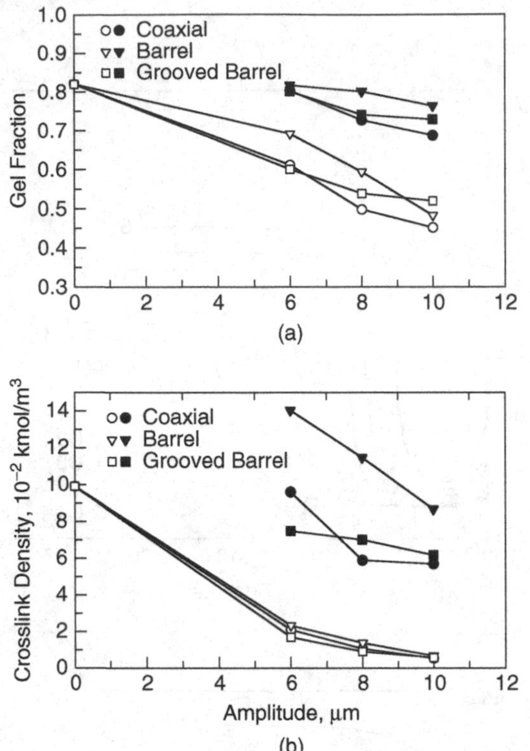

FIGURE 9.12
Gel fraction (a) and cross-link density (b) vs. ultrasound amplitude for DGRT (solid symbols) and RGRT (open symbols) prepared at a flow rate of 0.63 g/second, a gap of 2 mm, and a barrel temperature of 178°C using the coaxial, barrel, and grooved barrel reactors. (From Yun, J., Oh, J.S., and Isayev, A.I., *Rubber Chem. Technol.*, 74, 317–330, 2001. With permission.)

the coaxial reactor. This is also in agreement with the torque behavior depicted in Figure 9.8.

9.6.2 Various Types of Rubbers

GRT consists of various rubbers including NR, SBR, BR, and possibly EPDM. Therefore, its curing behavior is a combination of the co-curing of various rubbers. It is interesting to compare the gel fraction and cross-link density and their variation upon revulcanization of some of these individual devulcanized rubbers. Accordingly, Figure 9.13 represents the normalized gel fraction and normalized cross-link density of devulcanized and revulcanized NR, SBR, and EPDM rubber as a function of ultrasonic amplitude at a constant gap of 2.03 mm.[62] The normalized gel fraction and normalized cross-link density were calculated by the ratio of the corresponding measured values for devulcanized samples obtained at various conditions to that of the initial vulcanizate. Solid symbols at the ordinates indicated values of the

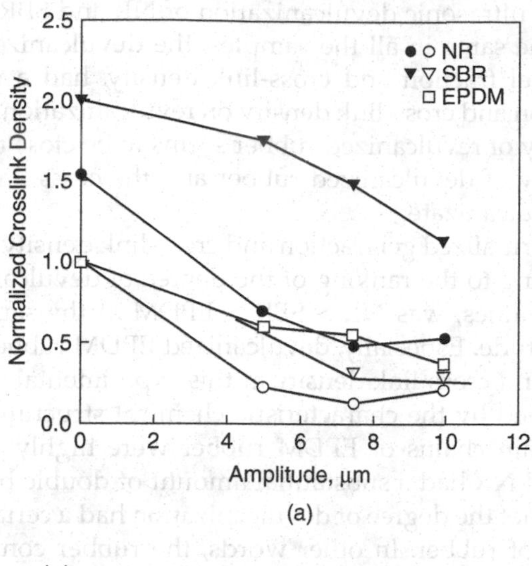

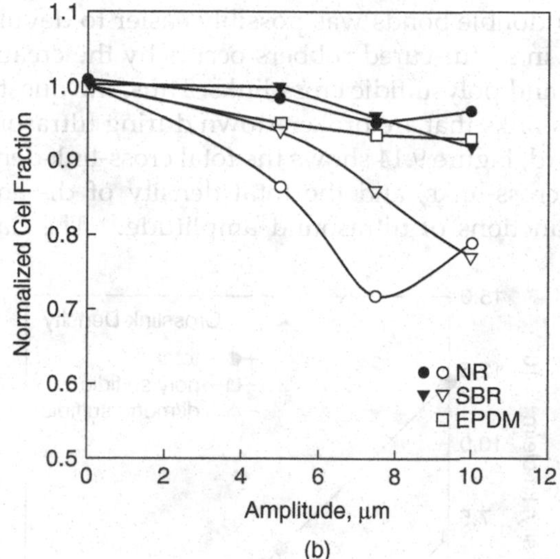

FIGURE 9.13
Normalized gel fraction vs. amplitude for unfilled devulcanized (open symbols) and revulcanized (solid symbols) NR, SBR, and EPDM rubbers obtained at a flow rate of 0.63 g/second, a gap of 2.03 mm, and a barrel temperature of 120°C using the coaxial reactor. (From Yun, J., Isayev, A.I., Kim, S.H., and Tapale, M., *J. Appl. Polym. Sci.*, 88, 434–441, 2003. With permission.)

gel fraction and cross-link density obtained using virgin rubber mixed with twice the amount of curatives, referred to as virgin 2 in Figure 9.9. Also, it was noted that the devulcanized EPDM rubbers obtained at a gap of 2.03 mm were not millable and reprocessable because of high gel content. Thus, revulcanized EPDM rubbers could not be obtained at this gap. For this reason, Figure 9.13 does not show data for revulcanized EPDM. Both the normalized gel fraction and normalized cross-link density decrease substantially during the ultrasonic devulcanization of NR and SBR. Since the curative level was the same in all the samples, the devulcanized SBR and NR, having higher gel fraction and cross-link density, had a correspondingly higher gel fraction and cross-link density on revulcanization. In fact, the total cross-link density of revulcanized rubber seems to be close to the sum of the cross-link density of devulcanized rubber and the cross-link density of the original virgin vulcanizate.

In addition, normalized gel fraction and cross-link density of devulcanized samples, according to the ranking of the degree of devulcanization defined by normalized values, was NR > SBR > EPDM at the same gap size and ultrasonic amplitude. Especially, devulcanized EPDM rubbers had the highest gel fraction and cross-link density at this experimental condition, which could be explained by the characteristic chemical structure of the rubbers. The polymer main chains of EPDM rubber were highly saturated, while those of SBR and NR had a substantial amount of double bonds. Therefore, it was believed that the degree of devulcanization had a certain order according to the type of rubber. In other words, the rubber containing a higher concentration of double bonds was possibly easier to devulcanize.

Cross–linking in sulfur-cured rubbers occurs by the creation of monosulfidic, disulfidic, and polysulfidic cross-links. Thus, the question arises about the type of cross-links that are broken down during ultrasonic devulcanization. In this regard, Figure 9.14 shows the total cross-link density, the density of polysulfidic cross-links, and the total density of di- and monosulfidic cross-links as functions of ultrasound amplitude.[44] The data indicate that

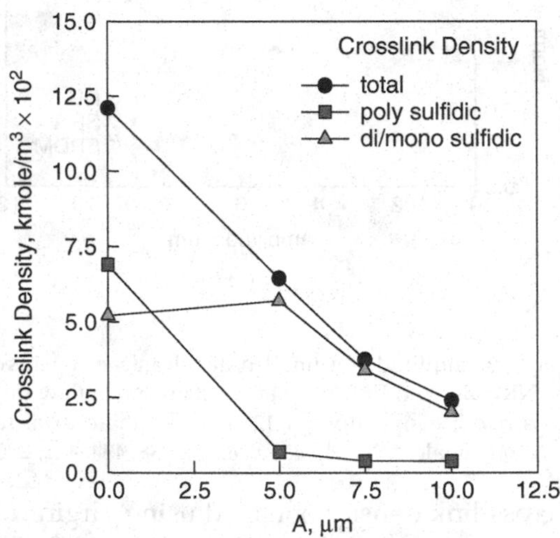

FIGURE 9.14
Total, polysulfidic, di-, and monosulfidic cross-link density in the unfilled devulcanized SBR vs. amplitude at a flow rate of 0.63 g/second, a gap of 1.52 mm, and a barrel temperature of 120°C using the coaxial reactor. (From Yu. Levin, V., Kim, S.H., and Isayev, A.I., *Rubber Chem. Technol.*, 70, 641–649, 1997. With permission.)

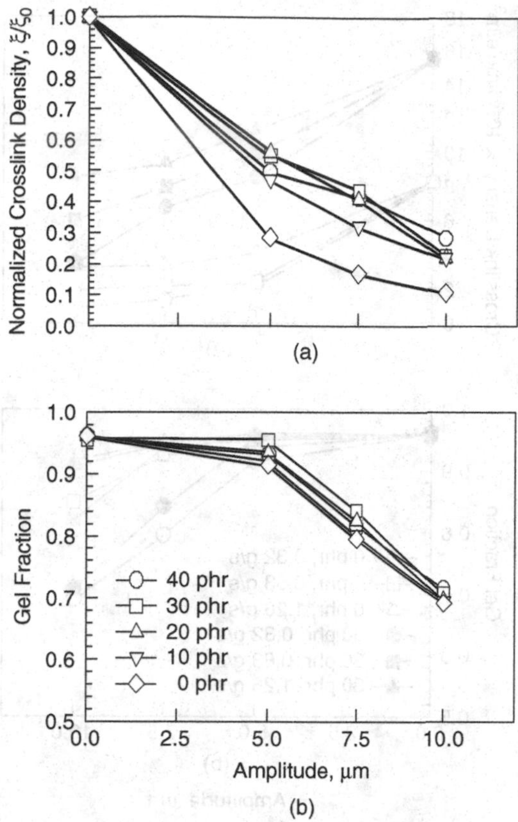

FIGURE 9.15
Normalized cross-link density (a) and gel fraction (b) vs. amplitude for silicone rubber compounds filled with various amounts of precipitated silica devulcanized at a flow rate of 0.32 g/second, a gap of 0.63 mm, and a barrel temperature of 180°C using the coaxial reactor. (From Shim, S.E. and Isayev, A.I., *Rubber Chem. Technol.*, 74, 303–316, 2001. With permission.)

ultrasound devulcanization is followed by a significant decrease in the number of polysulfidic cross-links. It is also seen that at low amplitudes, the number of di- and monosulfidic cross-links increases slightly and then decreases with a further increase of amplitude. Clearly, the decrease in the total cross-link density is governed predominantly by the breakup of the polysulfidic chains.

In case of rubber cured by peroxide, the cross-links are the C-C bonds with high bond energy. Thus, it is more difficult to break cross-links in such cases. PDMS is an example of such a system. The changes in normalized cross-link density and gel fraction of filled and unfilled PDMS, with amplitude and various filler concentrations, are depicted in Figure 9.15.[52] The rubbers are cured by peroxide. At constant amplitude, the unfilled system showed lower gel fraction and cross-link density than those of filled PDMS. The decrease in cross-link density with amplitude is more pronounced than that in gel

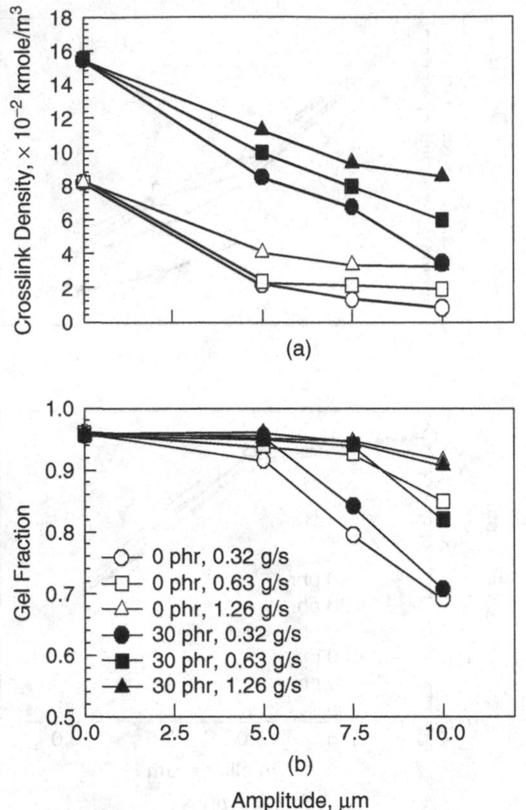

FIGURE 9.16
Cross-link density (a) and gel fraction (b) vs. amplitude for devulcanized unfilled and 30 phr precipitated silica–filled silicone rubber at various flow rates at a gap of 0.63 mm and a barrel temperature of 180°C using the coaxial reactor. (From Shim, S.E. and Isayev, A.I., *Rubber Chem. Technol.*, 74, 303–316, 2001. With permission.)

fraction. At an amplitude of 10 μm, the gel fraction is decreased by 30%, while the cross-link density is reduced by almost 90%. The degree of devulcanization depends not only on the amplitude, but also on the flow rate of material, as shown in Figure 9.16.[52] Upon increasing the flow rate from 0.32 to 1.26 g/second for both unfilled and filled PDMS, cross-link density increases. At higher flow rates, obviously less devulcanization takes place, as indicated by higher values of cross-link density. Also, gel fraction at higher flow rates changes slightly compared to cross-link density. When comparing cross-link densities as a function of amplitude for unfilled and filled rubbers, the cross-link density of the unfilled system reduces to a certain level followed by a slight decrease, whereas for the filled system, the cross-link density continuously decreases. More specifically, the cross-link density for unfilled PDMS at 10 μm amplitude decreases about an order of magnitude, while that in filled rubber reduces about fourfold. Taking into account the presence of two types of cross-links (chemical and physical) in filled silicone rubber, this may imply that ultrasound energy in filled systems is also used to break weak physical cross-links between rubber chains and fillers with increasing amplitudes, possibly leading to less breakage of chemical cross-links. It was shown that more breakage of chemical cross-links in PDMS can be achieved during ultrasonic devulcanization of vulcanizates in the presence of water.[63]

9.7 Rheological Properties

9.7.1 Experimental Observations

Rheological properties of devulcanized rubbers are important from a processability point of view. Thus, dynamic properties and flow curves of devulcanized rubbers are compared to those of the virgin compound. Figure 9.17 shows the absolute value of the complex viscosity, η^*, and tan δ vs. frequency at 100°C and a strain amplitude of ψ_0 = 0.07 for unfilled EPDM.[58,70] The complex viscosity and slope of the flow curves of devulcanized EPDM are higher than those of the virgin EPDM. This is generally true for various devulcanized rubbers, and it is possibly the result from the high gel content in the devulcanized EPDM. Also, the complex viscosity of devulcanized

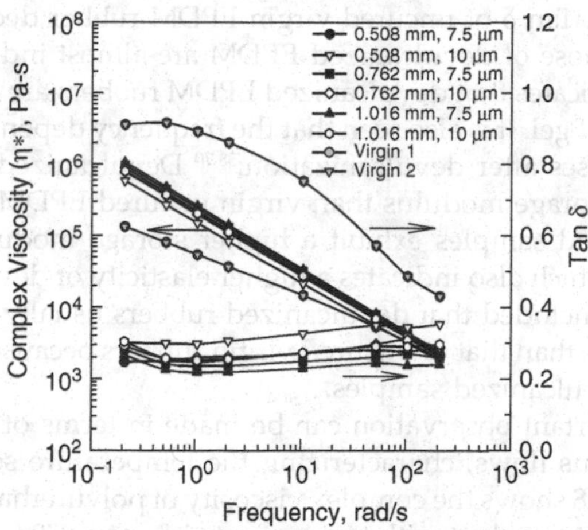

FIGURE 9.17
Complex viscosity and tan δ vs. frequency at an amplitude of strain γ_0 = 0.07 and 100°C for uncured virgin and devulcanized EPDM rubbers obtained at a flow rate of 0.63 g/second and a barrel temperature of 120°C, various gaps and amplitudes using the coaxial reactor. (Adapted from Yun, J. and Isayev, A.I., *Rubber Chem. Technol.*, 76, 253–270, 2003; and Yun, J., Ph.D. dissertation, University of Akron, 2003. With permission.)

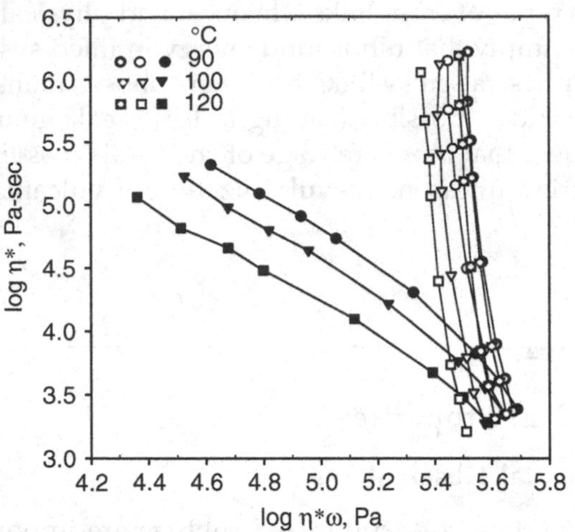

FIGURE 9.18
Complex viscosity, η*, vs. product of η*ω for virgin (solid symbols) and devulcanized samples of PU rubber obtained at a feed rate of 1.26 g/second, a gap size of 2.54 mm, a barrel temperature of 120°C, and amplitudes of 7.5 and 10 μm using the coaxial reactor. (From Ghose, S. and Isayev, A.I., *J. Appl. Polym. Sci.*, 88, 980–989, 2003. With permission.)

EPDM decreases with an increase of ultrasonic amplitude and a decrease of gap size. The viscosity is also an indirect indication of the degree of devulcanization. Namely, the lower the viscosity, the higher the degree of devulcanization. It should be noted that there is a change of tan δ after devulcanization. Tan δ of uncured virgin EPDM rubber decreases with frequency, while those of devulcanized EPDM are almost independent of frequency. This indicates that devulcanized EPDM rubbers are more elastic due to the presence of gel. It is also seen that the frequency dependency of storage modulus decreases after devulcanization.[58,70] Devulcanized EPDM rubbers have a higher storage modulus than virgin uncured EPDM. At all frequencies, devulcanized samples exhibit a higher storage modulus at the same loss modulus, which also indicates a higher elasticity of devulcanized material. It can be concluded that devulcanized rubbers usually exhibit a higher storage modulus than that of uncured virgin rubbers because of the presence of gel in the devulcanized samples.

Another important observation can be made in terms of activation energies, E, of viscous flows, characterizing the temperature sensitivity of viscosity. Figure 9.18 shows the complex viscosity of polyurethane rubber (PUR) as a function of its product with frequency ω, i.e., η*ω.[64] This plot shows that the slopes of the viscosity curves of the devulcanized samples are much higher compared to the slope of the viscosity curve of the virgin sample. This is an indication that the devulcanized samples exhibit higher shear thinning effect with respect to the original. This observation is also typical for all other ultrasonically devulcanized rubbers. The effect is more pronounced at lower shear stress values, indicating the possibility of the existence of yield stress in devulcanized rubbers. The value of the activation energy of viscous flow was calculated from these curves at a constant value of $\eta^*\omega$. The values of the activation energy, E, are 22 and 40 and 53 kJ/mole for virgin PU gum, devulcanized PUR at 7.5 and 10 μm, respectively. It is clear that devulcanized samples exhibit higher activation energies compared to the untreated gum sample.

It is thought that the activation energy is affected by factors such as the flexibility of the macromolecules and branching. It is also well known from comparison of the activation energies of viscous flow of linear and branched polymers that the linear polymers have a lower value of E than branched polymers.[74] Therefore, a higher value of activation energy in the case of devulcanized samples may be an indirect indication of branching taking place during devulcanization. When the ultrasonic waves act on the cross-linked rubber, not only are the sulfur bonds broken, but portions of the main chain are broken as well. These remnants of the network chain remain attached to the ruptured main chain leading to a branched structure. This reduces the flexibility (mobility) of the chains and consequently raises the activation energy.

Figure 9.19 displays the apparent viscosity of unfilled and 30 phr precipitated silica-filled PDMS at 130°C devulcanized at various amplitudes, in comparison with those of virgin silicone rubber.[71] As shown in Figure 9.19, devulcanization was achieved at a feed rate of 0.32 g/second, and a gap of 0.63 mm. It is seen that viscosity of unfilled and silica-filled devulcanized PDMS decreases as ultrasound amplitude increases. However, the decrease in viscosity for silica-filled devulcanized PDMS is insignificant compared to

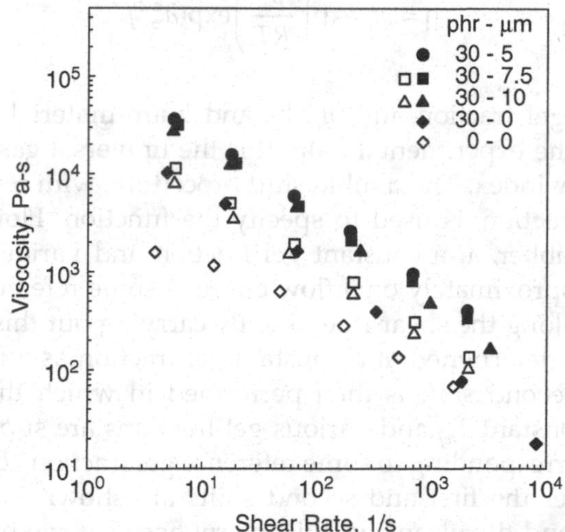

FIGURE 9.19
Viscosity of unfilled (open symbols) and 30 phr precipitated silica-filled (solid symbols) uncured virgin and devulcanized silicone rubber at various amplitudes at 130°C. Devulcanization was achieved at a feed rate of 0.32 g/second, a gap of 0.63 mm, and a barrel temperature of 180°C using the coaxial reactor. (From Shim, S.E., Ph.D. dissertation, University of Akron, 2001.)

that of unfilled rubber, due to less devulcanization at the presence of filler. As expected, viscosity for all specimens decreases with increasing shear rate, following a behavior slightly deviating from power law behavior since significant amount of unextractable gel exists in devulcanized samples. Also, the viscosity of devulcanized samples is higher than that of virgin samples.

9.7.2 Theoretical Description

An attempt was made to describe viscosity of devulcanized rubbers using a modified Cross model[75]:

$$\eta = \frac{\eta_o(T,\xi)}{1+\left[\dfrac{\eta_o(T,\xi)\dot{\gamma}}{\tau^*}\right]^{1-n}} \tag{9.1}$$

$$\eta_o(T,\xi) = m_o \exp\left(\frac{b_o}{RT}\right)\exp(b\xi^2) \tag{9.2}$$

or a modified power-law model.[37–39]:

$$\eta = m_o \exp\left(\frac{b_o}{RT}\right)\exp(b\xi^2) \tag{9.3}$$

where ξ is the gel fraction and m_o, b_o, and b are material constants to be specified from the experimental data. R is the universal gas constant and n is the power-law index. The double-shift procedure, with respect to temperature and gel fraction, is used to specify the function. Flow curves of the devulcanized rubber, at a constant gel fraction and various temperatures, superimpose approximately on a flow curve at some reference temperature T_{ref} by shifting along the shear rate axis. By carrying out this procedure, the shift factor was determined at a constant gel fraction as a function of temperature. The second shift is then performed in which the reduced flow curves at the constant T_{ref} and various gel fractions are superimposed on a master curve corresponding to some reference gel fraction, ξ_{ref}. The reduced flow curves after the first and second shifts are shown in Figure 9.20 for unfilled virgin and devulcanized silicone rubbers.[75] It can be seen that both devulcanized and virgin samples show non-Newtonian shear-thinning behavior. By comparing the two curves in Figure 9.20, it is apparent that the devulcanized samples have higher shear viscosity than that of the virgin sample. This is because the devulcanized samples contain significant amounts of gel inside. The higher the gel fraction, the higher the shear

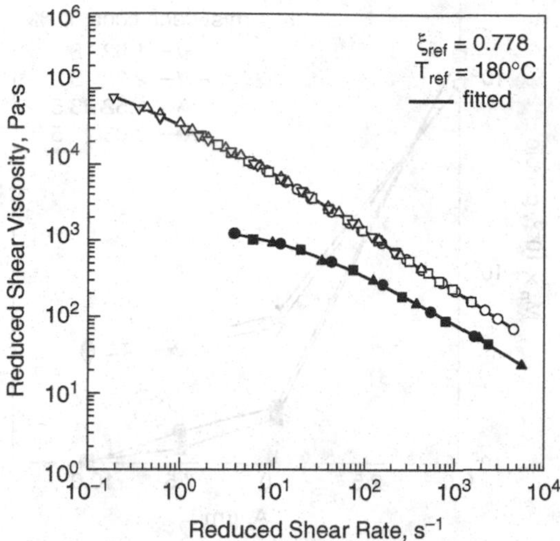

FIGURE 9.20
Master curves for reduced shear viscosity as a function of reduced shear rate for uncured virgin (solid symbols) and devulcanized (open symbols) unfilled silicone rubber at a reference temperature of 180°C. A reference gel fraction is 0.778. Devulcanization was carried out in the coaxial reactor at various conditions. (From Diao, B., M.S. thesis, University of Akron, 1997.)

viscosity. The lines in Figure 9.20 are fitted according to the modified Cross model described above (Equation 9.1 and Equation 9.2).[75]

9.8 Molecular Effect in Devulcanization

9.8.1 Sulfur-Cured Rubbers

Devulcanized rubber, as described in the previous section, consists of sol and gel. One can measure molecular characteristics such as number, M_n, and weight average, M_w, of the sol portion and compare them with those of the uncured virgin rubber. Such studies will help us to find the extent of changes in the molecular structure of rubber during devulcanization. So far, these studies have been carried out for various rubbers including SBR,[42,72] BR,[69,76] PUR cured by sulfur,[64] and PDMS cured by peroxide.[60,61] A study of the molecular weight of the soluble portion of the devulcanized rubbers is another way to judge the extent or degree of devulcanization and degradation. These studies reflect a reduction of M_w and M_n values in SBR, BR, and PUR, indicating degradation of molecular chains, the extent of which depends on the type of rubber.

Figure 9.21 denotes the number (a) and weight (b) average molecular weight of the sol part of devulcanized SBR.[72] SBRs containing 18% and 23.5% of styrene were used. As shown, the molecular weight of the sol component

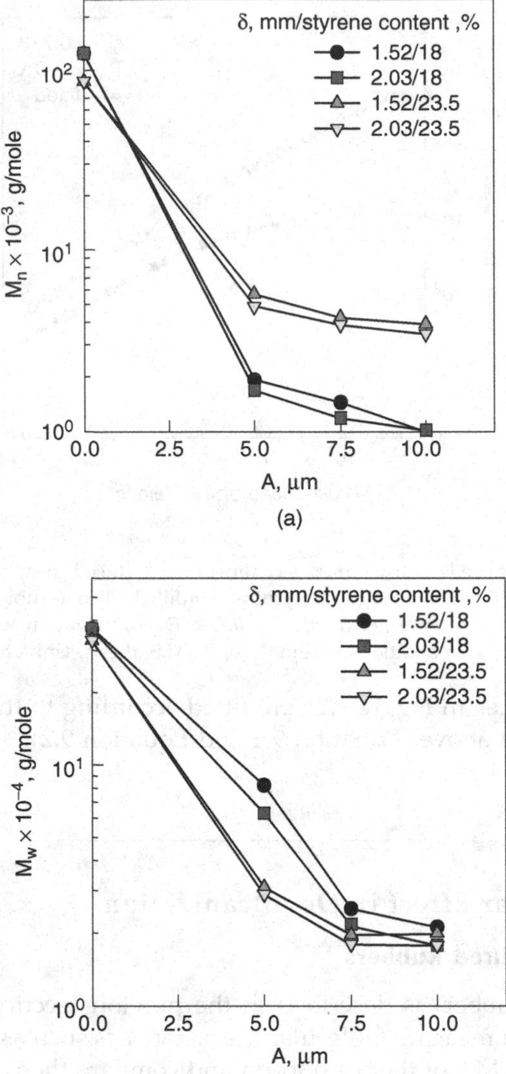

FIGURE 9.21
Number- (a) and weight- (b) average molecular weight of sol of the unfilled devulcanized SBR as a function of amplitude at various gaps and styrene contents obtained at a barrel temperature of 120°C and a flow rate of 0.63 g/second using the coaxial reactor. (From Kim, S.H., Ph.D. dissertation, University of Akron, 1998.)

is reduced as the amplitude of ultrasound is increased. Also, the molecular weight of the sol component is very low, more than an order of magnitude lower than the molecular weight of the virgin rubber. SBR with higher styrene content shows a higher number average molecular weight, indicating a higher degree of devulcanization, as shown by lower values of cross-link density and gel fraction. There are several possible interpretations for these experimental observations. The ultrasonic waves, in addition to severing chemical cross-links, also break up the short tails of some of the macromolecular chains. It is these broken fragments that are dissolved in the solvent during extraction. At the same time, the long macromolecular chains remain trapped in the gel and are apparently not removed from it by the solvent. This is a plausible explanation because devulcanized SBR of low styrene content has more double bonds available for cross-linking than SBR with high styrene content. Therefore, the opening of double bonds does not lead to degradation of macromolecular chains that remain entrapped in the gel because of the difficulty in removing them by the solvent. Another explanation is the possibility of the breakage of styrene in SBR with high styrene content. These styrene radicals lead to the formation of stable end groups in the presence of oxygen, inducing an increase of low molecular weight fragments.[72]

As indicated by Figure 9.21b, the weight average molecular weight of the sol of devulcanized SBR with lower styrene content is higher than that of the SBR with higher styrene content. This is because the SBR with low styrene content has more double bonds and generates more radicals, to allow branching by recombination.[77] Therefore, the molecular weight distribution (MWD) of the sol of the devulcanized SBR of low styrene content is broader than that of the other SBR. A comparison of Figure 9.21a and Figure 9.21b indicates that M_w reduces less than M_n, indicating significant broadening of the molecular weight distribution of sol in devulcanized SBR.

Similar significant reduction of M_w and M_n and broadening of MWD of sol was observed in BR[69] that has a similar or higher degree of unsaturation than SBR. However, the situation changes in the case of rubber having a lower degree of unsaturation. Less reduction of molecular weight is observed when the polymer has a lower degree of unsaturation, as with PUR.[64] Table 9.1 shows the molecular weight data obtained from gel permeation chroma-

TABLE 9.1
Number and Weight Average Molecular Mass of Virgin Gums and Sol Fractions of Unfilled Devulcanized Rubbers Obtained Using the Coaxial Reactor

Sample	Mn	Mw	Mw/Mn	Reference
SBR gum, styrene content 18%	124800	369400	2.96	72
SBR — 0.63 g/s, 1.52 mm, gap 10 μm	1000	21100	21.1	72
SBR gum, styrene content 23%	86800	330100	3.81	72
SBR — 0.63 g/s, 1.52 mm, gap, 10 μm	4000	20000	5.0	72
BR — gum	145600	191900	1.32	78
BR — 0.63 g/s, 1 mm gap, 10 μm	14100	34000	2.41	78
PDMS — gum	234000	414000	1.77	71
PDMS 0.32 g/s, 0.63 mm gap, 10 μm	110000	540000	4.91	71
PUR Gum	42300	65100	1.54	79
PUR — 1.26 g/s, 2.54mm gap, 10 μm	20400	27400	1.34	79

tography (GPC) measurements for various rubbers.[70,72,78,79]. As is evident from the M_n and M_w values, there is substantial molecular weight reduction in the case of the devulcanized samples. However, this reduction is more pronounced in SBR or BR and less pronounced in rubbers like PUR, which has a lower degree of unsaturation, or PDMS, which is a fully saturated rubber. Of note, the molecular weight of the devulcanized samples are obtained only from the extracted sol, which, in turn, comprises only 10 to 15% of the devulcanized samples. However, these molecular weight data further strengthen the results obtained from the lower maximum torque values in the cure curves of rubbers having lower degree of unsaturation (see Figure 9.9 and Figure 9.10) and measurements of mechanical properties, which will be discussed in the following section, that have shown devulcanization and degradation of the samples after the application of ultrasound.

9.8.2 Peroxide-Cured Rubber

The situation changes when the rubber is cured by peroxide, which creates much stronger carbon-carbon cross-links, than in the case of sulfur cross-linking, where the cross-links are weaker. Also, there is a difference in the devulcanization of organic and silicone rubbers. In particular, the difference in the ratio of the probability of bond cleavage in the main chain and in the cross-links in organic and silicone rubbers may lead to a difference in the weight and number average molecular weights of the sol fraction in the devulcanized unfilled silicone and organic elastomers. Figure 9.22 shows the molecular weight of the sol of the devulcanized silicone rubber.[48,75] Note that increasing the amplitude causes a significant decrease in M_n and an increase in M_w. Data for an amplitude of 5 μm are not available because at that amplitude, no appreciable change in sol fraction is detected in devulcanized PDMS, in comparison with the virgin vulcanizate. Apparently, the decrease of M_n corresponds to an increase in the amount of branched fragments with low molecular weight. The increase in M_w corresponds to the appearance of some branched fragments that have very high molecular weight. As a result, the polydispersity increases significantly. In the case of SBR, an increase in amplitude is accompanied by a significant decrease in both M_w and M_n of sol fraction.[72] The above facts clearly support our earlier hypothesis about the effect that the different bond energies of the main chain and the cross-links in SBR and PDMS have on their ultrasound devulcanization. Another observation significant in SBR is the change of the *cis-trans* ratio. The devulcanized and revulcanized samples show a higher *cis-trans* ratio compared to the virgin uncured and vulcanized samples.[72]

9.8.3 Molecular Mobility and Diffusion

In an effort to improve the understanding of the devulcanization process, complementary studies of the molecular mobility of the entanglement and

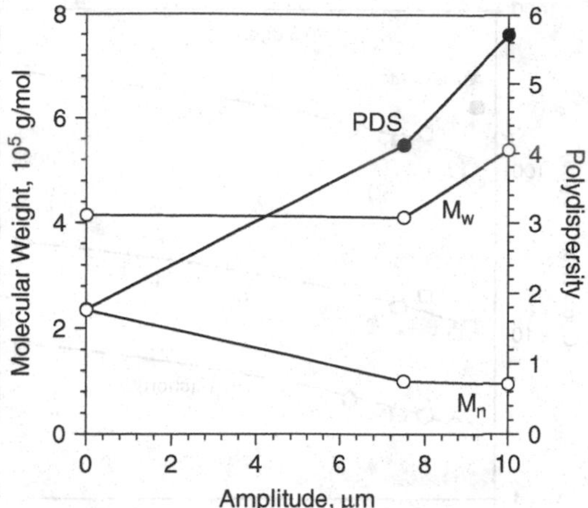

FIGURE 9.22
Molecular weights (open symbols) and polydispersity (filled symbols) of extracted sol of devulcanized PDMS as a function of ultrasound amplitude. The devulcanization samples were obtained using the coaxial reactor at a die gap of 0.63 mm, a flow rate of 0.32 g/second, and a barrel temperature of 180°C. (Adapted from Diao, B., Isayev, A.I., and Yu. Levin, V., *Rubber Chem. Technol.*, 72, 152–164, 1999; and Diao, B., M.S. thesis, University of Akron, 1997. With permission.)

cross-link network, and the production of the smaller sol molecules and their segmental mobility and diffusion rate were undertaken.[42,60,61,71,76,8,9] These investigations were conducted using proton pulsed NMR relaxation and pulse-gradient spin-echo (PGSE) diffusion measurements.

In the case of SBR, a two-component relaxation model was employed to explain the T_2 relaxation data.[42] These two major components were attributed, respectively, to the extractable sol and the unextractable gel. However, in later studies involving PDMS, a three-component model was used and was found to be successful in PDMS,[60,61,71] BR,[76,78] and PUR,[79] given by

$$\begin{aligned} \frac{A(2\tau)}{A(0)} &= f_S \exp(-2\tau / T_{2S})^E + f_L \exp(-2\tau / T_{2L}) \\ &+ (1 - f_S - f_L)\exp(-2\tau / T_{2M}) \end{aligned} \tag{9.4}$$

where T_{2S}, T_{2M}, and T_{2L} signify the short, medium, and long relaxation times, respectively; f_S, $(1\text{-}f_S\text{-}f_L)$, and f_L are their relative intensities; and E is the Weibull exponent, found to vary between about 1.0 (longest T_{2S}) and 1.6 (shortest T_{2S}). Figure 9.23 shows T_{2S}, T_{2M}, and T_{2L} plotted as functions of the fraction of chemically extractable sol for PDMS, the latter representing a measure of the extent of the devulcanization.[60,61,71] Data for the virgin uncured PDMS are included (100% "sol"). It is seen that the correlation is satisfactory, and that all component molecular mobilities are increased with

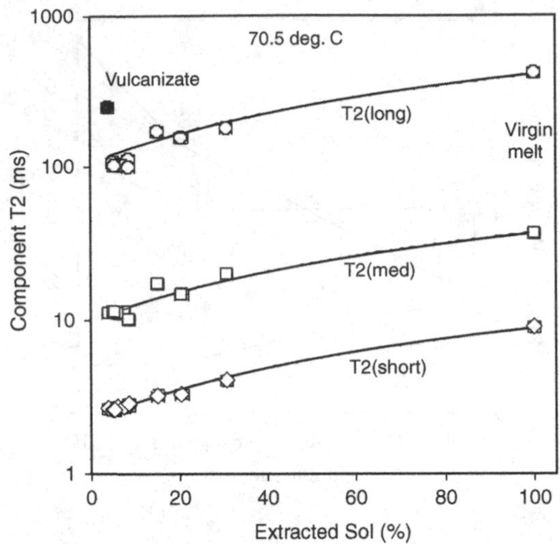

FIGURE 9.23
T_2 relaxation times of the short, medium, and long components for the series of virgin and devulcanized PDMS specimens as function of extractable sol content obtained at a die gap of 0.63 mm; various flow rates of 0.32, 0.63, and 1.26 g/second; various amplitudes of 5, 7.5, and 10 μm; and a barrel temperature of 180°C using the coaxial reactor. (From Shim, S.E., Parr, J.C., von Meerwall, E., and Isayev, A.I., *J. Phys. Chem.* B, 106, 12072–12078, 2002. With permission.)

increase of sol contents, as a result of ultrasound exposure. Interestingly, the factor of increase is the same for the three components, with the virgin melt having the highest mobilities. The relative contributions of the components to the echo, however, depend significantly on the ultrasound treatment.

This information, in the form of f_S, $f_M = (1\text{-}f_S\text{-}f_L)$, and f_L, is shown in Figure 9.24, again as function of the extractable sol fraction.[60,61,71] The presence of a significant f_S for the virgin melt clearly confirms the conclusion reached earlier for SBR[35,42,72] that, except for minor enhancements of segmental mobility (T_{2S}), physical cross-links in the form of entanglements are nearly indistinguishable in NMR relaxation from chemical crosslinks. As ultrasound treatment generates more sol, the fraction of long and intermediate components increases substantially at the expense of the short component. In contrast to the relaxation times, the results for the virgin melt more closely resemble the untreated vulcanizate rather than the most vigorously devulcanized specimens. This indicates that devulcanization reduces overall mobilities without significantly affecting the components' proportions, hence the scaled shape of the decay. The bold line in Figure 9.24 represents the extractable sol fraction itself, and its position well above f_L indicates that a substantial and increasing portion of the intermediate and perhaps short component arises from the chemically extractable material. Still, the fact that $f_L + f_M$ invariably exceeds the extractable sol fraction suggests that at least some of the intermediate (and, of course, short) component represents the unextractable material. In other words, the intermediate component must

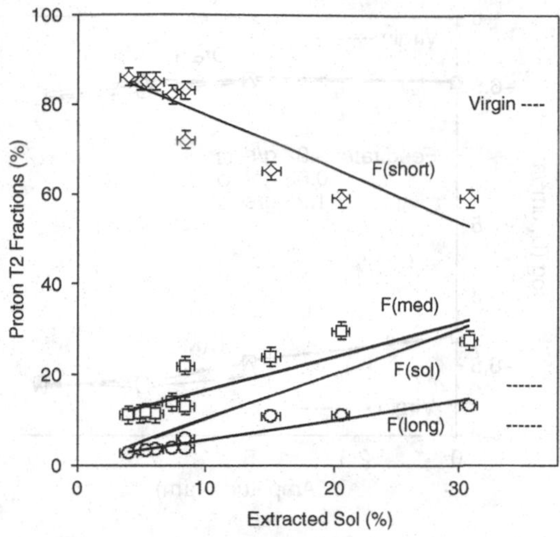

FIGURE 9.24
Dependence on sol content of the fraction of short, medium, and long T_2 relaxation time components for the series of devulcanized (solid lines) and virgin (dashed lines) PDMS specimens obtained at a die gap of 0.63 mm; various flow rates of 0.32, 0.63, and 1.26 g/second; various amplitudes of 5, 7.5, and 10 μm; and a barrel temperature of 180°C using the coaxial reactor. Bold line represents an equality of sol fraction. (From Shim, S.E., Parr, J.C., von Meerwall, E., and Isayev, A.I., *J. Phys. Chem.* B, 106, 12072–12078, 2002. With permission.)

divide its origins between detached and still attached network fragments of roughly equal segmental mobilities. On the basis of the component mobilities and relative contributions, and their dependences on sol fraction, it is now possible to assign the short component to the cross-linked and entangled network, increasingly loosened and diminished by ultrasound and penetrated by highly mobile and diffusing sol. Because of its distinct separation from the intermediate component, the long component must arise from the unreactive oligomers present in all specimens, irrespective of treatment, plus relatively small increments of very light, unbranched, material produced by ultrasound. It appears that ultrasound exposure of PDMS degrades the network by detaching molecular fragments of at least intermediate size, but leaves copious amounts of fragments dangling. Significantly, ultrasound seems inefficient at further tearing such fragments into oligomeric dimensions, or detaching oligomer-size material directly from the network. These conclusions were confirmed and further elaborated by the analysis of the diffusion data, as well as by chemical analysis of the extracted sol. The effect of ultrasound treatment on segmental mobility in PUR[79] is weaker than in other rubbers studied. Ultrasound exposure degrades the PUR network by detaching molecular fragments of substantial size that have molecular mobilities similar to the untreated network. But it is quite ineffective in generating short, unentangled species with higher mobilities.

Of interest in the diffusion experiments in PDMS,[60] BR,[76] and PUR[79] is the observation of a sharply bimodal diffusivity spectrum. The data suggest that

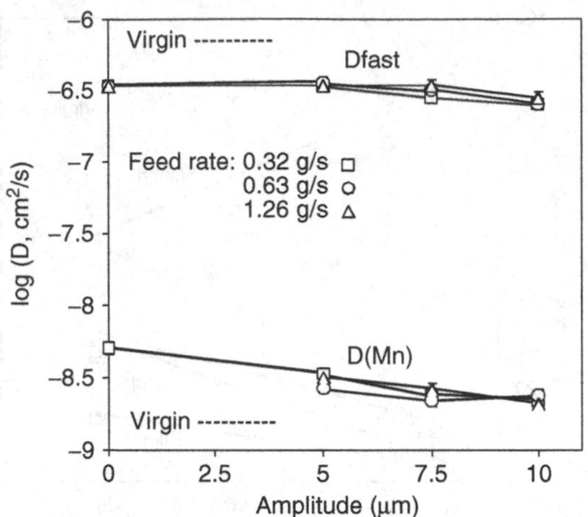

FIGURE 9.25
Diffusivity Dfast of the fast component and mean diffusivity D(Mn) of the slowly diffusing species in devulcanized (solid lines) and virgin (dashed lines) unfilled PDMS as a function of ultrasound amplitude obtained at a die gap of 0.63 mm and a barrel temperature of 180°C using the coaxial reactor. (From Shim, S.E., Parr, J.C., von Meerwall, E., and Isayev, A.I., *J. Phys. Chem.* B, 106, 12072–12078, 2002. With permission.)

the fast-diffusing portion arises from the oligomeric material relaxing at T_{2L}, while the lower-diffusivity portion originates with unentangled molecular fragments, i.e., nonoligomeric polydisperse light sol relaxing at T_{2M}. In Figure 9.25, the diffusivities D_{fast} and $D(M_n)$ in devulcanized PDMS samples are plotted as a function of ultrasound amplitude, showing that the two rates differ by almost two orders of magnitude.[60] The decrease of the diffusion rate with increase in ultrasound amplitude for the slow-moving species is greater than that of fast-moving molecules, indicating that as devulcanization conditions become more severe, copious additional unentangled sol molecules are created, with masses greater than those present in the virgin melt. This effect was also observed in SBR,[42] where it was more pronounced than in PDMS.

Devulcanization produces more sol, both entangled and unentangled (M M_c), likely of various degrees of branching, in copious amounts dependent on the condition of ultrasound treatment. It also greatly increases the amount of loosely attached network fragments. Devulcanization increases mobility (T_2) of the gel segments, of the added light sol, and of the unreacted oligomers, by equal factors, closely correlated with the amount of extractable sol. Three distinct T_2 components were found in virgin and devulcanized PDMS, BR, and PUR, comprised of entangled and cross-linked network; light sol and dangling network fragments; and a trace of unreactive light material (e.g., oligomers), respectively. As a result of the treatment, the molecular mobilities of all the three fractions increase by comparable factors, with sol diffusion responsible for enhancing the segmental mobility of the network

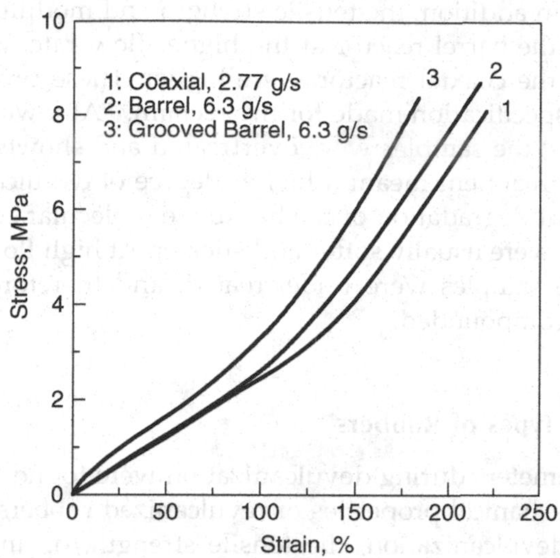

FIGURE 9.26
Stress-strain curve of RGRT obtained using the coaxial and barrel reactors at a gap of 2 mm and at an amplitude of 10 μm, and the grooved barrel reactor at a gap of 3.5 mm and an amplitude of 6 μm at a barrel temperature of 178°C. (From Yun, J., Ph.D. dissertation, University of Akron, 2003.)

of these rubbers. The corresponding diffusivity distribution in the unentangled portion of the sol was bimodal, separable into contributions from any fast-moving species and a slow-moving polydisperse species.[60,76,79]

9.9 Mechanical Performance of Revulcanized Rubbers

9.9.1 Tire Rubber

Vulcanizates obtained from devulcanized rubbers show a behavior that depends on processing conditions, type of reactor, type of rubber, and type of filler. Therefore, in this section, a comparison of the mechanical behavior of various rubbers will be presented. The comparison of stress-strain behavior of the vulcanizates prepared from devulcanized GRT produced by the three reactors at the maximum flow rate is shown in Figure 9.26.[70] The revulcanized sample obtained from the barrel reactor having a flow rate of 6.3 g/second shows a tensile strength of 8.7 MPa, elongation at break of 217%, and modulus at 100% elongation of 2.6 MPa. In addition, the revulcanized sample obtained from the grooved barrel reactor having a flow rate of 6.3 g/second shows a tensile strength of 8.3 MPa, an elongation at break of 184%, and a modulus at 100% elongation of 3.3 MPa. The maximum output of the barrel and grooved barrel reactors was higher than that of the coaxial reactor. In addition, the tensile strength and modulus of the sample obtained using the barrel reactor at the higher flow rate, which could not be achieved in the coaxial reactor, were higher. These properties met the higher level of specification made for tire reclaim.[73] At lower flow rates and high amplitudes, the samples were overtreated and showed inferior properties. The overtreatment meant a higher degree of devulcanization, along with a significant degradation of the backbone molecular chains. The overtreated samples were usually softer and stickier. At high flow rates and low amplitudes, the samples were undertreated, and therefore, could not be processed and compounded.

9.9.2 Various Types of Rubbers

Processing parameters during devulcanization were found to have a strong effect on the mechanical properties of revulcanized rubbers. Depending on the degree of devulcanization, the tensile strength, σ_b, and elongation at break, ε_b, of the revulcanized rubber varied. It was reported[43,72] that under some devulcanization conditions, the tensile strength of unfilled revulcanized SBR was found to be much higher than that of the original vulcanizate, with elongation at break being practically intact. In fact, it is true for various rubbers that are unable to crystallize under strain. In particular, Figure 9.27 shows the stress-strain curves of unfilled virgin vulcanizates and revulcanized NR, SBR, and EPDM obtained from devulcanized rubbers at various values of ultrasonic amplitudes.[62] The devulcanized rubbers were obtained by using the coaxial ultrasonic reactor depicted in Figure 9.3a at barrel temperature of 120°C, screw speed of 20 rpm, and a flow rate of 0.63 g/second. In contrast to usual findings that the mechanical properties of reclaimed rubber obtained by using different techniques are inferior to those of virgin vulcanizates,[73] the present data are rather unexpected. In the stress-strain curves of NR, stress-induced crystallization was observed in all the revulcanized samples. The sample of NR denoted as virgin 2 contained twice the amount of curatives compared to virgin 1. The best tensile properties of revulcanized NR obtained at an amplitude of 5 μm were the tensile strength, σ_b = 13.9 MPa and elongation at break, ε_b = 682%, compared to virgin 1, which had σ_b = 19.3 MPa and elongation at break, ε_b = 700%. Only 28% and 2.6% reduction in the tensile strength and the elongation at break, respectively, were seen. On the other hand, the revulcanized SBR and EPDM showed tensile properties superior to virgin vulcanizates. The best tensile properties of revulcanized SBR were obtained at an amplitude of 5 μm. For this sample, the tensile strength, σ_b = 1.94 MPa and elongation at break, ε_b = 199% compared to vulcanized virgin 1, σ_b = 1.23 MPa and elongation at break, ε_b = 217%. For revulcanized SBR, 58% increase in tensile strength at break and 8.3% decrease in elongation at break in comparison with virgin vulcanizate were observed. The best tensile properties of revulcanized EPDM were obtained at an amplitude of 10 μm. For this sample, the tensile

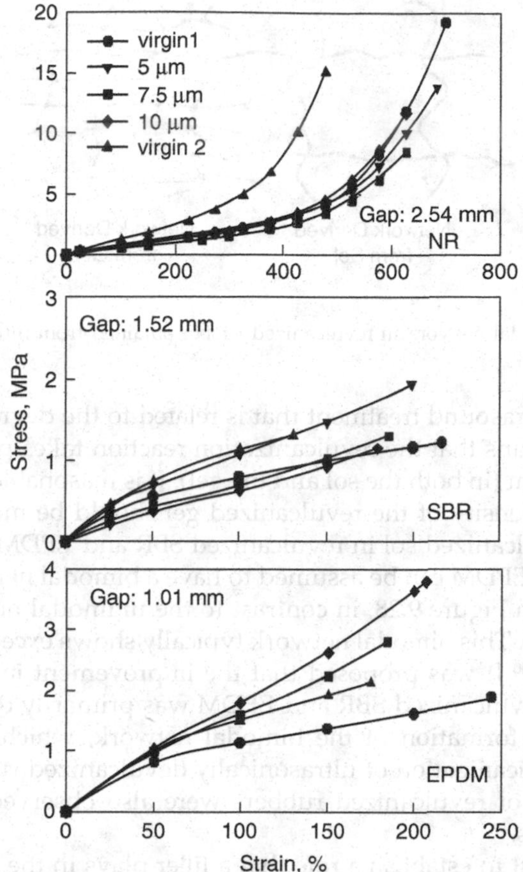

FIGURE 9.27
Stress-strain curves of virgin rubbers and revulcanized rubbers obtained at various amplitudes and gaps using the coaxial reactor, a flow rate of 0.63 g/second, and a barrel temperature of 120°C. (From Yun, J., Isayev, A.I., Kim, S.H., and Tapale, M., *J. Appl. Polym. Sci.*, 88, 434–441, 2003. With permission.)

strength, σ_b = 3.8 MPa, and the elongation at break, ε_b = 207% compared to virgin 1, σ_b =1.9 MPa and ε_b = 244%. The 100% increase in the tensile strength and 15% decrease in the elongation at break of revulcanized EPDM rubber were observed. Therefore, the tensile strength, $\sigma_{b,}$ of revulcanized SBR and EPDM rubbers exceeded that of virgin vulcanizates significantly, while the elongation at break, $\varepsilon_{b,}$ was practically intact. The stress-strain characteristics of the revulcanized samples exhibit an S-shaped curve, which is typical of a rubber crystallizable under strain. However, such a stress-strain behavior cannot be explained by the possibility of strain-induced crystallization in the SBR and EPDM networks.

The ultrasound treatment of rubber leads to an increase in sol fraction and a reduction in gel fraction. The sol and gel fraction strongly depend on the

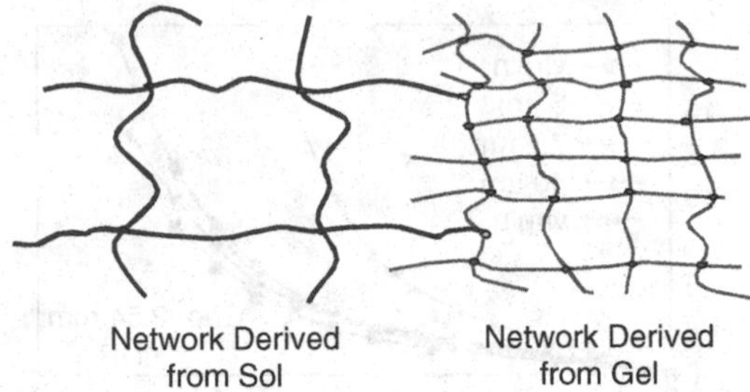

FIGURE 9.28
Schematics of bimodal network in revulcanized rubber obtained from ultrasonically devulcanized rubber.

condition of ultrasound treatment that is related to the degree of devulcanization. This means that the revulcanization reaction takes places in a heterogeneous system (in both the sol and the gel). It is reasonable to suggest that the cross-link density of the revulcanized gel should be much higher than that of the revulcanized sol in revulcanized SBR and EPDM. Thus, revulcanized SBR and EPDM can be assumed to have a bimodal network, as shown schematically in Figure 9.28, in contrast to the unimodal network in virgin SBR and EPDM. This bimodal network typically shows exceptional mechanical properties.[80] It was proposed that the improvement in the mechanical properties of revulcanized SBR and EPDM was primarily due to the extent of nonaffine deformation of the bimodal network, which appears in the process of revulcanization of ultrasonically devulcanized rubber. The superior properties of revulcanized rubbers were also observed in the case of unfilled EPDMs.[48]

It is of interest to establish a role that a filler plays in the devulcanization process and how it affects the mechanical properties of revulcanized rubbers. In this regard, Figure 9.29 shows the stress-strain curves for virgin and devulcanized 35 phr CB-filled NR vulcanizates.[51] Virgin vulcanizates were cured using 5 phr ZnO, 1 phr stearic acid, 1 phr CBS, and 2 phr sulfur. The revulcanization recipe contained 2.5 phr ZnO, 0.5 phr stearic acid, 0.5 CBS, and 2 phr sulfur. The experiments have shown that upon filling rubbers with CB, the mechanical properties of revulcanized rubbers typically deteriorate after devulcanization; the level of deterioration depends on devulcanization conditions and the amount and type of filler. This effect, due to the presence of filler, is clearly evident from Figure 9.29.

The tensile properties of revulcanized CB-filled EPDMs are compared to those of virgin vulcanizates in Figure 9.30.[81] For the unfilled rubber, no deterioration of properties is observed after revulcanization. In fact, unfilled revulcanizates show higher tensile strength, modulus, and energy at break than virgin vulcanizates. This is in agreement with the data reported in Figure 9.27. However, there is a significant deterioration in the tensile strength, elongation at break, and energy at break of revulcanized rubbers with an increase of filler contents. It is well known that CB surfaces contain

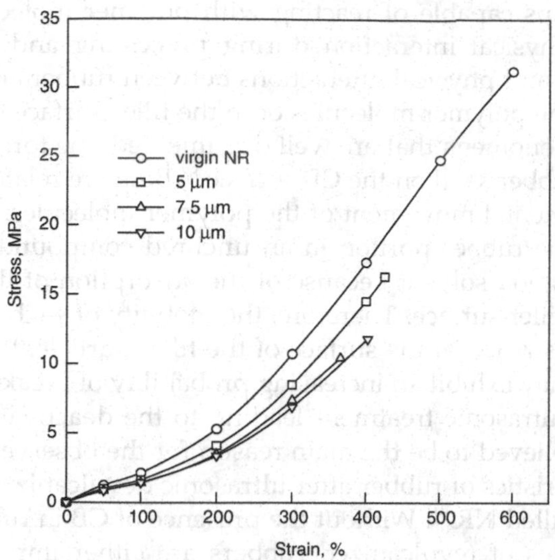

FIGURE 9.29
The stress-strain curves for 35 phr carbon black–filled virgin NR vulcanizate and revulcanized NR devulcanized in a coaxial reactor at a barrel temperature of 120°C, a gap of 2.54 mm, a flow rate of 0.63 g/second, and various ultrasonic amplitudes. (From Hong, C.K. and Isayev, A.I., *J. Appl. Polym. Sci.*, 79, 2340–2348, 2001. With permission.)

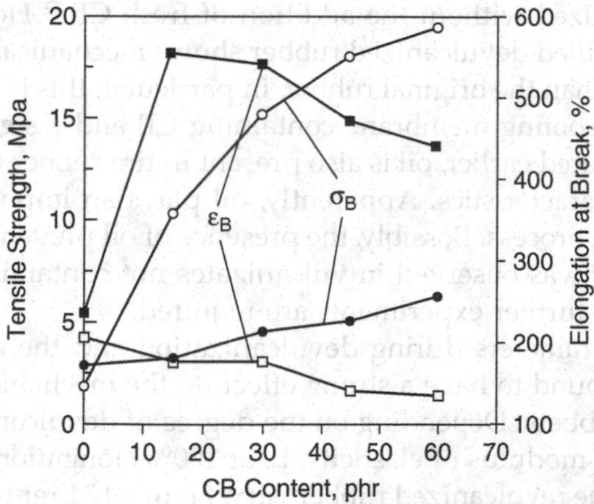

FIGURE 9.30
Tensile strength and elongation at break vs. CB content of virgin vulcanizates (open symbols) and revulcanized (filled symbols) EPDM rubbers devulcanized at a flow rate 0.63 g/second, a gap of 1.02 mm (for 60 phr a flow rate of 1.26 g/second and a gap of 2.03 mm), an amplitude of 10 μm, and a barrel temperature of 120°C using the grooved barrel reactor. (From Yun, J., Yashin, V.V., and Isayev, A.I., *J. Appl. Polym. Sci.*, 91, 1646–1656, 2004. With permission.)

functional groups capable of reacting with polymer molecules due to the reaction and physical interaction during processing and vulcanization.[82] These chemical and physical interactions between rubber and CB can cause adsorption of the polymer molecules onto the filler surface. This adsorption leads to two phenomena that are well documented: the formation of bound rubber and a rubber shell on the CB surface.[83] Both are related to the restriction of the segmental movement of the polymer molecules.[84] Bound rubber is defined as the rubber portion in an uncured compound that cannot be extracted by a good solvent because of the adsorption of the rubber molecules onto the filler surface. Therefore, the mobility of such rubber chains is considerably restricted at the surface of the filler particle. It is believed that filled rubbers may exhibit an increasing probability of breakup of the bound chains by the ultrasonic treatment leading to the deactivation of the filler. This effect is believed to be the main reason for the observed loss of performance characteristics of rubber after ultrasonic devulcanization in CB-filled SBR[72] and CB-filled NR.[51]. Without the presence of CB in rubber, the performance properties of revulcanized rubbers are either improved or do not deteriorate significantly.

It was suggested that ultrasonic devulcanization causes a partial deactivation of filler due to a breakup of the macromolecular chains attached to the surface of CB. In many cases, this effect leads to inferior properties of revulcanized CB-filled rubbers. Thus, ultrasonically devulcanized rubber was blended with virgin rubber.[85] The blend vulcanizates indicated a significant improvement in properties. Also, attempts were made to add a certain amount of a fresh CB into the devulcanized rubber. It was shown that the vulcanizates containing fresh CB also exhibited better properties than the rubber revulcanized without the addition of fresh CB.[72] However, in some cases, even CB-filled devulcanized rubber shows mechanical properties similar to or better than the original rubber. In particular, this is shown in Figure 9.31 for EPDM roofing membrane containing CB and a significant amount of oil.[65] As indicated earlier, oil is also present in tire rubber that shows good performance characteristics. Apparently, oil plays an important role in the devulcanization process. Possibly, the presence of oil prevents a deactivation of the filler that was observed in vulcanizates not containing oil. To prove this hypothesis, further experiments are required.

Processing parameters during devulcanization and the amount of curatives used are found to have a strong effect on the mechanical properties of revulcanized rubbers. Depending on the degree of devulcanization, the tensile strength, σ_b, modulus of elasticity, E, at 100% elongation and elongation at break, ε_b, of the revulcanized rubber vary. Figure 9.31 represents the stress-strain dependencies for samples of virgin vulcanizates and revulcanized roofing membrane obtained from devulcanized rubber prepared at various values of amplitudes and feed rates.[65] Tensile strength of the revulcanized rubber was found to be similar to that of the virgin vulcanizate, while the elongation at break decreased from about 450% to about 140 to 310%, depending on the process conditions during the devulcanization and the

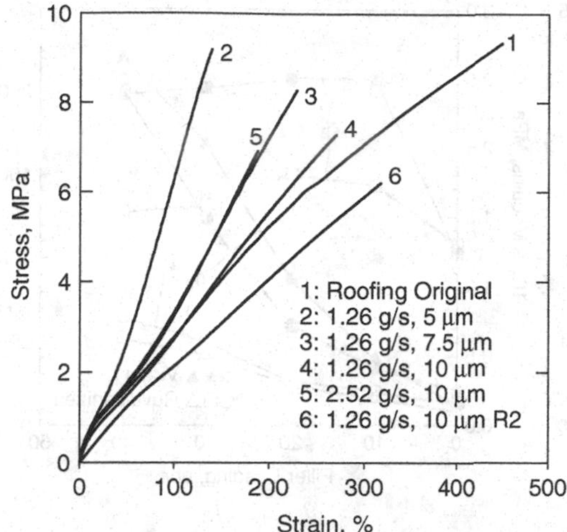

FIGURE 9.31
Stress-strain curves of original vulcanized and revulcanized EPDM roofing membranes obtained at a gap of 1 mm, a barrel temperature of 120°C, and various ultrasonic amplitudes and flow rates obtained using the grooved barrel reactor. R2 denotes reduced amount of curatives. (From Yun, J. and Isayev, A.I., *Polym. Eng. Sci.*, 43, 809–821, 2003. With permission.)

cure recipe. Note that at an ultrasonic amplitude of 5 μm, the tensile strength, σ_B, of a revulcanized sample is 9.2 MPa, that is, about the same as that of the original vulcanizate (9.3 MPa). However, the elongation at break, ε_B, of the revulcanized sample decreases to 140% from 450% of the original vulcanizate. In addition, the modulus at 100% of revulcanized roofing membrane obtained from the devulcanized sample at 5 μm shows an increase from 2.8 MPa to 6.3 MPa. Also, at an ultrasonic amplitude of 10 μm, the tensile strength, σ_B, and elongation at break, ε_B, of the revulcanized sample are 7.3 MPa and 270%, respectively. In this case, compared to the original vulcanizate, the tensile strength and elongation at break of revulcanized samples are respectively 80% and 60% of the original values.

Therefore, one can conclude that the tensile properties of recycled roofing membranes can be adjusted for different applications by varying the processing parameters during the devulcanization and the level of curatives during revulcanization. A higher output is achieved for roofing membrane (CB-filled EPDM compound) in the barrel reactor. The coaxial reactor was also used for devulcanization of unfilled EPDM, but the output of the coaxial reactor was limited, with the highest output being 0.63 g/second.

The mechanical properties of revulcanized silicone rubber samples are compared to those of virgin vulcanizates in Figure 9.32.[52,71] For the unfilled system, no deterioration of properties is observed. However, there is a significant deterioration in modulus, E, elongation at break, ε_b, and tensile strength, σ_b, of revulcanized rubbers with an increase of filler content. Hardness of revulcanized rubber is also lower than that of virgin vulcanizates.[52,71] There are two reasons for both reduced hardness and mechanical properties

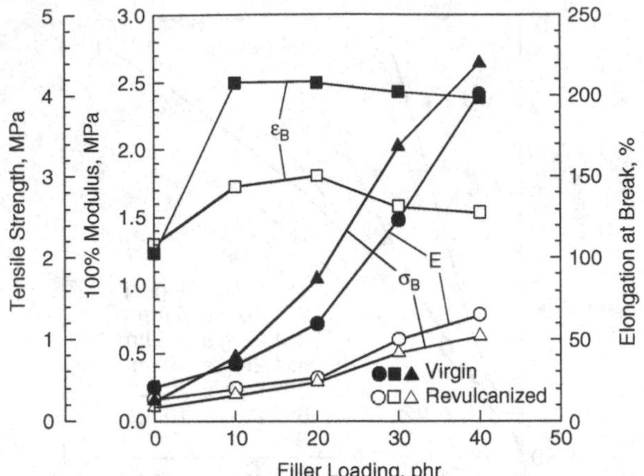

FIGURE 9.32
Comparison of mechanical properties of virgin (filled symbols) and revulcanized (open symbols) precipitated silica-filled silicone rubber devulcanized at a flow rate of 0.32 g/second, a gap of 0.63 mm, an amplitude of 10 μm, and a barrel temperature of 180°C obtained using the coaxial reactor. (From Shim, S.E. and Isayev, A.I., *Rubber Chem. Technol.*, 74, 303–316, 2001. With permission.)

of revulcanized samples. First is the possibility of a partial deactivation of the filler during ultrasonic devulcanization, as schematically shown in Figure 9.33. It is known that rubber molecules are bonded onto the filler surface and penetrate into the pores existing in silica filler, leading to bound rubber. The mobility of such chains is considerably restricted. This means that the chains of bound rubber can be more easily broken than those of the nonbound rubber during ultrasonic treatment. Accordingly, after ultrasonic treatment, a considerable amount of bound rubber chains are broken. This process may lead to a partial deactivation of filler, causing the loss of reinforcement. Another reason for decreased mechanical properties can be found in Figure 9.15 and Figure 9.16, where cross-link densities and gel fractions of virgin vulcanizates, devulcanized rubber, and revulcanized rubber were shown. It was shown in those figures that the cross-link density and gel fraction during devulcanization were significantly decreased. And they were not recovered after revulcanization, compared to those in virgin vulcanizates, although the same amount of curative as in virgin rubber was used. These results may also be related to the mechanical properties of the revulcanized samples. In the study of SBR, cross-link density of revulcanized specimens was much higher than that of virgin cured samples. In addition, it has been shown that the gel fraction of revulcanized rubber is close to those of virgin vulcanizates.[72] It should be noted that in the case of SBR, a twofold increase in the amount of curatives leads to a twofold increase in cross-link density. In contrast, in PDMS, a twofold increase in the amount of peroxide causes very little increase in cross-link density. Because peroxides decompose to generate free radicals when heated and form low molecular

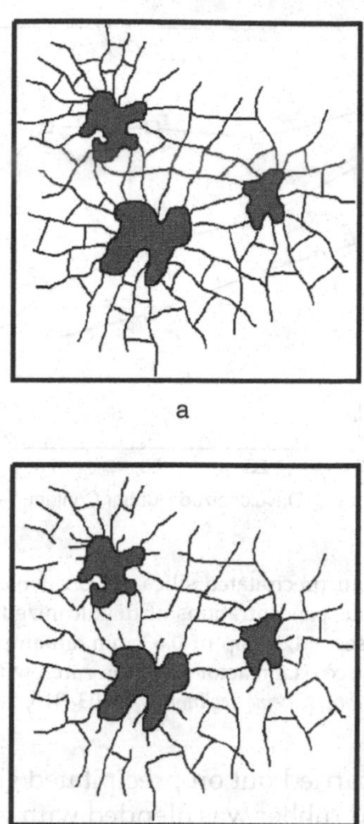

FIGURE 9.33
A schematic representation of deactivation of filler upon ultrasonic devulcanization (a) before ultrasonic devulcanization and (b) after ultrasonic devulcanization.

weight species removed by postcuring, the curatives do not remain in the vulcanizates after curing, in the case of peroxides.[86] In addition, because a cross-link reaction is achieved through the participation of vinyl groups in methyl vinyl silicone rubber,[87] further cross-linking reaction through the vinyl groups upon revulcanization is limited, since vinyl groups are consumed in the first stage of curing. This may be the reason for low cross-link density and gel fraction of revulcanized silicone rubber.

To increase the mechanical properties of recycled CB-filled rubber obtained by ultrasonic devulcanization, devulcanized CB-filled rubber was blended with virgin rubber.[81,85] In one study,[81] 30 phr CB-filled virgin EPDM compound was blended with 30 phr CB-filled devulcanized EPDM at various concentrations. It was observed that the modulus at 100% elongation increased with the content of devulcanized rubber, following values above the rule of mixture. The tensile strength and elongation at break of the blend is also improved, due to the adsorption of the polymer molecules onto the filler surface. With decrease in the content of devulcanized rubber in the blends, properties follow the values at or slightly below the rule of mixture.

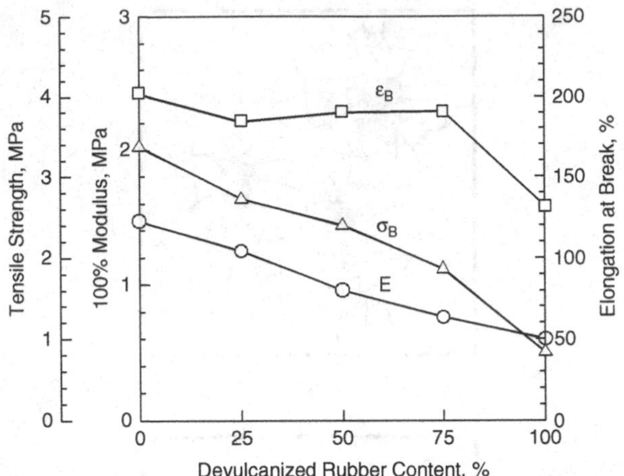

FIGURE 9.34
Mechanical properties of 30 phr precipitated silica–filled devulcanized/virgin silicone rubber blend vulcanizates with various concentrations of devulcanized rubber. Rubber was devulcanized at a flow rate of 32 g/second, a gap of 0.63 mm, an amplitude of 10 μm, and a barrel temperature of 180°C using the coaxial reactor and then cured with 0.5 phr DCP at 170°C. (From Shim, S.E. and Isayev, A.I., *Rubber Chem. Technol.*, 74, 303–316, 2001. With permission.)

Similar studies were carried out on precipitated silica-filled PDMS; 30 phr silica-filled devulcanized rubber was blended with virgin 30 phr filled PDMS at various concentrations.[52] Figure 9.34 represents the stress-strain behavior and modulus and ultimate values of the blend vulcanizates.[52] It is seen that 100% modulus, E, and tensile strength, σ_b, decrease with the content of devulcanized rubber following values close to or above the rule of mixture. The elongation at break, ε_b, of the blend vulcanizates containing up to 75% devulcanized rubber is close to that of the virgin cured samples.

9.9.3 Blends of Devulcanized and Virgin Rubbers

In recycling of rubbers, it is customary to add various proportions of ground rubbers in the virgin material. Therefore, blends of both filled and unfilled ground and ultrasonically devulcanized rubbers with virgin rubber have been prepared. The mechanical properties of blends of unfilled PUR virgin and devulcanized rubber have been measured.[88] Curatives were added in the blends of devulcanized rubber, based on the total rubber, while for blends of ground rubber, curatives were added based on virgin rubber content. Figure 9.35 shows the tensile properties of the two types of blends of PUR.[88] From a comparison of these two figures, it is quite apparent that the blends of the devulcanized samples (Figure 9.35a) have better tensile properties compared to the blends of ground samples (Figure 9.35b). For the blends of ground samples, only at a ground concentration of 25%, the tensile properties are similar to those of the original, whereas for the former, the properties

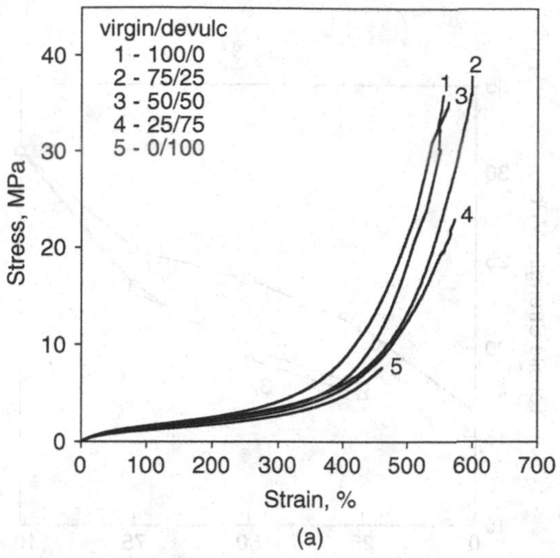

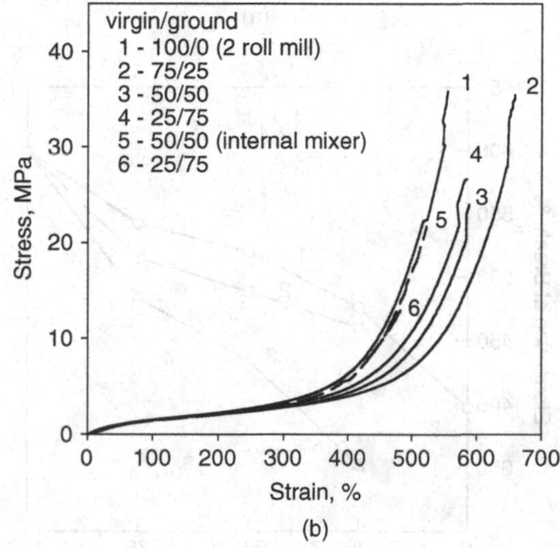

FIGURE 9.35
Stress vs. strain curves of vulcanized blends of virgin and devulcanized sample (a), and ground and virgin sample (b). The condition of devulcanization is a flow rate of 1.26 g/second, a gap size of 3 mm, an amplitude of 7.5 μm, and a barrel temperature of 120°C using the coaxial reactor. (From Ghose, S. and Isayev, A.I., *Polym. Eng. Sci.*, 44, 794–804, 2004. With permission.)

are superior to the original at this concentration and comparable at concentrations of 50%. It may be that ultrasonic devulcanization allows a better bonding of the devulcanized rubber to the virgin rubber in the blends in the case of ground-virgin blends.

Ultrasonically devulcanized and ground CB-filled NR was blended with virgin CB-filled material. Figure 9.36 gives the tensile strength of these NR/NR blends containing 35 phr CB.[85] The blends of devulcanized NR and virgin

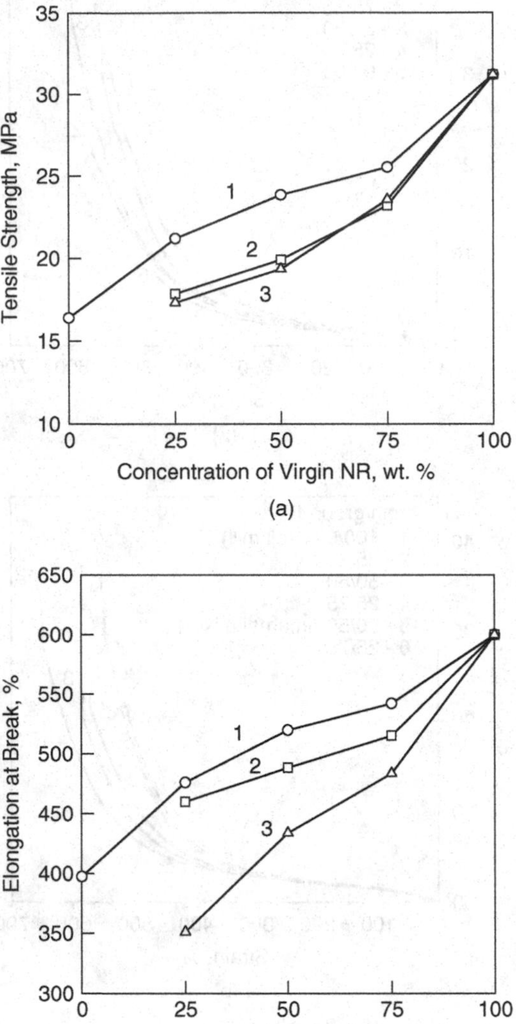

FIGURE 9.36
Tensile strength and elongation at break of 35 phr CB-filled virgin and devulcanized NR blends; (1) curatives were added to total rubber content; (2) blends of ground NR vulcanizates and virgin NR; curatives were added to virgin rubber, and (3) blends of ground NR vulcanizates and virgin NR; curatives were added to the total rubber content; NR was devulcanized at a flow rate of 0.63 g/second, a gap size of 2.54 mm, an amplitude of 5 μm, and a barrel temperature of 120°C using the coaxial reactor. (From Hong, C.K. and Isayev, A.I., *J. Mat. Sci.*, 37, 385–388, 2002. With permission.)

NR have much better tensile properties than blends with fully cured ground rubber. As the proportion of the virgin NR in the blends of devulcanized NR was increased, the mechanical properties progressively increased at or above the rule of mixture. Therefore, the mechanical properties of these kinds of blends can be improved significantly by ultrasonic devulcanization of ground vulcanizates. The modulus at 100% strain of NR/NR blends containing 35 phr CB was measured.[85] Blends of devulcanized and virgin NR showed a little drop in the modulus, while the modulus of ground and virgin NR was reduced. It is thought that this was due to the migration of curatives, and the blends of ground rubber and virgin rubber prepared using curatives for virgin rubber were likely to exhibit reduced modulus. The blends of ground rubber and virgin rubber, with curatives added based on total rubber content, showed higher modulus. However, they indicated lower tensile strength and elongation of break due to excess of curatives.

9.10 Mechanisms of Devulcanization

9.10.1 Cavitation

It is believed that the process of ultrasonic devulcanization is based on a phenomenon called cavitation. In this case, acoustic cavitation occurs in a solid body. This is in contrast to cavitation typically known to occur in liquids in the regions subjected to rapidly alternating pressures of high amplitude generated by high-power ultrasonics.[89] During the negative half of the pressure cycle, the liquid is subjected to a tensile stress, and during the positive half of the cycle, it experiences a compression. Any bubble present in the liquid will thus expand and contract alternately. The bubble can also collapse suddenly during the compression. This sudden collapse is known as cavitational collapse and can result in an almost instantaneous release of a comparatively large amount of energy. The magnitude of the energy released in this way depends on the value of the acoustic pressure amplitude and, hence, the acoustic intensity.

Although the presence of bubbles facilitates the onset of cavitation, it can also occur in gas-free liquids when the acoustic pressure amplitude exceeds the hydrostatic pressure in the liquid. For a part of the negative half of the pressure cycle, the liquid is in a state of tension. Where this occurs, the forces of cohesion between neighboring molecules are opposed and voids are formed at weak points in the structure of the liquid. These voids grow in size and then collapse in the same way as gas-filled bubbles. Cavitation may be induced in a gas-free liquid by introducing defects, such as impurities, in its lattice structure.

In the case of polymer solutions, it is well known that the irradiation of a solution by ultrasound waves produces cavitation of bubbles.[14,90] The formation and collapse of the bubbles play important roles in the degradation of polymers in solution. Most of the physical and chemical effects caused by ultrasound are usually attributed to cavitation — the growth and very rapid, explosive collapse of microbubbles as the ultrasound wave propagates through the solution. The intense shock wave radiated from a cavitating bubble at the final stage of the collapse is undoubtedly the cause of the most severe reactions. This shock wave is capable of causing the scission of macromolecules that lie in its path. The degradation arises as a result of the effect of the ultrasound on the solvent.

In any medium, cavities, voids, and density fluctuations exist. It is believed that these induce cavitation, leading to molecular rupture. In solid polymers, the microvoids present intrinsically are responsible for cavitation when they are subjected to a hydrostatic pressure in the manner of an impulse. One of the main causes of microvoid generation in polymer materials is the interatomic bond rupture when they are subjected to mechanical and thermal stresses. Extensive studies showing microvoid formation in stressed polymers have been carried out.[91]

When applied to rubbers, the cavitation usually corresponds to the effect of formation and unrestricted growth of voids in gas-saturated rubber samples after a sudden depressurization.[92] Cavitation usually has a broader sense and may be understood as the phenomena related to the formation and dynamics of cavities in continuous media.[36,49] In materials science, for example, it means a fracture mode characterized by formation of internal cavities.[93] In acoustics, the cavitation denotes the phenomena related to the dynamics of bubbles in sonically irradiated liquids.[94]

9.10.2 Breakup of Cross-Links and Main Chains

Structural studies of ultrasonically treated rubber show that the breakup of chemical cross-links is accompanied by the partial degradation of the rubber chain.[33,34,45,48,64,72,75,78,79] The mechanism of rubber devulcanization under ultrasonic treatment is presently not well understood, unlike the mechanism of the degradation of long-chain polymer in solutions irradiated with ultrasound.[90] Specifically, the mechanisms governing the conversion of mechanical ultrasonic energy to chemical energy are not clear. However, it has been shown that devulcanization of rubber under ultrasonic treatment requires local energy concentration, since uniformly distributed ultrasonic energy among all chemical bonds is not capable of rubber devulcanization.[36,38]

It is well known that some cavities or small bubbles are present in rubber during any type of rubber processing.[95] The formation of bubbles can be nucleated by precursor cavities of appropriate size.[92] The proposed models[36–38,45,49] were based upon a mechanism of rubber network breakdown caused by cavitation, which is created by high-intensity ultrasonic waves in the presence of pressure and heat. Driven by ultrasound, the cavities pulsate with amplitude, depending mostly upon the ratio between ambient and ultrasonic pressures (acoustic cavitation).

It is known that, in contrast to plastics, rubber chains break down only when they are fully stretched.[96,97] An ultrasonic field creates high-frequency extension-contraction stresses in cross-linked viscoelastic media. Therefore, the effects of rubber viscoelasticity have been incorporated into the description of dynamics of cavitation.[49,50] The devulcanization of the rubber network can occur primarily around pulsating cavities, due to the highest level of strain produced by the powerful ultrasound.[50] Generally, cleavage in polymer chains results in the production of macroradicals,[98,99] the existence of which have been confirmed spectroscopically by the use of radical scavengers such as diphenyl picrylhydracyl (DPPH). Obviously, in the absence of scavengers, the macroradicals are free to combine by either disproportionation or combination termination; disproportionation leads to smaller-sized macromolecules while combination termination gives a distribution dependent upon the size of the combining fragments.[100] Therefore, it is our present understanding that ultrasonic cavitation is the main mechanism governing the devulcanization process of rubbers.

9.11 Modeling of Ultrasonic Devulcanization Process

9.11.1 Cavitation Based on Elastic Model

The theoretical modeling of the devulcanization process involves, on a macromolecular level, material behavior during flow in the devulcanization process, and on a micromolecular level, molecular network degradation in ultrasonic fields. The rate of devulcanization is strongly dependent on the macroparameters of the media, e.g., temperature and pressure. These parameters depend on the viscosity of the rubber, which is constantly changing during the devulcanization process. Hence, the micro- and macromolecular aspects are inherently coupled. As a result, many individual processes taking place simultaneously should be considered, including nonisothermal flow of crumb rubber and devulcanized rubber in the devulcanization zone, acoustic cavitation of internal voids (bubbles) preexisting in the rubber, and the rate of devulcanization due to a breakup of overstressed molecular bonds, including the network chains and cross-links. In fact, the first simplified theoretical model of the ultrasonic devulcanization process that was proposed was based on these considerations.[36,101] To describe rubber flow, die filling modeling has been carried out with heat transfer that was affected by the rate of viscous dissipation due to flow, the rate of ultrasound dissipation due to oscillatory deformation, and the rate of energy consumption due to devulcanization. The rate of breakup of various bonds was modeled using the first-order kinetic equation, with time taken to be the current average residence time of rubber at a particular location in the devulcanization zone and bond energy dependent on the type of the bond. Rubber is considered to contain many voids, and the acoustic cavitation takes place around these voids. Volume concentration of voids is considered as an adjustable parameter. The parts of the molecular chains located near the bubble wall are affected by the highest tensile forces, which are assumed to be sufficiently high to break the molecular bonds. Owing to molecular chain conformation, the acting forces decrease with increasing distance from the bubble wall, and at some nearby distance are insufficient to break the molecular bonds. The maximum bubble diameter during its growth is determined by simulation of the cavitation process in cross-linked rubber. It is a function of the frequency and amplitude of the ultrasonic waves, the current hydrostatic pressure, and the elastic characteristics of the material that is represented by neo-Hookean models.[102] The acoustic pressure is the driving force causing the cavity to expand and collapse violently, resulting in the cleavage of chemical bonds. The presence of bubbles in rubber affects its acoustic properties, which differ considerably from those of the pure polymer, even if the void concentration is small. This is because the compressibility of the mixture is substantially altered, reducing the speed of sound. Therefore, the effect of these voids on the acoustic pressure was included. This type of modeling and simulation has been carried out for various devulcanization conditions such gap size, flow rate, temperature, frequency, and amplitude. The results of simulation allowed determination of the distributions of velocity, shear rate, temperature, hydrostatic and acoustic pressure, rate of breakage of various bonds (monosulfidic, disulfidic, polysulfidic, and carbon-carbon), and therefore gel fraction and cross-link density of devulcanized rubber. As an example, Figure 9.37 and Figure 9.38 show, respectively, comparisons of the predicted and experimental results for gel fraction and cross-link density of devulcanized SBR vs. ultrasonic amplitude, as obtained in the coaxial reactor.[45] As seen, the theory is qualitatively correct in predicting the variation of gel fraction and cross-link density of devulcanized SBR with increasing ultrasonic amplitude. However, the theory shows a lower degree of devulcanization at smaller ultrasonic amplitudes than observed from experiments. Also, the experiments showed that for SBR containing a high concentration of polysulfidic cross-links, concentration of di- and monosulfidic cross-links increased with ultrasonic amplitude, while theory predicts the opposite trend. This discrepancy can be explained by the transformation of polysulfidic cross-links into di- and monosulfidic cross-links, due to the formation of free radicals during the breakup of polysulfidic bonds. These complex chemical transformations during devulcanization are not accounted for in theoretical modeling.

The predicted and experimental entrance pressures during devulcanization of SBR in the coaxial reactor were also compared (Figure 9.39).[45] Viscosity, as a function of shear rate, temperature, and gel fraction, described by Equation 9.1 and Equation 9.2, was used in this simulation. The predicted pressure was found to be in fair agreement with the experimentally measured pressure, especially in view of high fluctuations in experimental data. Compared with the experimental data, the model shows correct trends in

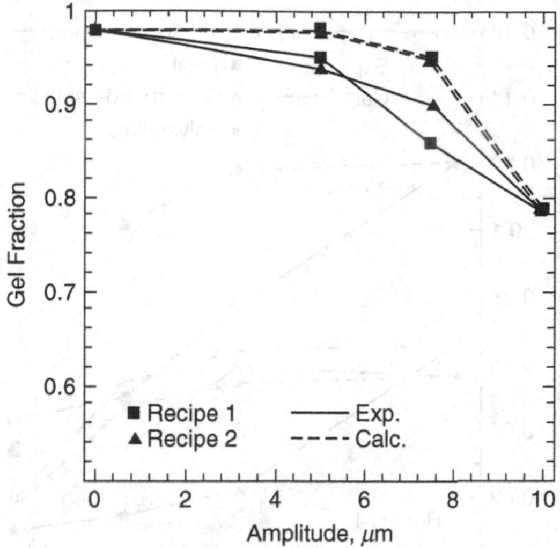

FIGURE 9.37
Experimental and calculated gel fraction of unfilled devulcanized SBR vs. ultrasonic amplitude obtained using the coaxial reactor at a barrel temperature of 120°C, a flow rate of 0.63 g/second, and a gap of 2.54 mm. (From Yushanov, S.P., Isayev A.I., and Kim, S.H., *Rubber Chem. Technol.*, 71, 168–190, 1998. With permission.)

the dependencies of the degree of devulcanization on the amplitudes of acoustic oscillations. Although the agreement with experimental results is only qualitative, the calculated results show that these modeling efforts are only the first step in developing the model for scale-up of the devulcanization process.

The numerical simulations of the continuous ultrasonic devulcanization process based on the cavitation collapse model[36] also revealed certain systematic qualitative disagreements between calculated results and available experimental data.[40,45,72] The theory predicts stronger dependencies of devulcanization on processing parameters than observed experimentally. Those disagreements are caused by the incapability of the cavitation collapse theory to describe devulcanization at high ambient pressure when no collapse occurs. Thus, the experimental data indirectly show that acoustic cavitation collapse plays a minor role in the degradation of vulcanized rubbers under ultrasonic treatment.

In the meantime, the experimental observations provide no arguments to reject the very idea about acoustic cavitation as the basic process in ultrasonic devulcanization. It is well known that, in contrast to plastics, rubber chains break down only when they are fully stretched.[96,97] Internal cavities are the sites in rubber where large displacements, and consequently highly stretched conformations of rubber chains, may take place. Voids in rubbers, for example, serve as crack initiation sites under dynamic fatigue tests.[95,103] Similarly, ultrasonic degradation of rubber should also occur primarily around pulsating bubbles due to the highest level of stress and strain pro-

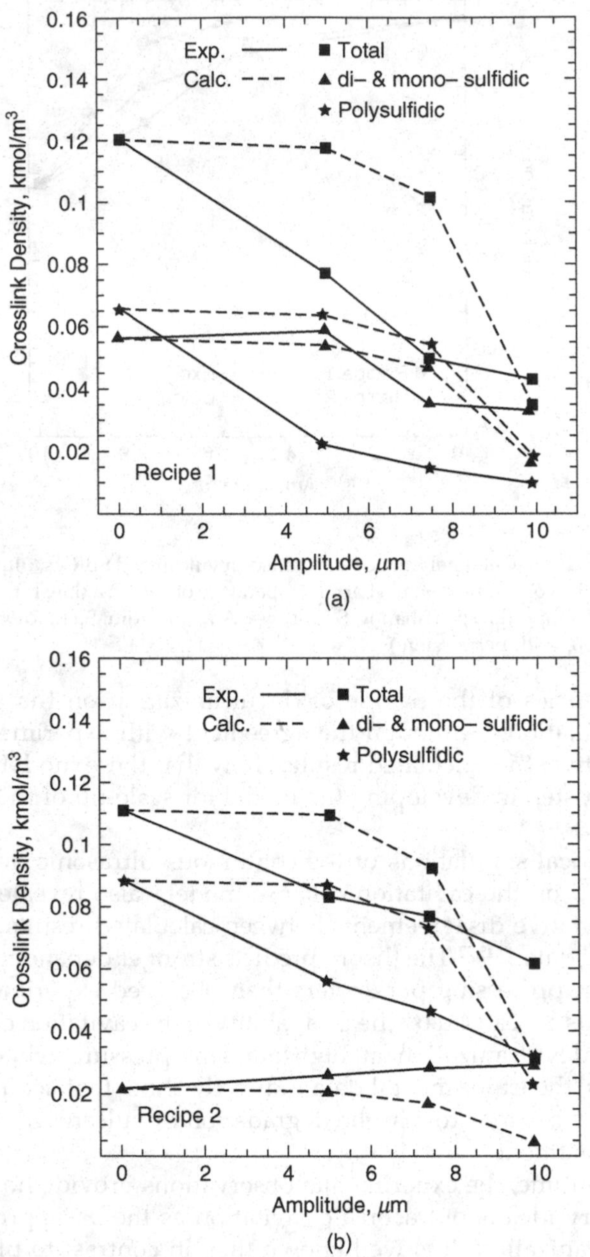

FIGURE 9.38
Experimental and calculated cross-link density of unfilled devulcanized SBR vs. ultrasonic amplitude obtained using the coaxial reactor at a barrel temperature of 120°C, a flow rate of 0.63 g/second, and a gap of 2.54 mm. (From Yushanov, S.P., Isayev A.I., and Kim, S.H., *Rubber Chem. Technol.*, 71, 168–190, 1998. With permission.)

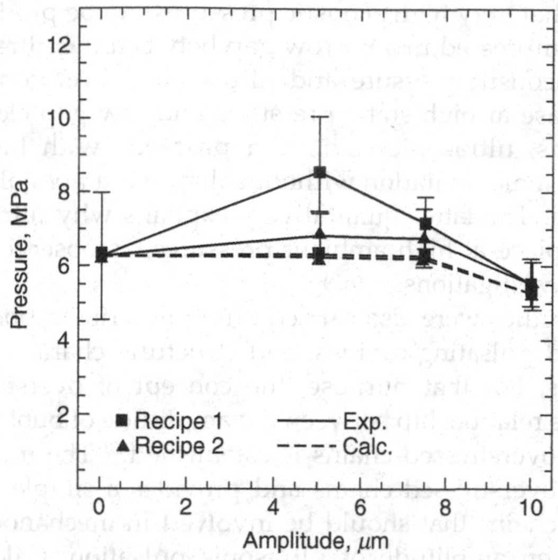

FIGURE 9.39
Experimental and calculated pressure vs. ultrasonic amplitude during devulcanization of unfilled SBR obtained using the coaxial reactor at a barrel temperature of 120°C, a flow rate of 0.63 g/second, and a gap of 2.54 mm. (From Yushanov, S.P., Isayev A.I., and Kim, S.H., *Rubber Chem. Technol.*, 71, 168–190, 1998. With permission.)

duced in this area. Cavitation collapse is obviously not a necessary condition for devulcanization.

9.11.2 Cavitation Based on Viscoelastic Model

The theoretical approach[36] based on the elastic model of rubber has been developed further using a viscoelastic model.[49] In order to do so, the effects of rubber viscoelasticity have been incorporated into the description of the dynamics of cavitation. Standard linear viscoelastic solids and Rouse viscoelastic solids were used for modeling rubber viscoelasticity in numerical simulations. The simulations demonstrated that viscoelasticity tremendously affects the dynamics of cavitation, acting to reduce the amplitude of oscillations and to decelerate the violent collapse contractions inherent in the pure elastic solid.[36] In the meantime, significant differences were observed between cavity pulsation in the standard viscoelastic solid and in the Rouse solid. No collapse-like effects were revealed during cavitation in the Rouse solid at low ambient pressures. High-amplitude stable oscillations around the dynamically swollen state were observed. It could be concluded that cavitation collapse is not the primary mechanism of ultrasonic devulcanization. Network degradation around pulsating cavities should be considered instead. The degradation is possible if acoustic pressure is high enough to cause high levels of strain during cavitation. A simple model of an ultrasonic reactor was used to estimate the acoustic pressure. The analysis demonstrated that very high acoustic pressures can be produced in a cured rubber layer compressed in a narrow gap between the ultrasonic horn and die. Both the acoustic pressure and ultrasonic power consumption were shown to increase at high static pressures and low gap clearances. Under these conditions, ultrasonic cavitation proceeds with high amplitudes. Therefore, ultrasonic cavitation without collapse is a possible mechanism of devulcanization. The latter qualitatively explains why ultrasonic devulcanization takes place at high ambient pressures, as observed in our earlier experimental investigations.

Theoretical studies were also carried out to describe rubber network degradation around pulsating cavities, and structural characteristics of devulcanized rubbers. For that purpose, the concept of overstressed chains is applied, and the relationship between the amplitude of bubble pulsation and the number of overstressed chains is established.[50] The model is based on the concept of overstressed chains and provides a simple tool to estimate the number of chains that should be involved in mechanochemical degradation at a given amplitude of ultrasonic pulsation. Calculations of the fraction of overstressed chains around pulsating cavities have indicated that cavitation shows a long-range effect on the rubber network. In particular, this results in a logarithmic dependence of the total fraction of overstressed chains on the concentration of the cavities. A simple estimation has been obtained demonstrating that high-amplitude cavitation with no collapse-like effects is capable of significant reduction of the rubber cross-link density. It has been also demonstrated that rubber degradation considerably affects elastic properties of rubber around a pulsating cavity, inducing certain irreversible effects including residual deformations.

The Dobson-Gordon theory of rubber network statistics has been employed to interpret the experimental data of the dependence of gel fraction in devulcanized rubbers on cross-link density in the gel. It has been demonstrated that the assumption of spatial-temporal randomness of rubber degradation allows one to achieve a fairly good agreement between experimental and theoretical data. Figure 9.40 presents the experimental data and the results of calculations for unfilled SBR, EPDM, and PDMS vulcanizates ultrasonically treated under the continuous (with flow) conditions.[104] It is seen that the simplest model based on random ruptures of chains and cross-links allows us to describe all of the shown data by fitting the parameter k_p/k_α for each kind of rubber, with k_p and k_α being the rate constants for the rupture of main chains and cross-links, respectively. Figure 9.40 also shows the values of the parameter found by the least-square method. The obtained values of the k_p/k_α ratio reveal that the rupture of rubber main chains is most significant in SBR and least significant in PDMS. These data correlate well with thermal stability of SBR, EPDM, and silicone polymer chains; PDMS is the most stable rubber and SBR is the least thermally stable rubber among them.

There are two distinct limiting cases in the considered model: $k_p = 0$ when only cross-links rupture, and $k_\alpha = 0$ when only polymer chains are subject

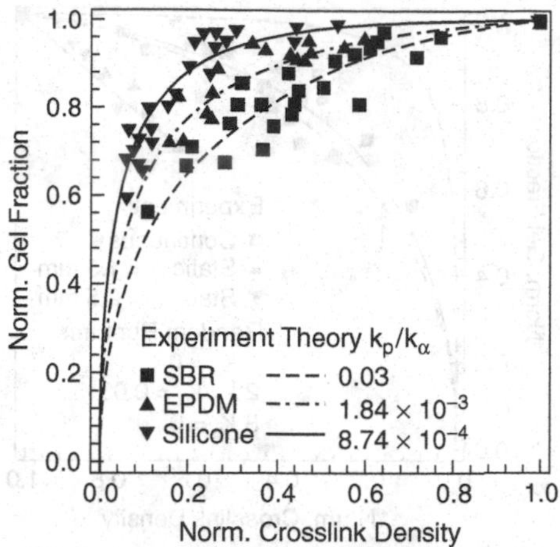

FIGURE 9.40
The experimental data and the results of simulations of structure of unfilled SBR, EPDM, and silicone rubber ultrasonically treated in the coaxial reactor. Symbols: experimental data. Curves: simulations based on a random rupture of chains and cross-links. (From Yashin, V.V., Isayev, A.I., Kim, S.H., Hong, C.K., Shim, S.E., and Yun, J., *SPE ANTEC Technical Papers*, 49, 2485–2489, 2003. With permission.)

to scission. The sets of experimental data on all rubbers treated in the continuous ultrasonic process lie in between the curves representing these two limiting cases, with values of k_p/k_α being different for different rubbers and/or for continuous ultrasonic reactors of different designs.

In the meantime, the experimental data on the treatment of SBR in a static ultrasonic device (no flow) does not follow this rule. Surprisingly, the obtained data points form a curve that lies outside the area bounded by the curves corresponding to $k_p = 0$ and $k_\alpha = 0$ as shown in Figure 9.41.[104] Namely, the experimental points are located lower than the curve with $k_\alpha = 0$. In other words, statically treated SBR samples exhibit lower gel fractions at a given cross-link density in gel than can be explained by the model based on random scission of polymer chains and cross-links alone. Some other process or processes must also be involved. It has been shown[105] that radical depropagation, which is triggered by a significant temperature buildup in the static device, may be responsible for this behavior. Depropagation acts toward elimination of dangling chains with radicalized ends created by main-chain rupture, thus resulting in a lower weight of insoluble fraction at the same concentration of elastically active rubber chains, i.e., at the same cross-link density.

Figure 9.42 demonstrates the effect of filler on the structure of EPDM vulcanizates ultrasonically treated under the continuous conditions.[104] It is seen that filler promotes rupture of polymer chains; the ratio k_p/k_α is greater for filled EPDM rubbers than for the unfilled vulcanizates, and it increases with an increase in the filler content ϕ. This effect is supposedly due to a decrease in mobility of EPDM chains caused by certain physical-chemical

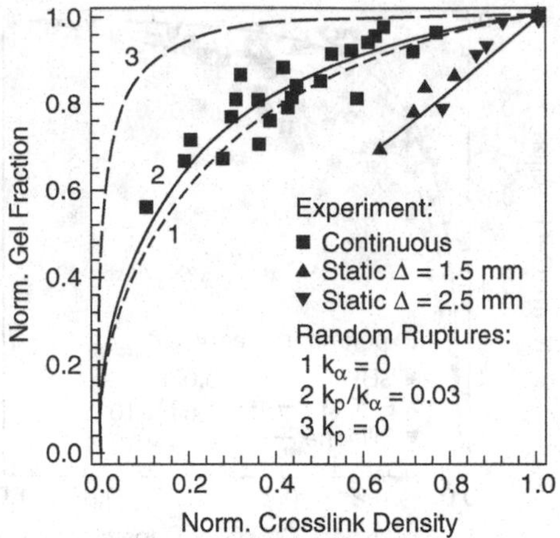

FIGURE 9.41
Gel fraction vs. cross-link density for unfilled SBR ultrasonically treated under the continuous (coaxial reactor) and the static conditions. Symbols: experimental data. Curves: simulations. Simulations of the static treatment take into account radical depropagation of broken chains. (From Yashin, V.V., Isayev, A.I., Kim, S.H., Hong, C.K., Shim, S.E., and Yun, J., *SPE ANTEC Technical Papers*, 49, 2485–2489, 2003. With permission.)

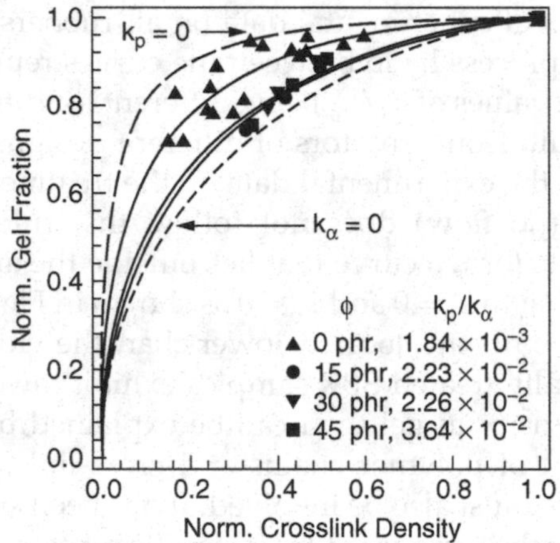

FIGURE 9.42
Gel fraction vs. cross-link density for the unfilled and CB-filled EPDM rubbers ultrasonically treated in the coaxial reactor. Symbols: experimental data. Curves: simulations based on a random rupture of chains and cross-links. (From Yashin, V.V., Isayev, A.I., Kim, S.H., Hong, C.K., Shim, S.E., and Yun, J., *SPE ANTEC Technical Papers*, 49, 2485–2489, 2003. With permission.)

interactions with the particles of CB.Also note that the data points for various filler contents ϕ in Figure 9.43 could seem to be described by a single value of k_p/k_α. However, this is not the case, since cross-link density of the virgin-filled vulcanizates depends on the filler content. A careful analysis of the data allowed us to show that the ratio k_p/k_α increases with an increase in ϕ as is demonstrated in Figure 9.42.

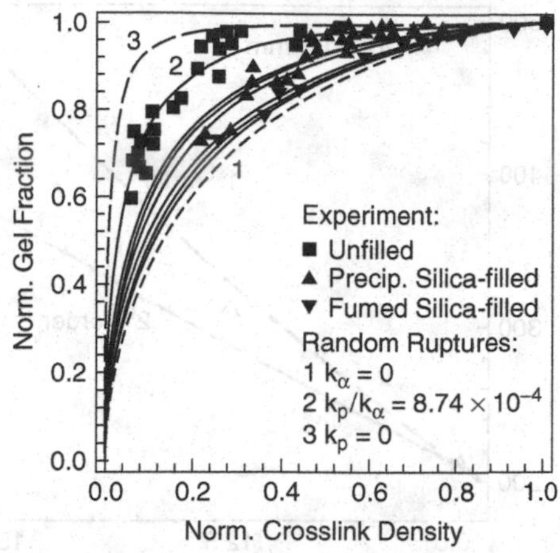

FIGURE 9.43
Gel fraction vs. cross-link density for the unfilled, precipitated silica–filled, and fumed silica–filled silicone rubbers ultrasonically treated in the coaxial reactor. Symbols: experimental data. Curves: simulations based on a random rupture of chains and cross-links. (From Yashin, V.V., Isayev, A.I., Kim, S.H., Hong, C.K., Shim, S.E., and Yun, J., *SPE ANTEC Technical Papers*, 49, 2485–2489, 2003. With permission.)

The effect of rubber-filler interactions on the structure of ultrasonically treated rubbers is further illustrated in Figure 9.43.[104] The latter shows the data on unfilled, precipitated silica–filled, and fumed silica–filled PDMS ultrasonically treated in the continuous reactor. Similar to what was observed in filled EPDM, silica-filled PDMS exhibits more polymer chain degradation during the treatment. As seen from Figure 9.43, fumed silica causes more polymer degradation than precipitated silica at the same degree of filling. Fumed silica is known to provide a better reinforcement to PDMS than precipitated silica. Fumed silica causes a greater reduction of mobility of polymer chains, which might enhance their rupture.

This simple approach involves only one adjustable parameter describing the ratio of rates at which rupture of main chain bonds and cross-link bonds take place. Application of the approach to devulcanized SBR and PDMS indicated that the rate of cross-link rupture is higher than that of main chain. A simple, qualitative explanation of this observation has been given, based on differences in their bond energies.

9.11.3 Ultrasonic Power Consumption

An attempt was also made to predict ultrasonic power consumption in the course of ultrasonic treatment of rubbers under the static conditions. Figure 9.44 presents the results of the calculation of power consumption at various ultrasonic amplitudes, A (a) and sample thickness, Δ_0 (b).[105] The numerically obtained data shown by the curves are compared with the values of ultrasonic power consumption measured at $t = 0$, which are denoted by symbols.

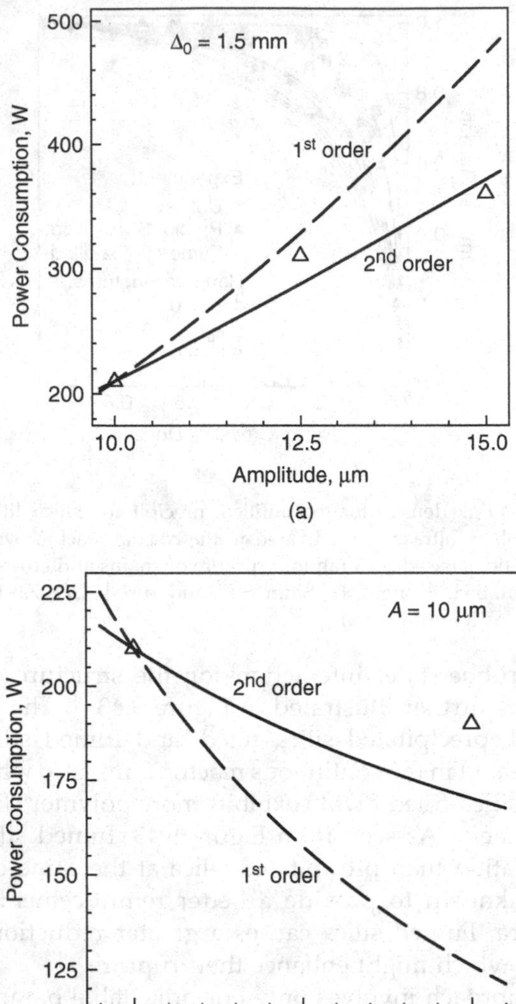

FIGURE 9.44
Ultrasonic power consumption as a function of amplitude in an unfilled SBR sample (a) with thickness of 1.5 mm, and (b) with thickness at an amplitude of 10 μm. Symbols: experimental data obtained under the static conditions. Dashed lines: linear approximation; solid lines: second-order approximation. (From Yashin, V.V., Hong, C.K., and Isayev, A.I., *Rubber Chem. Technol.*, 77, 50–77, 2004. With permission.)

The initial values of power consumption are chosen for comparison because they represent the samples of ground rubber before degradation. The volume fraction of voids, ϕ_0, is obtained from the condition that the calculated and measured values of power consumption coincide at A = 10 μm and Δ_0 = 1.5 mm. It should be noticed that the linear and second-order approximations for power consumption provide different values for ϕ_0. Namely, the first-order and second-order approximations give the values of $\phi_0 - 9.90 \times 10^{-3}$ and$\phi_0 - 1.24 \times 10^{-3}$, respectively. The corresponding values of ϕ_0 are used to calculate ultrasonic power consumption with and without nonlinear corrections, as shown in Figure 9.44 by the dashed and solid lines, respectively.

Figure 9.44 reveals that the model is qualitatively correct in the prediction of power consumption as a function of amplitude and sample thickness. It is seen, however, that the results of calculations based on the linear approximation exhibit steeper dependencies of power consumption on those parameters than the experimentally observed dependencies. Figure 9.44a also demonstrates that taking into account the nonlinear effects considerably improves the ability of the model to predict power consumption at various ultrasonic amplitudes. A noticeable improvement is also seen in Figure 9.44b, which represents power consumption as a function of sample thickness. There are not enough experimental data points in Figure 9.44, however, to draw a final conclusion.

To describe variations of power consumption in the course of ultrasonic treatment, the changes in dynamic properties of rubber due to degradation should be considered. Figure 9.45 shows the results of simulations of ultrasonic power consumption during the ultrasonic treatment of SBR at an amplitude of 10 μm and a sample thickness of 1.5 mm, assuming that the rubber properties do not change.[105] The calculations take into account the decrease in static pressure with exposure time, observed in the course of the experiment. Figure 9.45 indicates that the experimentally observed significant drop in power consumption could not be attributed just to a decrease

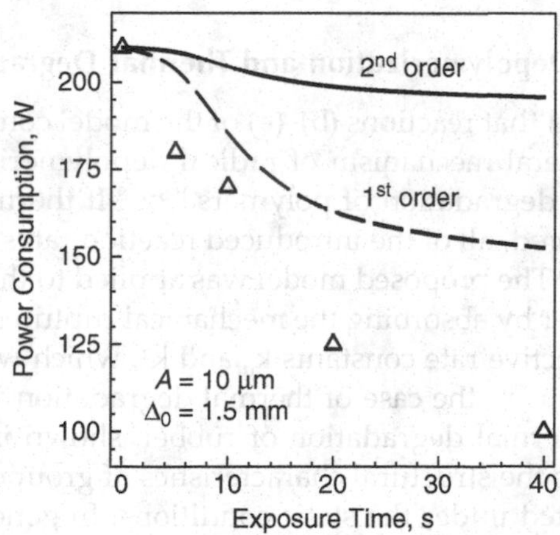

FIGURE 9.45
The results of simulations of ultrasonic power consumption vs. exposure time in the course of ultrasonic treatment of the unfilled SBR sample under the static conditions. The simulations take into account the decrease in static pressure during the experiment. The changes in dynamic properties of rubber due to degradation are not considered. Symbols: experimental data. Dashed line: linear approximation. Solid line: second-order approximation. (From Yashin, V.V., Hong, C.K., and Isayev, A.I., *Rubber Chem. Technol.*, 77, 50–77, 2004. With permission.)

in static pressure. The results of the calculations, accounting for the nonlinear effects, exhibit a minor decrease in power consumption under the reported pressure decay (solid line). The observed difference in Figure 9.45 between the results of simulations with and without consideration of the nonlinear effects is caused by the different pressure sensitivities of the radius of the voids predicted by these two approximations. Temperature buildup in the samples was simulated by considering the heat transfer, which included a heat-source determined by a time-dependent power consumption $W_{exp}(t)$ obtained from the experimental data. Figures 9.44 and 9.45 show power consumption $W_{exp}(t)$ for SBR samples of 1.5 mm thickness.

A model describing the contribution of thermal degradation of vulcanized rubbers under ultrasound treatment was also proposed.[105] Various reactions involved in this model are schematically shown in Figure 9.46. The model assumes that cross-linking does not alter the mechanism of degradation of polymer chains, and includes the following elementary reactions: (a) random scission of cross-links between macromolecules with the reaction rate constant k_α; (b) random scission of main chains with the reaction rate constant k_p that results in a formation of two shorter chains with radicalized ends; (c) depropagation reaction with the rate constant k_d in the course of which one monomeric unit is detached from an end-radicalized chain; (d) radical transfer reaction with the rate constant k_{tr} leading to a main chain scission with a transfer of the radical to one of the newly formed chain ends; and (e) first-order radical termination with the rate constant k_t. In Figure 9.46, the star-filled circles depict the location of radicals, and the indexes denote the number of monomeric units in chains before and after a correspondent degradation reaction.

9.11.4 Radical Depolymerization and Thermal Degradation

It should be noticed that reactions (b)–(e) of the model constitute a simplified version of the general mechanism of radical depolymerization that is well known in thermal degradation of polymers.[106–108] If thermomechanical degradation is considered, all of the introduced reaction rates are affected by the applied stress.[109,110] The proposed model was applied to the ultrasonic devulcanization of rubber by absorbing the mechanical ruptures of cross-links and chains into the effective rate constants k_α and k_p, which will not have a pure chemical meaning as in the case of thermal degradation without stress.

The model of thermal degradation of rubber, shown in Figure 9.46, was applied to describe the structural characteristics of ground SBR vulcanizates ultrasonically treated under the static conditions. In general, all of the processes shown in Figure 9.46 are thermally activated, with the rate constants being described by the well-known Arrhenius equation

$$k(T) = k_0 \exp\left(-\frac{E_a}{k_B T}\right) \tag{9.5}$$

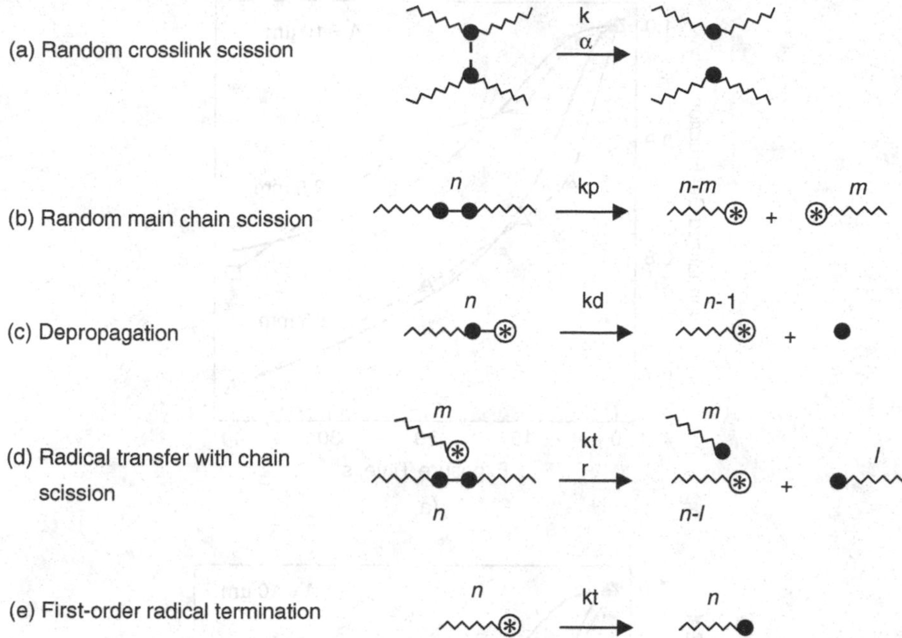

FIGURE 9.46
The model mechanism of thermal degradation of SBR. The star-filled circles: radicals. The indexes denote the number of monomeric units in chains before and after a degradation reaction.

with certain pre-exponential coefficient k_0 and activation energy E_a; k_B stands for the Boltzmann constant. Application of stress is known to significantly affect reactions in polymers.[109] According to the mechanochemistry of rubbers,[111] the first elementary act of hydrogen abstraction from the molecular chain up to the point of failure of rubber, the sequence, rate, and character of reactions are changed by application of stress. The latter makes identification of parameters of reactions affected by a mechanical action to be extremely difficult. Therefore, this problem has not been considered, since the primary goal was to investigate thermal degradation reactions; more specifically, radical depropagation might be responsible for the differences observed in structure of SBR treated under the continuous and the static conditions (see Figure 9.41).

The primary breakups of rubber chains and cross-links are initiated solely by the action of ultrasound and, therefore, the rate constants k_p and k_α do not depend on temperature T. It was also assumed that in the static mode of the ultrasonic treatment, their ratio is the same as that found for the continuous one, i.e., $k_p/k_\alpha = 0.03$. For the sake of simplicity, the temperature dependence of the termination rate constant k_t is also neglected.In the meantime, the depropagation and radical transfer rate constants k_d and k_{tr} are considered temperature dependent, so they retain the Arrhenius form.

Figure 9.47 and Figure 9.48 present the results of illustrative calculations, which indicate that the experimental data can be described by the proposed

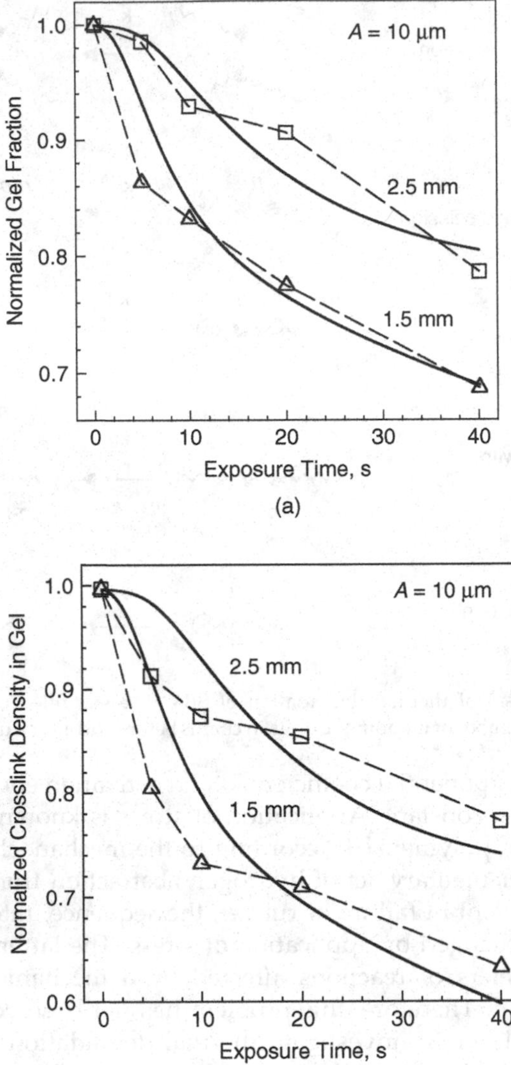

FIGURE 9.47
Normalized gel fraction (a) and normalized cross-link density in gel (b) in unfilled SBR samples with thickness of 1.5 and 2.5 mm, ultrasonically treated under the static conditions vs. exposure time. Symbols: experimental data. Dashed lines: connect experimental data for better visual clarity. Solid lines: the results of simulations. (From Yashin, V.V., Hong, C.K., and Isayev, A.I., *Rubber Chem. Technol.*, 77, 50–77, 2004. With permission.)

model based on thermal degradation of the rubber network using a certain set of reaction parameters.[105] Figure 9.47 shows the decrease in gel fractions (Figure 9.47a) and cross-link densities in gel (Figure 9.47b) of treated samples as a function of the time of exposure to ultrasound. The experimental data denoted by symbols are normalized and then connected by straight dashed

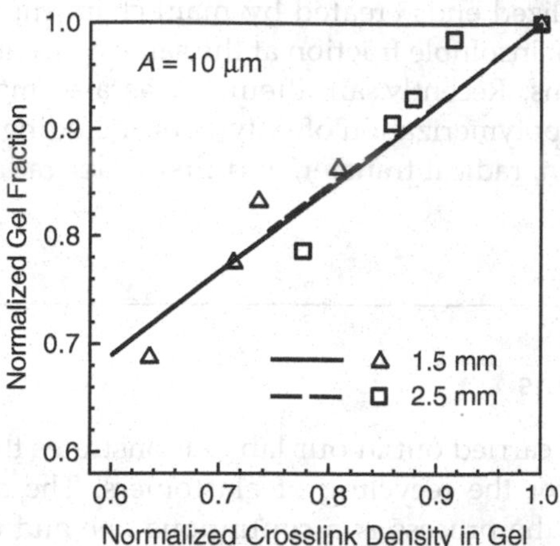

FIGURE 9.48
Normalized gel fraction vs. normalized cross-link density in gel for unfilled SBR samples ultrasonically treated under the static conditions. Symbols: experimental data; lines: simulations. (From Yashin, V.V., Hong, C.K. and Isayev, A.I., *Rubber Chem. Technol.*, 77, 50–77, 2004. With permission.)

lines for a better visual clarity. The solid lines represent the gel fraction $g(z,t)$ and cross-link density in gel $\xi(z,t)$ obtained by simulations, after averaging over sample thickness and normalization on their initial values.

The performed simulations of degradation of SBR samples of thickness 1.5 and 2.5 mm reveal that the experimental data on samples of these thicknesses can be described simultaneously only if the rate constant of chain scission, k_p, is taken to be greater for thinner samples. As the rate constant of cross-link scission, k_α, is bound to k_p by the fixed ratio k_p/k_α, k_α is also greater for thinner samples. This observation is not surprising since these two scission processes have been assumed to occur primarily due to the mechanical action of ultrasound. This action is more pronounced at higher strain amplitudes or, equivalently, at lower thickness at a given amplitude of ultrasound. Note that the Arrhenius parameters of other reactions are always taken to be the same.

In the experiments, the observed steep decrease in gel fractions and cross-link densities after exposure to ultrasound during 5 to 10 seconds is interpreted by the model as a result of acceleration of degradation by the radical depropagation and radical transfer processes triggered by temperature buildup.

Finally, Figure 9.48 demonstrates that the proposed model successfully describes the experimental data on the dependency of normalized gel fraction on normalized cross-link density in gel presented in Figure 9.40. The decrease in gel fraction at a given cross-link density in the statically treated samples in comparison with those treated under the continuous conditions can be explained by radical depropagation that occurs at high temperatures in the static ultrasonic device. Depropagation acts toward elimination of dangling chains with radicalized ends created by main-chain rupture, thus resulting in a lower weight of insoluble fraction at the same concentration of elastically active rubber chains. Recently, an attempt was also made to theoretically describe radical depolymerization of polymers, including random chain scission, depropagation, radical transfer, and first-order radical termination.[112]

9.12 Conclusions

The research work carried out in our lab demonstrates that ultrasound is an effective method for the recycling of elastomers. The advantage of using ultrasound is that the process is a continuous one and occurs in the order of seconds and does not require the use of any chemicals. The devulcanized rubber is soft and reprocessable and hence, meets the primary objective of recycling, i.e., reusing waste materials that would otherwise cause tremendous environmental pollution. Significant challenges exist in making this process viable to the industrial recycling of used tire and waste rubbers. Tire recycling requires a very high output of ultrasonic reactors (above 500 lbs/hour). At the present time, this challenge, due to the inability of the ultrasonic generators to deliver high power continuously, can be overcome by using multiple ultrasonic devices in one machine. However, the process is viable for use in recycling of factory rubber waste that does not require high output. Also, it is plausible that the rubber waste in a typical factory is of a known recipe and, therefore, it is easy to incorporate the recycled rubber into a similar compound. Obtained experimental results show that each rubber, due to its different molecular structure and compounding recipe, behaves differently during ultrasonic devulcanization. So far, information on ultrasonic devulcanization of only a few different types of rubbers has been obtained. Further work should be directed toward the study of various existing rubbers to find optimal conditions for their devulcanization. Moreover, the models of ultrasonic devulcanization that have been put forward so far use simplified approaches to very complex phenomena taking place during devulcanization. Therefore, significant challenges exist in the development of a suitable theory for the description of the devulcanization process, due to the complexity of the processes occurring simultaneously during ultrasonic devulcanization.

Acknowledgments

The financial support for this investigation has been provided by the grant DMI–0084740 from the National Science Foundation. The authors wish to express their appreciation to Professor E. D. von Meerwall and former graduate students and associates Drs. J. Chen, C. K. Hong, V. Yu. Levin, S. H. Kim, J. S. Oh, S. E. Shim, A. Tukachinsky, V. V. Yashin, J. S. Yun, S. P. Yushanov, and Mr. B. Diao, Mr. M. Tapale, and Mr. D. Schworm, whose contributions during the last decade made this work possible.

References

1. http://www.rma.org/scraptires/facts_figures.html
2. Klingensmith, W. and Baranwal, K., Recycling of rubbers, in *Handbook of Elastomers*, 2nd edition, Bhowmick, A. K. and Stephens H. L., eds., Marcel Dekker, New York, 835, 2001.
3. Schaefer, R. and Isringhaus, R.A., Reclaimed rubber, in *Rubber Technology*, 3rd edition;, Morton, M., ed., Van Nostrand Reinhold Company, New York, 505, 1987.
4. Nicholas, P.P., The scission of polysulfide crosslinks in scrap rubber particles through phase transfer catalysis, *Rubber Chem. Technol.*, 55, 1499–1515, 1982.
5. Klingensmith, B., Recycling, production and use of reprocessed rubbers, *Rubber World*, 203 (6), 16–21, 1991.
6. Phadke, A.A. and De, S.K., Vulcanization of cryo-ground reclaimed rubber, *Kautsch. Gummi Kusntst.*, 37, 776–779, 1984.
7. Fix, S.R., Microwave devulcanization of rubber, *Elastomerics*, 112 (6), 38–40, 1980.
8. Tsuchii, A., Suzuki, T., and Takeda, K., Microbial degradation of natural rubber vulcanizates, *Appl. Environ Microbiol*, 50, 965–970, 1985.
9. Bilgili, E., Arastoopour, H., and Bernstein, B., Analysis of Rubber Particles Produced by the Solid State Shear Extrusion Pulverization Process, *Rubber Chem. Technol.*, 73, 340–355, 2000.
10. Khait, K. and Torkelson, J.M., Solid-state shear pulverization of plastics: a green recycling process, *Polym. Plast. Technol. Eng.*, 38, 445–457, 1999.
11. Warner, W.C., Methods of Devulcanization, *Rubber Chem. Technol.*, 67, 559–566, 1994.
12. Isayev, A.I., Rubber recycling, *Rubber Technologist's Handbook*, White, R. and De, S.K., eds., Rapra Technology Ltd., UK, 2001.
13. Adhikari, B., De, D., and Maiti, S., Reclamation and recycling of waste rubber, *Prog. Polym. Sci.*, 25, 909–948, 2000.
14. Basedow, A.M. and Ebert, K., Ultrasonic degradation of polymers in solution, *Adv. Polym. Sci.*, 22, 83–148, 1977.
15. Price, G.J., *Advances in Sonochemistry*, 1, 231, 1990.
16. Price, G.J., Norris, D.J., and West, P.J., Polymerization of methyl methacrylate initiated by ultrasound, *Macromolecules*, 25, 6447–6454, 1992.
17. Schmid, G. and Rommel, O., The disintegration of macromolecules by ultrasonic waves, *Z. Elektrochem.*, 45, 659–661, 1939.
18. Schmid, G., The disruption of macromolecules: studies of the depolymerizing action of supersonic waves, *Physik. Z.*, 41, 326–337, 1940.
19. Jellinek, H.H.G. and White, G., The degradation of long-chain molecules by ultrasonic waves. III. Dependence of rate constant on chain length for polystyrene, *J. Polym. Sci.*, 7, 21–32, 1951.

20. Isayev, A.I., Wong, C., and Zeng, X., Flow of thermoplastics in dies with oscillating boundary, *SPE ANTEC Tech. Papers,* 33, 207–210, 1987.
21. Isayev, A.I., Wong, C., and Zeng, X., Effect of oscillations during extrusion on rheology and mechanical properties of polymers, *Adv. Polym. Technol.,* 10 (1), 31, 1990.
22. Isayev, A.I., *Proceedings,* 23rd Israel Conference on Mechanical Engineering, Technion, Haifa, 1990, Paper #5.2.3.
23. Isayev, A.I. and Mandelbaum, S., Effect of ultrasonic waves on foam extrusion, *Polym. Eng. Sci.,* 31, 1051–1056, 1991.
24. Peshkovsky, S.L., Friedman, M.L., Tukachinsky, A.I., Vinogradov, G.V., and Enikolopian, N.S., Acoustic cavitation and its effect on flow in polymers and filled systems, *Polym. Compos.,* 4, 126–134, 1983.
25. Garcia Ramirez, R. and Isayev, A.I., The effect of mobile boundary on extrusion of thermotropic LCPs, *SPE Technical Papers,* 37, 1084–1087, 1991.
26. Suslick, K.S., The chemical effects of ultrasound, *Sci. Am.,* 260, 80–86, 1989.
27. Suslick, K.S., Doktycz, S.J., and Flint, E.B., On the origin of sonoluminescence and sonochemistry, *Ultrasonics,* 28, 280–290, 1990.
28. Gooberman, G., Ultrasonic degradation of polystyrene. I. A proposed mechanism for degradation, *J. Polym. Sci.,* 42, 25–33, 1960.
29. Mangaraj, D. and Senapati, N., Ultrasonic vulcanization, U.S. Patent #4,548,771 (1986).
30. Okuda, M. and Hatano, Y., Devulcanization of rubber by ultrasonic waves, Japanese Patent Application #62,121,741, 1987.
31. Isayev, A.I., inventor, University of Akron, assignee, 1993, Continuous ultrasonic devulcanization of vulcanized elastomers, U.S. Patent #5,258,413.
32. Isayev, A.I. and Chen, J., inventors, University of Akron, assignee, 1994, "Continuous ultrasonic devulcanization of vulcanized elastomers and apparatus therefore, U.S. Patent #5,284,625.
33. Isayev, A.I., Chen, J., and Tukachinsky, A., Novel ultrasonic technology for devulcanization of waste rubbers, *Rubber Chem. Technol.,* 68, 267–280, 1995.
34. Tukachinsky, A., Schworm, D., and Isayev, A.I., Devulcanization of waste tire rubber by powerful ultrasound, *Rubber Chem. Technol.,* 69, 92–103, 1996.
35. Yu. Levin, V., Kim, S.H., Isayev, A.I., Massey, J., and von Meerwall, E., Ultrasound devulcanization of sulfur vulcanized SBR: crosslink density and molecular mobility, *Rubber Chem. Technol.,* 69, 104–114, 1996.
36. Isayev, A.I., Yushanov, S.P., and Chen, J., Ultrasonic devulcanization of rubber vulcanizates. I. Process model, *J. Appl. Polym. Sci.,* 59, 803–813, 1996.
37. Isayev, A.I., Yushanov, S.P., and Chen, J., Ultrasonic devulcanization of rubber vulcanizates. II. Simulation and experiment, *J. Appl. Polym. Sci.,* 59, 815–824, 1996.
38. Isayev, A.I., Yushanov, S.P., Schworm, D., and Tukachinsky, A., Modeling of ultrasonic devulcanization of tire rubbers and comparison with experiments, *Plast. Rubber and Compos. Process. Appl.,* 25, 1–12, 1996.
39. Yushanov, S.P., Isayev, A.I., and Levin, V.Y., Percolation simulation of the network degradation during ultrasonic devulcanization, *J. Polym. Sci.:* Part B: Polym. Phys., 34, 2409–2418, 1996.
40. Isayev, A.I., Yushanov, S.P., Kim, S.H., and Yu. Levin, V., Ultrasonic devulcanization of waste rubbers: experimentation and modeling, *Rheol. Acta,* 35, 616–630, 1996.
41. Yu. Levin, V., Kim, S.H., and Isayev, A.I., Vulcanization of ultrasonically devulcanized SBR elastomers, *Rubber Chem. Technol.,* 70, 120–128, 1997.

42. Johnston, S.T., Massey, J., von Meerwall, E., Kim, S.H., Yu. Levin, V., and Isayev, A.I., Ultrasound devulcanization of SBR: molecular mobility of gel and sol, *Rubber Chem. Technol.*, 70, 183–193, 1997.
43. Isayev, A.I., Kim, S.H., and Yu. Levin, V., Superior mechanical properties of reclaimed SBR with bimodal network, *Rubber Chem. Technol.*, 70, 194–201, 1997.
44. Yu. Levin, V., Kim, S.H., and Isayev, A.I., Effect of crosslink type on the ultrasound devulcanization of SBR vulcanizates, *Rubber Chem. Technol.*, 70, 641–649, 1997.
45. Yushanov, S.P., Isayev A.I., and Kim, S.H., Ultrasonic devulcanization of SBR rubber: experimentation and modeling based on cavitation and percolation theories, *Rubber Chem. Technol.*, 71, 168–190, 1998.
46. Diao, B., Isayev, A.I., Yu. Levin, V., and Kim, S.H., Surface behavior of blends of SBR with ultrasonically devulcanized silicone rubber, *J. Appl. Polym. Sci.*, 69, 2691–2696, 1998.
47. Tapale, M. and Isayev, A.I., Continuous ultrasonic devulcanization of unfilled NR vulcanizates, *J. Appl. Polym. Sci.*, 70, 2007–2019, 1998.
48. Diao, B., Isayev, A.I., and Yu. Levin, V., Basic study of continuous ultrasonic devulcanization of unfilled silicone rubber, *Rubber Chem. Technol.*, 72, 152–164, 1999.
49. Yashin, V.V. and Isayev, A.I., A model for rubber degradation under ultrasonic treatment: Part I. Acoustic cavitation in viscoelastic solid, *Rubber Chem. Technol.*, 72, 741–757, 1999.
50. Yashin, V.V. and Isayev, A.I., A model for rubber degradation under ultrasonic treatment: Part II. Rupture of rubber network and comparison with experiments, *Rubber Chem. Technol.*, 73, 325–339, 2000.
51. Hong, C.K. and Isayev, A.I., Continuous ultrasonic devulcanization of carbon black-filled NR vulcanizates, *J. Appl. Polym. Sci.*, 79, 2340–2348, 2001.
52. Shim, S.E. and Isayev, A.I., Ultrasonic devulcanization of precipitated silica-filled silicone rubber, *Rubber Chem. Technol.*, 74, 303–316, 2001.
53. Yun, J., Oh, J.S., and Isayev, A.I., Ultrasonic devulcanization reactors for recycling of GRT: comparative study, *Rubber Chem. Technol.*, 74, 317–330, 2001.
54. Isayev, A.I., Rubber recycling, in: *Rubber Technologist's Handbook,* White, J.R., and De, S.K., eds., RAPRA Technology Ltd., U. K, 511–547, 2001.
55. Isayev, A.I., Recycling of elastomers, in: *Encyclopedia of Materials: Science and Technology,* Buschow, K.H.J., ed., Elsevier, Amsterdam, (3), 2474–2477, 2001.
56. Hong, C.K. and Isayev, A.I., Continuous ultrasonic devulcanization of NR/SBR blends, *Rubber Chem. Technol.*, 75, 617–625, 2002.
57. Hong, C.K. and Isayev, A.I., Ultrasonic devulcanization of unfilled SBR under static and continuous conditions, *Rubber Chem. Technol.*, 75, 133–142, 2002.
58. Yun, J. and Isayev, A.I., Superior mechanical properties of ultrasonically recycled EPDM rubber, *Rubber Chem. Technol.*, 76, 253–270, 2003.
59. Shim, S.E., Ghose, S., and Isayev, A.I., Formation of bubbles during ultrasonic treatment of cured polydimethylsiloxane, *Polymer,* 43, 5535–5543, 2002.
60. Shim, S.E., Parr, J.C., von Meerwall, E., and Isayev, A.I., NMR relaxation and pulsed gradient NMR diffusion measurements of ultrasonically devulcanized poly(dimethylsiloxane), *J. Phys. Chem.* B, 106, 12072–12078, 2002.
61. Shim, S.E., Isayev, A.I., and von Meerwall, E., Molecular mobility of ultrasonically devulcanized silica-filled poly(dimethyl siloxane), *J. Polym. Sci.*, Part B: Polym. Phys., 41, 454–465, 2003.
62. Yun, J., Isayev, A.I., Kim, S.H., and Tapale, M., Comparative analysis of ultrasonically devulcanized unfilled SBR, NR and EPDM rubbers, *J. Appl. Polym. Sci.*, 88, 434–441, 2003.

63. Shim, S.E. and Isayev, A.I., Effects of the presence of water on ultrasonic devulcanization of polydimethylsiloxane, *J. Appl. Polym. Sci.*, 88, 2630–2638, 2003.
64. Ghose, S. and Isayev, A.I., Recycling of unfilled polyurethane rubber using high power ultrasound, *J. Appl. Polym. Sci.*, 88, 980–989, 2003.
65. Yun, J. and Isayev, A.I., Recycling of roofing membrane rubber by ultrasonic devulcanization, *Polym. Eng. Sci.*, 43, 809–821, 2003.
66. Boron, T., Roberson, P., and Klingensmith, W., Ultrasonic devulcanization of tire compounds, *Tire Technol. Int. '96*, 82–84, 1996.
67. Boron, T., Roberson, P., Forest, C., and Shringarpurey, Applied research on ultrasonic devulcanization of crumb rubber, 156th Meeting of the ACS Rubber Division, Orlando, Florida, Sep. 21–24, 1999, Paper No. 136.
68. Gonzalez de Los Santas, E.A., Sorieno-Corral, F., Lozano Gonzalez, Ma. J., and Cedillo-Garcia, R., Devulcanization of guayule rubber by ultrasound, *Rubber Chem. Technol.*, 72, 854–861, 1999.
69. Oh, J.S. and Isayev, A.I. Continuous ultrasonic devulcanization of unfilled butadiene rubber, J. *Appl. Polym. Sci.*, 93, 1166–1174, 2004.
70. Yun, J., Recycling of waste tires and EPDM rubber using high power ultrasonic devulcanization, Ph.D. dissertation, University of Akron, 2003.
71. Shim, S.E., Ultrasound devulcanization of silica filled silicone rubber: processing, property and molecular aspects, Ph.D. dissertation, University of Akron, 2001.
72. Kim, S.H., Continuous ultrasonic devulcanization of sulfur cured SBR vulcanizates, Ph.D. dissertation, University of Akron, 1998.
73. Makarov, V.M. and Drozdovski, V.F., *Reprocessing of Tires and Rubber Wastes*, Ellis Horwood, New York, chapter 2, 1991.
74. Vinogradov, G.V. and Malkin, A. Ya, *Rheology of Polymers*, Mir Publishers, (city), 116, 1980.
75. Diao, B., Basic study of continuous ultrasonic devulcanization of unfilled silicone rubber, M.S. thesis, University of Akron, 1997.
76. Oh, J.S., Isayev, A.I., and von Meerwall, E.D., Molecular mobility in ultrasonically devulcanized unfilled butadiene rubber, *Rubber Chem. Technol.*, 77, 745–758, 2004.
77. Basheer, R. and Dole, M. Kinetics of free radical decay reactions in butadiene-styrene block copolymers, *Radiat. Phys. Chem.*, 18, 1053–1060, 1981.
78. Oh, J.S., Applications of powerful ultrasound to thermoplastic elastomers and rubbers, Ph.D. dissertation, University of Akron, 2003.
79. Ghose, S., Isayev, A.I., and von Meerwall, E.D., Effect of ultrasound on thermoset polyurethane: NMR relaxation and diffusion measurements, *Polymer*, 45, 3709–3720, 2004.
80. Mark, J.E. and Erman, B., *Rubber Like Elasticity: A Molecular Primer*, John Wiley & Sons, New York, 1988.
81. Yun, J., Yashin, V.V., and Isayev, A.I., Ultrasonic devulcanization of carbon black filled ethylene propylene diene monomer rubber, *J. Appl. Polym. Sci.*, 91, 1646–1656, 2004.
82. Rivil, D., Surface properties of carbon, *Rubber Chem. Technol.*, 44, 307–343, 1971.
83. Stickney, P.B. and Falb, R.D., Carbon black-rubber interactions and bound rubber, *Rubber Chem. Technol.*, 37, 1299–1340, 1964.
84. Wolf, S. and Wang, M.J., in: *Carbon Black: Science and Technology*, Donnet, J.B., Bansal, R.C., Wang, M.J., eds., New York, Dekker, chapter 9, 1993.

85. Hong, C.K. and Isayev, A.I., Blends of ultrasonically devulcanized and virgin carbon black filled NR, *J. Mat. Sci.*, 37, 385–388, 2002.
86. Clarson, S.J. and Semlyen, J.A., *Siloxane Polymers*, Prentice Hall, New Jersey, 1993.
87. Caprino, J.C. and Macander, R.F., in: *Rubber Technology*, Morton, M., ed., Van Nostrand Reinhold, New York, pp. 375–409, 1987.
88. Ghose, S. and Isayev, A.I., Improved properties of blends of ultrasonically treated unfilled polyurethane rubber, *Polym. Eng. Sci.*, 44, 794–804, 2004.
89. Mason, T.J., *Sonochemistry*, Oxford University Press, New York, 1999.
90. Suslick, K.S., *Ultrasound: Its Chemical, Physical and Biological Effects*, VCH, New York, 1988.
91. Zhurkov, S.N., Zakrevskii, V.A., Korsukov, V.S., and Kuksenko, V.S., Mechanism of submicrocrack generation in stressed polymers, *J. Polym. Sci., Polym. Phys. Ed.*, 10, 1509–1520, 1972.
92. Gent, A.N., Cavitation rubber: a cautionary tale, *Rubber Chem. Technol.*, 63 (3), G 49–G53, 1990.
93. Bever, M.D., ed., *Encyclopedia of Materials Science and Engineering*, Pergamon Press, Oxford, (4), 2934, 1986.
94. Young, F.R., *Cavitation*, chapter 2, McGraw-Hill Co., London, 1989.
95. Kasner, A.I. and Meinecke, E.A., Porosity in rubber: a review, *Rubber Chem. Technol.*, 69, 424–443, 1996.
96. Kinloch, A.J. and Young, R.J., *Fracture Behavior of Polymers*, Applied Science Publishers, London, 1983.
97. Kausch, H.H., *Polymer Fracture*, Springer-Verlag, Berlin, 1987.
98. Tabata, M., Miyazawa, T., Kobayashi, O., and Sohma, J., Direct evidence of main-chain scissions induced by ultrasonic irradiation of benzene solutions of polymers, *Chem. Phys. Lett.*, 73 (1), 178–180, 1980.
99. Tabata, M. and Sohma, J., Spin trapping studies of poly(methyl methacrylate) degradation in solution, *Eur. Polym. J.*, 16, 589–595, 1980.
100. Lorimer, J.P., in *Chemistry with Ultrasound*, chapter 4, Mason, T.J., ed., SCI, New York, 1990.
101. Isayev, A.I., Chen, J., and Yushanov, S.P., *Simulation of Materials Processing: Theory, Methods and Applications*, Shen, S.F. and Dawson, P., eds., Balkema, Rotterdam, 77, 1995.
102. Gent, A.N. and Thomas, A.C., Forms for the stored energy function for vulcanized rubber, *J. Polym. Sci.*, 28, 625–628, 1958.
103. Kasner, A.I. and Meinecke, E.A., Effect of porosity on dynamic mechanical properties of rubber in compression, *Rubber Chem. Technol.*, 69, 223–233, 1996.
104. Yashin, V.V., Isayev, A.I., Kim, S.H., Hong, C.K., Shim, S.E., and Yun, J., Degradation of rubber network during the ultrasonic treatment, *SPE ANTEC Technical Papers*, 49, 2485–2489, 2003.
105. Yashin, V.V., Hong, C.K., and Isayev, A.I., Thermomechanical degradation of SBR during ultrasonic treatment under static condition, *Rubber Chem. Technol.*, 77, 50–77, 2004.
106. Reich, L. and Stivala, S.S., *Elements of Polymer Degradation*, McGraw-Hill, New York, 1971.
107. Madorsky, S.L., Thermal degradation of organic polymers, Interscience Publishers, New York, 1964.
108. Jellinek, H.H.G., Depolymerization, in: *Encyclopedia of Polymer Science and Technology: Plastics, Resins, Rubbers, Fibers*, Mark, H.F., Gaylord, N.G., Bikales, N.M., eds., Interscience Publishers, New York, (4), 740, 1964.

109. Casale, A. and Porter, R.S., Polymer stress reactions, Academic Press, New York, 1978.
110. Desai, A., The effect of accelerated heat aging on the strength and crosslink density of styrene-butadiene rubber, M.S. thesis, University of Akron, 1986.
111. Kuzmiski, A.Z., in *Developments in Polymer Stabilization*, Scott, G., ed., Applied Science Publishers, London, (4), 71, 1981.
112. Yashin, V.V. and Isayev, A.I., On the theory of radical depolymerization: a rigorous solution, *J. Polym. Sci.*, 41, 965–982, 2003.

10

Devulcanization by Chemical and Thermomechanical Means

Marvin Myhre

CONTENTS

10.1 Introduction

The reason for devulcanizing rubber or even discussing the devulcanization of rubber is the enormous stockpile of tires. Tires are piled up in fields, filling landfills, or used as reefs in the ocean offshore. As space becomes increasingly less available, uses for scrap rubber become increasingly more necessary. To be usable, the scrap rubber must be changed to a crumb so that a material with increased plasticity and an altered surface can be processed on rubber equipment and a new surface can be cross-linked to itself or virgin material. The unusual aspect of rubber reuse is that it has been pushed by environmental concerns and politics, rather than for scientific purposes.

Much work has been carried out and published on the mechanisms of vulcanization, but little on the mechanisms of devulcanization, which is the focus of this chapter.

A major difficulty in the regeneration of scrap rubber is maintaining the property levels of the virgin material when the regenerated rubber is used alone or is blended with virgin compound. The property levels are almost invariably reduced, or at least lowered. Another major problem is the change in compound viscosity, scorch, and cure characteristics, which for the most part, are negatively affected.

Ever since Charles Goodyear's patent[1] on vulcanization of rubber, efforts have been made to break cross-links that are formed during vulcanization, to obtain material the same as or similar to unvulcanized virgin compound. Many processes have been developed, used, and commercialized. All have produced a material of increased plasticity and have been identified as reclaim. There have been numerous publications[2–10] and patents[12–16] on the subject of reclaiming of scrap rubber. Reclaiming processes have used toxic chemicals, needed high heat, and caused environmental problems. A major negative, besides the environmental problems, has been that depolymerization of the polymer present in the material occurred during these processes. Since most everything was and is carried out in the presence of air, depolymerization is difficult to avoid. Since the early 1960s, few studies have been carried out on reclaiming or on producing products. Since the properties have always been reduced in the presence of the reclaim, the cost of overcoming the environmental problems alone have not been justifiable. Rubber materials are designed to last and are, therefore, difficult to reuse and to devulcanize.

When processes again began to emerge in the 1980s, they were different from early processes in that they were designed to be usable in a fabrication facility, they were basically dry processes (nonaqueous), or heat-only processes. The degree of devulcanization (breaking of cross-links) was, for the most part, less than when rubber was reclaimed. It is the author's belief that, unless complete or a high degree of devulcanization can be achieved without breaking or modifying the main chain, there are fewer uses for the rubber product. However, the best alternative is to regenerate the surface while maintaining the integrity of the rubber particles as much as possible.

10.2 Devulcanization — Definition

For the purpose of this chapter, the definition of devulcanization is as follows: Devulcanization of a sulfur-cured rubber is the process of cleaving, either totally or partially, of the sulfur cross-links formed during the vulcanization process. These cross-links can be poly-, di-, or monosulfidic in nature. It would be preferable for depolymerization (main chain cleavage) and main chain modification to be minimized during the process.

ASTM[17] currently defines vulcanization as an irreversible process during which a rubber compound, through a change in its chemical structure (e.g., cross-linking) becomes less plastic while the elastic properties are improved, and extended, over a greater temperature range. There is no definition of devulcanization. Earlier, in ASTM ST P 184A,[18] devulcanization was defined as a combination of depolymerization, oxidation, and increased plasticity. Over the years, many processes have been developed, and some commercialized, to allow for the reuse of vulcanized rubber. Almost all that have been commercialized have been reclaiming processes.

10.3 Early Methods of Devulcanization

As mentioned in the introduction, early methods of devulcanization were described as reclaim, since the intent was simply to reuse the material.[2–10] The original purpose of reclaiming was to break down the rubber and increase plasticization, in order to reuse the rubber by employing the normal rubber-processing equipment to manufacture new products. However, the problem with reclaimed rubber was that, since the chemicals and/or heat used for the process did not discriminate, all bonds, including carbon-carbon, were attacked, so a lot of main chain scission occurred. Because serious reduction in physical property levels always accompanies main chain scission, reclaimed rubber acted much like a process aid. Probably there was still some degree of cross-linking present, a degree of branching, and a reduced molecular weight. As shown by E.W.B. Owen,[2] using acetone and chloroform to extract the reclaimed material indicates attacks by oxygen, and subsequent molecular weight breakdown takes place. Therefore, in the reclaiming process, much main chain scission occurs due to free radicals producing hydroperoxides by the decomposition of oxygen. Methyl substituents attached to the carbon atom, connected by the double bond, enhance the hydroperoxidation at the alpha-methylenic carbon atom. Styrene-butadiene rubber (SBR) does not contain a substituent group; therefore, the hydroperoxidation proceeds at a different rate. Because oxygen is always present, hydroperoxides can and do form. The rate of attack, by the hydroperoxides, is different for different polymers. Therefore, if two or three different polymers are present, there could be polymer phases that would be at different levels of reclaiming, which would cause processing problems and quite likely have a negative effect on property values. This contributes to the difficulties encountered during reclaiming of, for example, a blend of natural rubber (NR) and SBR. There have been a number of good reviews on the reclaiming of rubber and the mechanisms involved during the early and mid part of the twentieth century.[19–21] Those mechanisms all involved high heat (or steam) only, heat and reclaiming agents, or high heat and alkali. These processes have also been recently reviewed.[22–25]

The major reclaiming processes are the "thermal," heater/pan, and alkali digester processes, as described in the following sections.[2,3,21] These worked well if the polymer was NR, but not if it was synthetic.

10.3.1 Thermal Process[2–4,7]

This is the oldest process and is carried out by heating rubber pieces in a pan, in an autoclave, with steam at 60 psi. The rubber is open to the steam and air. The full process is carried out over a period of 3 to 4 hours; the temperature is 260°C for 1 hour, after which the rubber mass is cooled with water, removed, and dried, then massed on a two-roll mill, producing sheeted rubber reclaim.

10.3.2 Heater or Pan Process[2–4,7]

This process is carried out using a single-shell steam autoclave at pressure of 100 to 300 psi. Reclaiming, charring, and wetting agents as well as plasticizer/oils are used. As an example, 2% ammonium persulfate or a 20% aqueous solution containing a wetting agent serving as a charring agent is mixed with the rubber pieces. Paraffinic oil (5% by weight) containing 2% each of coconut oil, fatty acid, and naphthenic oil is mixed with the waste. The material is heated in covered, -4-inch-deep pans in the autoclave for 3 hours and 15 minutes at 150 lbs. of steam, including a 15-minute warm-up. After the heating, the material is cooled, dried, and massed on a two-roll mill with 10% of a high-boiling, aromatic petroleum distillate, and strained through a 40 mesh screen.

10.3.3 Digester[2–4,7]

This process is similar to the heater process, but is carried out in a double-jacketed autoclave. The rubber waste pieces are submerged in an aqueous, caustic (sodium hydroxide) solution. Because the rubber is under water, the process is more or less done in the absence of air, other than what is already dissolved in the water. After the steam treatment, which is carried out at 150 psi steam pressure for 10 to 15 hours, the material is washed, dried, ground, massed, strained, and sheeted on a rubber mill. This was again fine, as long as the polymer was NR. However, if it was synthetic, then a metal chloride solution instead of a caustic soda solution had to be used to destroy the fibers.

Reclaiming has also been carried out using a continuous process called The Reclaimator Process,[7] which uses an extruder (cylinder and screw). Many patents have been granted for this process. A Banbury® was used in a process called the Lancaster-Banbury process.[7] It was also patented. This is a high-speed, high-temperature, high-steam Banbury,® mixing the ground rubber with oils, plasticizers, and powders.

The major effect on the pan, heater, and digester reclaiming processes is the surrounding medium/atmosphere. An open steam environment provides the greatest amount of oxidized products, and an alkaline medium, the least. An acid medium is somewhat in between and results in a slow, rather uniform rate of molecular breakdown. It was determined that only a very small amount of oxygen is necessary to accomplish this.

For synthetic rubber, reclaiming agents of the disulfide type were used early in the game, in particular for SBR, and are still employed today for processes described as devulcanization. Early on, pan heater or pan steam processes were used in the presence of the disulfides and swelling agents. The latter were used to help the reclaiming agents pass into the rubber more quickly. Materials that were, and still are, active reclaiming agents and catalysts, can be classified into several distinct groups[26]: (1) phenol alkyl sulfides and disulfides, (2) mercaptans, both aliphatic and aromatic, (3) amino compounds, and (4) unsaturated compounds. Numerous patents have been issued on each. Patent numbers are listed in this article.[26] Of these, most of the work has been carried out on the first group (sulfides and disulfides), which are the most effective reclaiming agents. Although these do not remove sulfur from the vulcanizates, they probably catalyze the oxidative breakdown of the polymer chain as well as the oxidative disruption of sulfur cross-links, thus restoring plasticity to the vulcanizate. In the presence of oxygen, their activity toward SBR is exceptionally good.

Mercaptans are not overly effective, in particular aliphatic mercaptans. Aromatic mercaptans are more effective, especially if the phenyl group is replaced with a naphthalene group. Amino compounds are effective, and their reactivity does not depend on oxygen. These agents are particularly effective for neoprenes and nitriles because of their extraordinary capacity to retard the heat hardening so predominant in these polymers.

The unsaturated compounds are for the most part indenes, dicyclopentadienes, coumarones, petroleum residues, and terpenes. All these act as plasticizers or softening agents. These probably work on the basis of swelling the material to the point of rupturing bonds.

Many of the processing advantages associated with reclaim are an indication of the structure of reclaim. Faster processing, greater tolerance for temperature, less heat buildup, better dimensional stability under heat, less shrinkage, reduced swelling, greater green strength, absence of nerve, and faster mixing all indicate a certain degree of a three-dimensional network (partly cross-linked).

10.4 Later Methods of Devulcanization

The methods of devulcanization developed within the last decade or so are designed to be relatively dry processes that can be done quickly, in a fabrication facility. These processes can be carried out on normal rubber-processing equipment. The devulcanization, in many of the processes, is only regeneration of the surface or close to the surface. Because the user would probably want to utilize the inherent properties of the ground rubber particles when blending them with virgin material, the following devulcanization processes will be discussed: thermal, chemical, mechanical, chemomechanical, thermomechanical, thermochemical, and biological.

10.4.1 Thermal Devulcanization

What was considered in the early days to be a thermal process[2–4,7] actually used steam, so it was still a wet process. The process used high temperatures 180° to 260°C (and sometimes higher) for lengthy periods of time. A strictly thermal process does not work well with SBR, because SBR hardens with heat, compared to NR, which degrades.

In the late 1970s, microwave devulcanization came on the scene.[11] It was felt that if the microwave energy could be controlled closely enough so that only sulfur cross-links would be broken, it would be a viable process. But microwaves cause the material to heat up, so care had to be taken to ensure the rubber did not burn. This process was used for a time, but has fallen into disuse, mainly due to cost.

10.4.2 Chemical Devulcanization

What is discussed here is a chemical treatment at temperatures of 100°C or lower. S. Yamashita and N. Kawabata[27] described a process that regenerates vulcanized rubber by treating the material at room temperature and atmospheric processes for several hours with ferrous chloride and phenyl hydrazine. Since this process was open to the air, the authors acknowledged that hydroperoxides formed by the reaction of the oxygen with the ferrous chloride–phenol hydrazine system could cause main chain scission.

A process for chemical devulcanization was patented in 1979[28] and then published in 1982.[29] This process used phase transfer catalysis to carry out the actual devulcanization at temperatures under 100°C for a period of 2 hours. According to the author, most of the cross-links were broken with minimal main chain scission. Because the reaction was carried out at the reflux temperature of the solvent, it was probably done in the absence, or minimal presence of, air.

The process of alkali reclaiming used alkali (NaOH) to do the work. The hydroxyl anion (OH^-) was the entity that actually attacked and broke the cross-links. Based on that, the phase transfer catalysis was designed to do the same thing. Using an OH^-, thereby abstracting a proton and forming water, made the process safe in that the disposal of dangerous by-products was not required.[29] If the OH^- anion could be used to selectively break the S–S bond without breaking the main chain, two reactions could take place. First, the disulfides could be converted to thiolate and sulfinates, as shown in Scheme 1.

$$2RS-SR + 2OH^- \rightleftharpoons 2RSOH + 2RS^-$$

$$2RSOH \rightleftharpoons R-\overset{\overset{\displaystyle O}{\|}}{S}-SR + H_2O$$

$$R-\overset{\overset{\displaystyle O}{\|}}{S}-S-R + 2OH^- \rightleftharpoons RSO_2^- + 2RS^- + H_2O$$

$$\overline{}$$

$$2RS\text{–}SR + 4OH^- \rightleftharpoons RSO_2^- + 3RS^- + 2H_2O$$

SCHEME 1
From Nicholas, P.P., *Rubb. Chem. Technol.*, 55, 1499, 1982. With permission.

The second possible step applies to the methylation of the monosulfide with methyl chloride.

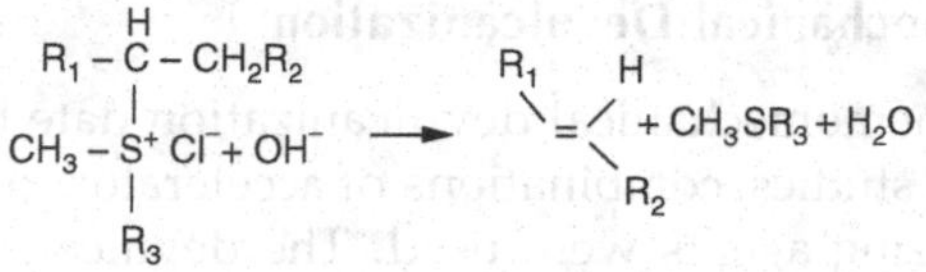

$$\begin{array}{c} H \\ R_1-\overset{|}{C}-CH_2R_2 \\ | \\ CH_3-S^+\,Cl^- + OH^- \\ | \\ R_3 \end{array} \longrightarrow \begin{array}{c} R_1 \quad\; H \\ \diagdown{=}\diagup \\ \diagdown \\ R_2 \end{array} + CH_3SR_3 + H_2O$$

SCHEME 2
From Nicholas, P.P., *Rubb. Chem. Technol.*, 55, 1499, 1982. With permission.

If it is at first methylated to methyl sulfonium chloride, then a beta elimination to an olefin and methyl sulfide may be the preferred reaction step.

When this process was compared to a phenyl lithium reaction with vulcanized rubber, which is known to selectively break polysulfide cross-links without breaking main chains, there is good agreement of the molecular weight distributions. Therefore, there may be a similar selectivity. In their reactions, Nicholas observed (Scheme 3) exclusive methyl transfer from the quaternary ammonium ion and no detectable R transfer.

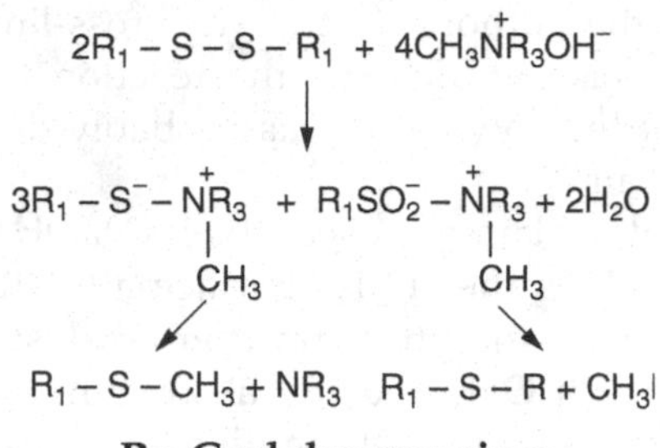

$$2R_1-S-S-R_1 + 4CH_3\overset{+}{N}R_3OH^-$$

$$\downarrow$$

$$3R_1-S^--\overset{+}{N}R_3(CH_3) + R_1SO_2^--\overset{+}{N}R_3(CH_3) + 2H_2O$$

$$R_1-S-CH_3 + NR_3 \qquad R_1-S-R + CH_3I$$

R_1: Cyclohexene ring

SCHEME 3
From Nicholas, P.P., *Rubb. Chem. Technol.*, 55, 1499, 1982. With permission.

The third treatment discussed here was presented at the Rubber Division, American Chemical Society (ACS) October meeting in Cleveland in 1995.[30] This process is carried out in a vessel equipped with mechanical stirring, with very little or no shear. At the beginning the temperature is ambient, but as the reaction proceeds it becomes exothermic, and therefore requires cooling.

10.4.3 Mechanical Devulcanization

In this section, only one process is discussed; it is performed at ambient temperatures, so heat is not a factor. This process is patented[31] and uses a rubber slurry (solvent swollen polymer) forced through screens of decreasing mesh size that are made from various metal alloys. These screens act as catalysts to reduce the cross-link density. Substantial devulcanization is claimed. A description of this process was published in 1998.[32]

10.4.4 Chemomechanical Devulcanization

Publications on chemomechanical devulcanization date back to World War II.[33,34] In these two studies, combinations of accelerator, zinc oxide, softening agents, and peptizing agents were used. The devulcanization was carried out on two-roll rubber mills at relatively low temperatures (< 80°C). Some of these basic rubber chemicals appear to attack the sulfur-sulfur bonds in combination with shear, reducing the cross-link density and increasing the plasticity. Since the process takes place on a two-roll mill in the presence of air, there would, in all likelihood, be a reduction in polymer molecular weight. This reduction in molecular weight would also increase plasticity. This was not discussed in the publications. Relatively good properties were obtained (see Table 10.1).

Some 30 or more years later, in 1979, work was published[35] on devulcanization using mechanical shear (i.e., a mill). The length of time of the process was quite long at 30 minutes; temperature was not discussed. Thiols and disulfides were used. Besides main chain breakdown, cross-links were opened and the viscosity was reduced (increased plasticity). This is more like reclaim than straight devulcanization because of the level of polymer chain breakdown.

In 1980, another patent[36] was issued for a process that used many chemicals in virtually the same manner, namely shear (mills and extruders), to chemomechanically devulcanize. Chemicals suggested were basically accelerators. As with the previous references, this was done in the presence of air; therefore, main chain scission could occur.

Between 1995 and 2000, STIK Polymers American published and heavily promoted their patented process for the devulcanization of scrap rubber.[37] Their process was also chemomechanical and was carried out on a refining mill, but it could also be done in an internal mixer. The company's Delink® comprised a number of basic rubber chemicals, including accelerators. If one considers what was published by F.B. Menadue,[33] and D.F. Twiss[34] in the 1940s, then STIK and some of the other processes are quite similar in the chemicals used and the manner in which they were carried out. What was different about the STIK process was the use of a devulcanizing masterbatch, and there was a specific ratio of one chemical to the other in the masterbatch. What STIK suggested was that through mechanical shear, in the presence of certain chemicals, cross-links were uncoupled, as shown below.

$$\{\text{—Sx—}\} + \text{Delink} \xrightarrow[<70°C]{\text{Shear}} \{\text{—Sx\_N} + \text{SN—}\}$$

SCHEME 4
From STIK Polymers America promotional material, 1995. With permission.

removing the sulfur and blocking the severed cross-links, then allowing the cross-links to reform with a lower sulfur rank at elevated temperature,[37] as shown in Scheme 5.

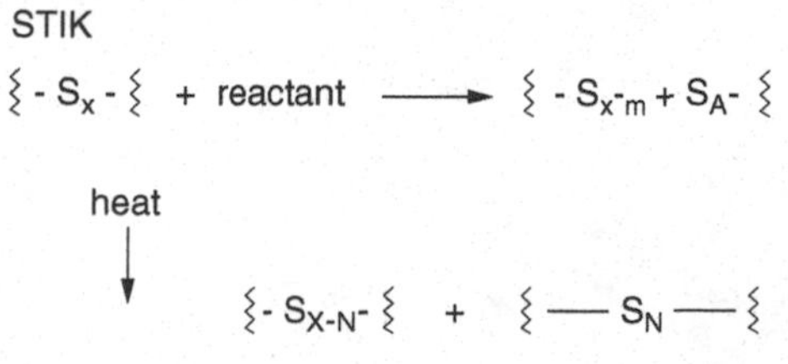

STIK

$$\wr\text{-}S_x\text{-}\wr + \text{reactant} \longrightarrow \wr\text{-}S_{x^-m} + S_A\text{-}\wr$$

$$\text{heat}\downarrow \quad \wr\text{-}S_{X\text{-}N}\text{-}\wr + \wr\text{—}S_N\text{—}\wr$$

SCHEME 5
From STIK Polymers America promotional material, 1995. With permission.

If this is so, then there could be free sulfur (from the original cross-links) present, and a higher cross-link density of the revulcanized scrap.

In their recent article, D. De et al.[38,39] suggest that when devulcanizing with disulfides (a frequent occurrence), heat and mechanical energy do two things. First, they split the disulfide into free radicals, as shown below.

$$\text{R-S-S-R} \xrightarrow{e} 2\text{RS}^{\cdot}$$

$$\xrightarrow{e} \text{RSS}\cdot + \text{R}^{\cdot}$$

SCHEME 6
From De, D., Adhikari, B., Maiti, S., *J. Polym. Mat.*, 333, 1997. With permission.

Second, the rubber does the same thing, so the main chain is severed into smaller units.

For example:

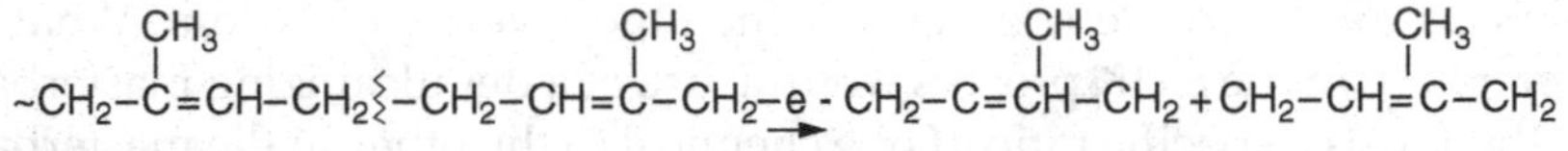

$$\sim\text{CH}_2\text{-C}(\text{CH}_3)\text{=CH-CH}_2 \wr \text{-CH}_2\text{-CH=C}(\text{CH}_3)\text{-CH}_2\text{-} \xrightarrow{e} \text{CH}_2\text{-C}(\text{CH}_3)\text{=CH-CH}_2 + \text{CH}_2\text{-CH=C}(\text{CH}_3)\text{-CH}_2$$

SCHEME 7
From De, D., Adhikari, B., Maiti, S., *J. Polym. Mat.*, 333, 1997. With permission.

and cross-link scission:

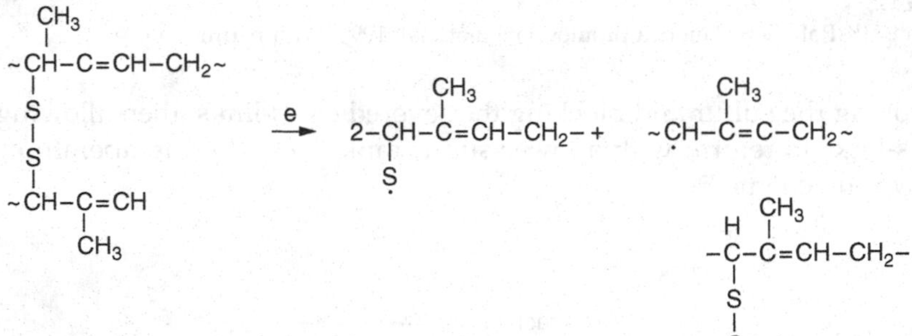

$$\begin{array}{c} \sim\text{CH}(\text{-C}(\text{CH}_3)\text{=CH-CH}_2\sim) \\ | \\ \text{S} \\ | \\ \text{S} \\ | \\ \sim\text{CH-C}(\text{CH}_3)\text{=CH} \end{array} \xrightarrow{e} 2\text{-CH}(\text{S}\cdot)\text{-C}(\text{CH}_3)\text{=CH-CH}_2\text{-} + \sim\dot{\text{C}}\text{H-C}(\text{CH}_3)\text{=C-CH}_2\sim \quad \text{-C(H)}(\text{S-S}\cdot)\text{-C}(\text{CH}_3)\text{=CH-CH}_2\text{-}$$

SCHEME 7
Continued.

With all the radicals that can form, many reactions (Scheme 8) with these radicals can also occur, such as:

Reaction 1:

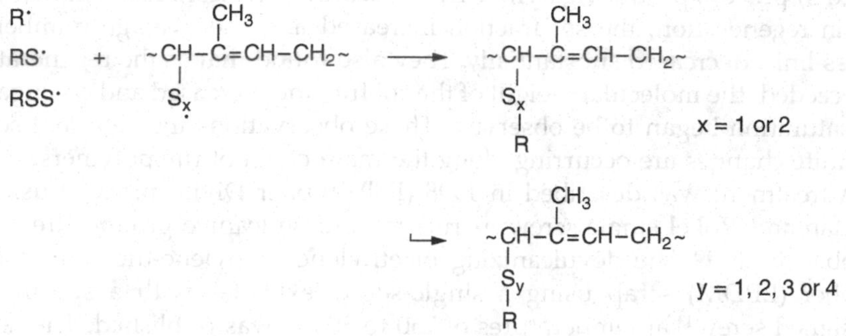

$$\left.\begin{matrix} R^{\cdot} \\ RS^{\cdot} \\ RSS^{\cdot} \end{matrix}\right. + \sim\!CH-\underset{\underset{\dot{S}_x}{|}}{\overset{\overset{CH_3}{|}}{C}}\!=\!CH-CH_2\!\sim \;\longrightarrow\; \sim\!CH-\overset{\overset{CH_3}{|}}{C}\!=\!CH-CH_2\!\sim \text{ (CH bearing } S_x\text{–R)} \quad x = 1 \text{ or } 2$$

$$\hookrightarrow \sim\!CH-\overset{\overset{CH_3}{|}}{C}\!=\!CH-CH_2\!\sim \text{ (CH bearing } S_y\text{–R)} \quad y = 1, 2, 3 \text{ or } 4$$

SCHEME 8
From De, D., Adhikari, B., Maiti, S., *J. Polym. Mat.*, 333, 1997. With permission.

Reaction 2:

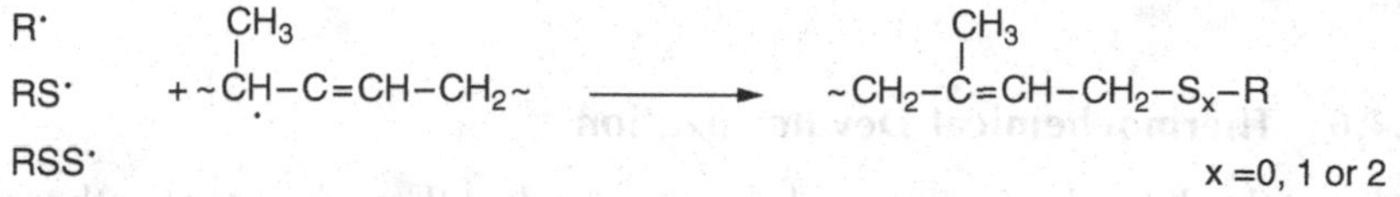

$$\left.\begin{matrix} R^{\cdot} \\ RS^{\cdot} \\ RSS^{\cdot} \end{matrix}\right. + \sim\!\dot{C}H-\overset{\overset{CH_3}{|}}{C}\!=\!CH-CH_2\!\sim \;\longrightarrow\; \sim\!CH_2-\overset{\overset{CH_3}{|}}{C}\!=\!CH-CH_2-S_x-R \quad x = 0, 1 \text{ or } 2$$

SCHEME 8
Continued.

When these reactions have taken place, the disulfide radicals can couple with the rubber radicals by acting as a peptizer, preventing recombination with itself and thereby maintaining the reduced molecular weight. The authors suggest that it is not the disulfide alone that is causing the polymer or cross-link scission. Because these processes are open to the air, the oxygen, in combination with the heat and mechanical shear, probably cause the scission of both the polymers and cross-links.

The unusual aspect of D. De's recent work[38, 39] is that the devulcanizing agent is from natural sources. Temperature is also involved, but is relatively low at 40 to 60°C. The natural material is not as efficient as the usual disulfides. More agent has to be used and the process time is rather lengthy, although it is shorter than the reclaim processes described earlier. However, it is probably more environmentally friendly and less toxic toward people.

10.4.5 Thermomechanical Devulcanization

A great deal of the recycled rubber crumb used, other than crumb itself, is obtained by thermomechanical treatment. No chemical agents are added;

the process involves just heat and/or shear. This kind of treatment has evolved into commercial processes such as National Rubber Industries' (NRI's) process producing Symar D.[40,41] NRI investigators have shown that upon regeneration, the sol fraction increased and the average number of cross-links decreased substantially. They also found that as the regeneration proceeded, the molecular weight of the sol fraction increased and conjugated unsaturation began to be observed. These observations indicate that some definite changes are occurring along the main chain of the polymers.

A treatment was described in 1998 (Fall Rubber Division, ACS) using a Rodan and Rokel single-screw extruder[42] to devulcanize ground tire scrap rubber. In 1999, the devulcanizing of ethylene-propylene-diene-monomer rubber (EPDM) scrap using a single-screw extruder with a specifically designed screw[43] at temperatures of 250 to 300°C was published. The negative aspect of thermomechanical regeneration of scrap rubber is that with all the shear that is used and the temperatures attained, there is much main chain scission that takes place. So there is considerable loss of physical properties when the recyclate is used by itself or when it is blended with virgin compound, even if the two are of the same formulation. As much as a 1% loss occurs per 1% recyclate added to the virgin material.

10.4.6 Thermochemical Devulcanization

These methods are basically a reclaim process, but they use agents other than alkali[38,39] and references therein. In the past, many used an autoclave, either steam or dry, or a kettle to carry out the process. Chemicals were used to soften or plasticize the crumb rubber. Temperatures ranged from 140 to 200°C. All were refined on a two-roll mill after the autoclaved treatment. The length of time was again long — up to 4 hours. How much of the reclaiming was due to the initial process and how much was due to the refining afterwards is not known. These processes were also rather messy and none seemed to have become commercially viable, at least not for long. All processes had to deal with the same problems; the chemicals used were highly toxic, so after the process was complete, disposal of the spent liquor or aqueous reclaiming solution was required, unless the process was a dry one.

10.5 Equipment

During the early times of reclaiming, almost everything was carried out in steam autoclaves so all that was necessary in terms of equipment was the autoclave and a mill for refining. Once removed from the autoclave, the rubber mass was then put through a refining process by using either a refining mill or a regular two-roll mill. Although the reclaiming was done under steam heat in the autoclave, there was a mechanical aspect to the process because of the refining under high shear in the presence of oxygen. It is quite likely that part of the actual reclaiming was due to this shear during the refining, which took place after the plasticization phase.

A special robust, high-shear Banbury® called the Lancaster-Banbury process was also used.[7] This process used high speeds and high heat (232 to 288°C), in conjunction with oil, plasticizer, and carbon black to reclaim the ground scrap in a short period of time. Refining mills (high shear) and two-roll mills were used in conjunction with plasticizing and devulcanizing agents to devulcanize ground rubber scrap. These processes, although relatively high shear, are carried out at a relatively low temperature compared to many of the other processes and equipment that were used.

One of the earliest continuous types of processes used was the Reclaimator process.[7] The Reclaimator was basically a cylinder containing a screw used to move the material through the tube, in the presence of chemicals, to produce shear between the wall and the screw. Temperatures ranged between 177 to 204°C at a residence time of 1 to 4 minutes.

Standard Banbury® mixers can also be used by first devulcanizing the scrap, then in the presence of virgin raw polymer, proceeding as if it were a normal mix by adding oils, fillers, etc.[37] The advantage of this approach would be the ability to do the devulcanizing and mixing in one step.

Most of the more modern processes use some type of two-roll mill or extruder, either alone or in conjunction with chemicals and heat. Modern reclaiming, with the use of different chemicals than those used earlier, is also carried out using not only steam autoclaves, but also dry autoclaves at high temperature in the presence of chemicals and air. The heat is either produced from the shear or it is external. Both single-screw and double-screw extruders have been used. Most of these, such as the Rodan and Rokel single-screw extruder,[42]use specially designed screws, with heated barrels, as opposed to a standard screw.[35,36,40,41] These types of extruders produce high shear at a relatively high screw speed and high barrel temperatures. These processes are designed for high throughput.

A different approach has recently been patented and described[31,32]; it uses screens made of metal alloys and iron and copper. Several screens of increasingly smaller mesh size are used through which a scrap rubber (polychloroprene and EPDM) slurry is pumped. This causes an apparent reduction in the cross-link density.

Microwave equipment[11] has also been used to devulcanize scrap rubber. Another approach has been to use an apparatus that is basically a drum equipped with a cooling jacket, and mechanical stirring (virtually no shear).[30] The material still has to be put through a mill if sheet material is desired or required.

The major problem with all these processes is not simply the high shear and high temperatures involved but rather, that the processes were carried out in the presence of air. Air causes major main chain scission and modification, which, in turn, cause a severe reduction in the physical property level of the material. Using equipment as described means that air will always be present. So radicals will be formed that attack not only the sulfur-sulfur cross-links, but also the polymer main chain; therefore, property values will always be reduced.

10.6 Materials and Chemicals

Much reclaiming of NR was carried out with only heat, but inorganic agents such as alkali were also used. Organic agents such as naphtha, aniline, reaction products of aryl-hydrazines with aldehydes and ketones, aromatic thiols, hydroxylamines, aliphatic amines and polyamines, and mercapto derivatives have been used. Aromatic thiols worked somewhat for SBR. Chemicals[4] such as beta-thionaphthiol, trichlorothiophenol, 9-thioanthrol (all sold under the name of Renacit) were used as reclaiming agents. However, the most effective chemicals are sulfides prepared from highly alkylated phenols. The most active sulfide is that prepared from 4,6–di-tert-butyl–3 methylphenol for synthetic polymer other than SBR. It is of interest that two, four, and six substituted compounds are of very low activity. The size of alkyl groups and their position on the aromatic ring greatly determined the activity of the disulfide toward reclaiming.

Agents such as naphthylamine and derivatives thereof have been used alone or with other agents.[43] Long chain saturated fatty acids (4-18 carbons), in the presence of steam and organic amines, have been used successfully to render a vulcanized rubber more processible.[12,13]

Dialkylarylmethyl hydroperoxide,[15] phenol sulfoxide $(\mathrm{AR{-}\overset{\overset{O}{\|}}{S}{-}AR})$, bis (trialkylphenol) sulfides,[14] aromatic sulfinic acids and their salts, and a multitude of accelerators have all been used to devulcanize scrap rubber.

It would seem plausible that since so many different materials render scrap rubber reprocessible, and increase plasticity, there may be a number of pathways to devulcanize.

10.6.1 Digester: Alkali

Sulfur is removed by reacting with the alkali, forming sodium sulfide. No further reaction of that sulfur with the polymer happens, even for long periods of time. This reaction happens quickly, in 20 to 30 minutes. The action of air is through the formation of a hydroperoxide attack of the olefinic systems and subsequent chain scission, most likely at the isoprene unit single carbon-carbon bond. Only very small amounts of oxygen are necessary to cause these reactions.[3,5]

It is well established that methods that work well[3] for NR do not always work well for synthetic rubbers, so other chemicals have become necessary.

Many of the reclaiming agents used and mentioned above can be classed into the following four major classes[26]:

1. Alkyl phenol sulfides and disulfides
2. Aliphatic and aromatic mercaptans
3. Amino compounds
4. Unsaturated compounds

For the most part, the unsaturated materials are oils and solvents such as terpenes, petroleum, naphthas, etc. The best amines are aliphatic ones containing 10 to 14 carbon atoms. Many of the agents used for reclaiming such as oils, terpenes, naphthas, and petroleum swell the rubber.

10.7 Compounding with Devulcanized Rubber

As stated earlier in the chapter, the processes now used to regenerate scrap rubber are different from the reclaiming processes. The chemomechanical process is the earliest process that the author considers to be similar to many of the modern-day processes in that chemicals were used in the presence of shear (a two-roll mill) at relatively low temperatures.[34] Physical property values of a blend of the regenerated material with a virgin tread compound at 25% are shown in Table 10.1 below. No values are given for the regenerated material itself, however. The regenerated material described here is from SBR tire tread. The virgin compound is the same formulation.

It seems that the devulcanizing system used, at least chemomechanical, does not matter a great deal in terms of property values in that the tensile strengths are of the same general level. However, some have been suggested to be much higher.

The author has carried out several hundred devulcanizations of a variety of rubber crumb of varying particle size with 50 to 100 different agents, and almost invariably, the tensile ranged between 8 and 14 MPa. Some examples are shown in Table 10.2.[30] The only devulcanized rubber crumb that had a much greater tensile strength was a low, filled, non-black NR compound that had a tensile strength of 20+ MPa (#5). Numbers 2 through 4 were

TABLE 10.1

Hardness (BS)	64
Tensile strength, MPa	16.6
Tear strength, N/cm	280

Source: Twiss, D.F. et al., British Patent #577,868, #578,482, 1946.

TABLE 10.2

Devulcanized Rubber Compounds

	1	2	3	4	5[e]
DVR[a]	100.0	100.0	100.0	100.0	100.0
Zinc Oxide	3.0	—	—	—	—
Stearic Acid	1.0	—	—	—	—
Sulfur	2.0	—	—	—	—
TBBS[b]	0.67	—	—	—	—
Rheometer Data at °C[c]	142	150	150	150	150
Min Torque, dN-m	7.7	63.8	65.1	37.3	26.0
MaxTorque, dN-m	104.0	104.4	106.7	79.1	70.1
Time to 2-Point Rise, min	2.0	2.1	2.1	1.3	1.8
Time to 90% Cure, min[d]	11.0	5.8	5.2	3.2	15.5
Cure Time, min[c]	11.0	10.8	10.2	8.2	20.5
Hardness, Shore A	68	63	64	61	47
100% Modulus, MPa	3.0	2.7	2.7	2.4	1.5
Tensile Strength, MPa	7.9	11.3	12.2	13.0	20.5
Elongation at Break, %	220	250	250	305	465
Tear Strength, Die C, N/cm	325	288	281	316	379

Note: See Section 10.9 for abbreviations.

[a] Number 1 is whole tire crumb; numbers 2, 3, and 4 are NR truck tire buffings; and number 5 is non-black, low, filled NR.

[b] N-tert Butyl-benzothiazole sulfenamide.

[c] Number 1 rheometer and cure was at 142°C; numbers 2 through 5 rheometers and cures were at 150°C.

[d] All were cured to 90% cure plus 5 minutes at the rheometer temperature.

[e] Unpublished data.

Source: From Myhre, M., and MacKillop, D.A., Modification of crumb rubber, Rubber Division, ACS, October 1995, Paper #21. With permission.

devulcanized crumb rubber that was revulcanized with no additional curative package added.

A major problem of devulcanized rubber, however, is the short scorch times and high viscosities observed. At present, these problems have not been solved to any large extent. Although the properties are not really good, they can be quite good and are good enough to be used for many applications. These materials can be processed and turned into products relatively easily. The results in Table 10.2 show that devulcanization of crumb rubber is probably not a straightforward process. The mechanisms may very well be different for different polymers, since four of the five shown did not appear to require the addition of curatives. This could mean that free sulfur is formed or that the cross-links are broken, then prevented from re-cross-linking by being blocked, and then released again at elevated temperatures. The cross-link sulfur rank is also very important.

Much of the work that has been carried out uses the devulcanized material as a compounding ingredient that can be a part of the rubber (Table 10.3) as a part of the formulation, or added on top of the formulation (Table 10.4).

TABLE 10.3

Natural Rubber/Devulcanized Rubber Blends

Formulation: Total Rubber 100.0; Zinc Oxide 5.0; Stearic Acid 1.5; Sulfur 2.5; TBBS 0.8.

	1	2	3	4
Natural Rubber	100.0	60.0	40.0	—
DVR	—	40.0	60.0	100.0
Rheometer Data at 142°C				
Minimum Torque, dN-m	10.5	7.9	6.0	27.0
Maximum Torque, dN-m	64.0	77.0	89.0	99.0
Time to 2-Point Rise, min	6.0	6.0	6.0	8.0
Time to 90% Cure, min	14.0	16.0	15.0	16.0
Cure Time at 142°C, min	14.0	16.0	15.0	16.0
Hardness, Shore A	40	52	57	70
100% Modulus, MPa	0.6	1.1	1.8	3.2
300% Modulus, MPa	1.9	4.8	6.2	—
Tensile Strength, MPa	26.9	19.5	12.9	7.0
Elongation at Break, %	725	555	390	205
Tear Strength, Die C, N/cm	360	385	370	230
Volume Swell, %	457	217	187	150

Note: See Section 10.9 for abbreviations.

Source: From Myhre, M., and MacKillop, D.A., Modification of crumb rubber, Rubber Division, ACS, October 1995, Paper #21. With permission.

Table 10.3 shows test results of a natural rubber/devulcanized rubber blend. The additives are based on 100 parts of rubber, which includes the devulcanized rubber (DVR). No DVR (#1) and no virgin material (#4) are the controls. The high viscosity of the DVR is obvious, but what is unusual in this case is the lower viscosity when combined with the natural rubber. It can be seen that the devulcanized material possesses much higher hardness and modulus, and much lower tensile properties and tear (#1 vs. #4). When DVR is blended with the NR, there is a reduction in tensile properties, but even at a level of 40 parts (#2), property values are quite good and can be used for many products. Tear strength is as good as the NR itself, even though the tear of the DVR alone is quite a bit lower. Also seen is the large reduction in volume swell (increased cross-link density) in the presence of the DVR.

Table 10.4 shows the test results of a blend of virgin compound and DVR. In this case, the DVR is added on top of the rubber at levels ranging from 10 to 25%, by weight. The DVR used in compounds two and three were devulcanized with different agents and used at a level of 25% of the total weight. Again, the scorch times were shorter and cure rates were faster than the virgin compound (#1). The modulus, tensile strength, and elongation differed little from the control. Tear strength was, however, noticeably lower (15 to 25%). Properties were still quite good, considering the DVR level was at 25%.

The devulcanized material in compounds four and five were different from two and three in that a different devulcanizing system was used. Also it was in the compounds at lower levels (10 and 15% by weight). The

TABLE 10.4

Natural Rubber Compounds with Devulcanized Truck Tire

	1	2	3	4	5
Virgin MB[a]	165.0	165.0	165.0	165.0	165.0
DVR[b]	—	55.0	55	18.5	29.0
TBBS	1.5	1.5	1.5	1.5	1.5
Sulfur	1.5	1.5	1.5	1.65	1.65
Rheometer Data at 150°C					
Minimum Torque dN-m	9.0	14.5	14.7	14.7	15.8
Maximum Torque, dN-m	74.0	74.4	74.4	70.5	70.5
Time to 2-Point Rise, min	6.2	5.2	5.0	4.5	5.0
Time to 90% Cure, min	11.8	8.9	8.4	8.5	8.5
Cure Time at 150°C	16.8	13.9	13.4	13.5	13.5
Hardness, Shore A	62	62	62	60	60
100% Modulus, MPa	2.1	2.0	2.2	1.8	1.9
300% Modulus, MPa	10.8	10.3	10.8	9.3	9.7
Tensile Strength, MPa	23.4	20.4	20.7	20.9	19.9
Elongation at Break, %	535	480	475	510	480
Tear Strength, Die C, N/cm	1018	870	763	1050	919
Volume Swell, %	279	264	263	225	257

Note: See Section 10.9 for abbreviations.

[a] NR/BR Tread Compound — NR 75.0; BR 25.0; Carbon Black 50.0; Aromatic Oil 4.0 Zinc Oxide 4.0; Stearic Acid 2.0; TMQ 2.0; 6PPD 2.0; Wax 1.5.

[b] Numbers 2 and 3 were devulcanized with different agents; Numbers 4 and 5 were devulcanized with an agent different from numbers 2 and 3.

Source: From Myhre, M., and MacKillop, D.A., Compounding with regenerated rubber, Rubber Division, ACS, October 2000, Paper #110. With permission.

property values, as with those of two and three, were quite good, especially tear strength. It is felt that all of these compounds, shown in Table 10.4, could be used for many applications.

Devulcanized rubber can be mixed just as any other rubber, whether it is part of a compound (Table 10.5 and Table 10.6) or by itself (Table 10.7). Both compounds (Table 10.5) were mixed like any other rubber compound in an internal mixer. Two different cure packages were used. The cure packages used in Table 10.5 and Table 10.6 are based on the NR/DVR total. The properties for both compounds were relatively good, including tear strength, considering there is more than twice the amount of DVR than virgin polymer. The compound shown in Table 10.6 is like that in Table 10.5 in that the DVR is part of a fully compounded material. The compound in Table 10.6 has more virgin polymer and black/oil, producing a higher elongation and a better tear strength. The material could be used in many applications.

The results in Table 10.7 show that a devulcanized material (whole car tire crumb rubber) can be mixed as a normal virgin compound, with rather low but surprisingly good properties. It is was very scorchy compound and had a high viscosity. However, the fact that this material had an elongation of 110% and could be mixed is rather surprising since there is no raw polymer present. An SBR tread compound with devulcanized whole tire crumb rub-

TABLE 10.5

Natural Rubber/Devulcanized Rubber[a] Compound

Base Formulation: NR 30; DVR 70; Carbon Black 15; Tall Oil 7.0; Zinc Oxide 3.0; Stearic Acid 2.0; Paraffin Wax 0.5; Antioxidant 0.5.

	1	2
CBS	1.0	—
TBSI	—	1.6
TBBS	—	0.4
Sulfur	2.0	1.2
Rheometer Data at 142°C		
Minimum Torque, dN-m	9.4	8.2
Maximum Torque, dN-m	77	75
Time to 2-Point Rise, min	4.5	5.5
Time to 90% Cure, min	13.5	25
Cure Time at 142°C	15	28
Hardness, Shore A	63	63
Modulus, MPa 100%	2.3	2.4
300%	9.9	10.7
Tensile Strength, MPa	16.7	15.8
Elongation at Break, %	425	385
Tear Strength, Die C, N/cm	515	395

[a] Whole car tire crumb.

Note: See Section 10.9 for abbreviations.

Source: From Myhre, M., and MacKillop, D.A., Modification of crumb rubber, Rubber Division, ACS, October 1995, Paper #21. With permission.

ber blend is shown in Table 10.8. The compound viscosity in the presence of the DVR increased and scorch time was reduced. There was not a significant reduction in vulcanizate properties, even though the DVR was present at a level of 25%.

Even though SBR is not capable of compensating for property losses by crystallizing, the author considers these properties (Table 10.8) to be good. Styrene-butadiene rubber is a less forgiving polymer compared to NR because it does not crystallize. The author feels that since NR crystallizes upon extension, it helps compensate for the reduction in properties when DVR is added.

Devulcanizing EPDM also seems to produce a good material, as shown in Table 10.9. When blended with a virgin compound and other compounding ingredients, the property values remained more or less unchanged, other than an increase in hardness and a marginal decrease in tear. The material became marginally more scorchy and was slightly slower curing. The virgin material (#1) was a final batch compound (curatives present). This blend (#2) was mixed in a laboratory internal mixer. Probably the best way to utilize devulcanized material is to put the same material back into the virgin compound. When that is done, the polymer is the same and filler

TABLE 10.6

Fully Compounded Natural Rubber/DVR Blend

Formulation: NR 70; DVR 65.0; Carbon Black 50; RP Oil (Witco) 12.0; Zinc Oxide 5.0; Stearic Acid 2.0; TBBS 1.2; Cryster Sulfur 2.67.

Rheometer Data at 145°C	
Minimum Torque, dN-m	9.0
Maximum Torque, dN-m	72.0
Time to 2-Point Rise, min	9.0
Time to 90% Cure, min	14.0
Cure Time at 145°C, min	15.0
Hardness, Shore A	62
100% Modulus, MPa	1.9
300% Modulus, MPa	8.4
Tensile Strength, MPa	15.4
Elongation at Break, %	455
Tear Strength, Die C, N/cm	607

Note: See Section 10.9 for abbreviations.

Source: From Myhre, M., and MacKillop, D.A., Modification of crumb rubber, Rubber Division, ACS, October 1995, Paper #21. With permission.

TABLE 10.7

Formulation: DVR 100.0; Carbon Black 35.0; Aromatic Oil 15.0; Zinc Oxide 5.0; Stearic Acid 1.5; TBBS 0.8; Sulfur 2.5.

Rheometer Data at 142°C	
Minimum Torque, dN-m	26.0
Maximum Torque, dN-m	121.0
Time to 2-Point Rise, min	3.0
Time to 90% Cure, min	16.0
Cure Time at 142°C, min	16.0
Hardness, Shore A	84
100% Modulus, MPa	8.1
Tensile Strength, MPa	9.2
Elongation at Break, %	110
Tear Strength, Die C, N/cm	183

Note: See Section 10.9 for abbreviations.

Source: From Myhre, M., and MacKillop, D.A., Modification of crumb rubber, Rubber Division, ACS, October 1995, Paper #21. With permission.

TABLE 10.8

SBR Compounds with Devulcanized Truck Tire

	1	2	3
Virgin MB	163.0	170.0	170.0
DVR	—	57.0	57.0
TBBS	1.5	1.5	1.5
Sulfur	1.5	1.5	1.5
Rheometer Data at 150°C	—	—	—
Minimum Torque, dN-m	16.4	22.9	22.5
Maximum Torque, dN-m	79.1	79.7	79.1
Time to 2-Point Rise, min	8.0	6.0	7.5
Time to 90% Cure, min	20.0	14.0	13.5
Cure Time at 150°C, min	25.0	19.0	18.5
Hardness, Shore A	66	67	67
100% Modulus, MPa	2.9	2.6	2.6
300% Modulus, MPa	14.8	12.9	13.2
Tensile Strength, MPa	20.5	17.4	18.2
Elongation at Break, %	405	385	370
Tear Strength, Die C, N/cm	493	423	426
Volume Swell, %	256	257	253

Note: See Section 10.9 for abbreviations.

Source: From Myhre, M., and MacKillop, D.A., Compounding with regenerated rubber, Rubber Division, ACS, October 2000, Paper #110. With permission.

TABLE 10.9

Virgin EPDM Compound/Devulcanized EPDM Blend

	1	2
Devulcanized EPDM	—	370.0
Virgin EPDM	100.0	450.0
Calcium Carbonate	—	72.0
TP888	—	8.0
Sulfur	—	2.25
Rheometer Data at 150°C		
Minimum Torque, dN-m	26.0	14.7
Maximum Torque, dN-m	43.0	62.2
Time to 2-Point Rise, min	2.5	1.9
Time to 90% Cure, min	10	9.5
Cure Time at 160°C, min	15.0	15.0
Hardness, Shore A	54	61
100% Modulus, MPa	1.6	1.9
300% Modulus, MPa	5.0	6.3
Tensile Strength, MPa	11.0	11.0
Elongation at Break, %	545	445
Tear Strength, Die C, N/cm	323	304

Note: See Section 10.9 for abbreviations.

Source: From Myhre, M., and MacKillop, D.A., Compounding with regenerated rubber, Rubber Division, ACS, October 2000, Paper #110. With permission.

TABLE 10.10

Base Formulation: NR 100; Carbon Black 50; Zinc Oxide 5.0; Stearic Acid 2.0.

	1	2	3	4	5
Virgin MB	157.0	157.0	157.0	157.0	157.0
DVR - 1	—	18.0	—	—	—
DVR - 2	—	—	18.0	—	—
DVR - 3	—	—	—	18.0	—
DVR - 4	—	—	—	—	18.0
Sulfur	2.5	2.9	2.9	2.9	2.9
TBBS	0.8	0.8	0.8	0.8	0.8
Rheometer Data at 150°C					
Min Torque, dN-m	5.1	6.4	7.1	7.2	7.9
Max Torque, dN-m	84.7	93.8	93.8	85.8	88.1
Time to 2-Point Rise, min	4.6	2.8	3.3	4.1	3.8
Time to 90% Cure, min	11.3	4.9	6.6	11.4	10.6
Cure Time at 150°C, min	16.3	9.9	11.6	16.11	15.6
Hardness, Shore A	64	64	62	62	64
Modulus, MPa					
100%	2.7	3.1	3.1	2.6	2.7
200%	7.4	8.0	7.8	6.6	7.2
300%	14.5	14.7	14.5	12.4	13.1
Tensile Strength, MPa	25.9	23.1	21.9	24.7	24.2
Elongation at Break, %	475	430	415	515	485
Tear Strength, Die C, N/cm	737	798	889	1138	1012
Volume Swell, %	227	223	228	248	242

Note: See Section 10.9 for abbreviations.

Source: From Myhre, M., Unpublished CRTDL data. With permission.

loading, etc. are the same. Adjustments would likely be needed because of the reduced scorch safety, increased cure rate, and viscosity. This information is shown in Table 10.10. Virgin compound was mixed, cured, cut up, and cryogenically ground into crumb rubber and was then devulcanized. The devulcanized material (chemomechanical) was then blended with the virgin stock at a level of 10% by weight, and curatives were added. Four differently devulcanized materials were evaluated. Chemical agents were used and the devulcanization was carried out on a two-roll laboratory mill (chemomechanical).

Hardness changed little, as did the modulus. Tensile strength reduction ranged from 5 to 15% compared to the virgin compound. Elongation varied from −12.5% to +8% and tear strength increased compared to the virgin material. It is felt that these compounds are very good and show that property values can be maintained. There did not appear to be obvious degradation of the polymer because of the devulcanization process.

10.8 Biotreatment

In the 1960s Saville and Watson studied extensively and published work on chemicals used as sulfur probes in vulcanized rubber.[49] These are, in some sense, devulcanizing by the removal of sulfur to determine the kind of cross-link (sulfur rank) present in a vulcanizate. The author feels that biotreatment may be similar in that sulfur, in this case, surface sulfur, is removed by microorganisms.

Quite a lot was done in the early part of the twentieth century on the effect of microorganisms, in particular, the process of fermentation by using yeasts on rubber.[50] What was realized was that certain organisms attacked NR. There were many conflicting reports in later publications on this attack by microorganisms. Much of the controversy could have been reconciled had the researchers considered the factors that earlier researchers were not aware of at the time of their work. The nature of the polymer, additives, types of organisms, and the physical and chemical state of the rubber were not always appreciated as being important.

Works carried out about 30 years later differed significantly from the earlier work; the more recent work was concerned with the effort to degrade rubber, while previous efforts dealt with prevention of failure due to microbial attack.

Organisms studied since the 1970s[51] were yeasts and fungi, microorganisms belonging to the Actinomycetes group, and Streptomycetes. (Yeasts appear to metabolize rubber without attaching itself to the rubber physically, whereas fungi becomes firmly attached.) It was determined that the organisms utilize rubber hydrocarbon as well as process oils, and that none of the microbial attacks were directed toward sulfur.

Factors that tended to suppress the growth of the microorganisms were the ingredients incorporated into rubber to activate and accelerate the curing, and to protect it. Strong antimetabolites were zinc oxide, mercaptobenzothiazole, dimethyl dithiocarbamate–based accelerators, and paraphenylenediamine antidegradants.

A number of these organisms produced rubber particles that became harder, drier, and more brittle than the untreated rubber. Although the cause was never really determined, the treated crumb had significant ion exchange activity. When mixed with soil, the fermented material produced a significant increase in water retention. Oxygen incorporated into the rubber at the surface as carboxyl groups rendered the surface hydrophilic, which may explain both activities. The microorganisms discussed so far all metabolized rubber carbon, but not sulfur.

A major problem with the use of microorganisms is the retardation of growth because of many of the compounding ingredients. There are three major factors that appear to retard the utilization of these organisms.First, rubber vulcanizates are hydrophobic, so are attacked only on the surface. The greater the available surface area, the more likely the microorganism attack. That means that the particle size needs to be small, probably 100 mesh and smaller. Smaller particle size rubber crumb requires more grinding, so it is more costly to produce the initial crumb. Second, rubber particles are highly cross-linked and branched materials. The latter is not conducive to biodegradation; therefore, polymeric materials of linear structure, or at least minimal branching, are preferable. Third, rubber vulcanizates contain many materials, most of which retard the activity of the microorganisms. To ensure the organism activity is not impeded, the ingredients that interfere would have to be removed. To do this, new cure systems and antidegradant packages would need to be developed, or a complete extraction of the rubber crumb would be necessary. Either approach would place intolerable economic pressures on recycling rubber if it was feasible. A likely use of this type of material would be as a soil conditioner.

During the last decade or so, microorganisms have been found that appear to break sulfur-sulfur and carbon-sulfur bonds at the surface. Several of the organisms used are chemolithotropes, which derive their energy from oxidation of sulfur compounds. Since only the particle surface is attacked, polar oxygen groups result on the surface. As opposed to just breaking the bonds, these organisms remove the sulfur (like chemical probes).

Romine, Romine, and Snowden-Swan[52] showed that the surface can be stripped of sulfur; however, the time frame is long (> 48 hours). The length of time would make the process less than appealing. Rubber chemicals that were most effective in retarding the microorganism growth were tetramethylthiuram, mono- and disulfides, and dimethylbutyl-p-phenylenediamine. Also, SBR alone affected the growth negatively.

10.9 Abbreviations

NR	Natural Rubber
SBR	Styrene Butadiene Copolymer
BR	Polybutadiene
EPDM	Ethylene Propylene Diene Terpolymer
DVR	Devulcanized Crumb Rubber
MB	Masterbatch
CBS	N-cyclohexyl-2-benzothiazole sulphenamide
TBBS	N-t-butyl-2-benzothiazole sulphenamide

Acknowledgments

The author would like to thank S. Maiti, et al. in allowing the use of data, as is, from their article in the *Journal of Polymer Materials,* December 1997.[38]

Also thanks to the Rubber Division, ACS, for allowing the use of material from *Rubber Chemistry & Technology* and for the use of information from its meeting publications, in particular the published study by P. Nicholas.[29] Thank you to Elaine, for all her computer work and for arranging data, tables, and other information for this chapter.

References

1. Goodyear, C., U.S. Patent #3,633, June 15, 1844.
2. Owen, E.W.B, Processes of reclaiming rubber and their relative merits, *Trans. Inst.Rub. Ind.*, 19, 111, 1943.
3. Le Beau, D.S., Basic reactions occurring during rubber reclaiming I, *India Rub. World*, 118, 59, 1948.
4. Cook, W. S., et al., Reclaiming agents for synthetic rubber, *Rubb. Chem.Technol.*, 22, 166, 1949.
5. Le Beau, D.S., Basic reactions occurring during rubber reclaiming II, *India Rub. World*, 118, 69, 1948; *Indust. Eng. Chem.*, 40, 1194, 1948.
6. Amberlang, J.C. and Smith, G.E.P., Behavior of reclaiming agents in sulfur and nonsulfur GR-S vulcanizates, *Indust. Eng. Chem.*, 46, 1716, 1954.
7. Ball, J.M., *Introduction to Rubber Technology*, 7th edition, van Nostrand Reinhold, New York, 1969, chapter 17.
8. Paul, J., *Rubber Reclaim*, Rubber Division, ACS, October, 1977, Paper #50.
9. Nourry, A., *Reclaim Rubber: Its Development, Application, and Future*, MacLaren and Sons, London, 1962.
10. Stern, H.J., *Rubber: Natural and Synthetic*, 2nd edition, MacLaren and Sons Ltd, London, page 302.
11. Novotney D.S., (to Goodyear Tire & Rubber Company), Microwave devulcanization of rubber, U.S. Patent #4,104,205 (1978).
12. Le Beau, R.V., Reclaiming synthetic rubber with an amine, U.S. Patent #2,423,032, 1947.
13. Le Beau, R.V., Reclaiming synthetic rubbers with a fatty acid and live steam, U.S. Patent #2,423,033, 1947.
14. Albert, H.E., Reclaiming rubber with bis (trialkyhphenol) sulfides, U.S. Patent #2,605,241, 1952.
15. Lewis, J.R., Process for the reclamation of rubber in the presence of alpha, alpha dialkylarylmethyl hydroperoxide, U.S. Patent #2,558,764, 1951.
16. Hensley, W.A., and Albert, H.E., Reclaiming rubbers, U.S. Patent #2,686,162, 1954.
17. Annual Book of ASTM Standards, 09.01 (1992), A 1566, *Standard Terminology Relating to Rubber.*
18. American Society for Testing and Materials, Special Technical Publication 184A, "Glossary of Terms Relative to Rubber and Rubber Technology."
19. Winkelmann, H.A., The present and future of reclaimed rubber, *Ind. Eng. Chem.*, 18, 1163, 1926.
20. Boiry, F., The reclaiming of rubber, *Rev. Gen. Caoutch.*, January, 205, 1927.
21. Stafford, W.G., and Wright, R.A., Fundamental aspects of reclaimed rubber, *Applied Science of Rubber*, Naunton, W.J.S., and Arnold, E., eds., London, 1961, chapter 4.

22. Myhre, M., and MacKillop, D.A., Rubber recycling, *Rubb. Chem.Technol.*, 75, 429, 2003.
23. Warner, W.C., Methods of devulcanization, Rubber Division, ACS, October 1993, Paper #6.
24. Adhikari, B., De, D., Maiti, S., Reclamation and recycling of waste rubber, *Prog. Polym. Sci.*, 25, 909, 2000.
25. Isayev, A.I., *Rubber Recycling, Rubber Technologist's Handbook*, De, S.K., and White, J.R., eds., RAPRA Technology Limited, Shawbury U.K., 2000, chapter 15.
26. Le Beau, D.S., Science and technology of reclaimed rubber, *Rubb. Chem. Technol.*, 40, 217, 1967.
27. Yamashita, S., Kawabata, N., Reclamation of vulcanized rubbers by chemical degradation, *Japan-USSR Polym. Symp. (Proc. Eng.)*, 355, 1976.
28. Nicholas, P.P. (to B.F. Goodrich), Devulcanized rubber composition and process for preparing the same, U.S. Patent #4,161,464, July 1979.
29. Nicholas, P.P., The scission of polysulfide crosslinks in scrap rubber particles through phase transfer catalyzing, *Rubb. Chem. Technol.*, 55, 1499, 1982.
30. Myhre, M., and MacKillop, D.A., Modification of crumb rubber, Rubber Division, ACS, October 1995, Paper #21.
31. Oliveira, L.C., Process for reclaiming cured or semi-cured rubber, U.S. Patent #5,677,354, 1977.
32. Baranwal, K., Rogers, J.W., Standley, P.M., Catalytic regeneration of rubber: part I, polychloroprene, Rubber Division, ACS, October 1998, Paper #39.
33. Menadue, F.B., Some technical aspects of rubber reclaiming, *Rubb. Age*, 56, 511,1945.
34. Twiss, D.F., Hughes, A.J., Amphlett, P.H., Improvements in or relating to the regeneration of vulcanized synthetic rubber-like materials, Brit. Patent #577,868, #578,482, 1946.
35. Okamoto, H., et al., Mechanochemical reclamation of comminated tyre scrap using vulcanization accelerators, *Int. Poly. Sci. Technol.*, 7, T/59 1980.
36. Watabe, Y., et al., Process for reclaiming scrap vulcanized rubber, U.S. Patent #4,211,676, 1980.
37. STIK Polymers America promotional material, 1995.
38. De, D., Adhikari, B., Maiti, S., Reclaiming of rubber by a renewable resource material, part I, *J. Polym. Mat.*, 333, 1997.
39. De, D., Maiti, S., Adhikari, B., Reclaiming of rubber by a renewable resource material, part II, *J. Appl. Polym. Sci.*, 73, 2951, 1999.
40. Kostanski, L.K., et al., Physico-chemical changes during re-activation (devulcanization) of tire-rubber crumb, Rubber Division, ACS, October 1999, Paper #58.
41. Kostanski, L.K., et al., Physico-chemical changes during re-activation of different types of vulcanized rubber, Rubber Division, ACS, October 1999, Paper #134.
42. Klingensmith, W., et al., Properties of synthetic rubber "devulcanized" using a Rodan and Rokel single screw devulcanizer, Rubber Division, ACS, October 1998, Paper #54.
43. Mouri, M., et al., New continuous recycling technology for vulcanized rubbers, Rubber Division, ACS, October 1999, Paper #84.
44. Cotton, F.H., and Gibbons, P.A., Process of reclaiming rubber, U.S. Patent #2,408,296, 1940.

45. Hildibrant, H. (Continental Gumi-Werke, A.G.), Regeneration of synthetic rubber vulcanizates, Ger. #1,244,390, 1967.
46. Stalinski, E., Reclaiming of natural or synthetic rubber, Fr. #1,517,694, 1968.
47. Myhre, M., and MacKillop, D.A., Compounding with regenerated rubber, Rubber Division, ACS, October 2000, Paper #110.
48. Myhre, M., Unpublished CRTDL data.
49. Saville, B., and Watson, A.A., Structural characterization of sulfur vulcanized rubber networks, *Rubb. Chem. Technol.*, 40, 100, 1967.
50. Beckman, J.A., et al., Scrap tire disposal, *Rubb. Chem. Technol.*, 47, 597, 1974.
51. Holst, O., Stenberg, B., Christianson, M., Biotechnological possibilities for waste tyre-rubber treatment, *Biodegradation*, 9, 301, 1998.
52. Romine, R.A., Romine, L., Snowden-Swan, Microbial processing of waste tire rubber, Rubber Division, ACS, 1995, Paper #56.

11

Conversion of Used Tires to Carbon Black and Oil by Pyrolysis

C. Roy, A. Chaala, H. Darmstadt, B. de Caumia, H. Pakdel, and J. Yang

CONTENTS

11.1 Abstract

The development of a tire vacuum pyrolysis process from the laboratory to the industrial scale is reported. This research included characterization of the pyrolysis products, the development of commercial applications for the products, and the engineering of a vacuum pyrolysis plant with a throughput capacity of 3 t/h. Fundamental data on the pyrolysis kinetics are presented. The heat of the pyrolysis reaction was determined in a pilot unit by measuring the temperature change of a bath of molten salts brought into indirect thermal contact with the pyrolysing tire particles. These data allowed us to model the heat transfer and to develop a new moving and stirred bed vacuum pyrolysis reactor with high heat-transfer capacity.

Vacuum pyrolysis is typically performed at a temperature of 500C and a total pressure of 20 kPa. At these conditions, the rubber portion of tires is transformed into oil and gas, whereas the carbon black filler is recovered as pyrolytic carbon black (CB_P). On a large scale, the process yields approximately 35 wt.% CB_P, 45 wt.% oils, and 20% gas on a tire steel–free basis. Several commercial applications for the different products have been investigated. The CB_P surface chemistry and activity are similar to those of some grades of commercial carbon blacks. Therefore, CB_P has the potential to replace commercial carbon black grades (e.g., N774 and N660) in certain rubber applications. CB_P was successfully tested as filler in road pavement. The total pyrolytic oil can be used as a heating fuel. It can also be distilled into different fractions: a light, a middle distillate, and a heavy fraction. The light fraction was positively tested as a gasoline additive. Furthermore, this fraction contains valuable chemicals such as D,L-limonene. The middle fraction was successfully tested as a plasticizer in rubbers. The heavy fraction represents a good-quality feedstock for the production of coke and can also be used in road pavements. The pyrolytic gas has a high calorimetric value. It can supply all of the heat required to achieve the pyrolysis reactions. Data on the process feasibility are presented for a 30,000 t/yr plant, using information obtained in a 3 t/h demonstration unit.

11.2 Introduction

In the U.S. and other industrialized countries, roughly one used tire is generated per capita per year.[1] A large portion of these tires are simply dumped in the open or in landfills. This practice is environmentally unacceptable. Tire piles can be the source of very toxic emissions in the case of a fire,[2] and may act as a breeding ground for mosquitoes[3]; therefore, governments limit this disposal route. For example, a directive of the European Union (EU) required its member states to stop all landfill of whole tires by July 2003.[4] In 2006, landfill of crumbed or shredded tires will be banned. In the EU, approximately one third of used tires are currently landfilled. New recycling routes must be developed.

Carcass retreading is one of the existing recycling methods.[5] Ground tires can be used in civil engineering applications, for example as an additive in road pavement.[6,7] A major outlet for scrap tires is their utilization as solid fuels,[8] especially in cement kilns.[9] The tire recycling or treatment processes mentioned above have some disadvantages. Retreading can only be performed when the carcass is not damaged. When tires are used as solid fuel, polycyclic aromatic hydrocarbons and soot are produced.[10–12] Therefore, expensive gas-cleaning devices are necessary for the removal of potentially hazardous compounds. Tire grinding is very expensive because it is performed at cryogenic temperatures and may require energy-intensive equipment.

A promising recycling route for used tires is their transformation into useful products by pyrolysis. Over the last 20 years, various pyrolysis processes have been developed at universities[13–16] and by the industry. Examples of industrial processes include: Kobe Steel in Kobe, Japan[17]; Onahama Smelting and Refining Co. in Iwaki City, Japan[18]; Metso Minerals in Tampere, Finland (formally Svedala)[19]; Titan Technologies Inc. in Albuquerque, N.M., U.S.; and LiGmbH, in Miltzow, Germany. Among these plants, to our knowledge Onahama is by far the largest tire pyrolysis plant (4 t/h) in the world; its current status is unknown. With the exception of the vacuum pyrolysis process developed at the Université Laval and by Pyrovac Int.,[16,20] very little information is available on the use and properties of the products obtained in large-scale tire pyrolysis processes. To provide additional information, a substantial portion of this chapter deals with vacuum pyrolysis process–related data and results.

Until recently, pyrolysis of tires has been practiced on a very limited scale due to the absence of an established market for the oil or especially for the pyrolytic carbon black (CBp) product. Therefore, the authors have emphasized CBp research and development activities to open up a market for the tire vacuum pyrolysis products, in particular CBp. In addition to tires, the vacuum pyrolysis of several other feedstock materials, such as biomass[20] and automobile shredder residues,[21] has been performed in our laboratory at Université Laval. The process has been scaled up from the laboratory (batch reactor) to a continuous-pilot unit with a capacity of 30 kg/h, and finally to an industrial-scale demonstration plant (3 t/h) now located in the city of Saguenay (Quebec, Canada). This unit was initially designed for pyrolysis of biomass, but it was also successfully tested with used tire feed.

Pyrolysis can be defined as the thermal decomposition of organic material in the absence of air (*i.e.,* a flow of nitrogen or under vacuum). The main components of tires are rubber, the carbon black filler, other organic materials, zinc oxide, and sulfur (see Table 11.1). Upon heating, the rubber and the other organic compounds decompose and are converted into oils and gases. The pyrolysis residue consists of the recovered carbon black filler, inorganic materials, and varying proportions of carbonaceous materials formed from the rubber decomposition products. In a fundamental study, the decomposition of extender oil and various rubbers used in tires was investigated by differential thermogravimetry (DTG).[22] Volatilization of the extender oil starts at a temperature as low as 100°C. The elastomers decompose between 300 and 500°C. Of the various elastomers, natural rubber is the least temperature-resistant compound, whereas for the decomposition of styrene-

TABLE 11.1

Typical Composition of Tires (Steel-Free Basis)

Compound	Weight (%)
Rubber (e.g., natural rubber (NR), styrene-butadiene rubber (SBR))	55–60
Carbon black filler	25–30
Extender oil	5–7
Zinc oxide	1–2
Sulfur	1–2
Others (e.g., fatty acids, phenolic resins, petroleum waxes)	4–8

Source: From Waddell, W.H., Bhakuni, R.S., Barbin, W.W., and Sandstrom, P.H., in *The Vanderbilt Rubber Handbook*, 13th ed., Ohm, R.F. (ed.), R.T. Vanderbilt Company, Norwalk, CT, U.S., 1990, 596. With permission.

butadiene rubber and polybutadiene rubber, higher temperatures are required. The thermal decomposition of tires can be simulated by linear addition of the DTG curves of the individual tire components (see Figure 11.1), indicating that the decomposition of the different rubbers occurs independently. These DTG data are useful when modeling the heat transfer during the pyrolysis reaction.[23]

The pyrolysis temperature and pressure have a significant influence on the pyrolysis reaction. For vacuum pyrolysis of tire sidewalls, it was observed that pyrolysis is completed at a temperature of approximately 420°C. A further increase of the pyrolysis temperature does not significantly change the pyrolysis yields (see upper graph in Figure 11.2).[24] However, as will be discussed later, a higher pyrolysis temperature is beneficial with respect to the CB_P quality. The situation is different when the pyrolysis is performed at atmospheric pressure.[13,15,17] With increasing pyrolysis temperature, the yield of oil decreases, whereas the gas yield increases (see lower

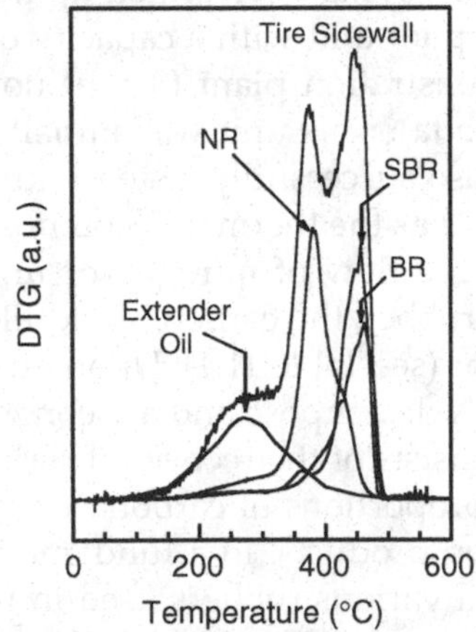

FIGURE 11.1

Experimental and simulated DTG curves of tire sidewall and components. (From Yang, J., Kaliaguine, S., and Roy, C., *Rubber Chem. Technol.*, 66 (1993) 213–229. With permission.)

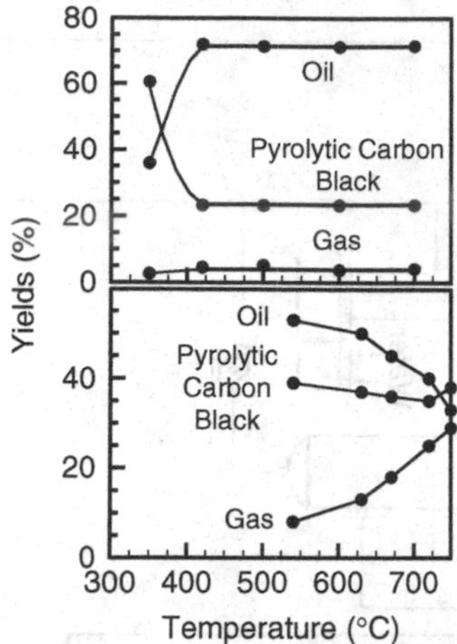

FIGURE 11.2
Yields of tire pyrolysis under vacuum pressure (upper graph) and at atmospheric pressure (lower graph).

graph in Figure 11.2).[17] This behavior is explained as follows: During vacuum pyrolysis, hydrocarbons originating from the decomposing elastomer are quickly removed from the hot reaction zone by the vacuum pump. Secondary reactions of these hydrocarbons are therefore limited. On the other hand, during atmospheric pyrolysis, the residence time of the gaseous hydrocarbons in the reactor is relatively long, with the consequence that secondary reactions occur to a considerable extent. These reactions include the cracking of large hydrocarbons to smaller fragments (increasing the gas yield and lowering the oil yield) and the formation of carbonaceous deposits on the recovered carbon black filler. These reactions are undesired because they reduce the commercial value of the pyrolysis products. Examples will be given later in the text.

11.3 Vacuum Pyrolysis Process

The vacuum pyrolysis process flow sheet is presented in Figure 11.3. From the used tire storage facility, separated from the plant building, tire crumb with a typical particle volume of approximately 10 mm^3 is loaded in a hopper and carried along by extracting screws into a pneumatic conveying system up to a cyclone, located on the roof of the plant building. The feedstock material flows down by gravity into the vacuum feeding system, consisting of two rotary valves set in series and between which an auxiliary liquid-ring vacuum pump suction line is connected. A positive pressure is maintained over the second rotary valve (between the vacuum pump suction and the

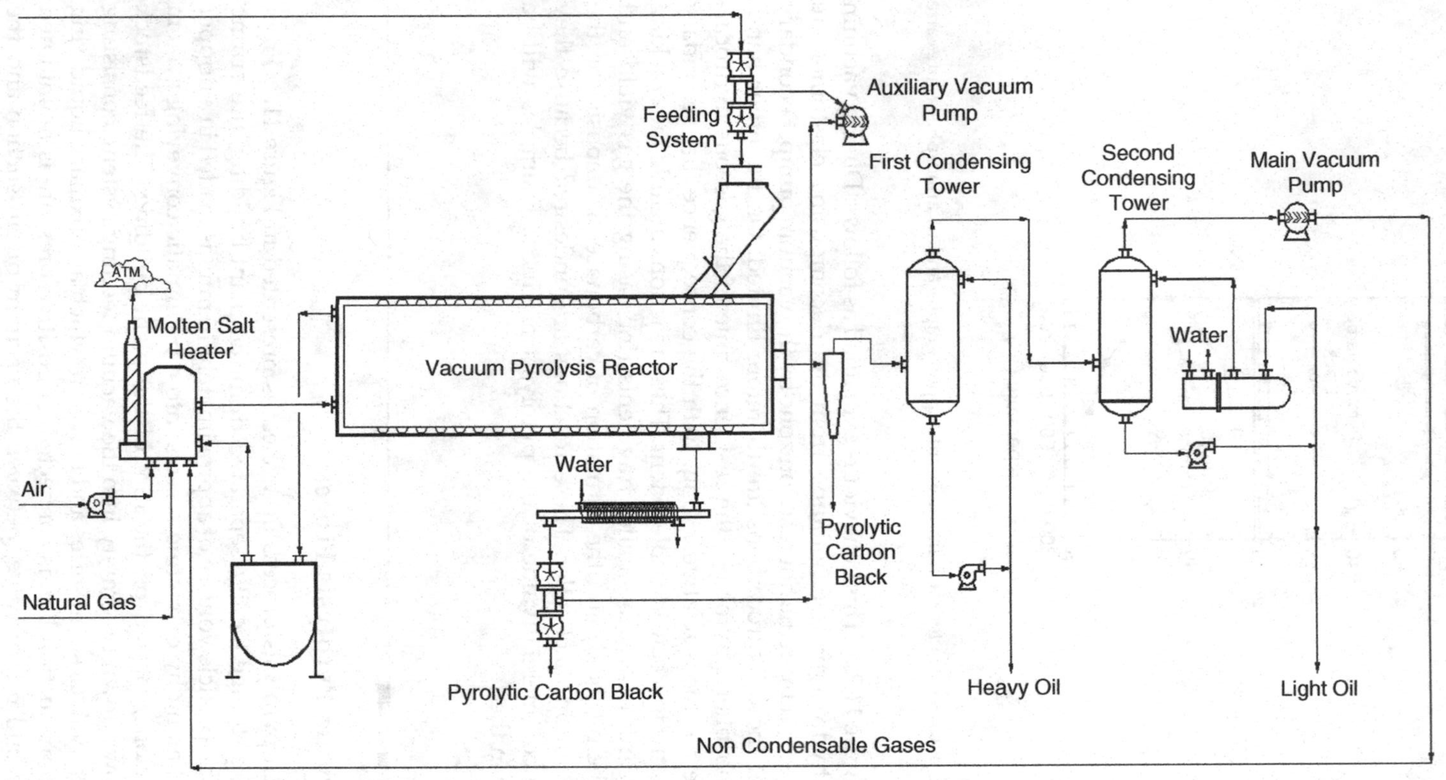

FIGURE 11.3
Tire vacuum pyrolysis, process flow sheet.

reactor) to avoid the escape of pyrolysis gas from the reactor. The feedstock is then fed by screw conveyors into the reactor, which has a length of 14.6 m and a diameter of 2.15 m. With this arrangement, mass flow rates of up to 3 t/h are reached. However, the test done with this unit and reported below was performed at a crumb feed rate of 0.5 t/h. Photographs of this plant are available in the literature.[20,25]

Inside the reactor, the tire particles are carried along and mixed by a raking system on two metal plates arranged one above the other. Molten salts circulate inside tubes below the plates, where they serve as a heating medium. The pyrolysing tire particles are first moved on the upper plate. At the end of the upper plate, they fall on the lower plate, where they are moved to the outlet. This arrangement ensures a very high heat transfer and efficient material mixing.[23] The residence time of the pyrolysing tire particles in the reactor is approximately 12 min. The temperature of the solid pyrolysis residue (CB_p) at the reactor outlet is set at approximately 530°C. The total pressure in the reactor is maintained at 15 to 20 kPa. The CB_p produced passes across the vacuum extraction system, which is composed of a pair of rotary valves. The cooled CBp is transported by belt conveyors and stored in 1.5 m^3 bags.

The pyrolytic vapors and gaseous products are withdrawn from the reactor via a 0.6 m header toward a two-stage (packed tower) condensation system, where the heavy oil is condensed in the first tower while the light oil is condensed in the second tower. The oil fractions from both packed towers are mixed together and stored.

The noncondensable gas drawn up by the main vacuum pump from the second packed tower is further cooled past the pump and compressed to a pressure of 170 kPa before being fired in the molten salt furnace. The heating value of the pyrolytic gas is sufficient to provide the entire heat required for the pyrolysis process. The excess gas is flared. From the molten salt 9 m^3 storage tank, which is equipped with heating elements and a piping loop circuit, the molten salt is pumped to the furnace (where it is heated), then circulates into the reactor heating plates, and finally flows back into the receiving tank.

11.3.1 Pyrolysis Yields

As mentioned earlier, during the process development over a 15-year period, tire vacuum pyrolysis was performed in several reactors: laboratory batch reactors, a pilot plant with a throughput capacity of 30 kg/h, and a demonstration plant with a design capacity of 3 t/h. The influence of the most important pyrolysis parameters (temperature and pressure) on the product distribution was studied in batch reactors.[24,26]. If the duration of the pyrolysis is long enough, a temperature of 450°C is sufficient to decompose the tire rubber. As shown in the upper graph of Figure 11.2 for tire sidewalls, a further increase of the pyrolysis temperature has a very small effect on the

TABLE 11.2

Yields on Crumb Rubber Vacuum Pyrolysis with Different Reactor Capacities

Reactor, Experiment #	Laboratory Reactor, G-821	Pilot Plant, H-70	Demonstration Plant, JT-1
Pyrolysis condition			
Temperature [°C]	500	450	530
Pressure [kPa]	0.4–0.8	20	15–20
Throughput [kg/h]	0.9[a]	20	540
Pyrolysis Yields [%]			
Pyrolytic oil	57.0	41.5	<50.7[b]
Pyrolytic carbon black	35.1	38.4	>30.0[b]
Gas	7.9	20.1	19.3
Total	100.0	100.0	100.0

[a] Batch experiment: the tire crumb was heated at a rate of 10°C/min to a final temperature of 500°C; the temperature was held constant for 3 h.

[b] The mass balance closure was not entirely satisfactory. The yield of CB_P is probably higher than 30%. The oil yield is probably overestimated.

yields. For large tire shreds, pyrolysis at 500°C in a batch reactor yielded approximately 57 wt.% oil, 35% CB_P, and 8% gas (see Table 11.2). Upon increase of the pyrolysis pressure, the yield of CB_P increases[24] as hydrocarbons originating from the decomposing tire rubber are adsorbed on the CB_P, where they form carbonaceous deposits. This reaction is limited at low pressures because the concentration of gaseous hydrocarbons is lower. As compared to vacuum pyrolysis, pyrolysis at atmospheric pressure yields less oil and more gas and CB_P at similar temperature.[27] The reason for the higher gas yield at atmospheric pressure is the cracking of large hydrocarbon molecules in the hot reactor. Under vacuum, this reaction is less important because the gaseous hydrocarbons are quickly removed from the reactor by the vacuum pump. As shown earlier, neither the tire size (i.e., whole tire vs. shreds of 20 mm × 30 mm) nor the kind of tire (i.e., truck vs. passenger) seems to significantly influence the pyrolysis product yields.[16]

Continuous tire vacuum pyrolysis yields considerably more gas (20%) as compared to batch pyrolysis (8%, Table 11.2). This is probably caused by the occurrence of severe secondary cracking reactions in the thick flowing bed of crumb particles in continuous feed reactors. The term "gas" is used here for the noncondensable compounds leaving the condensing system. There are important differences in the condensing units of the different setups. In the laboratory, the compounds leaving the batch reactor are condensed at much lower temperatures (typically at –78°C, using dry ice/limonene) as compared to continuous pyrolysis in a larger-scale reactor, where the condensation is in the range of 20 to 80°C. Moreover, in batch units, the pressure is much lower; hence more oil is produced in the reactor at the expense of gas. In the full-scale reactor, the pressure is higher, but the condensation temperature is also high. This leads to more hydrocarbon molecules in the vapor phase, which are eventually cracked to lighter molecules.

11.4 Properties of the Pyrolysis Products

11.4.1 Pyrolytic Carbon Black

11.4.1.1 Proximate and Elemental Analysis

The steel-free tire feedstock used for experiments at different scales consisted of approximately 67% wt. volatile matter (corresponding to the rubber and extender oil), 28% fixed carbon (main source is the carbon black), and 5% ash (e.g., zinc oxide, silica; Table 11.3). After pyrolysis, most of the rubber and extender oil is converted into oil and gas. The carbon black used as a reinforcing agent and the tire inorganic components find their way in the CB_P. However, the CB_P may also contain small amounts of not totally decomposed rubber and/or carbonaceous deposits. These deposits originate from hydrocarbons formed during the rubber thermal decomposition. The concentration of volatile matter in the CB_P indicates the degree of achievement of the pyrolysis reactions. For vacuum pyrolysis CB_P, values as low as 2.9% volatile matter were found (Table 11.3), indicating that (nearly) all of the tire rubber was transformed to oil and gas, and that the formation of carbonaceous deposits was limited. However, the concentration of volatile matter in CB_P is higher than that in commercial rubber–grade blacks (0.8%). The higher volatile content of the CB_P is also reflected by a higher concentration of hydrogen and nitrogen in the CB_P (Table 11.3). As mentioned earlier, the CB_P contains virtually all of the initial inorganics used to manufacture tires. Consequently, the ash concentration in CB_P is considerably higher (13 to 16 wt.%) as compared to that in commercial carbon black grades (0.5%). A significant reduction of the CB_P ash content can be achieved by an acid treatment, as discussed below.

The results of the proximate analysis of CB_P from the laboratory batch reactor and the demonstration plant are very similar, indicating that laboratory results might be used to predict properties of products obtained at a

TABLE 11.3

Proximate Analysis and Elemental Composition of the Tire Feedstock, Pyrolytic Carbon Blacks, and Commercial Carbon Blacks (wt.%)

Reactor, Experiment #	Tire Feedstock	Laboratory Batch Reactor, G-821	Pilot Plant, H-70	Demonstration Plant, JT-1	Commercial Carbon Black[a]
Volatile matters	67.2	3.3	7.7	2.9	0.8
Fixed carbon	27.9	83.8	79.2	81.4	98.7
Ash	4.9	12.9	13.1	15.7	0.5
Carbon	90.1	95.8	94.2	95.0	98.8
Hydrogen	7.7	0.5	1.2	0.6	0.4
Nitrogen	0.3	0.3	0.3	0.6	0.0
Sulfur	1.9	3.4	4.2	3.8	0.8

[a] Typical values from rubber tire–grade carbon blacks

TABLE 11.4

Surface Area and Structure of CB_P and Commercial Carbon Blacks

Reactor, Experiment #	Laboratory Batch Reactor, G-821	Pilot Plant, H-22[a]	Demonstration Plant, JT-1	Carbon Black Grades Used in Tires			
				N115	N330	N539	N660
Surface area [m²/g]	80.1	95.0	77.3	145	82	43	36
Structure, DBP No. [cm³/100 g]	—	79.2	95.0	113	102	110	90

[a] Pyrolysis of truck tire particles.

much larger scale. However, for the CB_P obtained in the pilot plant, the concentration of volatile matter was approximately twice as large as the other CB_P samples. This could indicate a lack of optimization of the operating conditions during the pyrolysis run.

11.4.1.2 Specific Surface Area and Structure

The specific surface area (inversely proportional to the particle size) and the structure are considered to be the two most important carbon black properties. The primary carbon black particles have a spherical shape. These primary particles are fused together in aggregates (looking similar to grapes). In carbon black science, the degree of aggregation is designated by the term "structure" and is measured by absorption of di-butyl phthalate (DBP) in the voids between the primary particles of the same aggregate. A high DBP number indicates a high-structure carbon black with aggregates consisting of many primary particles. In low-structure carbon blacks, only a few particles are fused together.

In tires, many different carbon black grades are used.[28] The surface area and structure of these blacks range from 35 to 150 m^2/g and 65 to 125 $cm^3/100$ g, respectively. The recovered CB_P contains a mixture of all these grades. One might therefore expect the CB_P surface area and structure values to be close to the values of a mixture of the different grades used in tires. However, during service in the tire and pyrolysis, carbon black aggregates might break down, lowering the structure. Ash components and carbonaceous deposits "dilute" the recovered carbon black filler in the CB_P, leading to lower structure and surface area. Furthermore, carbonaceous deposits might fill the voids between the primary particles, further reducing the surface area. The surface area and structure of the vacuum CB_P from the demonstration plant were, indeed, close to the average values of the commercial grades used in tires (see Table 11.4). With respect to its surface area and structure, the CB_P under vacuum conditions is close to commercial blacks of the N300 series. This suggests that the mechanisms leading to lower surface area and structure discussed above have little influence on vacuum pyrolysis CB_P.

TABLE 11.5
Acid-Base Demineralization of CB_P

Pilot Plant, Run # H03[a]	CB_P Nondemineralized	CB_P Demineralized[b]	
Ash content [wt. %]	14.6	6.3	
Surface area [m²/g]	42.1	53	
Relative Concentration of Individual Ash Component [%]			**Reduction Ratio, %**
ZnO	29.6	13.4	55
SiO_2	27.0	14.0	48
Al_2O_3	11.5	4.4	62
CaO	10.2	1.2	88
Na_2O	2.4	1.3	46
Fe_2O_3	2.7	0.7	74
MgO	1.0	0.1	90
K_2O	0.1	0.1	0
Others	15.5	7.8	50
Total	100.0	43.0	57

[a] Pyrolysis of tire sidewall shreds.[29]
[b] One single treatment with 1 N H_2SO_4 and 1 N NaOH.

Source: From Chaala, A. et al., *Fuel Proc. Technol.*, 46 (1996), 1–15. With permission.

11.4.1.3 CB_P Demineralization

As mentioned earlier, CB_P contains nearly all of the original inorganic tire additives. Consequently, the ash content of CB_P is considerably higher as compared to that of commercial carbon blacks (Table 11.3). Since a high ash content prevents the use of CB_P in certain applications, it is desirable to decrease the CBp mineral content. This can be done by successive treatment with an acid and a base[29] or strictly with an acid.[30] The effects of such a demineralization treatment for a CB_P produced in the pilot plant for tire sidewall are summarized in Table 11.5 for a single acid/base treatment. Treatment with H_2SO_4 followed by treatment with NaOH reduced the ash content from 14.6 to 6.3 wt.%. However, if the acid/base treatment is repeated, ash concentrations lower than 1% can be obtained. After a one-step acid/base treatment, the surface area increased significantly from 42 to 53 m²/g, whereas after repeated treatments a value of 65 m²/g was obtained. It should be mentioned that the tire sidewalls used as pyrolysis feedstock are low-surface-area carbon blacks.[28] Thus, the surface areas of the sidewall-derived CB_P were lower as compared to CB_P from whole tires (Table 11.4 and Table 11.5).

CB_P contains a variety of ash compounds. In the case of the CB_P from tire sidewalls, the main compounds are ZnO, SiO_2, Al_2O_3, and CaO (Table 11.5). In addition to carbon black, some tires contain silica used as filler.[31] Thus, the CB_P recovered from these tires should contain more SiO_2 than the CB_P investigated above. By the demineralization treatment, the concentrations of

Al_2O_3 and CaO were much more reduced as compared to concentrations of ZnO and SiO_2. This indicates that scrap tires should be thoroughly washed to remove any dirt prior to being introduced into the pyrolysis reactor. The changes of the organic CB_P portion upon demineralization were studied by surface spectroscopic methods.[29] It was shown that the demineralization treatment affected the organic portion only slightly. It can be concluded that from a technical standpoint, the acid/base or the single acid treatment is a suitable method to reduce the ash concentration of CB_P. The process economics, however, is not established yet.

11.4.1.4 CB_P Surface Chemistry

11.4.1.4.1 *Elemental Composition*

The most important (90%) commercial application of carbon black is rubber reinforcement.[32] Therefore, it is logical to investigate whether CB_P can replace some commercial carbon black grades as semireinforcing filler in rubbers. For such an application, the surface chemistry and activity are very important because both properties determine the strength of the carbon black – rubber interaction. The surface chemistry of CB_P and commercial carbon blacks has been studied in this laboratory by x-ray photoelectron spectroscopy (XPS) and secondary ion mass spectroscopy (SIMS).

The first information of the XPS experiment is the chemical composition of the surface. On the surface of commercial rubber–grade carbon blacks, only oxygen and sulfur are found in addition to carbon (hydrogen cannot be detected by XPS). Carbon is the major element on the surface; the combined concentration of oxygen and sulfur is usually smaller than 2 atom%.[33–35] On CB_P, in addition to the elements found on commercial blacks, nitrogen and zinc are detected.[33,34] The nitrogen most probably originates from the vulcanization accelerators and the antioxidants used during tire manufacture, whereas zinc oxide is added as a vulcanization catalyst.[28] The elemental surface composition of the CB_P depends on the pyrolysis conditions. With increasing pyrolysis pressure and temperature, the surface concentration of sulfur and zinc decreases (see Table 11.6). During pyrolysis, zinc oxide reacts with sulfur to form zinc sulfide.[36] The zinc sulfide is partly covered by carbonaceous deposits. Thus, with increasing coverage, the surface concentration of zinc and sulfur decreases. As mentioned earlier, the concentration of these deposits increases with the pyrolysis pressure. The increasing coverage of the zinc sulfide with increasing pyrolysis temperature might be due to an increased mobility of the deposits at higher temperature.

11.4.1.4.2 *Chemical Nature of the Carbon on the CB_P Surface by XPS*

X-ray photoelectron spectroscopy (XPS) also yields information on the chemical nature of the different elements present on the surface. The carbon XPS spectra of commercial carbon blacks were fitted to five peaks: an asymmetrical so-called graphite peak for carbon atoms in graphite-like structures (C_0);

TABLE 11.6
Elemental Composition of the Surface of CB_P Obtained under Various Pyrolysis Conditions and a Commercial Rubber-Grade Carbon Black

Sample		Element [atom %]				
CB_P Pyrolysis Temperature and Pressure		Carbon	Oxygen	Nitrogen	Sulfur	Zinc
500°C	0.3 kPa	91.3	3.3	0.7	2.7	2.0
500°C	10.0 kPa	91.5	3.3	0.7	2.5	2.0
500°C	20.0 kPa	92.3	3.0	1.0	2.0	1.7
600°C	0.3 kPa	91.9	3.4	0.9	2.0	1.8
700°C	0.3 kPa	93.4	3.4	0.6	1.4	1.2
Commercial black, N539		98.5	1.0	[a]	0.5	[a]

[a] Element not detected

Source: From Darmstadt, H., Roy, C., and Kaliaguine, S., *Carbon*, 32 (1994) 1399–1406. With permission.

three peaks for carbon atoms with one, two, or three bonds to carbon and/or sulfur atoms (*e.g.*, C-OH, C = O, and –COOH groups, C_2, C_3, and C_4); and finally, a $\pi \rightarrow \pi^*$ peak (C_5), which is also due to carbon atoms in graphite-like structures. The carbon black spectra are dominated by the graphite peak. Because of the low concentration of noncarbon elements on the surface (Table 11.6) the peaks for carbon atoms with other elements ($C_2 - C_4$) are very weak (see Figure 11.4). In addition to the peaks discussed above, the spectra of the CB_P showed a peak assigned to carbon atoms in small aromatic compounds or aliphatic structures (C_1). These compounds are part of the carbonaceous deposits. The area of this peak can be used as a measure of the amount of carbonaceous deposits formed at different pyrolysis conditions. This is shown in Figure 11.5 for CB_P obtained by pyrolysis of particles from tire sidewalls in a laboratory batch reactor.[33,34] The amount of carbonaceous deposits strongly decreased with decreasing pyrolysis pressure (100 to 0.3 kPa) and increasing temperature (420 to 700°C). Higher temperatures and lower pyrolysis pressures limit the adsorption of hydrocarbons on the carbon black surface. Thus, fewer carbonaceous deposits are formed from these adsorbed hydrocarbons. To preserve the integral properties of the virgin carbon black as much as possible, the pyrolysis should be performed at low pressure and at the highest possible temperature.

11.4.1.4.3 Chemical Nature of the Carbon on the CB_P Surface by SIMS

Secondary ion mass spectroscopy (SIMS) was used as a second surface spectroscopic analysis method. In the SIMS experiment, the sample surface is bombarded with high-energy ions, causing the ejection of charged and neutral fragments from the surface. Charged fragments (ions) are detected in a mass spectrometer. Which fragments are ejected depends on the chemical nature of the surface. The yields of certain ions can be correlated to the amount of carbonaceous deposits on carbon blacks. The SIMS spectra of carbon blacks show intense peaks of C_2^- and C_2H^- ions.[35,37] The surface of

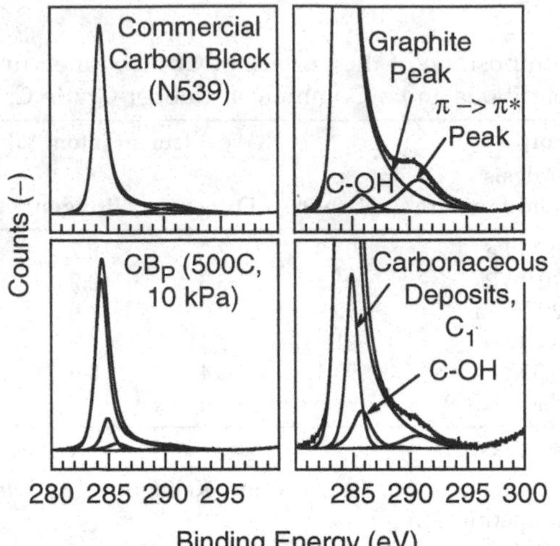

FIGURE 11.4
XPS carbon spectra of a commercial carbon black and a CB_P; spectra on right enlarged to 20% of maximum intensity.

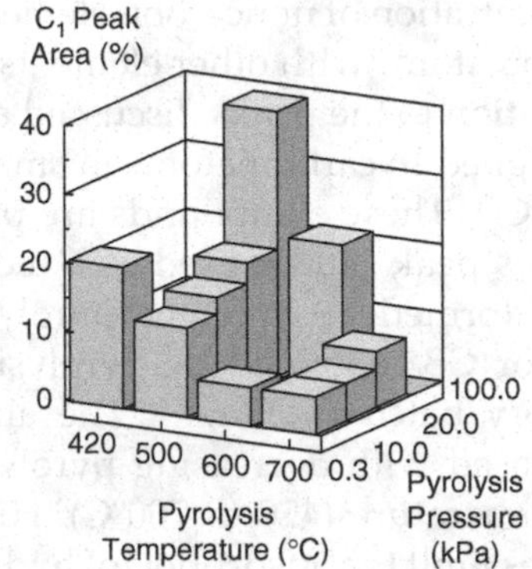

FIGURE 11.5
Dependence of the area of the CB_P XPS C_1 peak (measure for the amount of carbonaceous deposits) on pyrolysis temperature and pressure. (From Roy, C., Chaala, A., and Darmstadt, H., *J. Anal. Appl. Pyrolysis*, 51 (1999) 201–221. With permission.)

carbon blacks mainly consists of graphite (graphene) layers with many defects. The most important contribution to the C_2^- peak arises from the interior of these layers. The carbonaceous deposits consist of small, aromatic structures and aliphatic groups. The concentration of C-H groups at the edges of the small, aromatic structures is much higher as compared to a carbon black surface without deposits. Most of the C_2H^- ions originate from the edges of the small, aromatic structures. The ratio of the C_2H^-/C_2^- SIMS peaks is therefore a measure of the concentration of carbonaceous deposits on the CB_p surface.

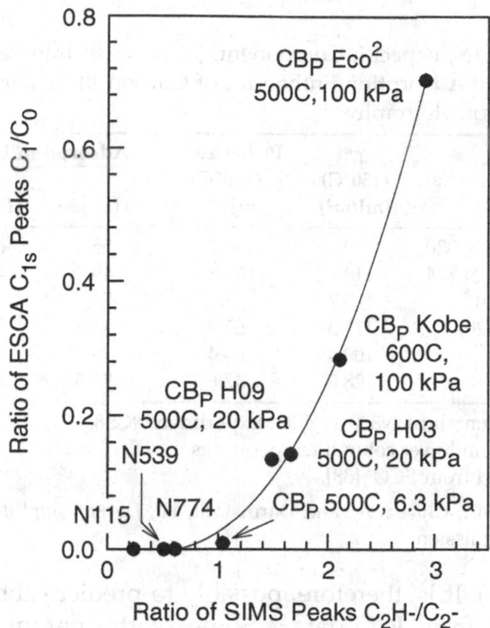

FIGURE 11.6
Correlation between two surface spectroscopic measures (XPS and SIMS) for the concentration of carbonaceous deposits on carbon blacks. (From Darmstadt, H., Sümmchen, L., Roland, U., Roy, C., Kaliaguine, S., and Adnot, A., *Surf. Interface Anal.*, 25, 1997, 245–253. With permission.)

In the previous section, the relative area of the XPS C_1 peak was discussed as another measure for the concentration of carbonaceous deposits. The correlation between the XPS and SIMS parameters is shown in Figure 11.6 for a variety of CB_P and commercial carbon blacks. The CB_P were produced by atmospheric pyrolysis[17,38] and under vacuum,[34] respectively. The excellent agreement between the two surface spectroscopic techniques confirms that the chemical nature of the surface of CB_P from vacuum pyrolysis is much closer to commercial grades as compared to CB_P from atmospheric pyrolysis.

11.4.1.4.4 Carbon Black Surface Activity by Inverse Gas Chromatography

The interaction of the carbon black surface with the rubber is of critical importance for the reinforcement behavior in rubbers. A suitable method for the determination of the carbon black surface activity with respect to rubber molecules is inverse gas chromatography (IGC). In this technique, probe molecules are injected into a gas chromatograph equipped with a carbon black–filled column. The probe molecules can be selected so that they represent a small portion of a rubber. The residence time of the probe molecules in the column depends on their interaction with the carbon black. A strong interaction causes a long residence time; if there is no interaction, the probe molecule will pass across the column as fast as an inert marker

TABLE 11.7

Surface Energies (Nonspecific Component, γ_s^d), Specific Interaction of Benzene (I^{sp} benzene), and Adsorption Enthalpies of Carbon Blacks Determined by Inverse Gas Chromatography

Sample	γ_s^d (150°C) [mJ/m²]	I^{sp} Benzene (150°C) [mJ/m²]	Adsorption Enthalpy [kJ/mol]		
			Hexane	Hexene	Benzene
Commercial black N 330	150.4	80.2	55.6	61.8	65.2
Commercial black N 774	118.1	63.8	49.5	49.6	49.6
CBP (polyisoprene)[a]	103.7	65.5	56.8	63.9	66.6
CBP passenger (H20)	113.3	65.0	61.3	58.3	57.7
CBP truck (H22)	109.4	69.4	79.2	74.5	67.9
CBP (passenger, 1atm)[c]	28.0[b]	37.6[b]	51.5	52.5	53.3

[a] The feedstock contained N330 as filler. Experiment #G58.
[b] Extrapolated from lower column temperatures.
[c] Sample obtained from ECO^2 [38].

Source: From Roy, C., Chaala, A., and Darmstadt, H., *J. Anal. Appl. Pyrolysis*, 51 (1999) 201–221. With permission.

(usually methane). It is, therefore, possible to predict rubber reinforcement of carbon blacks from IGC data.[39] Some of the parameters that can be obtained by IGC are: the nonspecific portion of the carbon black surface energy (γ_s^d), the specific interaction with molecules having a certain functional group (I^{sp}), and the adsorption enthalpy (ΔH). Of these parameters, ΔH is a measure of the strength of the active sites on the carbon black surface, whereas γ_s^d and I^{sp} are related to the number and strength of active sites on the carbon black surface.

To compare the surface activity of a carbon black before and after pyrolysis, two samples were studied[40]: an N330 commercial carbon black and a sample obtained by vacuum pyrolysis at 500°C and 6 kPa (absolute) of a polyisoprene rubber containing N330 (CBp (polyisoprene)). For these two carbon blacks, the adsorption enthalpies of hexane, hexene, and benzene were comparable (see Table 11.7), indicating that the strength of the active sites was preserved during the pyrolysis. However, γ_s^d and I^{sp} of CB_P (polyisoprene) were approximately one third lower than the corresponding values of the original commercial N330. This decrease can be explained by a loss of one third of the active sites on the carbon black surface during the pyrolysis. It was assumed that these "lost" sites are covered by carbonaceous deposits on the carbon black surface. Since more carbonaceous deposits are formed during atmospheric pyrolysis than during vacuum pyrolysis, one can expect that the values of γ_s^d and I^{sp} for the CB_P obtained under atmospheric pyrolysis will be lower than the corresponding values for the CB_P obtained after vacuum pyrolysis. Comparing the surface energy values of the CB_P (passenger) and the CB_P (passenger, 1atm) samples showed that this was indeed the case (Table 11.7). In spite of the loss of some active sites, γ_s^d and I^{sp} for the CB_P obtained by vacuum pyrolysis of passenger and truck tires, respectively, the values were comparable to the γ_s^d and I^{sp} values for the low-surface-area commercial blacks, such as N774.

11.4.1.5 CB_P Applications

11.4.1.5.1 Use of CB_P as Filler in Polymers

Based on the various characterization studies that were conducted, it can be concluded that the surface activity of CB_P from vacuum pyrolysis is high enough to replace low-surface-area commercial grades (such as N774) in some rubber applications, such as conveyer belts and rubber boots. It is known that vacuum pyrolysis CB_P can also be used in nonrubber applications. The mechanical properties of CB_P-filled polyvinyl chloride (PVC) were found to be comparable to the compound filled with a low-surface commercial black.[41] An example is the elongation at break, which decreases for a laboratory CB_P and a low-surface-area commercial black in a very similar fashion (see Table 11.8). The other viscoelastic properties of CB_P-filled PVC are in-between PVC filled with low- and high-surface-area commercial blacks.[41] Researchers from the Indian Institute of Technology recently reported that the reinforcing properties of postpyrolysis heat-treated, demineralized CB_P were as good as commercial grades.[42]

As mentioned earlier, during pyrolysis carbonaceous deposits are formed. These deposits "bake" a portion of the carbon black aggregates together. Upon incorporation into rubber, these agglomerates are not broken up, which leads to poor reinforcing behavior of the CB_P. However, the agglomerations can be broken up by physical agitation into carbon black aggregates and carbonaceous deposits. The coke-like carbonaceous deposits have a higher density than the carbon black aggregates. It is possible, therefore, to remove by air classification a large portion of the undesirable carbonaceous deposits from the desirable carbon black filler.[43] The mechanical properties of rubbers filled with this treated CB_P were found to be comparable to low-surface commercial black of the N774 series.[44,45] Upon replacement of N774 in a nitrile rubber (NBR) formulation, the tensile strength slightly decreased

TABLE 11.8
Mechanical Properties of CB_p-Filled PVC

Sample	Surface Area (m^2/g)	Structure, DBP No. ($cm^3/100$ g)	Ultimate Elongation (%) at Break with Various Carbon Black Loadings			
			0% CB	1%	5%	10%
Commercial black, Sterling R	25	70	140	76	22	9
CBP (Laboratory)[a]	60	85	140	73	22	18

[a] Sample produced in a laboratory reactor at 550°C and 4.6 kPa.

Source: From Dufeu, J.B., Roy, C., Ajji, A., and Choplin, L., *J. Appl. Polym. Sci.*, 46 (1992), 2159–2167. With permission.

TABLE 11.9

Mechanical Properties of CB_P-Filled NBR

Sample	Tensile Strength (MPa)	Ultimate Elongation (%)	Durometer Hardness (Points)
Commercial black, N774	10.6	289	74
CBP [a]	10.0	299	74

[a] Sample produced by pyrolysis at atmospheric pyrolysis,[44] the concentration of carbonaceous deposits was reduced by physical agitation and classification.[43]

Source: Fader, J.H., Upgraded Pyrolyzed Carbon Black (CB_p) from Waste Tires and Scrap Rubber as Reinforcing Filler in Rubber Compounds, *Proceedings*, Rubber Division, American Chemical Society, Cincinnati, OH, U.S., October 17–20, 2000. With permission.

from 10.6 to 10.0 MPa, whereas ultimate elongation slightly increased from 289 to 299% (see Table 11.9). However, this process, which has been patented, does not substantially reduce the amount of ash remaining in the CB_P.[43]

The electrical properties of carbon black fillers are very important in applications such as cable insulations and antistatic plastics. Carbon black imparts electrical conductivity to normally resistive plastics. In addition to the structure (DBP adsorption) and surface area, the surface chemistry has an important influence on the electrical conductivity of carbon black and carbon black–filled plastics. In the case of CB_P, the electrical conductivity is lower when compared to virgin commercial carbon black. However, Pantea and co-workers[46] recently showed that a postpyrolysis heat treatment of the CB_P can increase the electrical conductivity to values higher than those obtained with virgin blacks.[47,48] The CB_P electrical conductivity can be controlled for a given application.

11.4.1.5.2 *Use of CB_P for Reinforcement of Bitumen*

Another contemplated commercial application involves the use of pyrolytic carbon black as filler in road bitumen. Extensive testing of this application has been performed in the authors' laboratory in collaboration with the Ministère des Transports du Québec (MTQ), Québec City, Canada.[49–52] Commercial carbon black was first used as a bitumen modifier about 30 years ago by the Australian Carbon Black Pty. Ltd. in collaboration with the Australian Road Research Board.[53] The idea for such an application was based on the concept that commercial carbon black is known to reinforce rubber polymers, improve their durability, and increase ultraviolet (UV) protection. Scientists have demonstrated that the use of this modifier as a reinforcing agent for bitumen cement may lead to a similar improvement.[54] In fact, to attain high road performance and to be convenient for specific climate and traffic conditions, conventional road bitumen binders must be modified.

The objective of the studies cited above was to investigate the possibility of incorporating CB_P and commercial carbon black (CB_c) in conventional bitumens to produce a binder with improved properties in comparison to unmodified bitumen concrete. A comparison between the influence of CB_P

and CB_c on mechanical properties of the concrete included compaction and shear properties, rutting, resistance to water, and thermal stress–restrained specimen tensile strength.

11.4.1.5.2.1 Compaction and Shear Properties of Bituminous Concrete — This test method is used for guidance in selection of the optimum bitumen content and to determine void requirements in the bituminous concrete sample. Details concerning the specimen preparation, equipment, and experiments are described in the corresponding standard test method (ASTM D 3387). The U.S. Corps of Engineers Gyratory Testing Machine (GTM) was used to determine compaction and shear properties of concrete.

The results showed that among the four concrete samples prepared with the different bitumen concentrations (5%, 5.1%, 5.2%, and 5.3 wt.%), only the sample with a 5.3 wt.% bitumen content had the percentage of voids required by the GTM standard.[55] This can be seen in Table 11.10, which compares the measured voids of the 5.3% sample and the requirements, at different gyrations. No significant differences between the asphalt samples with and without CB_p were observed. Thus, the addition of CB_p to the bitumen only slightly affects the concentration of voids in the asphalt.

The data indicate that voids filled with bitumen (VFB) increases with the CB_p concentration. The modifier filled the intergranular pores. This can also prove that the integration of the CB_p in the bitumen-aggregate mixtures does not alter the cohesion specifications of the straight binder.

11.4.1.5.2.2 Rutting Test — This test consists of repetitive crossing of a regulated pneumatic wheel, which exerts a vertical force on a standard parallepipedic paving mixture sample. The wheel involves a relative rut, depending on the concrete nature. The result is expressed as a percent number, which links the rut thickness to the sample thickness. According to the SUPERPAVE method used by the MTQ, the test is performed at 60°C with a 50 mm thickness plaque, and compacted at 6% of voids. The standard test method has been described in greater detail by Lelièvre.[56] The rutting must be < 10% for 1,000 cycles and > 20% for 3,000 cycles.

The results describing the rutting phenomenon are summarized in Figure 11.7. The results indicate that the CB_p-modified bitumen met all the requirements. In addition, the histograms clearly show an improvement in the rutting potential as a result of the CB_p reinforcement. These results confirm the rheological and the classical mechanical tests,[51] which showed that at high temperatures, the CB_p blends considerably reduced the thermal susceptibility by increasing the rigidity and the viscosity of the binder.

11.4.1.5.2.3 Effect of Water on Cohesion of Compacted Bituminous Mixtures — The temperature variations and the aging factor have a definite effect on road behavior. The technique used for measurement of the loss of cohesion, resulting from the action of water on compacted bituminous concrete, is method ASTM D 1075.

TABLE 11.10
Results of the Gyratory Testing Machine

Binder		Gyrations	Voids (%)	Required Voids (%)	VFB[a]
Straight Shell					
		10	15.1	>11	38.5
Binder content	5.3 (%)	20	12.4		44.0
Maximum density[b]:	2.604	40	9.7		50.9
Aggregates:	Ray-car	60	7.9	4–8	55.5
Bulk density aggregates[b]:	2.775	80	7.2		58.8
Test temperature:	135°C	100	6.5		61.6
		120	6.0		63.8
		150	5.3		66.6
		200	4.8	>2	68.8
		230	4.2		71.8
Shell + 10% CB_p					
		10	14.7		39.2
Binder content	5.3 (%)	20	12.0		44.9
Maximum density[b]:	2.604	40	9.4		51.8
Aggregates:	Ray-car	60	7.6	4–8	56.4
Bulk density aggregates[b]:	2.775	80	6.9		59.9
Test temperature:	135°C	100	6.3		62.5
		120	5.7		64.7
		150	5.1		67.3
		200	4.7	>2	70.1
		230	4.1		72.2

[a] Voids filled with bitumen (%).
[b] The maximum and bulk densities were determined according to ASTM D 2041.

Source: Chebil, S., Chaala, A., and Roy, C., *Polym. Recycling*, 2 (1996) 257–269. With permission.

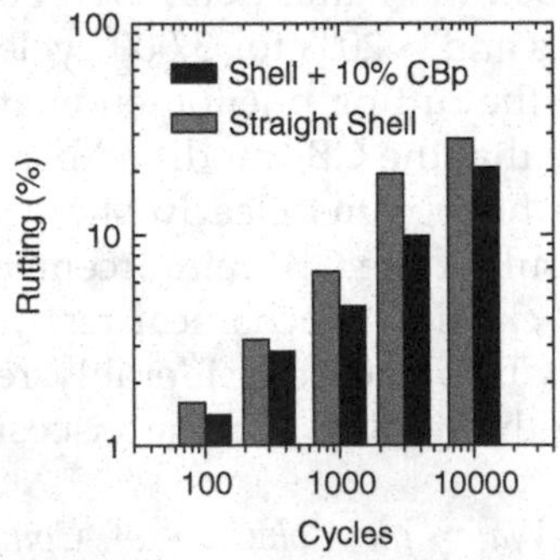

FIGURE 11.7
Rutting test of CB_P-reinforced bitumen and nonreinforced bitumen. (From Chebil, S., Chaala, A., and Roy, C., *Polym. Recycling*, 2 (1996) 257–269. With permission.)

First, two lots of three samples with equivalent bulk density were chosen (ASTM D 2726). One of the two lots was immersed in water under prescribed conditions (ASTM D 1075). The compressive strength was compared between the fresh (group 1) and aged (group 2) specimens. The index of retained strength (effect of water on cohesion) and the loss of coating (percentage indicating the loss of concrete material involved by the effect of water) can then be deduced as follows:

$$\text{The index of retained strength}\,(\%) = \frac{S_2}{S_1} \times 100 \tag{10.1}$$

$$\text{Loss of coating}\,(\%) = \frac{S_1 - S_2}{S_1} \times 100 \tag{10.2}$$

where S_1: compressive strength of dry specimens (group 1); and S_2: compressive strength of immersed specimens (group 2).

The CB_p has a significant effect on the cohesive properties of the binder. The index of retained strength and the loss of coating differ by 18.4% when the CB_p is added.

This pronounced improvement, which is clearly due to the presence of the modifier, indicates that the CB_p reinforces the compacted bitumen-aggregate mixture against the effect of water and aging. As proven with the rheological tests and the Strategic Highway Research Program, U.S. (SHRP) specifications, it is clear that the CB_p has, at high temperature, a positive influence on the cohesion and aging phenomena.

11.4.1.5.2.4 Thermal Stress Restrained Specimen Tensile Strength — The method of the American Association of State Highway and Transportation Officials (AASHTO TP10-93) determines the tensile strength and temperature at fracture of field on laboratory-compacted bituminous mixtures by measuring the tensile load in a specimen, which is cooled at a constant rate while being restrained from contraction. The information obtained from this experiment gives indications about thermal cracking and road life. Figure 11.8 illustrates the effect of the CB_p on the straight bitumen behavior under the described specific conditions.

The results indicate that the presence of CB_p decreases the slope of the curve describing the variation of the constraint as a function of temperature. The slope decreases from –42.9 PSI/°C (slope "a") to –35.5 PSI/°C (slope "b"). This proves the decrease of the modified concrete susceptibility at low temperature, which leads to constraints and deformation resistance. The present results obtained for these compacted bituminous mixtures confirm the results attained with the Bending Beam Rheometer (BBR) method and the SHRP specifications described above. As already described for the Fraass

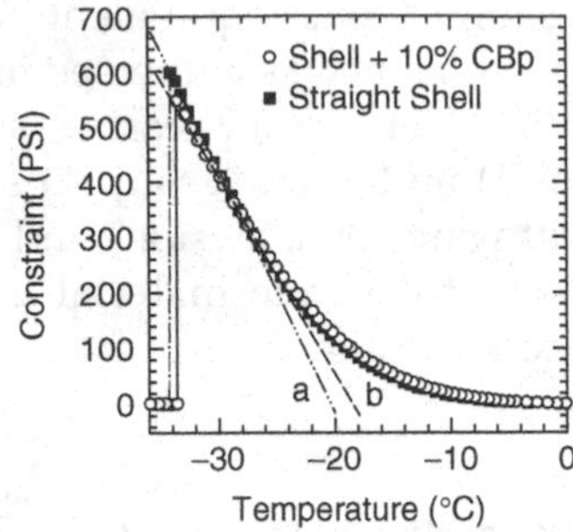

FIGURE 11.8
Constraint as a function of temperature for CB_P-reinforced bitumen and nonreinforced bitumen. (From Chebil, S., Chaala, A., and Roy, C., *Polym. Recycling*, 2 (1996) 257–269. With permission.)

point results,[57] this test confirms the resistance of the modified concrete subjected to gradual contraction and freezing.

The surface activity of CB_p is very important for this type of application (for road use). Strong interactions occur between asphaltene bitumen and CB_p particles, causing the formation of networks. It has been found that the addition of 5 to 30 wt.% of CB_p in road bitumen increased the high service-temperature by reducing the thermal susceptibility of the pure bitumen and increasing the rutting resistance and the cracking resistance. To prepare high-performance binders without exceeding the processing viscosity, the addition of CB_p of 5 to 15 wt.% has been suggested.[51]

The integration of CB_p into the bitumen matrix improves the road bitumen quality, particularly its resistance to rutting. This effect will be appreciated in warm climates, where the difference between extreme season temperatures is not very high. In cold countries such as Canada, where this difference is around 60°C, CB_p added to road bitumen is only effective in summer. During winter, the bituminous concrete formulated with CB_p will exhibit neither significant improvement, nor deterioration.

11.4.1.5.3 Use of CB_P in Other Applications

Another reported potential application of CB_P is its use as feedstock for the production of activated carbon.[58–60] Surface areas above 1000 m^2/g were obtained for some CB_P-derived activated carbons.[59] CB_P was also tested as feedstock material for the production of fluorinated carbon electrodes for lithium batteries. Due to the presence of carbonaceous deposits on its surface, CB_P reacts at mild conditions with fluorine. Preliminary tests indicated that CB_P is, indeed, a suitable material for this application.[61] Other industrial uses reported are in the ink industry[62] and in the copper industry as a carbon-reducing material.[18]

Despite the promising applications discussed above, in reality, CB_P has found only limited penetration in the user sector. One of the resistances expressed is the CB_P-suspected heterogeneity, because this recycled material is a blend of various grades of commercial carbon blacks. Until it is pro-

TABLE 11.11
Physicochemical Properties of the Pyrolytic Oils

Property	Laboratory Retort G-821	Pilot Plant H-70	Demo Plant JT-1
Density, kg/m^3 25°C	941	956	970
Density, °API	18.8	15.8	14.3
Viscosity, cSt: 25°C and 50°C	84 and 20	11.5 and 5.0	14.6 and 6.1
Flash point, °C	39	41	—
Sediments, wt.%	0.0	0.7	0.0
Water, wt.%	2.0	0.1	0.2
Carbon, wt.%	85.2	85.8	86.1
Hydrogen, wt.%	10.9	9.8	9.6
Nitrogen, wt.%	0.8	0.8	0.9
Sulfur, wt.%	1.1	1.1	1.2
Ash, wt.%	0.02	0.03	0.01
Oxygen*, wt.%	2.0	2.5	2.2
CCR, wt.%	1.5	4.1	4.5
Gross calorific value**, MJ/kg (BTU/lb)	44.8 (19280)	43.6 (18763)	43.8 (18849)
Refractive index 25°C	1.5166	1.5368	1.5442
pH	7.4	6.3	6.7

* Determined by difference.
** Anhydrous basis.

duced and made available in bulky volumes through a reliable and steady pyrolysis process, the user sector will be hesitant to pay the right price for this valuable product.

11.4.2 Pyrolysis Oil

11.4.2.1 Whole Oil

The physicochemical properties of the pyrolytic oils obtained using a laboratory retort, a pilot plant, and a large-scale demo plant, are presented in Table 11.11. The pyrolytic oils G-821, H-70, and JT-1 were characterized. The gross calorific values of the oils were high at 44.8 MJ/kg, 43.6 MJ/kg, and 43.8 MJ/kg for G-821, H-70, and JT-1, respectively. These values are higher than the heating value of the used tires from which the oils are recovered (33 MJ/kg).[5] The gross calorific value of tire pyrolysis oils exceeds by far the heating value of bituminous coal (28 MJ/kg)[63] and even wood charcoal (30 MJ/kg).[64] Their relatively low ash (< 0.03 wt.%), sulfur (1.1 to 1.2 wt.%), and Conradson carbon residue (CCR) (1.5 to 4.5 wt.%) content are noticeable (Table 11.11).

Because G-821 oil was produced in a batch reactor, it was relatively free of sediments. H-70 oil contained a relatively small amount of sediments (0.7 wt.%), mostly consisting of carbon black particles. The JT-1 oil freshly collected from the condensing towers contained 4.8 wt.% sediments. These undesirable compounds can be removed from the oil by decantation and

TABLE 11.12

Elemental Composition of the Pyrolytic Oils

Property	Laboratory Retort G-821	Pilot Plant H-70	Demo Plant JT-1
Carbon, wt.%	85.2	85.8	86.1
Hydrogen, wt.%	10.9	9.8	9.6
Nitrogen, wt.%	0.8	0.8	0.9
Sulfur, wt.%	1.1	1.1	1.2
Ash, wt.%	0.02	0.03	0.01
Oxygen [a], wt.%	2.0	2.5	2.2
C/H atomic ratio	1.53	1.37	1.34

[a] By difference.

filtration. Other physicochemical characteristics of the crude oils are given in Table 11.12.

According to the C/H ratio, one can suggest that G-821 is rich in aliphatic hydrocarbons. The viscosity of this oil is higher than the other oils. Based on the refractive index, H-70 and JT-1 oils are high in aromatics. This is confirmed by their high density, their low H/C atomic ratio, and their high carbon residue. The aromatization process, which occurs through a Diels-Alder mechanism, i.e., alkanes → alkenes → aromatics → polyaromatics, is enhanced by the high pyrolysis temperatures. These oils represent a valuable raw feedstock material for the production of high-quality electrode coke and carbon black.

Their high content in carbon makes the pyrolytic oils a desirable feedstock for the fabrication of high-value carbon materials that could, in turn, be used in various industries (e.g., production of calcium carbide, steelworks, etc.). The absence of vanadium and the low content in heteroatoms such as sulfur are definite assets for these oils. The chlorine content was reported to range between 100 and 180 ppm.[16]

The chemical composition of tire pyrolysis oil in terms of unsaturated hydrocarbons is basically similar to that of the oils obtained by coking of various petroleum residues, by thermal cracking of gas/oil fractions, and by steam cracking of gasoline.[65] Therefore, the pyrolytic oil can be blended with these condensates, then subjected to the usual posttreatments at the refinery site.

The metal composition presented in Table 11.13 was determined for pyrolytic oil that was obtained by vacuum pyrolysis of used tire in the pilot plant (run #H20). The details concerning the feedstock characteristics and the operating conditions of the pyrolysis test are described elsewhere.[16] The high concentrations in iron, aluminum, zinc, and sodium possibly originate from the carbon steel pyrolysis equipment (Fe); the dust- and aluminum-based additives (Al); the zinc oxide added in the formulation of the tire (Zn); and the carbon black quenching water, the binding agent, and the sodium-based additives for Na. It is worth noticing that the amount of aluminum reported

TABLE 11.13
Mineral Composition of the Pyrolytic Oil (Run #H03) [weight ppm]

V	Mn	P	Mg	Na	Ba	As	Ti	Ni
< d.l	0.006	0.142	0.134	1.280	0.077	< d.l	0.198	0.073
Fe	**Cu**	**Al**	**Zn**	**Pb**	**Ca**	**Cr**	**Cd**	**Co**
5.585	0.104	4.030	2.044	0.918	0.458	0.093	0.024	< d.l

Note: d.l = detection limit of the measurement method.

Source: Roy, C., Chaala, A., and Darmstadt, H., *J. Anal. Appl. Pyrolysis*, 51 (1999) 201–221. With permission.

in Table 11.13 is probably underestimated, because the ash recovered from the oil was not totally digested in the *aqua regia*.

The important requirements for heavy fuels used as diesel fuels are their ignition quality, viscosity, water, sediment, and sulfur contents. The ignition property of diesel fuels is evaluated by the cetane number. A good indication of the ignition quality is given by the diesel index, which is related to the aniline point and the density of the fuel. The viscosity required at the fuel pump to provide better fuel injection characteristics is typically 10 to 15 cSt.[66] The temperature can be adjusted to provide the right viscosity at the fuel pump. The water and sediment content must be reduced to the levels that are specified for distillate fuel, before being supplied to the engine fuel filters. During combustion, the sulfur in the fuel is transformed into sulfur oxides. Since SO_2 is corrosive, the sulfur content of the fuel must be low. Furthermore, the presence of SO_2 in the exhaust gas may affect the lubricating oil.

Using the whole pyrolytic oil as a fuel would require preliminary treatments such as decantation, centrifugation, and filtration. The treated pyrolytic oil might then be used alone or blended with other fuels such as CIMAK-B10 diesel fuel, which has specifications that are given in Table 11.14.

TABLE 11.14
Characteristics of Pyrolytic Oils and Commercial CIMAK-B10 Diesel Fuel

Properties	Laboratory Retort G-821	Pilot Plant H-70	Demo Plant JT-1	CIMK-B10 (a)
Density, kg/m^3 25°C	941	956	970	975
Viscosity, cSt 50°C	20	5.0	6.1	9.7
Flash point, °C	39	41	35	50[b]
Water, wt.%	2.0	0.1	0.2	0.5
Sulfur, wt.%	1.1	1.1	1.2	3.5
Ash, wt.%	0.02	0.03	0.01	0.1
CCR, wt.%	1.5	4.1	4.5	10
Vanadium (ppm)	Nil	Nil	Nil	300

[a] Maximum value allowed.
[b] Minimum value allowed.

The addition of pyrolytic oil to this kind of diesel fuel would reduce the viscosity of the resulting blend. Consequently, the atomization would be improved, ensuring a more thorough combustion of the fuel. Based on its fuel properties, tire-derived pyrolytic oil can be considered to be a valuable fuel for blending with conventional heating fuels.

11.4.2.2 Oil Fractions

In addition to be globally used as a heating fuel, the pyrolytic oil can be fractionated into different cuts, which could be further blended with the corresponding petroleum fractions. Several schemes of fractionation of the oil can be performed. The configuration depends on the application and throughput capacity. In the present work, the following, sometimes overlapping, oil fractions were studied: initial boiling point (ibp) – 204°C, > 204°C, 240 to 450°C, > 350°C, and > 400°C. The pyrolytic oil from pilot plant run #H03 was fractionated into: i.b.p. – 204°C, 27 wt.%, 204 to 350°C, 17 wt.%, and a residue > 350°C, 56 wt.%.

11.4.2.2.1 Pyrolytic Light Fraction (i.b.p. –204°C)

The fraction (i.b.p. – 204°C) was distilled up to 160°C to obtain a light naphtha fraction (i.b.p. – 160°C) similar to that produced from petroleum. The chromatographic analysis indicated that the fraction (i.b.p. – 160°C) is composed of 45 wt.% aromatics, 22 wt.% olefins, 15 wt.% isoparaffins, 1 wt.% n-paraffins, 7 wt.% naphthenes, and 10 wt.% high-molecular-weight hydrocarbons and heterocyclic compounds.[67] Compared to typical petroleum naphtha,[67] the pyrolytic fraction is rich in aromatics including BTX and alkyl benzene. Due to their high octane number, aromatic hydrocarbons are desirable in the composition of gasoline. The catalytic reforming process increases the gasoline octane number by converting n-paraffin hydrocarbons into isoparaffin and aromatic hydrocarbons. However, due to more stringent environmental regulations, refiners tend to partially substitute the aromatic hydrocarbons with isoparaffinic hydrocarbons. It has been found that under regular engine operating conditions, aromatic hydrocarbons condense, leading to the formation of hazardous polycyclic aromatic hydrocarbons (PAH).[68] PAH formed during the combustion reach the ambient air with the flue gas. Because the pyrolytic naphtha fraction has a chemical composition similar to that of petroleum naphthas produced in destructive processes such as coking and thermal cracking processes, it should be subjected to a hydrotreatment prior to be used in the refinery as a distillate gasoline pool.

Work performed at the research laboratory of Imperial Oil in Sarnia, Ontario, showed that the addition of 2 vol.% of a tire-derived pyrolytic light naphtha fraction to petroleum naphtha (i.b.p. – 152°C) and to regular unleaded gasoline (Mogas GF 711) increased the aromaticity of the resulting blends, which enhanced the octane number.[67] The increased sulfur, olefin, and nitrogen contents of the resulting blend are still below the hydrofining process specification requirements. Further study will be necessary to evaluate the hydrofining efficiency for the saturation of olefins and the reduction of metals as well as heteroatoms.

It has been found by Pakdel, et al.[69] that the light fraction (i.b.p – 204°C) contains 20 to 25 wt.% *dl*-limonene, a high-value fine chemical that may be used in industrial applications as an environmentally friendly solvent, a component for the production of resins and adhesives, and a dispersing agent for pigments.[69,70] The authors have studied the reaction mechanism of the *dl*-limonene formation during vacuum pyrolysis and have determined that: 1) *dl*-limonene is formed by the dimerization of isoprene units following a low-energy reaction mechanism (intramolecular cyclization to form *dl*-limonene is also possible); 2) a pyrolysis temperature higher than 500°C tends to crack the limonene molecules to trimethylbenzene, *m*-cymene, and indane, which have boiling points similar to *dl*-limonene; 3) *dl*-limonene yield increases as the pyrolysis pressure decreases; 4) any heat and mass transfer limitations during the pyrolysis hamper *dl*-limonene formation and favor a secondary degradation of the pyrolysis products; and 5) a proven and economical method to deodorize the *dl*-limonene–enriched fractions is needed.

11.4.2.2.2 *Pyrolytic Oil Fractions (>204°C) and (240 to 450°C).*

The commercial oil Sundex 790 was substituted with a deasphaltized pyrolytic oil fraction (> 204°C) in a rubber formulation used for the fabrication of floor carpets.[71] The main common physicochemical property of these oils is the aromatic content. The plasticizing effect of the pyrolytic oil has been evaluated by a rubber manufacturing company, Mondo Rubber Intl., Laval, Québec, Canada. The plasticizing properties of both oils are presented in Table 11.15 and indicate that the tire-derived pyrolytic oil exhibits a lower Mooney viscosity (40.5) than the Sundex 790 (42.0). The cure time is reduced with the pyrolytic oil, more so than with the commercial oil. Thus, the vulcanization process occurs more rapidly when pyrolytic oil is used. The tensile strength, the elongation at break, and the hardness of the rubber compounds obtained indicated that this vacuum pyrolytic oil is a good plasticizer.

Since the pyrolytic oil contains more volatile components than the Sundex 790 oil, it was decided to prepare a heavier oil fraction (240 to 450°C) to test this fraction as a plasticizer and to compare it with the commercial aromatic oil Dutrex R 729. The Mooney viscosity and the elastic and viscous moduli were found to decrease with increasing oil content of pyrolytic oil (240 to 450°C). The oil affects the curing characteristics as expected, i.e., by a decrease in both the highest and lowest torque and some delaying of the scorch and cure times. The tensile properties were also examined. In terms of moduli, the pyrolytic oil fraction (240 to 450°C) seems to be a more efficient softener than the aromatic oil Dutrex R 729. However, the ultimate properties evolved in completely different manners. At the lower levels up to 10 phr, the pyrolytic oil gave a larger drop in stress at break than the commercial oil; at 10 phr, the effects are equal. At higher levels (> 10 phr), the commercial oil significantly increases the elongation at break, while the pyrolytic oil does

TABLE 11.15

Evaluation of Pyrolytic Oil (>204°C) as a Plasticizer in the Floor Carpet Formulation

	Pyrolytic Oil (>201°C)	Sundex Oil 790
Processing Properties		
Viscosity ML $_{(1+4)}$ 100°C (ASTM D-1646)	40.5	42.0
Mooney Scorch (121°C)		
Minimum viscosity	18	17
Scorch time, min	17.0	34.5
Cure index	4	10
Rheological Properties 168.5°C		
Minimum viscosity	10.0	10.2
Scorch time t02, min	2.1	3.1
Modulus at 12 min	54.5	48.0
90% cure time (tc90), min	6.2	9.7
Vulcanized Material Properties, Cure 16 min 160°C		
Hardness, shore-A (ASTM D2240)	64	62
Tensile, MPa (ASTM D412)	15.18	12.14
Elongation, % (ASTM D 412)	810	890
300% modulus, MPA (ASTM D412)	3.17	2.62
Tear Die –C, KN/m (ASTM D624)	31.16	30.29
Specific gravity, g/cm^3	1.39	1.39

Source: From Roy, C., Chaala, A., and Darmstadt, H., The vacuum pyrolysis of used tires. end-uses for the oil and carbon black products, *J. Anal. Appl. Pyrolysis*, 51 (1999) 201–221. With permission.

not produce much change. Thus, the pyrolytic and commercial oils seem to interfere with the vulcanization network in different pathways. This was expected considering this kind of oil, its origin, and its specific chemical composition. The compression set data are in line with the tensile properties, i.e., there is a marginally higher loss with the pyrolytic oil. The above results indicate that the heavy pyrolytic oil has good mechanical and lubricating properties, very similar to those of commercial processing oil.

11.4.2.2.3 Pyrolytic Residue (> 350°C)

The pyrolysis residue (> 350°C) represents 56 wt.% of the total oil that was obtained by vacuum pyrolysis of used tires during pilot run #H03 (Table 11.16).[72] This fraction has a high density, a high Conradson carbon residue, and a high content in asphaltenes. The latter are polyaromatic hydrocarbons, which are known to be the main precursor of coke.

The heavy oil fraction was subjected to a laboratory-scale delayed coking experiment under conditions similar to those used in industrial coking plants. Low sulfur (0.7 wt.%) and low ash content (0.24 wt.%) characterized the cokes obtained (Table 11.17).[72]

TABLE 11.16

Characteristics of the Pyrolysis Fraction (>350°C)

Density 20°C, kg/m³	948.1
Viscosity 50°C, cSt	6.8
Asphaltenes, wt.%	12.4
Toluene insolubles, wt.%	0.27
Refractive index, n_D^{20}	1.523
Sulfur, wt.%	0.8

Source: From Chaala, A. and Roy, C., *Fuel Proc. Technol.*, 46 (1996), 227–239. With permission.

TABLE 11.17

Coke Characteristics

Properties	Pyrolytic Coke	Petroleum Coke
Proximate Analysis, wt.%:		
Volatile matters	8.7	8.1
Ash	0.24	1.9
Fixed carbon	91.06	90.0
Elemental Analysis, wt.%:		
Carbon	91.6	90.5
Hydrogen	4.4	4.1
Nitrogen	1.0	0.71
Oxygen [a]	2.3	0.69
Total sulfur	0.7	4.00
Pycnometric density, Kg/m³	1800	1850
Interlayer Spacing, d_{002},Å	3.4985	3.5697

[a] Calculated by difference.

Source: From Chaala, A. and Roy, C., *Fuel Proc. Technol.*, 46 (1996), 227–239. With permission.

The major metals found in the cokes (zinc, 33.4 ppm and silicon, 15.30 ppm) find their source in the original tire. The absence of vanadium, an undesirable contaminant of commercial cokes, is noticeable. The x-ray diffractogram showed that the coke obtained is characterized by a lower interlayer spacing (d_{002} = 3.498 Å), indicating that this coke is graphitizable.[73] The coking experiment also yielded a highly aromatic naphtha fraction (i.b.p. –205°C), a light gas/oil (205 to 350°C) rich in aromatic hydrocarbons, a heavy gas/oil (> 350°C) that can be recycled in the coke manufacturing process, and finally, a gas with a high heating value.

11.4.2.2.4 Residual Fraction > 400°C Pyrolytic Bitumen

The residual fraction i.p.b. > 400°C, herein called pyrolytic bitumen (PB), was obtained by vacuum distillation of the bottom residue (> 350°C) at a

TABLE 11.18

Physicochemical Properties of Pyrolytic Bitumen PB (H18) and Petroleum Bitumen

Characteristics	Pyrolytic Bitumen PB	Petroleum Bitumen PC
Classical Tests:		
Penetrability 25°C, 1/10 mm	n.d.	113
Softening point, °C	29.0	48.5
Fraas point, °C	–15.0	–8.5
Dynamic viscosity 50°C, cP	436	n.d.
Generic Composition, wt.%:		
NC5-asphaltenes	26.6	19.0
Maltenes	71.6	80.6
Toluene-insolubles	1.8	0.4
Elemental Composition, wt. %:		
Carbon	88.4	85.3
Hydrogen	10.2	10.4
Nitrogen	0.6	0.8
Sulfur	0.6	3.3
Oxygen[a]	0.2	0.2
Atomic ratio C/H	0.72	0.68

[a] Determined by difference.

Source: From Chaala, A., Ciochina, O.G., Roy, C., Bousmina, M., *Polym. Recycling*, 3 (1997/98), 1–15. With permission.

yield of 29 wt.% based on the total oil produced in pilot plant run #H18.[74] It was characterized and tested as a binder or a modifier for road bitumens. Compared with petroleum bitumens, which are largely used in road pavements, PB has a high penetrability, a low softening point, and a low dynamic viscosity (see Table 11.18). The PB content of toluene-insolubles, which consists of carbonaceous materials, is higher than that found with petroleum bitumens (1.8 wt.% vs. 0.4 wt.% on average). These fines were physically carried out of the reactor by the gas and vapor streams and were recovered in the condensed oil product.

The elemental composition of the pyrolytic bitumen presented in Table 11.18 showed that it is rich in carbon and has a low hydrogen content. Also, the sulfur content is rather low, while the nitrogen content is similar to that of petroleum bitumens. The atomic carbon/hydrogen ratio (C/H) suggests that PB contains more aromatic hydrocarbons than petroleum bitumens. It is known that bitumens suitable for styrene-butadiene-styrene (SBS) blends should have high aromatic and low asphaltene contents.[75] In certain cases, special petroleum aromatic fractions are added to the bitumens to improve their compatibility. It has been found that the addition of aromatic oils (e.g., furfural extracts) enhances the peptization effect of the SBS-petroleum bitumen blends.[76] The addition of PB to an SBS blend may have the same peptization effect on the blend components as the addition of petroleum aromatic fractions.[76]

11.4.2.2.4.1 Rheological Properties of PB and Petroleum Bitumen PC

11.4.2.2.4.1.1 Rheological Functions G′(ω), G″(ω), and η(ω)* — Mechanical properties of the pyrolytic bitumen such as storage modulus G′, loss modulus G″, and complex viscosity η* were determined by rheological measurements under oscillating conditions.[74] Storage and loss moduli are components of the complex modulus G*, which is defined by the equation $G^* = G' + iG''$ and which reflects the total stiffness. The storage modulus G′ represents the elastic portion of the complex modulus, while the loss modulus G″ represents the viscous contribution of the bitumen. In the case of bitumen, a linear zone was previously determined by a strain sweep. In this zone, the stress is related to the strain by a linear equation, and the rheological behavior of the material is called viscoelastic.

Rheometers are not able to measure the dynamic functions at very low (10^{-3} Hz) and very high (500 Hz) test frequencies. Therefore, a time-temperature superposition principle is required to extrapolate the results obtained in restricted intervals of temperatures and frequencies to a wider range of conditions, allowing the behavior to be predicted. The curve obtained by superimposing curves plotted at various temperatures into a single function is called a "master curve." The construction of master curves has been described elsewhere.[74]

The variations of the storage modulus G′, loss modulus G″, and η* for PB as a function of reduced frequency are plotted at the reference temperature 30°C (Figure 11.9). At low temperatures, or at high frequencies, PB approaches the elastic state in which the bitumen can resist the deformation caused by loading. The stress developed during deformation is lower and the bitumen is more capable of recovering from its preloading condition. At high temperatures, or at low frequencies, G′ decreases continuously and approaches a complete viscous behavior or a complete dissipation of energy in viscous flow. The ability of the bitumen to limit the dissipation of energy is represented by G″. The shape of the G″(ω) curve is similar to that of G′(ω).

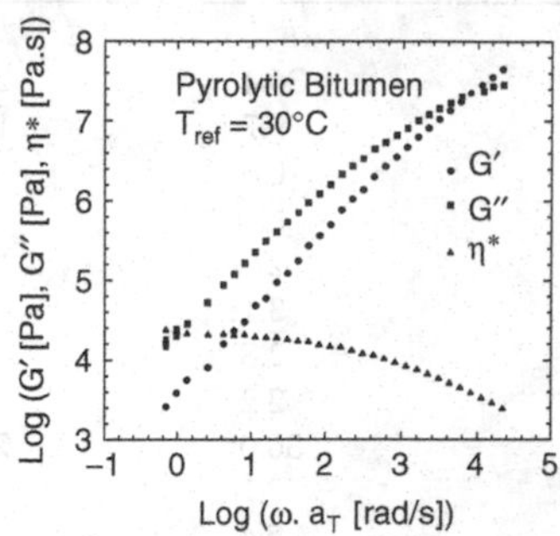

FIGURE 11.9
Master curves of G′, G″, and η* for pyrolytic bitumen. (From Chaala, A., Ciochina, O.G., Roy, C., and Bousmina, M., *Polym. Recycling*, 3 (1997/1998) 1–15. With permission.)

The viscosity of PB was very low throughout the frequency range tested. The increase and the decrease of the viscosity may be directed by the equilibrium existing between the components of the PB and the external stresses, such as temperature. It has been found earlier that the asphaltenes extracted by n-pentane from pyrolytic residue have a higher concentration of carbonyl groups and oxygenated sulfur than the asphaltenes of the petroleum bitumens.[50] These functional groups enable the products to form weak polar-polar interactions. At low temperature, the polar components contained in the PB matrix yield weak networks of polar-polar associations, providing elasticity to the bitumen. At high temperature, the networks become weaker, leading to the formation of a Newtonian fluid.

Based on these rheological tests, one can conclude that the addition of pyrolytic bitumen to a petroleum bituminous matrix causes the same effect as the integration of CB_p described earlier. The pyrolytic residue enhanced the high service-temperature by reducing the thermal susceptibility of the pure bitumen, i.e., the blend exhibits a significant ability to withstand severe thermal stress when subjected to gradual contraction and freezing.

11.4.3 Pyrolysis Gases

An analysis of the noncondensable gases is provided in Table 11.19. The gases are rich in hydrogen, methane, propane, and butane fractions. These gases have a high calorific value (46 MJ/kg), which makes them a valuable fuel. There is enough gas to supply all of the pyrolysis reactions. The excess gas can be burned to raise steam or generate electricity in turbines. A rough calculation indicated that combustion of the excess gas in a gas turbine and

TABLE 11.19
Non–Condensable Gas Composition

Reactor, Experiment #	Laboratory Batch Reactor, G-821	Pilot Plant, H-70	Demonstration Plant, JT-1
Gas [Vol. %]			
H_2	36.4	21.8	19.0
CH_4	9.7	14.7	23.1
CO	0.6	2.1	3.0
CO_2	6.6	4.9	5.1
C_2H_4	2.5	6.2	9.1
C_2H_6	3.1	7.1	7.4
C_3H_6	2.8	6.1	8.2
C_3H_8	2.2	5.5	3.2
C_4H_8	30.0	25.0	16.3
C_4H_{10}	0.7	0.0	0.0
Others	5.4	6.6	5.6
Total	100.0	100.0	100.0
Molecular weight [g/mol]	30.2	32.8	29.9
Gross calorific value [MJ/kg]	46.9	46.7	46.0

of all the pyrolysis oil in a diesel engine, for a 30,000 t/yr used tire vacuum pyrolysis plant, would generate 6 MWe net.

11.4.4 Data on Economics

The profitability of a scrap tire pyrolysis plant is, of course, process specific. The larger the plant capacity, the higher is the profitability. Continuous feed plants with capacity of less than 2 t/h will hardly enable positive cash flows. Because only very few privately owned large-scale plants for tire pyrolysis are currently in operation, it is impossible to report reliable figures about capital and operating costs. Figures of between $11 and 16 million and $125 to $200 (U.S.)/t are cited for capital costs and operating costs of 30,000 t/yr plants.[77] There is a consensus, however, about the desirability of a tipping fee to help kick off projects. There is also an agreement on the role of the CB_p as a key contributor to the process economics. A market value of $500 (U.S.)/t and more for the CB_p certainly improve the plant profitability. This fact pinpoints the need to enhance CB_p R&D activities in the direction of wide, rewarding areas of applications for this new material.

11.5 Conclusions

Pyrolysis has several advantages over other alternative tire recycling methods. No toxic substances are emitted, and various commercial applications for all of the products obtained are possible. An important difference between atmospheric and vacuum pyrolysis performed at 500°C is that the residence time of the hydrocarbons formed during the rubber cracking process is considerably shorter in the vacuum process. Thus, undesirable reactions such as the formation of carbonaceous deposits on the pyrolytic carbon black and the secondary decomposition reactions of valuable products such as limonene are limited. The oil yield is also increased due to the low pressure.

The carbon black surface area and structure (aggregation of the carbon black particles) changes only slightly during the vacuum pyrolysis. However, a small amount of carbonaceous deposits is formed on CB_P and deactivates approximately one third of the active sites on the CB_P surface. The strength of the remaining sites is the same as in the carbon black initially present in the tire. Thus, the tire CB_P has the same surface activity as that of a low-surface commercial carbon black such as grade N774. Tests in rubber and plastic compounds confirmed, indeed, that CB_P can replace certain commercial grades in these applications. Another potentially important end use of the CB_P is as an additive for road bitumen. For this end-use application, the raw CB_P must simply be crushed and sieved. Upgraded forms of CB_P

and hence, higher-value products, are also possible if the raw pyrolytic carbon black is demineralized, which is technically possible. CB_p is key to any pyrolysis process profitability.

The whole pyrolytic oil may be directly used as fuel or subjected to distillation. Various schemes of distillation have been performed for this oil and each fraction was tested in various applications. The lighter fractions can be used as a source of high-value chemicals such as BTX and limonene, and as extender oil in rubber formulation. The heavier fractions have been tested as additives in road bitumen and as feedstock for the production of coke. The PB obtained from vacuum pyrolysis of used tires can be profitably used as a compatibilizing agent for SBS-modified bitumen. PB enhances the aromaticity of petroleum bitumen, which enables the peptization of asphaltene and rubber.

Potential end uses for the pyrolytic oil in the petroleum refining industry include the following:

- Feedstock of the fluid catalytic cracking (FCC) process to produce light products (enhancement of resources in high-target-value fuels)
- Feedstock of the thermal cracking process to produce light products and olefin-rich gases
- Feedstock of the delayed coking process to produce light products and coke

Pyrolysis processes have been developed that enable the conversion of tires into oil, carbon black, and gas raw products. To enable such plants to reach the marketplace, focus in the future should be on the development of strong sectors of applications for the CB_p that reflect the genuine commercial value of this novel but overlooked recycled material.

Acknowledgment

The authors express their gratitude to Elsevier ACS Rubber Division, Wiley, R.T. Vanderbilt Co. and Rapra Technology Ltd. for authorizing the reprint of several tables and figures presented in this chapter.

References

1. Blumenthal, M., Scrap tire recycling markets update, *BioCycle*, 43 (2002) 58–62.
2. Steer, P.J., Tashiro, C.H.M., McIlveen, W.D., and Clement, R.E., PCDD and PCDF in air, soil, vegetation and oily runoff from a tire fire, *Water Air Soil Pollut.*, 82 (1995) 659–674.

3. Lampman, R., Hansonand, S., and Novak, R., Seasonal abundance and distribution of mosquitoes at a rural waste tire site in Illinois, *J. Am. Mosquito Control Assoc.*, 13 (1997) 193–200.
4. Anonymous, The changing face of tire recycling in Europe, *Scrap Tire News*, 17, 1 (2003) 1–14.
5. Ferrer, G., The economics of tire remanufacturing, *Resour., Conserv. Recycling*, 19 (1997) 221–255.
6. Myhre, M. and MacKillop, D.A., Rubber recycling, *Rubber Chem. Technol.*, 75 (2002) 429–474.
7. Savas, B.Z., Ahmad, Sh., and Fedroff, D., "Freeze-thaw Durability of Concrete with Ground Waste Tire Rubber," Transportation Research Record No. 1574, Nov. 1996, National Research Council, Washington, DC, U.S., (1996) 80–88.
8. Blumenthal, M., Use of scrap tires as a supplemental fuel, in: *Advances in Instrumentation and Control: International Conference and Exhibition Proceedings of the 1996 International Conference on Advances in Instrumentation and Control, ISA: 96.* Part 1 (of 2), Oct. 6–11, 1996, 51, 1, Chicago, IL, U.S.
9. Siuru, B., Tire fuels benefit cement making process, *Scrap Tire News*, 11 (1997) 10.
10. Adalevendis, Y.A., Atal, A., Carlson, J., Dunayevskiy, Y., and Voutros, P., Comparative-study on the combustion and emissions of waste tire crumb and pulverized coal, *Environ. Sci. Technol.*, 30 (1996) 2742–2754.
11. Levendis, Y.A., Atal, A., and Carlson, J.B., On the correlation of CO and PAH emissions from the combustion of pulverized coal and waste tires, *Environ. Sci. Technol.*, 32 (1998) 3767–3777.
12. Mastral, A.M., Callen, M.S., and Garcia, T., Polyaromatic environmental impact in coal-tire blend atmospheric fluidized bed (AFB) combustion, *Energ. Fuels*, 14 (2000) 164–168.
13. William, P.T. and Brindle, A.J., Aromatic chemicals from the catalytic pyrolysis of scrap tyres, *J. Anal. Appl. Pyrolysis*, 67 (2003) 143–164.
14. Williams, P.T., Black magic? High value products from scrap tyres, *Chem. Rev.*, 12 (2002) 17–19.
15. Kaminsky, W. and Mennerich, C., Pyrolysis of synthetic tire rubber in a fluidised-bed reactor to yield 1,3-butadiene, styrene and carbon black, *J. Anal. Appl. Pyrolysis*, 58 (2001) 803–811.
16. Roy, C., Chaala, A., and Darmstadt, H., The vacuum pyrolysis of used tires. end-uses for the oil and carbon black products, *J. Anal. Appl. Pyrolysis*, 51 (1999) 201–221.
17. Kawakami, S., Inoue, K., Tanaka, H., Sakai, T., Pyrolysis process for scrap tires, in: Jones, J.L., Radding, S.B. (eds.), *Thermal Conversion of Solid Wastes and Biomass*, ACS Symposium Series 130, ACS Publishers, Washington, DC, U.S., (1980) 557.
18. Kono, H., Onahama Smelting and Refining Co., Ltd. Iwaki City, Fukushima, Japan, private communication (1987).
19. Faulkner, P. and Weinecke, W., Carbon black production from waste tires, *Miner. Metall. Process.*, 18 (2001) 215.
20. Roy, C., Blanchette, D., deCaumia, B., Dubé, B., Pinault, J., Bélanger, É., and Laprise, P., "Industry Scale Demonstration of the Pyrocycling™ Process for the Conversion of Biomass to Biofuels and Chemicals," in: Kyritsis, S., Beenackers, A.A.C.M., Helm, P., Grassi, A., Chiaramonti, D. (eds.), *Proceedings*, First World Conference and Exhibition on Biomass for Energy and Industry, June 5–9, 2000, Sevilla, Spain, (2), James & James Science Publishers, London, U.K., (2001) 1032–1035.

21. Roy, C. Chaala, A., Recycling of automobile shredder residues by vacuum pyrolysis, *Resour. Conserv. Recycling,* 32 (2001) 1–27.
22. Yang, J., Kaliaguine, S., and Roy, C., Improved quantitative determination of elastomers in tire rubber by kinetic simulation of DTG curves, *Rubber Chem. Technol.,* 66 (1993) 213–229.
23. Yang, J., Blanchette, D., de Caumia B., and Roy, C., Modelling, scale-up and demonstration of a vacuum pyrolysis reactor, *Prog. Thermochem. Biomass Conversion,* Bridgwater, A. V. (ed.), Tyrol, Austria, Sept. 17–22, 2000, Blackwell Science Ltd., Oxford, U.K., (2001) 1296–1311.
24. Roy, C., Rastegar, C., Kaliaguine, S., Darmstadt, H., and Tochev, V., Physicochemical properties of carbon blacks from vacuum pyrolysis of used tires, *Plast. Rubber Composites Appl.,* 23 (1995) 21–30.
25. Roy, C., Dubé, F., Blanchette, D., de Caumia, B., and Plante, P., Progress in the Demonstration of the Biomass Vacuum Pyrolysis Pyrocycling Process, *Proceedings,* Twelfth European Biomass Conference, Amsterdam, The Netherlands, June 17–21, 2002.
26. Roy, C., Labrecque, B., and de Caumia, B., Recycling of scrap tires in oil and carbon black by vacuum pyrolysis, *Resour. Conserv. Recycling,* 4 (1990) 203–213.
27. Cinliffe, A.M. and Williams, P.T., Composition of Oil Derived from the Batch Pyrolysis of Tyres, *J. Anal. Appl. Pyrolysis,* 44 (1998)131–152.
28. Waddell, W.H., Bhakuni, R.S., Barbin, W.W., and Sandstrom, P.H., Pneumatic tire compounds, *The Vanderbilt Rubber Handbook,* 13th ed., Ohm, R.F. (ed.), R.T. Vanderbilt Company, Norwalk, CT, U.S., (1990) 596.
29. Chaala, A., Darmstadt, H., and Roy, C., Acid-base method for the demineralization of pyrolytic carbon black, *Fuel Process. Technol.,* 46 (1996) 1–15.
30. Mahramanlioglu, M., Adsorption of uranium on adsorbents produced from used tires, J. *Radioanal. Nucl. Chem.,* 256 (2003) 99–105.
31. Wang, M.J., Kutsovsky, Y., Zhang, P., Mehos, G., Murphy, L.J., and Mahmud, K., Using carbon-silica dual phase filler — improve global compromise between rolling resistance, wear resistance and wet skid resistance for tires, *Kautschuk Gummi Kunststoffe,* 55 (2002) 33 – 40.
32. Kühner, G. and Voll, M., Manufacture of carbon black, in: *Carbon Black Science and Technology,* 2nd ed., Donnet, J.-B., Bansal, R.C., and Wang, M.J. (eds.), Marcel Dekker, New York, U.S., (1993) 1–65.
33. Darmstadt, H., Roy, C., and Kaliaguine, S., ESCA characterization of commercial carbon blacks and of carbon blacks from vacuum pyrolysis of used tires, *Carbon,* 32 (1994) 1399–1406.
34. Darmstadt, H., Roy, C., and Kaliaguine, S., Characterization of carbon blacks from commercial tire pyrolysis plants, *Carbon,* 33 (1995) 1449–1455.
35. Darmstadt, H., Cao, N.-Z., Pantea, D., Roy, C., Sümmchen, L., Roland, U., Donnet, J.-B., Wang, T.K., Peng, C.H., and Donnelly, P.J., Surface activity and chemistry of thermal carbon blacks, *Rubber Chem. Technol.,* 73 (2000) 293–309.
36. Darmstadt, H., Roy, C., and Kaliaguine, S., Inorganic components and sulfur compounds in carbon blacks from vacuum pyrolysis of used tires, *Kautschuk Gummi Kunststoffe,* 47 (1994) 891–895.
37. Darmstadt, H., Sümmchen, L., Roland, U., Roy, C., Kaliaguine, S., and Adnot, A., Surface chemistry of pyrolytic carbon black by SIMS and Raman spectroscopy, *Surf. Interface Anal.,* 25 (1997) 245–253.
38. Ledford, C.D., "Process for Conveying Old Rubber Tires into Oil and a Useful Residue," U.S. Patent #5,095,040, March 10 (1992).

39. Donnet, J.-B., and Lansinger, C.M., Characterization of surface energy of carbon black surfaces and relationship to elastomer reinforcement, *Kautschuk Gummi Kunststoffe*, 45 (1992) 459–468.
40. Darmstadt, H., Roy, C., Kaliaguine, S., and Cormier, H., Surface energy of commercial and pyrolytic carbon blacks by inverse gas chromatography, *Rubber Chem. Technol.*, 70 (1997) 759–768.
41. Dufeu, J.B., Roy, C., Ajji, A., and Choplin, L., PVC filled with vacuum pyrolysis scrap tires — derived carbon blacks: an investigation on rheological, mechanical and electrical properties, *J. Appl. Polym. Sci.*, 46 (1992) 2159–2167.
42. Bhadra, S., De, P.P., Mondal, N., Mukhapadhyaya, R., and Das Gupta, S., Regeneration of carbon black from waste automobile tires, *J. Appl. Polym. Sci.*, 89 (2003) 465–473.
43. Fader, J.H., Method for reclaiming carbonaceous material from a waste material, U.S. Patent #5,037,628, August 6 (1991).
44. Fader, J.H., "Upgraded Pyrolyzed Carbon Black (CB_p) from Waste Tires and Scrap Rubber as Reinforcing Filler in Rubber Compounds," *Proceedings*, Rubber Division, American Chemical Society, Cincinnati, OH, U.S., October 17 – 20 (2000).
45. Fader, J.H., "Manufacturing Reinforcing Fillers from Scrap Tyres," *Proceedings*, 10th ETRA Conference, Brussels, Belgium, March 26 – 29, 2003.
46. Pantea, D., Darmstadt, H., Kaliaguine, S., and Roy, C., Heat-treatment of carbon blacks obtained by pyrolysis of used tires. effect on the surface chemistry, morphology and electrical conductivity, *J. Anal. Appl. Pyrolysis*, 67 (2003) 55–76.
47. Pantea, D., Darmstadt, H., Kaliaguine, S., Blacher, S., and Roy, C., Surface morphology of thermal, furnace and pyrolytic carbon blacks by nitrogen adsorption — relation to the electrical conductivity, *Rubber Chem. Technol.*, 75 (2002) 691–700.
48. Pantea, D., Darmstadt, H., Kaliaguine, S., Sümmchen, L., and Roy, C., Electrical conductivity of thermal carbon blacks. influence of surface chemistry, *Carbon*, 39 (2001) 1147–1158.
49. Chaala, A., Roy, C., and Ait-Kadi, A., Rheological properties of bitumen modified with pyrolytic carbon black, *Fuel*, 75 (1996) 1575–1583.
50. Darmstadt, H., Chaala, A., Roy, C., and Kaliaguine, S., SIMS and ESCA characterization of pyrolytic carbon black reinforced bitumen, *Fuel*, 75 (1996) 125–132.
51. Chebil, S., Chaala, A., and Roy, C., Modification of bitumen with scrap tire pyrolytic carbon black. Comparison with commercial carbon black. Part I: mechanical and rheological properties, *Polym. Recycling*, 2 (1996) 257–269.
52. Chebil, S., Chaala, A., Darmstadt, H., and Roy, C., Modification of bitumen with scrap tire pyrolytic carbon black. Comparison with commercial carbon black. Part II: microscopic and surface spectroscopic investigation, *Polym. Recycling*, 3 (1997/98) 17–28.
53. Bahia, H.U., The use of crumb rubber and carbon black for modification of paving-grade asphalt binders, *Carbon Black World '94*, Houston, TX, U.S., (1994).
54. Alliotti, A.G., Carbon Black — Its Nature and Possible Effect on the Characteristics of Bituminous Road Binders, *Proceedings*, Australian Road Research Board, (1), Part 1, (1962), 912–918.
55. Heukelom, W., Une methode amelioree de caracterisation des bitumes par leurs proprietes mecaniques, *Bulletin de Liaison, Laboratoire des Ponts et Chaussees*, 76 (1975) 55–64.

56. Lelièvre, A., *Les Enrobés Bitumineux,* Le griffon d'argile, Québec, Canada, (1994) 36–40.
57. Rostler, F.S., White, R.M., and Cass, P.J., "Modification of Asphalt Cements for Improvement of Wear Resistance of Pavement Surface," *Report N° FHWA-RD,* (1977) 72–24.
58. Cunliffe, A.M. and Williams, P.T., Influence of process conditions on the rate of activation of chars derived from pyrolysis of used tires, *Energ. Fuels,* 13 (1999) 166–175.
59. Lehmann, C.M.B., Rostam-Abadi, M., Rood, M.J., and Sun, J., Reprocessing and reuse of waste tire rubber to solve air-quality related problems, *Energ. Fuels,* 12 (1998) 1095–1099.
60. Ariyadejwanich, P., Tanthapanichakoon, W., Nakagawa, K., Mukai, S.R., and Tamon, H., Preparation and characterization of mesoporous activated carbon from waste tires, *Carbon,* 41, (2003) 157–164.
61. Nedelec, J.M., "Valorisation des Noirs de Carbone Pyrolytiques," Report Ecole Nationale Supérieure de Chimie de Clermont-Ferrand, Clermont-Ferrand, France (2002).
62. Bridges, P., Conrad Industries Inc., Centralia, WA, private communication (1987).
63. Lin, H., The Combustion of Anthracites and Low Grade Bituminous Coals, *Proceedings,* International Conference on Coal Science, Pittsburgh, PA (1983).
64. Roy, C., de Caumia, B., Pakdel, H., Plante, P., Blanchette, D., and Labrecque, B., Vacuum pyrolysis of used tires, petroleum sludges and forestry wastes: technological development and implementation perspectives, *Biomass Thermal Processing, Proceedings,* First Canada/European Community R&D Contractors Meeting, Ottawa (1990).
65. Petroleum (Refinery Process, Survey), *Encyclopaedia of Chemical Technology,* 3rd ed., 17 (1982), 210–218.
66. Anonymous, Heavy Fuel Specification, Report of MAN B&W Diesel AG, Augsburg, Germany (1995).
67. Benallal, B., Pakdel, H., Chabot, S., Roy, C., and Poirier, M.A., Characterization of pyrolytic light naphtha derived from vacuum pyrolysis of used tires — comparison with petroleum naphthas, *Fuel,* 74, (1995), 1589–1594.
68. Anonymous, Environmental Protection and Industrial Safety 1, *Ullmann's Encyclopaedia of Industrial Chemistry,* 5th edition, (B-7), 1995, 438–440.
69. Pakdel, H., Roy, C., Aubin, H., Jean, G., and Coulombe, S., Formation of dl-limonene in used tire vacuum pyrolysis oils, *Environ. Sci. Technol.,* 25 (1991) 1646–1649.
70. Roy, C., U.S. Patent #4,740,270 (1988); Canadian Patent #1,271,151, July 3, 1990; U.S. Patent #5,087,436, February 11, 1992; U.S. Patent #5,099,086, March 24, 1992; Australia Patent #636,350, April 29, 1993; U.S. Patent #5,208,401, May 1993; U.S. Patent #5,229,099, July 20, 1993; Australia Patent #651,029, July 7, 1994; Canadian Patent #1,334,433, February 14, 1995; European Patent Application #94 100 421.0, September 29, 1997.
71. Leblanc, J.L., Roy, C., Mirmiran, S., Benallal, B., and Schwerdtfeger, A., The plasticizing properties of heavy oils obtained from the vacuum pyrolysis of used tires, *Kautsch. Gummi Kunstst.,* 49, (1996) 194–199.
72. Chaala, A. and Roy, C., Production of coke from scrap tire vacuum pyrolysis oils, *Fuel Proc. Technol.,* 46 (1996) 227–239.

73. Ciochina, O., Darmstadt, H., Chaala, A., Roy, C., Neau, L., and Monthioux, M., Coke prepared from heavy pyrolytic oils, *Eurocarbon 98,* July 5–9, 1998, Strasbourg, France, 3–4.
74. Chaala, A., Ciochina, O.G., Roy, C., and Bousmina, M., Rheological properties of bitumen modified with used tire-derived pyrolytic oil residue, *Polymer Recycling,* 3 (1997/1998) 1–15.
75. Mancini, G., Del Manso, F., and Bocchi, L., Correlation Between Chemical Type of Bitumen Fractions and their Interactions with SBS Copolymers, Paper G-6, Symp. on Chemistry of Bitumens, Roma (1991).
76. Breadael, P., Andriolo, P., and Killens, E., A Structural Study of the Hot Storage Stability of SBS-Modified Bitumens, Paper G-5, Symp. on Chemistry of Bitumens, Roma (1991).
77. Roy, C., Chaala, A., and Darmstadt, H., The Potential Uses of Products from Vacuum Pyrolysis of Scrap Tires, 2è Conférence maghrébine de génie des procédés, April 22–25, 1996, Gabes, Jerba, Tunisia.

12

Markets for Scrap Tires and Recycled Rubber

Tjaart P. Venter

CONTENTS

12.1 Introduction

It is not the endeavor in this chapter to teach the reader about every single possible use of scrap tires and recycled rubber, but rather, to offer a way of observing and analyzing the existing and potential markets. Without a market, there is no recycler. "Market" can mean that products are sold or that a service is rendered, for instance, in shredding scrap tires for landfill disposal. Whether you are already in the business or want to start out in this intriguing field, the market is the place to begin your analysis. By analyzing the market, you can plan your strategies, financial performance, and profits for ensuing years. This chapter does not provide the wisdom of a marketing guru. There are many excellent books on this subject by renowned educators and practitioners.[1]

Here we suggest the concept of a value tier system to segment the market for yourself in a practical way. Normally "value" will mean the economic value one can add or create in the product or service. Such value is appreciated by your customers, and they pay you for the product/service according to how much they appreciate the value and how much your competitors would charge for the same value. Value can also include solutions to environmental issues. At the end of this chapter you will find a description of how to obtain market and product information. Some uses of scrap tires and rubber are also listed. Please refer to previous chapters in this book for additional information on products derived from scrap tires and their present and potential markets.

As discussed later, longer-term viability in the scrap tire recycling industry (around the world) has been under suspicion. A number of ventures, some on a rather grand scale, have failed. To this end liberties were taken and a recap offered of how to distinguish financially between good and bad projects. In the author's opinion, endless more projects would have been started and would have failed had the financiers not had the vision to make sure the markets were there and available.

12.2 Background

In evaluating markets for recycled rubber, the variety of rubbers sold in virgin form has to be kept in mind, including: natural rubber (NR), styrene-butadiene-rubber (SBR), ethylene-propylene-diene-methylene rubber, butyl rubber (BR), polyurethane rubber, thermoplastic elastomers, and a long list of others. Although reclaimed (recycled) BR is important in the tire industry, it is the large volume of NR and SBR used in tire compounds that warrants attention. Hence, the markets mainly discussed here are those for scrap tires and recycled tire rubber.

The use of scrap tires and scrap rubber evolved over time, away from their literal burial place: the landfill site. Markets were developed not only on a purely economical basis, but also with government intervention, for example, by way of banning tires from landfills and instituting compulsory levies or gate fees.

What can scrap tires and their rubber be used for? Applications are legion and vary widely. Scrap tires from Japan are sold as secondhand, usable tires in other (sometimes less-developed) countries. In India, treads for steel-banded, wooden wheel, two-wheel carts are made from scrap tire treads. A couple in Mexico built their house with scrap tires and wrote a book about how to do that. All over the world, scrap tires are used as fuel in cement kilns. In some countries such as Australia, recycled rubber crumb is used as flexible filler in ceramic tile adhesives. In Brazil, a company uses the tire rubber in devulcanized form for manufacturing durable pallets, and also incorporates the steel from the tires into their product. Scrap tires are excellent silage cover anchors all over Europe, and in Africa scrap tires are the only source of raw material for a special shoe-manufacturing industry. Scrap tires are raw material in pyrolysis processes to produce oils, gas, and char, while recycled crumb rubber is the major modifying ingredient in asphalt-rubber in the U.S., Canada, Europe, South Africa, Australia, and various parts of Asia. It is impossible to describe all uses of scrap tires and derived products in one chapter; therefore, the large, well-known markets in the U.S. and Europe are first described with appropriate detail. Then a system is proposed as a tool to analyze the market for existing and new ventures.

12.3 The Market in the U.S.

For developed countries, the rule of thumb is that one scrap tire is generated per inhabitant. That proves true for the U.S., where it was estimated that the number of scrap tires in 2003 totaled 300 million, close to the U.S. population count of 290.8 million estimated by the 2003 census.[2] Table 12.1, extracted from the Rubber Manufacturers Association's report "U.S. Scrap Tire Markets 2003,"[2] provides a good overview of the U.S. market.

Table 12.1 provides information on the U.S. market published in July 2004, but what was the situation some years back? At the May 6–9, 1997 meeting of the Rubber Division of the American Chemical Society held in Anaheim, California, John R. Serumgard of the Scrap Tire Management Council pre-

TABLE 12.1

Estimated Total U.S. Scrap Tire Market: 2003 (millions of tires)

Tire-derived fuel (TDF)		
Cement kilns	53	18.3% of total generation
Pulp/paper mills	26	9%
Dedicated tires to energy	10	3.4%
Electric utilities	23.7	8.2%
Industrial boilers	17	5.9%
Total fuel use	129.7	44.7%
Products	34.7	12%
Ground rubber	28.2	10.1%
Cut/punched/stamped	6.5	2.2%
Civil engineering	56.4	19.4%
Miscellaneous/agriculture	3	1%
Electric arc furnaces	0.5	0.2%
Export	9	3.1%
Total use	233.3	80.4%
Total generation	290.2	100%

sented the following information [3]: In 1990 the Environmental Protection Agency estimated the total markets for scrap tires at less than 11% of annual generation. At the end of 1994 the market stood at 55.4%, which in 1996 increased by a whopping 37% to 202 million, or 75.9% of the total scrap generation of 266 million. Table 12.2, extracted from Serumgard's published paper,[3] provides a summary of the U.S. market position in 1996.

Let us analyze the growth in the various market segments. Total market growth was from 202 million tires in 1996 to 233 million in 2003, an increase of only 15% over 7 years. But compare that with an increase in scrap tire generation from 266 to 290 million over the same period, an increase of 9%. However, there was a major shift in the actual consumption pattern for scrap tires; TDF use dropped from 152.5 million tires in 1996 to 130 million in 2003. One reason for that may be because TDF provides too low a revenue. It should not be forgotten, however, that TDF is the largest single use of scrap tires. Products almost doubled in the same 7-year period; tires used for

TABLE 12.2

Estimated Scrap Tire Market Capacity at January 1, 1997 (millions of tires)

Fuel	152.5	57.3% of total generation
Products	20.5	7.7%
Ground rubber	12.5	4.7%
Cut/punched/stamped	8	3%
Civil engineering	10	3.8%
Export	15	5.6%
Miscellaneous/agriculture	4	1.5%
Total use	202	75.9%
Total generation	266	100%

TABLE 12.3
Crumb Rubber Markets in North America 2003 (Tons)

Asphalt modifications	54,545	20.5%
Molded products	95.455	35.9
Sport surfacing	68,182	25.6
Tires/automotive	22,727	8.5
Surface modified/reclaim	4,545	1.7
Animal bedding	11,364	4.2
Horticultural	9,545	3.6
Total	**266,363**	100

ground rubber increased from 12.5 to 28 million, to show average growth of over 17% per year. Please take heed of the extracts from the source publications quoted hereunder.[3,4] Civil engineering use jumped from 10 to 56 million tires, an astounding 65% per year, and today it is the second largest market for scrap tires. Large volumes of material are used in a variety of applications, mostly replacing other materials previously used in construction such as soil, clean fill, drainage aggregate, lightweight fill, expanded polystyrene, etc. Miscellaneous and agricultural use went from 4 to 3 million scrap tires, while exports remained constant.

Details are overlooked in the information above, so the reader should be aware of that when forming conclusions. A case in point is the ground rubber market. Serumgard's paper delivered in May 1997[3] included the following quote:

> Even if the market demand for ground rubber would double, the impact on the industry would be minimal. Currently, many companies are operating at less than half their production capacity, which drives up costs. These increased production costs cannot be passed along because: 1) the market for recyclable rubber is not elastic; 2) ground rubber supply greatly exceeds demand; and 3) the market forces are placing a downward pressure on prices. The result of all these factors is that many of the marginal producers have been or will soon be forced out of business.

Table 12.3, extracted from the RMA[2] 2004 report, outlines the ground rubber market in North America in 2003.

At the October 8–11, 2002, meeting of the Rubber Division of the American Chemical Society held in Pittsburgh, Pennsylvania, Michael Blumenthal of the Rubber Manufacturers Association quoted the statistics published in December 2002.[2] Notice the caution dispensed in his paper:

> At present, there is market demand for an equivalent of 77% of the annually generated scrap tires in the U.S. (218 million of the 281 million generated). Still an entrepreneur could perceive that there is ample market demand for ground rubber products since a significant number of states are willing to fund the purchase of ground rubber products or

provide grants for processing equipment. There are some would-be investors that are under the impression that there still remain untold millions of scrap tires without markets, as well as stockpiles of scrap tires waiting around to be processed and sold as ground rubber. Given that some states are offering substantial grant packages (i.e., California, Maryland, North Carolina, Georgia, South Carolina), there are some entrepreneurs who are preparing to enter the ground rubber production arena. Yet, the existence of easy financial terms and apparently unlimited supply of scrap tires does not guarantee long term success.

Case in point is the Santee River Rubber Company, which received upwards of $36 million in bonds and raised another $20 million, had most modern processing equipment available and claimed to have effective markets for all of the six million tires they were to process annually. This largest of all ground rubber operations in the United States remained in business a mere six months.

If anything can illustrate the main objective of this chapter in the book, it is the experience of Santee River — no market, no revenue, no profit, no sustainability, no company!

As recently as the Fall Technical Meeting and Rubber Expo '03 held in Cleveland, Ohio, mid-October 2003, and afterwards, Michael Blumenthal of the Rubber Manufacturers Association shared his insight and experience of how markets for scrap tires are developing since the last study in 2001. The next study report is expected around March 2004. Mr. Blumenthal's summary is quoted with thanks in the next paragraphs.[6]

Trends in the U.S. Scrap Tire Markets

From the end of 1998 through the end of 2001, the total number of scrap tires going to a market increased from 177.5 million tires (66% of the 270 million generated) to 218 million (77.6% of the 281 million generated). There were increases in all three of the major markets: fuel, civil engineering and ground rubber applications. In the period 2001 to 2003 there was a continuation in the advancements of these major markets.

In the tire-derived fuel (TDF) market, the increases that have occurred in the past two years are functions of same three factors that contributed to earlier increases: a decreased demand for cement, elevated cost of energy (natural gas) and an improvement in the quality and consistency of TDF. Over the past two years there has been a significant increase in the amount of TDF going into pulp and paper mill boilers, especially in the Southeastern portion of the country.

The use of scrap tires in civil engineering applications continues to increase, although not at the rate experienced from 1998 – 2001. In 1998, some 20 million scrap tires were used. The 2001 data indicated that 40 million scrap tires were used. In 2003 it is anticipated that some 50 million scrap tires will be used in this market application. Three large-scale applications for tire shreds accounted for most of the markets: landfill applications, septic field drain medium and road construction. The increase is a function of three factors: 1) cost competitiveness of the shreds, 2) increased acceptance by regulatory agencies and 3) market responsiveness by the industry.

In the ground rubber market there are two classes of particle sizes: "ground rubber" (also referred to as crumb rubber—10 mesh and smaller) and "coarse" rubber (one quarter inch pieces and larger, with a maximum size of one-half inch). Each of these size ranges has distinct market applications. Over the course of the last two years the greater growth in market share has been with the "coarse" sized particles. This particle range is used in playground surfacing, running track material, soil amendments and some bound rubber products. The smaller particle sizes are used for the more traditional applications (asphalt rubber and molded and extruded rubber products).

Overall, there are a series of trends in the U.S. scrap tire industry:

Larger, regionally based tire processors are established which happens at the expense of the smaller, locally based processors.

There are still companies attempting to establish a nation-wide network of processing operations, but this effort continues to [be] elusive.

Processing technologies for ground rubber continue to improve, as does the quality of the tire-derived material.

There appears to be an abundance of the 20 – 40 mesh ground rubber, while the production of coarse rubber and ultra-fine mesh rubber appear to be below market demand.

Recently a new market application for tires has emerged. Tires are being used as a source of carbon, steel and energy for electric arc furnaces (EAFs). Tires are placed into the raw materials used to make high carbon steel products. At present there is only one EAF permitted to use tires, but several more are in the process of applying for their permits. Should this use of tires become wide spread, there could be a significant increase in the number of tires consumed.

Markets are dynamic and during the time it takes to publish a book, markets change. Please obtain updated information from the Rubber Manufacturers Association's website.[2]

TABLE 12.4

Postconsumer Tire Arisings by Member State (thousand tons)

Member State	1996	1998	2000	2002 Estimate
Austria	40	41	50	50
Belgium	65	70	70	70
Denmark	38	38.5	37.5	39.5
Finland	30	30	30	30
France	480	380	370	401
Germany	650	650	650	650
Greece	58	58.5	58.5	58.5
Ireland	7.64	7.64	32	32
Italy	360	360	350	350
Luxembourg	2	2	2.75	3
Netherlands	65	65	67	67.5
Portugal	19.82	45	52	52
Spain	115	330	244	280
Sweden	65	65	60	60
UK	400	380	435	435
Total EU	2,430.640	2,522.640	2,508.750	2,570.500

Note: Revisions based upon 2002 questionnaire responses.

12.4 The Market in Europe

In the European Union (EU), the market is measured differently. Table 12.4 contains data extracted from the proceedings of the European Tyre Recycling Association's (ETRA) conference on March 26, 2003. It shows where and what quantities of scrap tires are generated.[7]

The estimate of 2,570,500 tons of total scrap tires generated during 2002 in the EU is recalculated in passenger car equivalent tires at 7 kg or 15.4 lb and equates to a total of 367 million tires scrapped by a total of 373 million inhabitants. Remember the rule of thumb?

So where are all these scrap tires going? That information is shown in Table 12.5, which contains data derived from the same ETRA conference proceedings.

TABLE 12.5

European Union Market Evolution 1992–2002 (Estimated) (% of total tire arisings)

	1992	1994	1996	1998	2000	2002 est.
Landfill	62	56	49	40	39	35
Reuse/export	6	8	8	11	10	10
Retreading	13	12	12	11	11	11
Recycling	5	6	11	18	19	21
Energy	14	18	20	20	21	23

TABLE 12.6
Material Recycling in the EU for 1995 and 2001

Market	1995	2001	2001
Civil engineering	12%	19%	14.6 million tires/yr
Sports and play surfaces	39%	33%	25.4 million
Consumer/industrial products	21%	23%	17.7 million
Construction	7%	14%	10.8 million
Other (incl. TDF)	21%	11%	14.3 million
Total	100%	100%	82.8 million

Over the 10 years, there was a drastic downward trend in landfilling, and this will continue due to environmental laws. Reuse and export essentially represents what happens to secondhand tires with enough tread to still be used. There was a reasonable gain shown in this category, which means tires are more properly utilized. Retreading sagged a bit, which could be the effect of cheaper tires being imported into Europe from Asia — a trend worth noting because it affects the whole western world.

Because the EU is so varied in composition between member states, it would be unfair to compare its performance with that of the U.S. However, let us look at the progress of recycling and energy uses (see Table 12.6).

Once again, the civil engineering and construction uses seem to be the up and coming markets.

12.5 Markets Elsewhere in the World

Canada has a tire recycling market similar to that of the U.S. in character. Of course it is much smaller, as one would expect, with a population around 20 million compared to a population of approximately 280 million in the U.S. One great difference, though, is that Canadians can feasibly market their recycled rubber products in the U.S. because of their particular state incentive or levy systems, which require a recycler of tires to invoice the goods he actually produced from scrap tires. Only then can he claim the government incentive. Such incentive schemes are not typically available to U.S. producers.

In Japan, the rules are very stringent with regard to tread depth on a tire; as a result, reuse and export of tires are important. From the Japan Tire Recycling Association Inc. statistics,[8] the recycling rate for tires in Japan varied over the years from 91% in 1986 down to 85% in 1988, and up again to 93% for the years 1993 to 1995. The last figure was 88% in 2000. In Australia, the population of approximately 20 million people will scrap some 20 million tires annually if we use our rule of thumb. Besides the particle rubber markets, there are no special processes applied, although over the last number of years research has been done on rubber "renewal" formulas.

The particle market is estimated at 15,000 tons/yr. Some tires are used as fuel, notably in a cement kiln. This last application has been questioned because of suspicious air emissions. The usual landfilling is done with excess tires, but only after shredding.

Markets in the rest of Asia, China, the other parts of the Eurasia continent, Africa, and South America will be determined by the wealth of each country's economy. In South Africa, the scrap tire generation is estimated at 11 million per year, and the market is estimated to be 20% of that. The South African population of 42 million does not follow our rule of thumb for scrap tire generation! India needs special mention because no usable material of any nature, including all parts of scrap tires, is discarded or landfilled. Once again, this indicates the relative wealth of the country.

12.6 Market Analysis: The Practical Approach

12.6.1 Marketing's Four Ps

An existing recycler of tires and rubber should analyze the available markets from time to time to test his or her position. Market analysis should also be the starting point for a new venture. The most important characteristics of markets can be grouped by using the well-known Four Ps: Product, Price, Place, and Promotion. Unfortunately, rather than being mutually exclusive, these characteristics are mutually dependent. In other words, you must review the 4 Ps quite a few times to reach your final decision of what to produce and offer in the market, at what price, where, and how you will tell potential customers about these products.

As a first step, you decide which Products will most likely give you the largest profit. Then the question is raised whether you can already produce these products or whether you want to produce them. What will be the costs, both capital and operational? Remember that selling price less total cost is your profit, and do not forget the taxman!

Next is the task of defining Place. You have to decide where you want to sell your products, through which distribution channels, to which specific customers, and in which geographic areas.

The next decision to make is at what Price you will sell your product. Whenever one thinks price, one is inadvertently compelled to think about competitors' prices. Note that, with new products or a completely new recycling plant, you are only determining, at best, your competitor's price; that is not necessarily the price your prospective customer will pay for your product. You may achieve a better price for specific reasons, but typically, you must sacrifice price to break into a market. In other words, unless your product is truly unique and you can demand a better price than your competition, you will find yourself cutting price to get a part of the business.

Tires and rubber have been recycled for a while, so you will find competition everywhere.

Promotion, the last P, is about telling the customers how wonderful your products are and why they need them! Because recycled tire and rubber products mostly fall into categories that are classed as commodities, it is very difficult to make the products unique, to distinguish your product from the rest. One way would be to have a secret process, e.g., to produce very fine rubber crumb, as indeed some producers in the U.S. do. Another way might be if your process requires incorporation of very expensive equipment to produce your unique product. Then the cost of entry into your market becomes too high for the next man to take the risk. When you have a unique product, you can sell on an almost monopolistic basis, and then you have plenty to tell your customers about.

12.6.2 The Concept of Product/Service Value

When tires on a vehicle are replaced, the old ones can be sold as secondhand tires, they can be retreaded, or they can be scrapped. The scrap tires and their rubber can be put to use in a tier of value related to use of the original rubber polymers and/or to market values of the recycled materials. The value added to the scrap by the recycler is important, but it is the value to the recycler's customer that is the pivot of his market. Although mainly rubber produced from scrap tires is discussed here, it is recognized that the steel recovered and the textile fibers separated from the rubber can make decent contributions to the profits in the tire recycling process. Every market as defined by the Four Ps will be unique; in other words, the set of circumstances and market characteristics will likely never be the same elsewhere. For this reason, consider each customer to be unique. How do we then make the correct choices of products, places, prices, and promotion? We look for value — for the best possible profit. Let's take apart this maze (the recycled tire and rubber market) in a structured way, by segmenting it, to discover the best value for us.

12.6.3 The Value Tiers of Recycled Tires and Rubber

The suggestions of the value tier made here are by no means absolute, but serve as a guideline only. Remember that we attribute the highest value to the product closest to the virgin rubber or to a product that has intrinsic value not achievable by another product (uniqueness).

12.6.3.1 Tier 1: Substitute Virgin Rubber

Recycled rubber can be used as a substitute for virgin rubber in the rubber compounds in tires and industrial rubber products, in the following forms: as reclaimed or devulcanized rubber; as ultra finely ground rubber modifiers;

as inert, low-cost fillers with process-enhancing characteristics; and as replacement for virgin rubber, e.g., replacement of block copolymers (SBS type) as rubberizing material in rubberized asphalt by recycled rubber crumb.

12.6.3.2 Tier 2: Use Recycled Rubber for Its Rubber Character in Bound Systems

Rubber particles of various sizes and shapes are used in combination with binders, such as polyurethane, or virgin rubbers in the production of articles that consist predominantly of recycled rubber, specifically because of the rubber properties required in the end. Examples include: solid tires and wheels, harbor bumpers, welcome mats, safety surfacing in play parks, the shock-absorbent layer under artificial grass on sports fields, and a host of others.

12.6.3.3 Tier 3: Flexible and Low-Cost Fillers

Rubber particles, normally sizes below 0.75 mm, are used in nonrubber formulations as fillers, where they impart flexibility to the end product and deliver cost advantages. The rubber crumb has a low bulk density; thus, the cost per volume is low compared to much heavier fillers such as sand, talc, dolomite, etc. Cases in point are latex-based sealants, cement-based ceramic tile adhesive, and flexible screeds where the volume of product in the application determines the cost. Flexibility may be a sought-after characteristic in the end product.

12.6.3.4 Tier 4: Use of Larger-Size Particles of Rubber

The use of tire rubber chips, or granules, and other types of tire and tire rubber particles brings us to civil engineering applications, such as lightweight fill material in embankments, roads, etc. Granules are used as a receptor for bullets at shooting ranges; the metal can be screened out efficiently afterwards. Smaller-size granules of 0.8 to 1.6 mm (crumb) is used on sports fields as an in-fill in the artificial grass. Larger granules are used as a soil enhancer under natural grass, such as on putting greens on golf courses. In most of these cases, the recycled rubber is a replacement for another material and is required for these applications because it does the job better.

12.6.3.5 Tier 5: Whole Tires or Parts of Tires

The next tier of value includes the use of whole tires and parts of tires, but excludes retreading of tires and salvaging of casings for retreading. Whole tires are used as harbor bumpers. On farms, they anchor silage covers and are used to stop soil erosion by physically restricting erosion in ditches. Tires provide a low-cost material for crash barriers at motor race courses. They are cut into strips and woven into blast mats for mines and other blasting.

Strips of tires are woven into truck-loading bay mats. Who has not seen or used a child swing made from a tire?

12.6.3.6 Tier 6: Energy Value

Lastly, energy use of scrap tires amounts to the largest single consumption, accounting for more than half of the total use of scrap tires "recycled" in the U.S., and some 23% in the EU. How much value the material has in this use is determined by cost factors such as gate fee (negative cost!), cost of transport of tires and TDF, cost of processing to TDF, delivered prices of alternative energy such as coal, desirability of tires as the specific fuel for a particular plant, air emission laws, etc. A small deviation here is provided by the age-old process of pyrolysis, whereby the tires' energy is converted into more easily used materials such as oil, gas, and char. The char can also be used as is for producing smokeless fuel, and with further refinement, as a pigment or as a type of "carbon black" reinforcer in rubber compounds, notably not high-performance tires.

12.6.4 Conclusion

It is important to be broad-minded in your own market analysis and segmentation. You can use any such value tier system applicable to your own situation or area or even financial position. You may have certain restrictions such as capital, or you may already be operating some machinery and want to expand your market only within the product range that you are able to manufacture. The easiest start-up is a sharp knife and some scrap textile automobile tires from which you can produce child swings. Or get tractor tires and sell them as sand pits for children's playgrounds. Then again, you may want to shred tires for TDF. There is no substitute for initiative and imagination in setting up the best value tier set for your market. Be sure of one thing: Marketing scrap tires and derived products is an ongoing war. Using the above guideline will bring you closer to a decision about where and how you want to pitch your troops and war equipment (company and its employees) in the war (marketplace). The guideline will also help define the enemy and their positions (the competition with their strategies). Sometimes you win and sometimes you lose. But doing proper market segmentation and strategic marketing, by using the Four Ps and the above guideline, will help you achieve a much better strike rate.

May you also be reminded of the simplest management process already described many years ago: Plan, Organize, Lead, and Control (POLC). The best way is to plan in money terms, i.e., prepare a budget. A budget always starts with the market and your marketing plan; then you organize whatever it takes to achieve that plan. An organization is run by people who need to be led and motivated toward achieving their goals. To ensure smooth accomplishment of what you set out to do, you will control the results and your employees' actions toward achieving the profits you set out to make. Please do not be fooled by the simplicity of the process outlined. It is very effective, especially so if your competition does not follow such a structured approach. Again, heed the one warning that, unfollowed, has led to the demise of many in the tire and rubber recycling market (see examples quoted earlier in this chapter): Do not first produce a product and then go out and look for a market. Start with the market!

12.7 Financial Evaluation: a Quick Reminder

So, you would like to build a tire recycling plant? Or add equipment to your existing plant because you want to manufacture new products? You may find price to be one of your best and most important competitive tools. Because price minus cost equals profit, the total cost of your product determines the price you have to obtain and the success of your venture. This section will help you evaluate your project in financial terms. The three methods reviewed here are the payback method, the net present value method, and the internal rate of return method. An accountant, net present value tables, and a financial calculator are indispensable aids. Also, there are many books available on managerial finance.[9]

12.7.1 Opportunity Cost

Let's assume that the shareholders have $1 million to invest in a new project and that tire recycling can offer them a "return" of, say, $100,000 per year, or 10% after taxes. If their next best option is a bakery, which returns $80,000, or 8% after taxes, their opportunity cost for investing in the tire recycling plant would be $80,000. In other words, the yield on the best alternative is the opportunity cost of your project because you passed up the opportunity to invest in that "second-best" alternative project. An opportunity that is always available of course, is for the investor to put his money in a bank and earn interest.

12.7.2 Payback Period

This is an easy method to use for short-term or small projects: How many years or months will it take to recover your original investment? If the profit on the above $1 million project were $400,000 per year after taxes, it would take 2 1/2 years to recover the capital outlay, excluding the cost of capital and the time value depreciation of money. But what about the income derived from the project after 2 1/2 years? When comparing various investment opportunities, this method may prove unsatisfactory.

TABLE 12.7
Net Present Value Method

	Tire Recycling			Project Bakery		
Year	**Net Cash Inflow**	**Interest Factor**	**Present Value**	**Net Cash Inflow**	**Interest Factor**	**Present Value**
1	200,000	0.91	182,000	200,000	0.91	182,000
2	250,000	0.83	207,500	200,000	0.83	166,000
3	300,000	0.75	225,000	200,000	0.75	150,000
4	350,000	0.68	238,000	200,000	0.68	136,000
5	350,000	0.62	217,000	200,000	0.62	124,000
			$1,069,500			**$758,000**

12.7.3 Net Present Value Method

This is also called a discounted cash flow technique, by which the time value of money is taken into account. You can, of course, have fun if you are in business, but the fun stops suddenly if a business is liquidated! Therefore, it is recommended that you calculate the viability of the business by calculating the net cash flows for the future. Let's take a 5-year period for the tire recycling plant and the bakery, where the recycling plant shows profit growth to maturity and the bakery remains at the same profit level throughout the period. With an interest factor of 10% applied to both for comparison purposes, the calculation of net present value is as shown in Table 12.7.

The net present value (NPV) of a project is the sum of the present values of the cash flows, inflows, and outflows. This is calculated by discounting the cash flows at the rate of the cost of the capital, e.g., the interest rate at which the capital is borrowed and which, in our example, is 10%. From the sum of the cash flows, the original capital cost has to be deducted. Note that the capital is spent now, so the NPV is exactly that capital cost. If the NPV is positive, you can invest. If it is negative, stay away. In the example:

Tire recycling project NPV = $1,069,500 – $1,000,000
= **$69,500**

Bakery NPV = $758,000 – $1,000,000
= **–$242,000**

Conclusion: DO NOT start with the bakery. DO START with the tire recycling plant.

12.7.4 Internal Rate of Return Method

The internal rate of return (IRR) is defined as that interest rate which equates the cost of the capital investment to the present value of all the cash flows, i.e., the NPV. In the example above, the interest rate was 10%. For the tire recycling project, the NPV will equate to the $1 million capital cost at a higher rate than 10%. For the bakery, the NPV equates to the $1 million capital at a much lower rate than 10%. The last two deductions mean that the IRR for the tire recycling plant is higher than 10% and for the bakery, much lower than 10%. When the IRR is exactly the same as the cost of the capital (the interest you pay on the capital), then there is a break-even point, i.e., the project will not produce a profit. However, the larger the margin by which the IRR exceeds the cost of capital, the more profitable the venture will be, and vice versa. To calculate the IRR, you can use a trial-and-error method or let the computer do all the trial and error for you. Commonly used spreadsheet programs will do these tricks.

12.8 Where to Get Market Information

A variety of market information sources are available. The most important sources are mentioned below.

The Internet is by far the widest-ranging research tool for finding market information. Use search engines such as Google or Alta Vista and key words such as "scrap tires" or "tire recycling." The problem you are bound to have is the glut of possible information.Topical or industry websites are invaluable, such as those of the Rubber Manufacturers Association, USA at rma.org or the Rubber Division, American Chemical Society at rubber.org[2] and the European Tyre Recyclers Association (ETRA) at etra.eu.com. For Japanese information, try jinjapan.or.jp and jinjapan.org/stat.[8] Some information is available at trade magazine websites, and of course, the magazines are just about worth their weight in gold for keeping you up to date with competitive movements in the industry. Refer to *Scrap Tire News* in the U.S.[5] and *European Rubber Journal* in Europe.[10]

One publication that stands out and needs special mention is *The Scrap Tire Users Directory,* available through scraptirenews.com.[5] The claim that it is "The Business Reference Book of the Scrap Tire & Rubber Industry" is absolutely true. The information contained in this directory covers subjects as diverse as applications, quantities, prices, suppliers to the industry, equipment manufacturers, and people involved in the tire and rubber recycling industry inside and outside of North America.

Books can point you in a direction and provide a good basis of information.[11] Refer to a section at the end of this chapter on Further Suggested Reading. Joining industry associations will be extremely advantageous, especially if you go to the meetings and visit the expos. Once again, the Rubber Division of the American Chemical Society and ETRA spring to mind. In Canada, there is the Canadian Rubber Manufacturers Association and in Japan, the Japan Tire Recycling Association Inc.

However, you will need information of the market specific to your own situation. Talk to suppliers and to your prospective customers. Scan the local commercial telephone and business directories. Talk to local and federal government agencies and officials. In this way, you will build up your knowledge library of your competitors and their products and strategies. Learn from your customers what the competition does wrong and, of course, what they do right.

12.9 A Handy List: Uses of Scrap Tires and Recycled Rubber

12.9.1 Tier 1: Substitute Virgin Rubber

Asphalt/rubber (bitumen/rubber)
Porous asphalt mix
Roadway filter drains
Tire compounds (noncritical)
Reclaim rubber
Bituminous crack sealants
TPE-type rubber/plastics mixtures (automotive parts)

12.9.2 Tier 2: Bound Systems Utilizing Rubber Characteristics of Recycled Rubber

Solid tires and wheels
Harbor bumpers
Welcome mats
Safety surfacing
Artificial grass underlay
Carpet underlay
Vehicle load bay mats
Animal transport floors
Horse stalls
Breeding shed floors — ostriches, emus, pigs
Cow mats
Antifatigue matting
Runners for shops and factories
Commercial doormats
Paving bricks and blocks
Swimming pool surrounds
Pedestrian and railroad crossings
Cricket pitch covers
Golf walkways and clubhouse flooring
Golf driving tees
Gymnasium mats
Athletic tracks
Traffic islands
Speed humps and rumble strips
Curbs
Roadside demarcators
Nonslip surfaces, e.g., step covers
Roofing materials
Brake linings
Sound barriers
Tennis courts
Shock and noise proofing of floors

12.9.3 Tier 3: Flexible and Low-Cost Fillers

Latex sealants
Non-water-based sealants
Ceramic tile adhesive
Flexible screeds
Explosives — rubber supplies combustible material as well

12.9.4 Tier 4: Use of Larger-Size Particles of Rubber

Lightweight fills in embankments
Artificial grass in-fill
Loose fill on playgrounds
Equestrian training arenas
Shooting ranges
Soil enhancer, e.g,. in putting greens
Loose fill on parking lots
Filter media

12.9.5 Tier 5: Whole Tires or Parts of Tires

Sand pit — tractor tire
Silage cover anchors/clamps
Crash and other barriers
Truck loading bay mats — woven strips
Cat/dog "baskets"
Landfill drainage and leachate layers
Road furniture and sign posts
Harbor bumpers and boat fenders
Soil erosion control
Blasting mats
Child swings
Artificial reefs
Shoes/shoe soles
Septic tank fill

12.9.6 Tier 6: Energy Use

To produce TDF
Supplemental coal fuel
Conversion by pyrolysis — oil, gas, char
Cement kilns
Sole source electricity generation

References

1. Kotler, Philip, Marketing Management; Kotler, Philip and Armstrong, Gary, *Principles of Marketing;* McDonald, Malcolm H.B., *Marketing Plans, How to Prepare Them: How to Use Them.*
2. Website of the Rubber Manufacturers Association: rma.org
3. Serumgard, John R., Expanding Markets for Scrap Tires and Rubber, Paper No. 12 presented at the meeting of the Rubber Division, ACS, Anaheim, CA, May 6–9, 1997.
4. Blumenthal, Michael, Changes Impacting the Ground Rubber Industry, Paper No. 115 presented at the meeting of the Rubber Division, ACS, Pittsburgh, PA, October 8–11, 2002.
5. scraptirenews.com, the industry magazine *Scrap Tire News* and their publication, *The Scrap Tire & Rubber Users Directory 2005.*
6. Blumenthal, Michael, private communication, November 2003.
7. *Proceedings,* ETRA conference, Brussels, Belgium, March 25–27, 2003, and thereafter.
8. Japan Tire Recycling Association Inc.: jinjapan.org/stats and jatma.or.jp
9. Weston, J. Fred and Brigham, Eugene F., *Essentials of Managerial Finance;* Brigham, Eugene F. and Houston, Joel F., *Fundamentals of Financial Management.*
10. crain.co.uk
11. amazon.com and barnesandnoble.com

Further Suggested Reading and Contacts

Contact the Library of the Rubber Division, ACS through rubber.org.

Papers presented at the spring and fall meetings of the Rubber Division, ACS, are obtainable from their library.

Snyder, Robert H., *Scrap Tires: Disposal and Reuse.*

Rader, Charles P., et al., *Plastics, Rubber and Paper Recycling: A Pragmatic Approach.*

U.S. Environmental Protection Agency and Charlotte Clark, et al., *Scrap Tire Technology and Markets.*

Dufton, P.W., *Scrap Tyres A Wasting Asset?*

Scheirs, John, *Polymer Recycling: Science, Technology and Applications.*

Paschich, Ed and Hendricks, Paula, *The Tire House Book.*

The magazine *Rubber World* and rubberworld.com as well as their publication *2002 Rubber Red Book,* A Lippincott & Peto, Inc. publication.

Rapra, United Kingdom, *Rapra Review report No. 99, Recycling Of Rubber,* 1997.

Index

B

C

D

E

F

H

I

N

O

S

T

U

W

X

Y

Z

Printed in the United States
by Baker & Taylor Publisher Services

Printed in the United States
by Baker & Taylor Publisher Services